Wild Mushrooms and Health

Medicinal mushrooms have been used since ancient times. Certain mushrooms can be used to treat numerous conditions, including those related to cardiovascular health, obesity, cholesterol balance, bone health, diabetes, and cancer. *Wild Mushrooms and Health: Diversity, Phytochemistry, Medicinal Benefits, and Cultivation* presents reports on numerous species of wild medicinal mushrooms with discussion of drug-discovery implications, analysis of bioactive substances, and prospects for cultivation.

FEATURES

- Comprehensive review of medicinal mushrooms as sources of promising bioactive molecules and prospective compounds for drug discovery
- Information on diversity, distribution, ethnomycology, ecology, cultivation, descriptions of specific species, and folk medicinal uses of mushrooms throughout the world
- Emphasis on identification, documentation, bioactive substances, and the nature of mushroom bioactivity
- Discussion of the nutraceutical properties of wild mushrooms, including high protein content comparable to that of meat, and low fat content, which make them a complete dietary food source
- Exploration of methods used in the collection, identification, documentation, cultivation, analysis, and conservation of mushrooms for drug discovery

An installment in the *Exploring Medicinal Plants* series, this volume is a comprehensive resource for medical researchers, scientists, and pharmaceutical companies. In addition, this resource is appropriate for mycologists and botanists interested in pharmacognosy.

Exploring Medicinal Plants

Series Editor

Azamal Husen

Wolaita Sodo University, Ethiopia

Medicinal plants render a rich source of bioactive compounds used in drug formulation and development; they play a key role in traditional or indigenous health systems. As the demand for herbal medicines increases worldwide, supply is declining as most of the harvest is derived from naturally growing vegetation. Considering global interests and covering several important aspects associated with medicinal plants, the Exploring Medicinal Plants series comprises volumes valuable to academia, practitioners, and researchers interested in medicinal plants. Topics provide information on a range of subjects including diversity, conservation, propagation, cultivation, physiology, molecular biology, growth response under extreme environment, handling, storage, bioactive compounds, secondary metabolites, extraction, therapeutics, mode of action, and healthcare practices.

Led by Azamal Husen, PhD, this series is directed to a broad range of researchers and professionals consisting of topical books exploring information related to medicinal plants. It includes edited volumes, references, and textbooks available for individual print and electronic purchases.

Plants as Medicine and Aromatics: Conservation, Ecology, and Pharmacognosy
Mohd Kafeel Ahmad Ansari, Bengu Turkyilmaz Unal, Munir Ozturk and Gary Owens

Sustainable Uses of Medicinal Plants
Learnmore Kambizi and Callistus Bvenura

Medicinal Plant Responses to Stressful Conditions
Arafat Abdel Hamed Abdel Latef

Aromatic and Medicinal Plants of Drylands and Deserts: Ecology, Ethnobiology and Potential Uses
David Ramiro Aguillón Gutiérrez, Cristian Torres León, and Jorge Alejandro Aguirre Joya

Secondary Metabolites from Medicinal Plants: Nanoparticles Synthesis and their Applications
Rakesh Kumar Bachheti, Archana Bachheti

Aquatic Medicinal Plants
Archana Bachheti, Rakesh Kumar Bachheti, and Azamal Husen

Antidiabetic Medicinal Plants and Herbal Treatments
Azamal Husen

Ethnobotany and Ethnopharmacology of Medicinal and Aromatic Plants: Steps Towards Drugs Discovery
Adnan Mohd, Mitesh Patel and Mejdi Snoussi

Wild Mushrooms and Health: Diversity, Phytochemistry, Medicinal Benefits, and Cultivation
Kamal Ch. Semwal, Steven L. Stephenson, and Azamal Husen

For more information, please visit: www.routledge.com/Exploring-Medicinal-Plants/book-series/CRCEMP

Wild Mushrooms and Health: Diversity, Phytochemistry, Medicinal Benefits, and Cultivation

Edited by Kamal Ch. Semwal, Steven L. Stephenson, and Azamal Husen

CRC Press
Taylor & Francis Group
Boca Raton London New York

CRC Press is an imprint of the
Taylor & Francis Group, an **informa** business

First edition published 2024
by CRC Press
6000 Broken Sound Parkway NW, Suite 300, Boca Raton, FL 33487–2742

and by CRC Press
4 Park Square, Milton Park, Abingdon, Oxon, OX14 4RN

CRC Press is an imprint of Taylor & Francis Group, LLC

ISBN: 978-1-032-37219-8 (hbk)
ISBN: 978-1-032-54248-5 (pbk)
ISBN: 978-1-003-33593-1 (ebk)

DOI: 10.1201/b23190

Contents

About the Editors

Kamal Ch. Semwal is an Associate Professor in the Department of Biology of the College of Sciences at the Eritrea Institute of Technology in Asmara, where he teaches interdisciplinary subjects in the botanical sciences. He received his Ph.D. from HNB Garhwal University (a Central University), BGR Campus Pauri, Srinagar Garhwal, Uttarakhand, India, in 2004. He began his career as a research associate at the Directorate of Mushrooms in Solan (Himachal Pradesh) and later taught at Gondar University and Mekelle University in Ethiopia (East Africa). His research on mushrooms is well recognized and has been cited numerous times in various research journals. He is the author or coauthor of several dozen papers published in peer-reviewed journals, as well as book chapters and review articles. He also served as a co-editor of a book entitled *Ecosystem and Environmental Pollution*, published by Discovery Publishing House (P) Ltd., India, and as co-investigator of the study *Collection, identification, documentation of wild edible and medicinal mushrooms of Garhwal Himalaya of Uttarakhand* funded by UCOST, Dehradun, India. His papers have considered a number of topics within botany, particularly regarding the diversity, taxonomy, and ecology of wild mushrooms in the Northwestern Himalayas of India. He has more than 18 years of experience in fungal taxonomy and ecology. He has described several species of wild mushrooms new to science and is currently serving as a reviewer for many highly reputed international scientific journals throughout the world.

Steven L. Stephenson is currently a research professor in the Department of Biological Sciences at the University of Arkansas, USA, where he teaches courses in plant biology, forest ecology, and plant ecology. He received his Ph.D. from Virginia Polytechnic Institute and State University (Virginia Tech) in 1977 and taught at Fairmont State College (now University) for 27 years before moving to the University of Arkansas in 2003. He was a Senior Fulbright Scholar at Himachal Pradesh University, India in 1987, Visiting Scientist at the Australian Antarctic Division in 1995, and William Evens Fellow at the University of Otago in New Zealand in 2003. His studies of myxomycetes (slime molds) and fungi have taken him to all seven continents and every major type of terrestrial ecosystem. He is a Fellow of the Mycological Society of America and the author or coauthor of 18 books and more than 480 book chapters and papers published in peer-reviewed journals. His notable works include *The Kingdom Fungi: The Biology of Mushrooms, Molds, and Lichens* (2010), *Mushrooms of the Southeastern United States* (2017), and *Secretive Slime Moulds: Myxomycetes of Australia* (2021).

Azamal Husen served as Professor and Head of the Department of Biology at the University of Gondar, Ethiopia, and is a foreign delegate at Wolaita Sodo University, Wolaita, Ethiopia. Previously, he was a visiting faculty member of the Forest Research Institute and the Doon College of Agriculture and Forest in Dehra Dun, India. His research and teaching experience of 20 years involves studies of biogenic nanomaterial fabrication and application; plant responses to environmental stresses; nanomaterials at the physiological, biochemical, and molecular levels; herbal medicine; and clonal propagation for the improvement of tree species. He has conducted several research projects sponsored by various funding agencies, including the World Bank (FREEP), the National Agricultural Technology Project (NATP), the Indian Council of Agriculture Research (ICAR), the Indian Council of Forest Research Education (ICFRE), and the Japan Bank for International

Cooperation (JBIC). He received four fellowships from India and a recognition award from the University of Gondar, Ethiopia, for excellent teaching, research, and community service. Husen has been on the editorial board and the panel of reviewers of several reputed journals published by Elsevier, Frontiers Media, Taylor & Francis, Springer Nature, RSC, Oxford University Press, Sciendo, The Royal Society, CSIRO, PLOS, MDPI, John Wiley & Sons, and UPM Journals. He is on the advisory board of Cambridge Scholars Publishing, UK. He is a fellow of the Plantae group of the American Society of Plant Biologists and a member of the International Society of Root Research, the Asian Council of Science Editors, and INPST. To his credit are over 200 publications, and he is Editor-in-Chief of the *American Journal of Plant Physiology.* He is also working as Series Editor for *Exploring Medicinal Plants*, published by Taylor & Francis Group, USA; *Plant Biology, Sustainability, and Climate Change*, published by Elsevier, USA; and *Smart Nanomaterials Technology*, published by Springer Nature Singapore Pte Ltd., Singapore.

Preface

We are living in a world with as many as one and a half to six million different species of fungus, only about 144,000 of which have been formally described thus far. Why is there a need to know these yet undescribed fungi, more commonly referred to as mushrooms? First, mushrooms are rich in many bioactive substances, such as alkaloids, flavonoids, polyphenols, saponins, sterols, tannins, triterpenoids, p-coumaric acids, and many more. These bioactive substances are prospective agents for the discovery of new drugs that will benefit mankind. For example, clinical studies have demonstrated that some substances derived from mushrooms have the potential to treat various ailments. Although an unknown number of wild mushrooms have yet to be discovered and examined for the presence of new bioactive substances that could yield new drugs, mushrooms are valuable for other reasons. Many species of wild mushrooms are collected and consumed as food in countries throughout the world. Only a few of these have been domesticated and cultivated to the extent that they have considerable economic value. There is little question that if more of these edible wild mushrooms were systematically identified and cultivated, it would be possible to overcome food security in certain areas of the world.

Among the species of wild edible mushrooms, there are many examples that are ectomycorrhizal and thus important components of forest ecosystems. As such, there are some technical and ecological constraints in collecting and cultivating their fruiting bodies. However, accounts of the successful cultivation of wild ectomycorrhizal species provide evidence that there should be a greater effort to increase the number of species that can be cultivated and exploited as a source of food. Although there have been some success stories, there are very large differences in the extent to which mushroom cultivation occurs between different countries. A few countries grow appreciable amounts of edible or medicinal species, but many cultivate only a few species, such as the oyster mushroom (*Pleurotus* spp.) or the button mushroom (*Agaricus* spp.), in very low numbers.

Our knowledge of the diversity, distribution, and taxonomy of wild mushrooms is incomplete because of the information gap that exists in many countries. For some countries and geographical areas, we do not yet have anything approaching a comprehensive list of the mushroom varieties present. This means that there is no basis for carrying out studies to determine what potentially useful bioactive substances might be discovered, especially with respect to wild mushroom species found only in a particular part of the world. There is little question that additional exploration and documentation of wild mushrooms would lead to new drug discovery opportunities.

This book addresses various aspects of the diversity, distribution, ecology, culinary and medicinal uses, ethnomycology, cultivation, and biological activity of wild mushrooms in countries all over the world. In addition, the bioprospection of mushrooms and the economics of commercial mushroom cultivation and marketing are considered. Throughout the book, emphasis is placed on the role of field exploration for wild mushrooms as the first step towards the documentation, identification, and research into their pharmaceutical properties, some of which may have value for treating different ailments, including some currently untreated diseases.

The purpose of this book is to disperse the knowledge of wild mushrooms among researchers, scientists, students, and citizen scientists (generally including those interested in wild mushroom consumption or foraging). The editors are grateful to all the individuals who contributed chapters to this book for their time and effort. We also extend our gratitude to Randy Brehm, Tom Connelly, and all associates at Taylor & Francis Group, LLC / CRC Press for their active cooperation. We shall be happy to receive comments and criticism, if any, from subject experts and general readers of this book.

Kamal Ch. Semwal
Steven L. Stephenson
Azamal Husen

Contributors

Gopalappa Abhisheka
Department of Biology
Jnanasudha Pre-University College
India

Ashok Aggarwal
Botany Department
Kurukshetra University
India

Uzma Altaf
Department of Botany
University of Jammu
India

F. M. Aminuzzaman
Department of Plant Pathology, Faculty of Agriculture
Sher-e-Bangla Agricultural University
Bangladesh

Ignacio Arroyo
Finca Boryana
San Gerardo de Dota, Costa Rica

Jaica Jane C. Balcruz
Department of Biological Sciences, College of Science
Central Luzon State University
Philippines

Rajeev Bhat
Estonian Literary Museum
Estonia

Rajendra Prasad Bhatt
Department of Botany & Microbiology
HNB Garhwal University
India

Anuj Chauhan
Forest Pathology Section, Forest Protection Division,
Forest Research Institute
India

K. Das
Department of Plant Pathology, Faculty of Agriculture
Sher-e-Bangla Agricultural University
Bangladesh

Angeles M. De Leon
Department of Biological Sciences, College of Science
Central Luzon State University
Philippines

Tatek Dejene
Ethiopian Forest Development (EFD)
Ethiopia

Robin G. Doss
Maialu S.A.
Santo Domingo de Heredia, Costa Rica

Rich Milton R. Dulay
Address 1- Department of Biological Sciences, College of Science
Central Luzon State University, Science City of Munoz
Nueva Ecija, Philippines 3120
Address 2- Center for Tropical Mushroom Research and Development, College of Science
Central Luzon State University Science City of Munoz
Nueva Ecija, Philippines 3120

Mahesh C. A. Galappaththi
Postgraduate Institute of Science (PGIS)
University of Peradeniya
Peradeniya 20400 Sri Lanka

N. S. K. Harsh
Former Scientist, Forest Research Institute
India

Gabbalkaje Harish Jithesh
Department of Biosciences
Mangalore University
India

Aivar Jürgenson
School of Humanities
Tallinn University
Estonia

Ranjna Kaundal
Forest Pathology Section, Forest Protection Division
Forest Research Institute
India

Samantha C. Karunarathna
Center for Yunnan Plateau Biological Resources Protection and Utilization, College of Biological Resource and Food Engineering
Qujing Normal University
China

Yu Li
Engineering Research Center of Edible and Medicinal Fungi
Ministry of Education, Jilin Agricultural University
Changchun 130118, Jilin Province, P. R. China

Pu Liu
Engineering Research Center of Edible and Medicinal Fungi
Ministry of Education, Jilin Agricultural University
Changchun 130118, Jilin Province, P. R. China

Shivannegowda Mahadevakumar
Forest Pathology Department
Division of Forest Protection
India

A. Manoj Kumar
Post Graduate and Research Department of Botany
Government College for Women
India

Pablo Martín-Pinto
University of Valladolid
Avda. Madrid 44, 34071 Palencia, Spain

Justine Reine T. Matillano
Department of Biological Sciences, College of Science
Central Luzon State University
Philippines

Tahir Mehmood
Department of Botany
University of Jammu
India

Charlyn M. Miguel
Department of Biological Sciences, College of Science
Central Luzon State University
Philippines

Steven L. Miller
Department of Botany
University of Wyoming
USA

John Paul R. Orevillo
Department of Biological Sciences, College of Science
Central Luzon State University, Science City of Munoz
Nueva Ecija, Philippines 3120

Vipin Parkash
Forest Pathology Section, Forest Protection Division
Forest Research Institute
India

Supriti Paul
Forest Pathology Section, Forest Protection Division
Forest Research Institute
India

C. K. Pradeep
Microbiology Division
Jawaharlal Nehru Tropical Botanic Garden and Research Institute
India

Ain Raal
Institute of Pharmacy, Faculty of Medicine
University of Tartu
Estonia

Venugopalan Ravikrishnan
Department of Biosciences
Mangalore University
Mangalagangotri, Karnataka, India

Carlos Rojas
Department of Biosystems Engineering and Engineering Research Institute
University of Costa Rica
Costa Rica

Sanjeev Kumar
Department of Botany
University of Jammu
India

Adam Sayner
GroCycle Courses
UK

Yash Pal Sharma
Department of Botany
University of Jammu
India

Kodandoor Sharthchandra
Department of Biosciences
Mangalore University
India

Upendra Singh
Department of Botany & Microbiology
HNB Garhwal University
India

Kandikere Ramaiah Sridhar
Department of Biosciences
Mangalore University
India

Surya Sudheer
ERA-Chair for Food (By-) Products Valorisation Technologies (VALORTECH)
Estonian University of Life Sciences
Estonia

Tahir Syed
Department of Biosciences
Mangalore University
India

Wubalem Tadesse
Ethiopian Forest Development (EFD)
Ethiopia

J. F. Tanni
Department of Plant Pathology, Faculty of Agriculture
Sher-e-Bangla Agricultural University
Bangladesh

Busha Teshome
Ethiopian Forest Development (EFD)
Ethiopia

Priyanka Uniyal
Department of Botany
Govt. P. G. College
India

Komal Verma
Department of Botany
University of Jammu
India

1 Biological Activity of Six Wild Edible Mushrooms of the Western Ghats, India

Venugopalan Ravikrishnan, Kandikere Ramaiah Sridhar

1.1 INTRODUCTION

Cultivated and wild edible mushrooms serve as attractive functional foods throughout the world owing to their nutritional value, texture, taste, and flavor. In addition, they possess adequate nutritional components (protein, indispensable amino acids, carbohydrates, fiber, and vitamins) and low quantities of fat, calories, sodium, and cholesterol (Mattila et al. 2000). Moreover, cultivated and wild mushrooms serve as a repository of therapeutic bioactive compounds (polyketides, phenolics, steroids, and terpenes [Kues and Liu 2000]).

Damage by free radicals is linked to many human diseases (cancer, diabetes, atherosclerosis, and cirrhosis) as well as aging (Halliwell and Gutteridge 1984). Synthetic antioxidants like butylated hydroxyanisole (BHA), butylated hydroxytoluene (BHT), and tertiary-butylated hydroxyquinone (TBHQ) exhibit protective activity against reactive oxygen, but they are known for carcinogenesis (BHA and BHT) and may possess tumor-producing/tumor-promoting properties (BHA) (Botterweck et al. 2000). Hence, wild and cultivated mushrooms serve as a potential natural source of antioxidants (Ferreira et al. 2009; Boonsong et al. 2016; Sánchez 2017). Among the antioxidants in mushrooms, phenolics have gained more attention owing to their excellent reactive oxygen quenching potential, anti-inflammatory activities, and anti-tumor potential (Puttaraju et al. 2006; Palacios et al. 2011). Mammalian cells possess intracellular defense molecules such as catalase (CAT), glutathione peroxidase (GP), and superoxide dismutase (SOD) to protect themselves against excessive free radicals. Although many organisms have protective mechanisms against damage caused by reactive oxygen, such support is not enough to prevent such damage, and thus supplementary aids are needed (Simic 1988; Turkoglu et al. 2007). Additional air to protect the oxidative damage comes from several compounds, especially minerals (Se and Zn), vitamins (A, E, and β-carotene), and proteins (albumin, ceruloplasmin, and transferrin) (Ostrovidov et al. 2000). Such aids could be further supplemented by cultivated and wild mushrooms in the human diet. Antioxidant components in mushrooms are found in fruit bodies, mycelia, and culture broth (polysaccharides, tocopherols, phenolics, carotenoids, ergosterol, and ascorbic acid) (Sánchez 2017). Increased knowledge of the harmful reactive oxygen species, synthetic antioxidants, and antibiotic resistance by microorganisms have tremendously promoted an interest in exploring natural bioactive compounds with antioxidant and antimicrobial activities.

Mushrooms are a potential source of remediation for several cancers (stomach, esophagus, and lungs) prevalent in Canada, China, Japan, Korea, Russia, and the United States (Ferreira et al. 2007). In addition to antioxidant potential, mushrooms also possess antimicrobial activity (Heleno et al. 2013; Khan et al. 2016). There exists a number of synthetic antimicrobial agents that have been built up to control or kill microorganisms without affecting human metabolism. However, uncontrolled applications of such antibiotics result in the emergence of drug-resistant pathogenic bacteria, which are threatening human life. Many mushrooms serve as natural antibiotics against several

DOI: 10.1201/b23190-1

bacteria, including multi-drug-resistant bacteria (e.g., Gao et al. 2005; Turkoglu et al. 2007; Barros et al. 2007a, b, 2008a, b; Alves et al. 2012). Extracts of fruit bodies and mycelia from the culture of over 2000 mushroom species have been examined and the successful isolation of diterpene, a pleuromutilin useful in the treatment of mycoplasma infections in animals, has been achieved, leading to the development of the first commercial antibiotic of basidiomycete origin (Kavanagh et al. 1950; Brizuela et al. 1998; Sandven 2000).

Domesticated edible mushrooms are available throughout the year due to cultivation, harvest, postharvest, processing, and storage practices (Palacios et al. 2011). Edible wild mushrooms are gaining importance owing to their nutritional, medicinal, and anti-pathogenic potential in the gastrointestinal tract. In view of the utilization of mushrooms as a nutraceutical source, this chapter provides details of the antioxidant and antimicrobial potential of six edible wild mushrooms [*Amanita hemibapha* (Berk. and Broome) Sacc., *Boletus edulis* Bull., *Hygrocybe alwisii* (Berk. and Broome) Pegler., *Lentinus polychrous* Lév., *L. squarrosulus* Mont. and *Termitomyces schimperi* (Pat.) R. Heim] occurring in the foothills of the Western Ghats.

1.2 MUSHROOMS AND PROCESSING

Tender edible wild mushrooms (*Amanita hemibapha, Boletus edulis, Hygrocybe alwisii, Lentinus polychrous, L. squarrosulus*, and *Termitomyces schimperi*) were collected from three locations in the foothills of the Western Ghats region. Replicate samples were oven-dried (58±2°C) to constant mass. The dried mushrooms were milled and preserved in air-tight containers in the refrigerator.

1.3 ANTIOXIDANT ANALYSES

1.3.1 Total Antioxidant Activity

Evaluation of the total antioxidant activity (TAA) of mushrooms was performed according to Prieto et al. (1999). The mushroom flour was extracted with methanol (1 mg/ml) (0.1 ml) mixed with sodium phosphate (28 mM) and ammonium molybdate (4 mM in 0.6 M sulfuric acid). Samples were then incubated (95°C for 90 min) and cooled to room temperature. Absorbance (695 nm) was assessed using a phosphomolybdenum complex with a methanol blank. The TAA was expressed as µM ascorbic acid equivalents per gram of mushroom flour (µM AAEs/g).

1.3.2 Ferrous Ion-Chelating Capacity

The Fe^{2+} chelating capacity of methanolic extracts from mushroom flour was assessed using the method by Hsu et al. (2003). A volume of 2 mM $FeCl_2$ (0.1 ml) and 5 mM ferrozine (0.2 ml) were added to methanol extract (1 ml), and fresh methanol was added to a volume of 5 ml. Samples were incubated at room temperature (10 min) and the absorbance of the Fe^{2+}-ferrozine complex was assessed (562 nm). Samples without methanolic extract served as controls to evaluate the capacity of ferrous ion chelation.

$$\text{Ferrous ion-chelating capacity}\,(\%) = \left(1 \frac{A_{s562}}{A_{c562}}\right)100$$

(Where A_c is absorbance by the control and A_s is absorbance by the sample.)

1.3.3 DPPH Free Radical-Scavenging Activity

The free radical-scavenging activity of the methanol extract of mushroom flour was determined using the method proposed by Singh et al. (2002). To 1 ml of the extracts, 0.01 mM 2,2-diphenyl-1-picrylhydrazyl (DPPH) (4 ml) was mixed and allowed to react for up to 20 min at room temperature. The reagents devoid of methanolic extracts served as controls and their absorbance (517 nm) was assessed to determine free radical–scavenging activity.

$$\text{Free radical-scavenging activity (\%)} = \left(\frac{A_{c517} \; A_{s517}}{A_{c517}} \right) 100$$

(Where A_c is absorbance by the control and A_s is absorbance by the sample.)

1.3.4 Reducing Power

The reducing power of mushroom flour was measured using the technique by Oyaizu (1986) with a minor alteration. Varied concentrations (200–1,000 μg) of methanolic mushroom flour extracts were mixed with 2.5 ml phosphate buffer (0.2 M, pH 6.6) followed by 2.5 ml potassium ferricyanide (1%). After mixing, samples were incubated (at 50°C, 20 min). Later, 2.5 ml trichloroacetic acid (10% w/v) was added and samples were mixed and centrifuged (402×*g*, 10 min). The supernatant (2.5 ml) was blended with distilled water (2.5 ml). To this mixture, 0.5 ml ferric chloride (0.1%) was added, and the absorbance was measured (700 nm). Increased absorbance of the reaction mixture indicated increased reducing power.

1.4 ANTIMICROBIAL ASSAYS

1.4.1 Microbial Strains

Strains of Gram-positive bacteria (*Bacillus cereus*, MTCC430; *Bacillus subtilis*, MTCC441; *Staphylococcus aureus*, MTCC96; and *Streptococcus pneumoniae*, MTCC655), Gram-negative bacteria (*Enterobacter aerogenes*, MTCC2823; *Escherichia coli*, MTCC443; *Haemophilus influenzae*, MTCC3826; *Proteus vulgaris*, MTCC1771; and *Pseudomonas aeruginosa*, MTCC424) and yeast (*Candida albicans*, MTCC183) were obtained from the Microbial Type Culture Collection (MTCC) in Chandigarh, India. Bacteria were maintained on Muller-Hinton agar, whereas the yeast was cultured on Sabouraud dextrose agar. Bacteria were cultured overnight and 200 μl aliquots were transferred into 20 ml of sterile Muller-Hinton broth. Yeast was incubated in Sabouraud dextrose broth (37°C) to stabilize the culture and obtain a concentration of 10^5 CFU/ml.

1.4.2 Mushroom Extraction

Whole tender mushroom powder (5 g) each from triplicate sampling was extracted in 25 ml of methanol using a rotary shaker (120 rpm) for up to 24 h at room temperature. The extract was dried in an oven (50±2°C). The dried methanol extract was dissolved in 20% dimethyl sulfoxide (DMSO) to prepare a stock solution (2 mg/ml). DMSO served as a solvent control and its impact on the growth of microorganisms was examined.

1.4.3 Cup Diffusion Method

The antimicrobial activities of the mushroom extracts were assessed using the cup diffusion method (Bauer et al. 1966). A 24 h bacterial culture (0.1 ml; 10^5 CFU/ml) was spread on Muller-Hinton agar, while the yeast culture was spread on Sabouraud dextrose agar. Sterile porcelain cups (6 × 6 mm; inner tunnel, 4 × 6 mm) were slightly pressed into the seeded plates on the respective agar medium and then loaded with mushroom extract (200 µg/cup). Each plate comprised four cups (three with mushroom extract as triplicate samples and one with DMSO as a control) with one standard antibiotic disc. The standard antibiotic discs contained vancomycin (30 µg/disc) against bacteria and fluconazole (25 µg/disc) against yeast. The plate cultures were incubated (at 37±1°C, 18–24 h) for the growth of bacteria and for the growth of yeast at 26±1°C (48–72 h). The plates were assessed periodically for growth inhibition. Wherever inhibition was observed, the diameter of the inhibition zone was measured using a dial caliper.

1.5 DATA ANALYSIS

The total antioxidant, ferrous ion-chelation, DPPH free radical-scavenging, and reducing activities of the six edible mushrooms were compared by one-way analysis of variance (ANOVA) (SigmaPlot #11; Systat Software Inc., USA).

1.6 ANTIOXIDANT ACTIVITIES

All the mushrooms showed considerable total antioxidant activity, with the lowest levels observed in *H. alwisii* (84 µM AAEs/g) and the highest in *L. polychrous* (394 µM AAEs/g) (Figure 1.1a). The total antioxidant activity, in increasing order, showed a trend of: *H. alwisii* < *T. schimperi* < *L. squarrosulus* < *B. edulis* < *A. hemibapha* < *L. polychrous*. The ferrous ion-chelating capacity was lowest in *L. polychrous* (32.6%) and highest in *A. hemibapha* (62.1%) (Figure 1.1b). Ferrous ion-chelating capacity increased in the following order: *L. polychrous* < *B. edulis* < *H. alwisii* < *T. schimperi* < *L. squarrosulus* < *A. hemibapha*. DPPH radical-scavenging was lowest in *L. polychrous* (9%) and highest in *T. schimperi* (28%) (Figure 1.2a). The DPPH radical-scavenging increased in the following order: *L. polychrous* < *L. squarrosulus* < *A. hemibapha* < *H. alwisii* < *B. edulis* < *T. schimperi* (Figure 1.2b). Reducing power was lowest in *H. alwisii* (1 mg: absorbance, 0.044) and highest in *B. edulis* (1 mg: absorbance, 0.294). One-way ANOVA revealed that the total antioxidant activity was significant among the wild mushrooms ($p<0.001$) except for *B. edulis* vs. *L. squarrosulus*. The ferrous ion-chelating capacity was significantly differed among the mushrooms studied ($p<0.001$). The DPPH radical-scavenging was significant among all comparisons (*H. alwisii* vs. *L. polychrous*: $p<0.05$; *H. alwisii* vs. *L. squarrosulus*: $p<0.01$; other combinations: $p<0.001$) except for *A. hemibapha* vs. *H. alwisii*, *L. polychrous*, and *L. squarrosulus*.

The ferrous ion-chelating capacity of the crude extract of *L. polychrous* was comparable to that of water and ethanol extracts of polysaccharides (Ravikrishnan 2019). Interestingly, the ferrous ion-chelating capacity of the methanol extract of *L. polychrous* was comparable to that of its aqueous and ethanol extracts of crude polysaccharides, indicating that the polysaccharide fraction of the mushrooms studied possesses potential antioxidant capacity. However, such ability was much lower in crude extract than in water and ethanol extracts of polysaccharides of *L. squarrosulus*. The DPPH radical-scavenging capacity was much higher in the water and alcohol extract of polysaccharides of *L. polychrous* than in the crude extract, while it was comparable in *L. squarrosulus*. The reducing power of polysaccharides in the aqueous extracts of *L. polychrous* and *L. squarrosulus* was comparable with that in the crude extracts, while polysaccharides in the ethanol extract showed extremely low reducing power.

The biological activity, especially the antioxidant and antimicrobial activity, of edible mushrooms is valuable for human health protection. The impairment caused by free radicals is correlated to

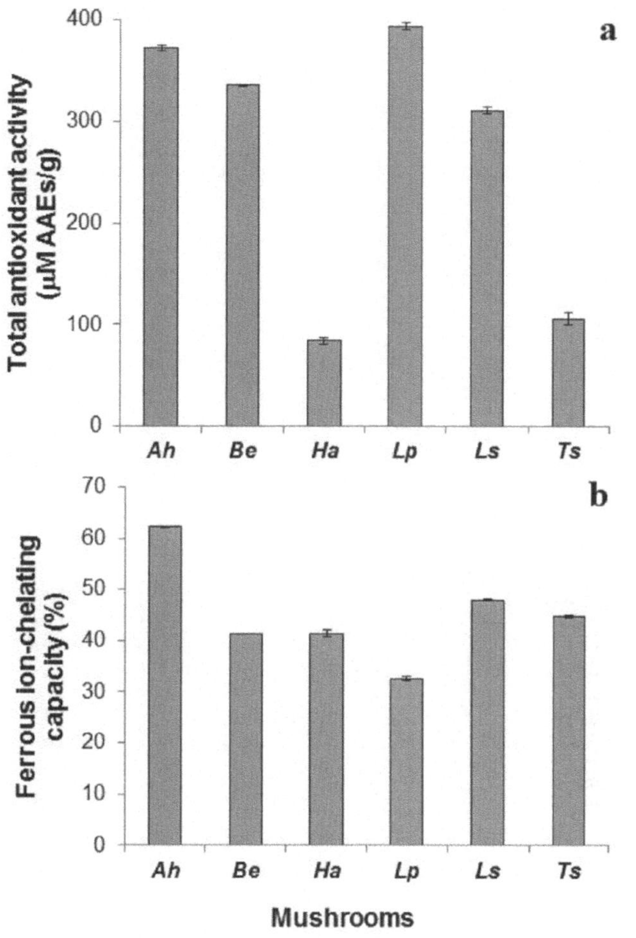

FIGURE 1.1 Total antioxidant activity (a) and ferrous ion-chelating capacity (b) of six mushrooms (n=3±SD) (Mushrooms: *Ah, Amanita hemibapha; Be, Boletus edulis; Ha, Hygrocybe alwisii; Lp, Lentinus polychrous; Ls, L. squarrosulus; Ts, Termitomyces schimperi*).

many diseases like cancer, diabetes, atherosclerosis, cirrhosis, and enhancement of aging (Halliwell and Gutteridge 1984). In addition, synthetic antioxidants are not valued due to their cancer-causing or cancer-promoting qualities (Botterweck et al. 2000). Thus, the natural antioxidants of edible mushroom origin will be of immense value. As mentioned earlier, mushrooms have already been demonstrated to possess cancer-preventing properties (stomach, esophagus, and lungs) in many countries (Canada. China, Japan, Korea, Russia, and the United States) (Ferreira et al. 2007).

The total antioxidant capacity of *A. hemibapha* was much higher than that of edible *Amanita* spp. found in scrub jungles on the west coast of India (Greeshma et al. 2018). The rest of the antioxidant activities of *A. hemibapha* were comparable with those of *Amanita* spp. All four antioxidant potentials of *L. squarrosulus* are very low compared with those in the same species grown in the scrub jungles of southwest India (Ghate and Sridhar 2017). All four antioxidant parameters of *T. schimperi* were extremely low compared to those in *Termitomyces clypeatus* grown in the scrub jungles in southwest India (Ghate and Sridhar 2017). Similarly, all activities except for DPPH radical-scavenging activity were very high in *T. umkowaan* grown in the scrub jungles of southwest India (Karun et al. 2016). Similarly, the free radical-scavenging activity of *L. squarrosulus* (Kerala) and *T. schimperi* (Himachal Pradesh) was higher than that of the same species assessed in this study (Puttaraju et al. 2006). Such differences in free

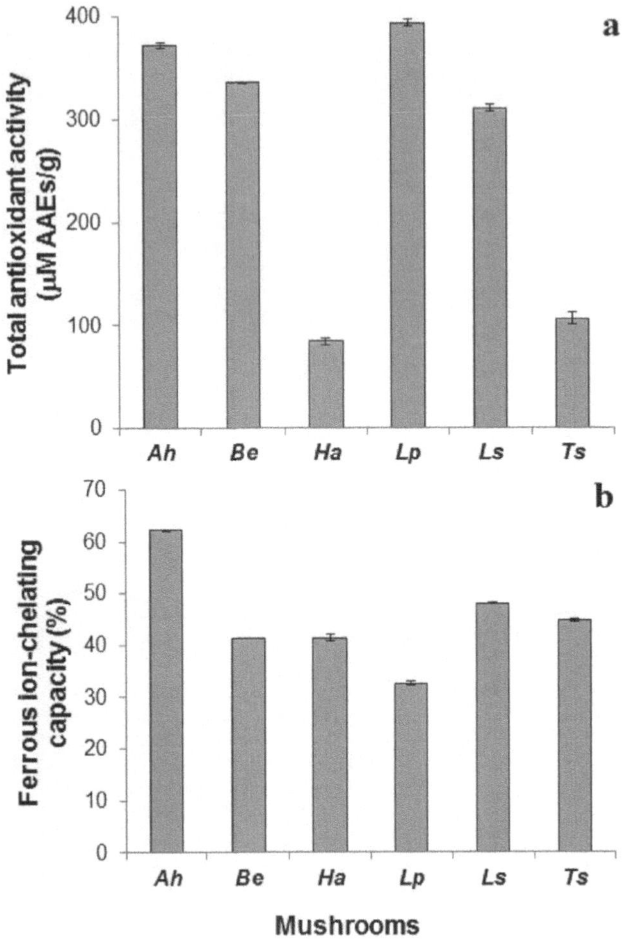

FIGURE 1.2 DPPH radical-scavenging activity (a) and reducing power (b) of the edible wild mushrooms (n=3±SD) (Mushrooms: *Ah, Amanita hemibapha; Be, Boletus edulis; Ha, Hygrocybe alwisii; Lp, Lentinus polychrous; Ls, L. squarrosulus; Ts, Termitomyces schimperi*).

radical-scavenging were also seen between *L. polychrous* in this study and the same species collected from Similipal (Orissa, India) (Tripathy et al. 2014). It is likely that such variation in antioxidant activity among the same mushroom species depends on its geographic location and the quantity of antioxidant compounds.

1.7 ANTIMICROBIAL ACTIVITIES

None of the bacteria or yeast was inhibited by the control cups containing DMSO. Crude mushroom extracts showed variable results against Gram-positive bacteria (Table 1.1). Six mushroom extracts showed inhibition of Gram-positive bacteria between 9.7 mm (*L. squarrosulus* against *S. aureus*) and 18 mm (*L. squarrosulus* against *B. subtilis*). The highest inhibition of *B. cereus* (15.3 mm), *S. aureus* (12 mm), and *S. pneumoniae* (15 mm) was shown by *S. schimperi*, *A. hemibapha/H. alwisii*, and *A. hemibapha*, respectively. None of the mushrooms showed greater inhibition of Gram-positive bacteria than the standard antibiotic vancomycin.

The impact of mushroom extracts against Gram-negative bacteria also showed variable results (Table 1.1). Mushroom extracts yielded 7–12 mm inhibition against *E. aerogenes* with the highest

TABLE 1.1
Antimicrobial Activity (inhibition diameter, mm) of Six Edible Wild Mushrooms (200 µg) and Standard Antibiotics (n=3±SD) (NI, no inhibition; –, not applied).

	Gram-positive bacteria				Gram-negative bacteria					Yeast
	Bacillus cereus	Bacillus subtilis	Staphylococcus aureus	Streptococcus pneumoniae	Enterobacter aerogenes	Escherichia coli	Haemophilus influenzae	Proteus vulgaris	Pseudomonas aeruginosa	Candida albicans
Amanita hemibapha	12.0±0.00	15.7±0.58	12.0±1.00	15.0±0.00	8.0±0.00	7.0±0.00	15.0±1.73	8.0±0.00	12.0±0.00	10.7±0.58
Boletus edulis	10.7±0.58	13.0±0.00	10.0±0.00	10.0±0.00	10.7±0.58	9.0±0.00	13.0±0.00	8.0±0.00	11.0±0.00	12.0±0.00
Hygrocybe alwisii	12.3±0.58	16.0±1.00	12.0±0.00	14.3±0.58	7.0±0.00	8.0±0.00	11.3±0.58	9.3±0.58	9.3±1.15	10.0±0.00
Lentinus polychrous	10.0±0.00	12.7±0.58	10.7±0.58	14.3±1.15	8.3±0.58	NI	13.3±0.58	NI	9.7±0.58	11.7±0.58
Lentinus squarrosulus	11.0±1.00	18.0±0.00	09.7±0.58	13.0±0.00	7.0±0.00	NI	13.0±0.00	6.7±0.58	10.3±1.53	12.7±1.15
Termitomyces schimperi	15.3±1.15	13.0±0.00	10.3±0.58	14.0±0.00	12.0±0.00	11.0±0.00	10.3±0.58	13.7±0.58	9.7±0.58	11.3±0.58
Vancomycin (30 µg)	19.7±0.58	25.7±0.58	22.0±0.00	26.0±0.00	20.0±0.00	22.0±0.00	21.7±0.58	18.7±0.58	31.0±0.00	–
Fluconazol (25 µg)	–	–	–	–	–	–	–	–	–	26.0±0.00

inhibition by *T. schimperi*; *E. coli* showed inhibition of 0–11 mm with the highest inhibition by *T. schimperi*; *H. influenzae* showed inhibition of 10.3–15 mm with the highest inhibition by *A. hemibapha*; *P. vulgaris* showed inhibition 0–13.7 mm with the highest inhibition by *T. schimperi*; and *P. aeruginosa* showed inhibition 9.3–12 mm with the highest inhibition by *B. edulis*. *Lentinus polychrous* did not inhibit *E. coli* or *P. vulgaris*, and *E. coli* was not inhibited by *L. squarrosulus*. Among the mushrooms, *A. hemibapha* showed the highest inhibition of *H. influenzae* (15 mm). None of the mushrooms showed stronger inhibitory effects against Gram-negative bacteria than the standard antibiotic vancomycin. All the test mushrooms inhibited the yeast *Candida albicans* at an inhibition range of 10–12.7 mm, with the highest inhibition by *L. squarrosulus*. However, its inhibitory effect was lower than that of the standard antibiotic fluconazole.

Similar to synthetic antioxidant agents, conventional antibiotics pose problems like multi-drug resistance in several threatening microbes (Alves et al. 2012). Fruiting body and mycelial extracts of more than 2,000 mushrooms possess diterpene pleuromutilin, which is useful for treating mycoplasma infections (Kavanagh et al. 1950; Sandven 2000). It seems to be that wild mushrooms are a possible source of antimicrobial compounds valuable for protecting human health against pathogenic microbes.

According to Lindequist et al. (2005), wild mushrooms must possess antimicrobial metabolites to survive in their natural habitat; thus wild mushrooms are considered rich sources of natural antibiotics. It is well known that the control of Gram-negative bacteria is more difficult than that of Gram-positive bacteria, especially by cell wall-inhibiting antibiotics. The current study showed that all mushrooms studied were capable of inhibiting one or the other type of bacteria, along with the yeast (*C. albicans*). Thus, the wild mushrooms studied possess broad-spectrum antibiotics. Although the inhibition of bacteria and yeast did not reach the levels of standard antibiotics, these wild mushrooms are capable of inhibiting a broad range of microbial pathogens.

Suay et al. (2000) reported that the extracts of more than 75% of polypore mushrooms showed antimicrobial properties. Iacobellis et al. (2005) found that terpenoids are responsible for the antibacterial action of mushrooms (*Cuminum cyminum* and *Carum carvi*). As described by Cowan (1999) the phenol and tannin contents of *Pleurotus ostreatus* showed antibacterial activity similar to that of medicinal plants, following a mechanism of action characterized by cell membrane lysis, protein synthesis inhibition, proteolytic enzymes activity, and microbial adhesin activation. Giri et al. (2012) suggested that the basidiocarps of *L. squarrosulus* are a source of pharmacologically active metabolites having diverse curative applications including inhibitory activity against bacteria (*Bacillus cereus*, *B. subtilis*, *Escherichia coli*, *Proteus vulgaris*, *Pseudomonas aeruginosa*, and *S. aureus*) and yeast (*C. albicans*), as seen in the present study. Another study by Alves et al. (2012) evaluated the effects of 13 wild mushrooms (collected from different habitats) against several multi-drug-resistant and pathogenic clinical isolates of bacteria in Portugal. They suggested that wild mushrooms like *Fistulina hepatica*, *Ramaria botrytis*, and *R. delicata* possessed the most prominent antimicrobial agents. A study by Santoyo et al. (2009) showed considerable antimicrobial activity in the aqueous extract of edible wild and cultivated mushrooms.

1.8 CONCLUSIONS AND OUTLOOK

Wild mushrooms in natural habitats are becoming more valuable in our diet than cultivated mushrooms for their nutritional, medicinal, and pharmacological potential. They are major sources of antioxidants as well as antimicrobial compounds. Six wild mushrooms of the foothills of the Western Ghats showed considerable antioxidant potential (total antioxidant, ferrous ion-chelation, DPPH radical-scavenging, and reducing activities). Comparison of wild mushrooms of the same species or allied species from the same or different geographic regions showed variation in antioxidant activity. It is expected that such variation depends on the geographic location of the specimens and the quantity of antioxidant compounds present in the mushrooms. Wild mushrooms are known to possess antibacterial and antifungal metabolites for their survival in natural habitats (Lindequist et al. 2005), making them valuable for human nutrition and for managing pathogenic microbes in the gastrointestinal tract. The present study showed that mushroom extracts inhibited several pathogenic bacteria and yeast. It is expected that the wild mushrooms studied possess broad-spectrum antibiotic capabilities. Despite the inhibition of pathogenic bacteria and yeast, the inhibitory effect was not as strong as that of standard antibiotics evaluated in the present study. There is a further need to assess these wild edible mushrooms for nutraceuticals, such as valuable metabolites for human nutrition and health promotion.

1.9 ACKNOWLEDGMENTS

The first author (VR) appreciates the facilities provided by the Department of Biosciences, Mangalore University. He is thankful to the research supervisor Prof. M. Rajashekhar and the Board of Research in Nuclear Sciences (BRNS), Bhabha Atomic Research Centre, Mumbai, India for funding this research.

REFERENCES

Alves, M.J., Ferreira, I.C.F.R., Martins, A. & Pintado, M. 2012. Antimicrobial activity of wild mushroom extracts against clinical isolates resistant to different antibiotics. *Journal of Applied Microbiology* 113: 466–475.

Barros, L., Baptista, P., Estevinho, L.M. & Ferreira, I.C.F.R. 2007b. Effect of fruiting body maturity stage on chemical composition and antimicrobial activity of *Lactarius* sp. mushrooms. *Journal of Agricultural and Food Chemistry* 55: 8766–8771.

Barros, L., Calhelha, R.C., Vaz, J.A., Ferreira, I.C.F.R. & Eslevinho, B.P. 2007a. Antimicrobial activity and bioactive compounds of Portuguese wild edible mushrooms. *European Food Research and Technology* 225: 151–156.

Barros, L., Cruz, T., Baptista, P., Estevinho, L.M. & Ferreira, I.C.F.R. 2008a. Wild and commercial mushrooms as source of nutrients and nutraceuticals. *Food Chemistry and Toxicology* 46: 2742–2747.

Barros, L., Venturini, B.A., Baptista, P., Estevinho, M. & Ferreira, I.C.F.R. 2008b. Chemical composition and biological properties of Portuguese wild mushrooms: a comprehensive study. *Journal of Agricultural and Food Chemistry* 56: 3856–3862.

Bauer, A.W., Kirby, W.M.M., Sherris, J.C. & Turck, M. 1966. Antibiotic susceptibility testing by a standardized single disk method. *American Journal of Clinical Pathology* 45: 493–496.

Boonsong, S., Kalypradit, W. & Wilaipun, P. 2016. Antioxidant activities of extracts from five edible mushrooms using different extractants. *Agriculture and Natural Resources* 50: 89–97.

Botterweck, A.A.M., Verhagen, H., Goldbohm, R.A., Kleinjans, J. & van den Brandt, P.A. 2000. Intake of butylated hydroxyanisole and butylated hydroxytoluene and stomach cancer risk: results from analyses in the Netherlands cohort study. *Food Chemistry and Toxicology* 38: 599–605.

Brizuela, M.A., Garcíam, L., Pérez, L. & Mansur, M. 1998. Basidiomicetos: nueva fuente de metabolites secundários. *Revista Iberoamericana de Micología* 15: 69–74.

Cowan, M.M. 1999. Plant products as antimicrobial agents. *Clinical Microbiology Reviews* 12: 564–582.

Ferreira, I.C.F.R., Baptista, P., Vilas-Boas, M. & Barros, L. 2007. Free-radical scavenging capacity and reducing power of wild edible mushrooms from northeast Portugal: individual cap and stipe activity. *Food Chemistry* 100: 1511–1515.

Ferreira, I.C.F.R., Barros, L. & Abreu, R.M.V. 2009. Antioxidant in wild mushrooms. *Current Medicinal Chemistry* 16: 1543–1560.

Gao, Y.H., Tang, W.B., Gao, H.E., Chan, E., Lan, J., et al. 2005. Antimicrobial activity of the medicinal mushroom *Ganoderma*. *Food Reviews International* 21: 211–229.

Ghate, S.D. & Sridhar, K.R. 2017. Bioactive potential of *Lentinus squarrosulus* and *Termitomyces clypeatus* from the southwestern region of India. *Indian Journal of Natural Products and Resources* 8: 120–131.

Giri, S., Biswas, G., Pradhan, P., Mandal, S.C. & Acharya, K. 2012. Antimicrobial activities of basidiocarps of wild edible mushrooms of West Bengal, India. *International Journal of PharmTech Research* 4: 1554–1560.

Greeshma, A.A., Sridhar, K.R., Pavithra, M. & Tomita-Yokotani, K. 2018. Bioactive potential of nonconventional edible wild mushroom *Amanita*. In: Gehlot, P. & Singh, J. (Eds.), *Fungi and Their Role in Sustainable Development: Current Perspectives*. Springer Nature, Singapore, pp. 719–738.

Halliwell, B. & Gutteridge, J.M.C. 1984. Oxygen toxicity, oxygen radicals, transition metals and disease. *Biochemistry Journal* 219: 1–4.

Heleno, S.A., Ferreira, I.C.F.R., Esteves, A.P., Ćirić, A., Glamočlija, J., et al. 2013. Antimicrobial and demelanizing activity of *Ganoderma lucidum* extract, p-hydroxybenzoic and cinnamic acids and their synthetic acetylated glucuronide methyl esters. *Food Chemistry and Toxicology* 58: 95–100.

Hsu, C.L., Chen, W., Weng, Y.M. & Tseng, C.Y. 2003. Chemical composition, physical properties and antioxidant activities of yam flours as affected by different drying methods. *Food Chemistry* 83: 85–92.

Iacobellis, N.S., LoCantore, P., Capasso, F. & Senatore, F. 2005. Antibacterial activity of *Cuminum cyminum* L. and *Carum carvi* L. essential oils. *Journal of Agricultural and Food Chemistry* 53: 57–61.

Karun, N.C., Sridhar, K.R., Niveditha, V.R. & Ghate, S.D. 2016. Bioactive potential of two wild edible mushrooms of the Western Ghats of India. In: Watson, R.R. & Preedy, V.R. (Eds.), *Fruits, Vegetables and Herbs: Bioactive Foods in Health Promotion*. Elsevier Inc., Oxford, pp. 344–362.

Kavanagh, F., Hervey, A. & Robbins, W.J. 1950. Antibiotic substances from basidiomycetes. VI. *Agrocybe dura*. *Proceedings of the National Academy of Sciences* 36: 102–106.

Khan, A.A., Gani, A., Ahmad, M., Masoodi, F.A., Amin, F. & Kousar, S. 2016. Mushroom varieties found in the Himalayan regions of India: Antioxidant, antimicrobial, and antiproliferative activities. *Food Science and Biotechnology* 25: 1095–1100.

Kües, U. & Liu, Y. 2000. Fruiting body production in basidiomycetes. *Applied Microbiology and Biotechnology* 54: 141–152.

Lindequist, U., Niedermeyer, T.H.J. & Julich, W.-D. 2005. The pharmacological potential of mushrooms—Review. *Evidence-Based Complementary and Alternative Medicine* 2: 285–299.

Mattila, P., Suonpää, K. & Piironen, V. 2000. Functional properties of edible mushrooms. *Journal of Nutrition* 16: 694–696.

Ostrovidov, G., Franck, P., Joseph, D., Martarello, L., Kirsch, G., et al. 2000. Screening of new antioxidant molecules using flow cytometry. *Journal of Medicinal Chemistry* 43: 1762–1769.

Oyaizu, M. 1986. Studies on products of the browning reaction. *Japanese Journal of Nutrition and Dietetics* 44:307–315.

Palacios, I., Lozano, M., Moro, C., A'Arrigo, M., Rostagno, M.A., et al. 2011. Antioxidant properties of phenolic compounds occurring in edible mushrooms. *Food Chemistry* 128: 674–678.

Prieto, P., Pineda, M. & Aguilar, M. 1999. Spectrophotometric quantitation of antioxidant capacity through the formation of a phosphomolybdenum complex: specific application to the determination of vitamin E. *Analytical Biochemistry* 269: 337–341.

Puttaraju, N.G., Venkateshaiah, S.U., Dharmesh, S.M., Dharmesh, S.M., Urs, S.M.N. & Somasundaram, R. 2006. Antioxidant activity of indigenous edible mushrooms. *Journal of Agricultural and Food Chemistry* 54: 9764–9772.

Ravikrishnan, V. 2019. *Nutritional and Bioactive Properties of Wild Mushrooms from the Foot-Hills of Agumbe Ghat, Karnataka*. Ph.D. Dissertation, Department of Biosciences, Mangalore University, Karnataka, India, p. 181.

Sánchez, C. 2017. Reactive oxygen species and antioxidant properties from mushrooms. *Synthetic and Systems Biotechnology* 2: 13–22.

Sandven, P. 2000. Epidemiology of candidemia. *Revista Iberoamericana de Micología* 17: 73–81.

Santoyo, S., Ramírez-Anguiano, A.C., Reglero, G. & Soler-Rivas, C. 2009. Improvement of the antimicrobial activity of edible mushroom extracts by inhibition of oxidative enzymes. *International Journal of Food Science and Technology* 44: 1057–1064.

Simic, M.G. 1988. Mechanisms of inhibition of free-radical processed in mutagenesis and carcinogenesis. *Mutation Research* 202: 377–386.

Singh, R.P., Murthy, C.K.N. & Jayaprakasha, G.K. 2002. Studies on antioxidant activity of pomegranate (*Punica granatum*) peel and seed extracts using *in vitro* methods. *Journal of Agricultural and Food Chemistry* 50: 81–86.

Suay, I., Arenal, F., Asenio, F., Basilio, A., Cabello, M.A., et al. 2000. Screening of basidiomycetes for antimicrobial activities. *Antonie van Leeuwenhoek* 78: 129–139.

Tripathy, S.S., Rajoriya, A. & Gupta, N. 2014. Wild mushrooms of Odisha: Prospective candidates of antioxidant sources. *Advances in Plant & Agriculture Research* 1: 129–133.

Turkoglu, A., Duru, M.E., Dogan, N.M., Kivrak, I. & Gezer, K. 2007. Antioxidant and antimicrobial activities of *Laetiporus sulphureus* (Bull.) Murrill. *Food Chemistry* 101: 267–273.

2 Distribution, Ecology, and Taxonomy of Medicinal Mushrooms

Steven L. Stephenson, Kamal Ch. Semwal

2.1 INTRODUCTION

The fungi (kingdom Fungi) represent a large and diverse group of organisms that are found in every ecosystem on the earth and are particularly common and conspicuous in forests. Microfungi, which do not produce a fruiting body easily visible to the naked eye, are the most widespread and abundant members of the kingdom Fungi, but macrofungi, which produce a visible fruiting body, are much better known. These macrofungi are often referred to as "mushrooms" in a very broad sense. Some mycologists restrict the use of the term mushroom to the gilled mushrooms (or agarics), but the concept used in this chapter reflects the title of the book of which the chapter is part.

2.2 TAXONOMY

2.2.1 Classification

In the traditional view of the kingdom Fungi, four phyla were recognized. These were the Chytridiomycota, Zygomycota, Ascomycota, and Basidiomycota (Alexopoulos et al. 1996). However, recent phylogenetic studies have revealed that certain fungi once placed in Chytridiomycota and Zygomycota should be removed and placed elsewhere, with some members of the latter now included in the "new" phylum Glomeromycota. In addition, numerous classes and orders within the traditional four phyla have been redefined (Hibbett et al. 2007). The taxonomic treatment used in this chapter considers these recent changes where appropriate, but its primary approach is general when referring to the members of a particular phylum, where the more traditional concept is followed.

2.2.2 Microfungi

The vast majority of all fungi are microfungi, including all members of the Chytridiomycota and Zygomycota as well as most of the Ascomycota and some of the Basidiomycota. Although microfungi, by definition, are too small to be readily observed in nature, it is certainly possible to notice their mycelia on certain substrates, where they have the appearance of a "fuzzy" external covering.

There is little question that some microfungi have medicinal properties. The discovery by Alexander Fleming in 1928 that a substance (later named penicillin) produced by a species of *Penicillium* inhibited the growth of certain types of bacteria is considered one of the most noteworthy events of the 20th century (Stephenson 2010). Some microfungi are mycoparasites, which means that they are parasites of mushrooms. This is the case for a number of species in the genus *Hypomyces*, including *Hypomyces lactifluorum*, which infects mushrooms

DOI: 10.1201/b23190-2

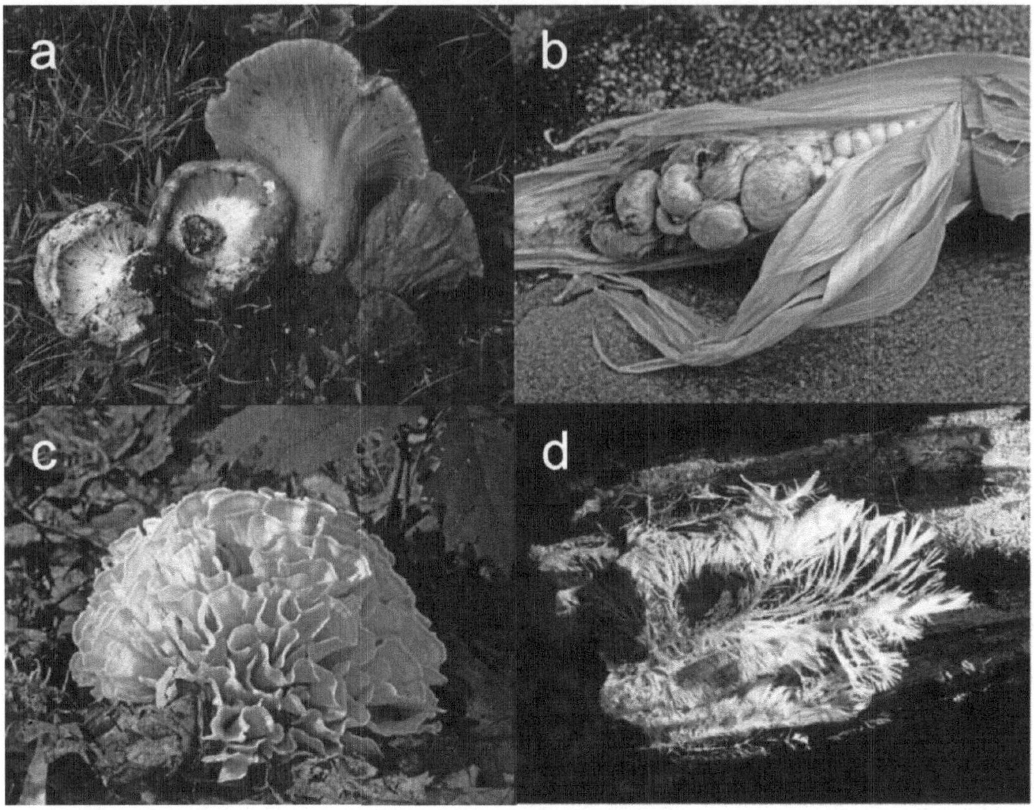

FIGURE 2.1 (a) *Hypomces lactiflorum*, a mushroom that is parasitic on other mushrooms [Emily Johnson]. (b) *Ustilago maydis*, a mushroom that forms gall-like growths on an ear of corn [Emily Johnson]. (c) *Sparrasis spatulata*, the *cauliflower mushroom [Emily Johnson].* (d) Mycelium of a mushroom that decays wood [Emily Johnson].

in the genera *Russula* and *Lactarius*. *Hypomyces lactifluorum* causes the host mushroom to become distorted and appear bright orange-red (Elliott and Stephenson 2018). Surprisingly, the resulting "lobster mushroom" (so named because of its color) is considered a choice edible (Figure 2.1A).

As a group, the microfungi are metabolically active organisms. Many produce secondary metabolites that have been exploited for applications in the pharmaceutical and food industries (Archer et al. 2008). *Penicillium* is not the only microfungus of medicinal importance, and microfungi of this and other genera play important roles in the production of certain types of food, including a variety of cheeses. The most famous are the blue cheeses such as Roquefort, in which *Penicillium roquefori* is the principal microfungus involved (Alexopoulos et al. 1996). Although microfungi constitute a fascinating and understudied group of organisms, the primary focus of this chapter is on macrofungi (mushrooms).

2.2.3 Ascomycetes

Relatively few members of the Ascomycota (or "Ascomycetes") produce fruiting bodies large enough to be considered mushrooms. The most conspicuous examples are the morels and truffles. The truffles are an exceedingly valuable group of hypogeous (i.e., they produce subterranean fruiting bodies) mushrooms characterized by unique flavors and high nutritional value. In addition to being used for food, morels are known to have anti-tumor, antitoxic, antibacterial, and anti-inflammatory properties (Tietel and Masaphy 2018). The best-known truffles are members of the genus *Tuber*, which are primarily found in Europe, North America, and Asia. *Tuber* belongs to a different clade from the "desert truffles" (*Terfezia* and *Tirmania*) which are primarily found in arid and semi-arid areas of northern Africa and the Middle East. The medicinal properties of desert truffles have been described by Mandeel and Al-Laith (2007), Shavit and Shavit (2014), and Veeraraghavan et al. (2021)

Morels are greatly valued as food and have the advantage of occurring above ground, where they can be more easily collected than truffles. In addition, morels are not likely to be confused with other, non-edible mushrooms. Long thought to have health-related benefits, recent studies have demonstrated that morels have anti-oxidative and anti-inflammatory properties (Tietel and Masaphy 2018). The biology of morels is not completely known, but some species are thought to form mycorrhizal associations. Other ascomycetes well known for their medicinal properties are members of the genera *Ophiocordyceps* and *Tolypocladium* (the caterpillar fungi). These mushrooms have long been used in traditional medicine and gained notoriety some years ago when a number of female Chinese distance runners attributed their improved performances, which included world records at that time, to a supplement called CordyMax made from *Ophiocordyceps sinensis* (Stephenson 2010).

2.2.4 Basidiomycetes

The majority of mushrooms are produced by members of the Basidiomycota (commonly referred to as "basidiomycetes"). These are the mushrooms frequently encountered in nature. Mushrooms produced by basidiomycetes are usually categorized on the basis of their overall morphology, which may or may not correspond with their phylogenetic relationships, as noted herein. Some of the best-known categories are agarics, jelly fungi, boletes, chanterelles, coral fungi, polypores, tooth fungi, puffballs, and stinkhorns. These categories represent starting points for identification in field guides and other similar books on mushrooms (e.g., Lincoff 1989; Desjardin and Hemmes 2002; Hall et al. 2003; Halling and Mueller 2004; Young 2004; Miller and Miller 2006; Roody 2003; Elliott and Stephenson 2018). Mushrooms with real or reputed medicinal properties are not confined to just one category but instead occur across a wide range of categories, including agarics (*Agaricus subrufescens* [almond mushroom]), coral fungi (*Artomyces pyxidatus* [crown-tipped coral mushroom]), jelly fungi (*Tremella fuciformis* [snow mushroom]), puffballs (*Calvatia gigantea* [giant puffball]), stinkhorns (*Phallus impudicus* [common stinkhorn]), and tooth fungi (*Hericium erinaceus* [lion's mane]). At least some medicinal mushrooms can be found anywhere in the world. For example, *Tremella fuciformis*, despite its common name, occurs in the subtropics and tropics, and the same is true for numerous polypores.

The Basidiomycota is made up of three major clades, each of which is recognized as a subphylum. These are the Agariomycotina, the Pucciniomycotina, and the Ustilaginomycotina (Hibbett et al. 2007). The Agaricomycotina is the largest subphylum and contains most of the mushrooms considered to have medicinal properties. The Pucciniomycotina (commonly referred to as rusts) and the Ustilaginomyota (commonly known as smuts) are plant parasites of considerable economic but little medicinal importance. One of the few exceptions is *Ustilago maydis* (corn smut), which infects the kernels on an ear of corn, producing a gall-like growth (Figure 2.1B). These structures, highly valued as food in Latin America, are suspected to possess some medicinal properties (Valdez-Morales et al. 2016; Aydoğdu and Gölükçü 2017).

Although various systems of classification have been proposed for the subphylum Agaricomycotina in an effort to group phylogenetically related forms together (Hibbett et al. 2014), a practical approach is to consider the assemblages of taxa that have been placed in the various orders of the most important (with respect to medicinal mushrooms) class, Agaricomycetes. Among the orders with particular relevance to this chapter are the Agaricales (the agarics) and the Polyporales (the polypores).

2.2.5 Polypores

Numerous medicinal mushrooms are polypores, while other examples are members of different orders of the Agaricomycetes or, in some instances, entirely different groups of fungi. This is the case for morels, truffles, and caterpillar fungi (*Cordyceps*) and members of related genera, including species such as *Tolypocladium inflatum*, which is a source of the drug cyclosporine (used to prevent organ rejection after transplants [Patocka et al. 2021]). All of these are ascomycetes.

The order Polyporales is well accepted as a distinct but related phylogenetic assemblage (clade) of the class Agaricomycetes. Justo et al. (2017) proposed a family-level classification that recognized 18 different clades, which were assigned family names. Three of these were described as new. Among the families with mushrooms considered to have medicinal properties are the Fomitopsidaceae (*Fomes fomentarius* [horsehoof mushroom], Grifolaceae (*Grifola frondosa* [hen of the woods]), Polyporaceae (*Ganoderma applanata* [artist's conk], *G. tsugae* [hemlock polypore], *Pycnoporus cinnabarinus* [cinnabar polypore], and *Trametes versicolor* (turkey tail)], and Sparassidaceae (*Sparassis spathulata* [cauliflower mushroom, Figure 2.1C]).

2.3 ECOLOGY OF FUNGI

2.3.1 Roles in Terrestrial Ecosystems

Mushrooms have three major roles in terrestrial ecosystems. Numerous mushrooms are saprotrophs and break down (decompose) various types of organic matter, including dead logs and stumps, dead leaves, twigs, cones, old fruits and inflorescences, and dung, as well other fungi. In dead logs, the mycelium of the saprotrophic fungus is often readily apparent (Figure 2.1D). By breaking down this organic material, fungi release the nutrients that other organisms (especially plants and ultimately animals) need to survive. As a result of their role as decomposers, mushrooms are nature's major recyclers. Some mycologists have suggested that this function could be recognized as separate and distinct from their role as decomposers, but the two are so closely linked that they are essentially inseparable.

2.3.2 Mycorrhizal Mushrooms

Some mushrooms are symbionts and form mycorrhizal associations with plants. These associations are of two basic types—vesicular arbuscular mycorrhizae (often abbreviated as VAM) and ectomycorrhizae. In the former, some fungal hyphae penetrate the cell walls of a root of the host plant and form distinctive intracellular structures called vesicles and arbuscules. It is estimated that 85% of all vascular plants form VAM associations with microfungi in the phylum Glomeromycota. Ectomycorrhizal associations are much less common than VAM associations but do involve mushrooms, mostly basidiomycetes but also a few ascomycetes. In an ectomycorrhizal association, the fungal hyphae do not penetrate the individual cells of a root but do form a layer (called a sheath or mantle) around the outside (Figure 2.2A). Both types of mycorrhizal associations are highly beneficial to the host plant because they increase the effective surface area of the root, thus enhancing the uptake of water and nutrients. In return, the fungal partner obtains photosynthates from the plant. Trees are among the major groups of plants that form ectomycorrhizal associations.

Basidiomycete mushrooms form ectomycorrhizal associations with the members of a number of plant families, such as the Betulaceae (birches [*Betula*] and alders [*Alnus*]), Fagaceae (oaks [*Quercus*],

FIGURE 2.2 (a) Ectomycorrhizal roots of the American chestnut tree (*Castanea dentata*). The fungi present on the roots (identified from DNA sequences) included species of *Cortinarius*, *Lactarius*, and *Russula* [*Mourad Ben Hassine Ben Ali*]. (b) *Amanita hemibapha*, an ectomycorrhizal mushroom found in the tropical forests of Southeast Asia [Dennis Desjardin]. (c) *Laetiporus sulphureus*, a saprotrophic mushroom that decomposes wood [Emily Johnson]. (d) *Gymnopus dryophilus*, a saprotrophic mushroom that decomposes dead leaves [Todd Elliott].

beeches [*Fagus*], chestnuts [*Castanea*], lithocarps [*Lithocarpus*], chinkapins [*Castanopsis*]), and Pinaceae (pines [*Pinus*], spruces [*Picea*], and firs [*Abies*]) in the Northern Hemisphere, as well as the Leptospermoideae (tea trees [*Leptospermum*], eucalyps [*Eucalypus*]), and Nothofagaceae (southern beeches [*Nothofagus*]) in the Southern Hemisphere (Warcup 1980; Steidinger et al. 2019). These mushrooms do not occur in regions of the world where their host trees are lacking but can be introduced to new areas along with their host trees. Such is the case in New Zealand and Australia, where *Amanita muscaria* (fly agaric), formerly limited to the Northern Hemisphere, now commonly occurs (Robinson 2010).

Ectomycorrhizal mushrooms are exceedingly common in the boreal forests and mid-latitude temperate forests of the Northern Hemisphere, where their host trees are often dominant. Examples include the forests that occur over northern portions of North America, Europe, and Asia. At comparable latitudes in the Southern Hemisphere, relatively few hosts (e.g., *Eucalyptus* and *Nothofagus*) are present. Although ectomycorrhiza-forming trees do occur in tropical forests (Buyck et al. 1996; Henkel et al. 2002; Smith et al. 2011), the majority of trees exhibit VAM relationships with fungi. Consequently, ectomycorrhizal mushrooms are much less common in some tropical forests (Figure 2.2B). The largest mushrooms found are mostly saprotrophs, including species that decompose wood and other types of dead plant material. As such, the general pattern is for saprotrophs to be common in all types of forests, whereas ectomycorrhizal mushrooms are limited largely to temperate and boreal forests. Among the most common and widespread genera are *Amanita*, *Boletus*, *Cortinarius*, *Hebeloma*, *Laccaria*, *Lactarius*, *Rhizopogon*, *Russula*, *Scleroderma*, *Suillus*, and *Tricholoma*.

In the treeless areas of both the Arctic and the alpine tundra in the Northern Hemisphere, ectomycorrhizal hosts are often dwarf willows. In some localities, *Dryas octopetala* (mountain avens), an unusual ectomycorrhizal herbaceous plant and a member of the Rosaceae, is quite common (Ryberg et al. 2009). The assemblages of ectomycorrhizal mushrooms associated with these communities are surprisingly diverse, including species in such genera as *Cortinarius*, *Hebeloma*, *Laccaria*, *Lactarius*, and *Russula*. Because of the severe climatic conditions, many of these species do not produce fruiting bodies most years. Based on DNA extracted from the root tips of the potential host plants, there are fungal sequences present that cannot be identified at the level of species, which suggests that the diversity of mushrooms is even higher than field observations would indicate.

The dominant plant communities of the Arctic and alpine tundra are generally low-growing, with many of the dwarf willows no more than ten centimeters tall. The ectomycorrhizal mushrooms associated with these willows can be nearly as tall as or even taller than their host plants. As such, the general appearance of these communities is that of "giant mushrooms" in a miniature forest. This was noted for *Boletus edulis* (king bolete) in the Swiss Alps by Treindl and Leuchtmann (2019) and has been observed for other species in areas of subalpine tundra in central Alaska in the United States (Stephenson, per. obs.).

Most ectomycorrhizal trees display a low level of specificity for the mushrooms with which they can form associations. However, there are exceptions. For example, almost all members of the group of mushrooms that includes the genera *Suillus* and *Rhizopogon* are invariably associated with trees belonging to the family Pinaceae.

2.3.3 Mushrooms as Saprotrophs

Saprotrophic mushrooms exhibit a simply incredible diversity of forms, occur in every imaginable environment, and are collectively capable of utilizing virtually every type of substrate derived from plants and other fungi. However, any given mushroom is limited to certain types of substrate or is almost invariably associated with a particular habitat (Ye et al. 2019). Noteworthy in this respect are *Ganoderma tsugae* (hemlock polypore), which is almost entirely restricted to the decaying wood of *Tsuga* (hemlocks); *Xylaria magnoliae* (magnolia cone xylaria), which occurs only on old inflorescences of magnolia (*Magnolia*); and *Xylaria liquidambaeis* (*Liquidamber xylaria*), which is found exclusively on old, decaying fruits of the sweetgum tree (*Liquidamber styraciflua*).

Some mushrooms decompose only wood, and these wood-decay mushrooms are not the same species associated with the decomposition of dead leaves and other types of non-woody plant material. It has long been known that when a log undergoes decomposition, the mushrooms associated with the early stages of the process are not necessarily the same as those associated with later stages. In other words, there is a successional pattern that must be taken into consideration when searching for a particular species of mushroom.

Mushrooms that decompose wood are often assigned to two categories based on whether or not they can degrade both cellulose and lignin or just cellulose alone. The members of the first category (called white-rot mushrooms) produce the enzymes necessary to degrade both cellulose and lignin. The material left behind has a somewhat fibrous appearance and is very pale white. In contrast, wood degraded by members of the second group (called brown-rot mushrooms) is brown in color and tends to be broken up into somewhat cubical fragments that quickly disintegrate into a powdery brown residue. Brown-rot mushrooms are especially common on conifers and are most abundant in cool temperate and boreal regions of the Northern Hemisphere, though few are found in the tropics or in temperate regions of the Southern Hemisphere. In marked contrast, white-rot mushrooms are found in all parts of the world.

Among the more common examples of brown-rot mushrooms are *Laetiporus sulphureus* (chicken of the woods), *Piptoporus betulinus* (birch polypore), *Fomitopsis pinicola* (red-belted polypore), and *Schizophyllum commune* (common split-gill mushroom). Examples of white-rot mushrooms are *Trametes versicolor* (turkey tail mushroom), *Pleurotus ostreatus* (oyster mushroom), *Lentinula edodes* (shiitake), and *Phellinus pini* (red root rot). There are many more species of white-rot mushrooms than of brown-rot mushrooms.

The distinction between saprotrophs and parasites is not clearly drawn for some of the mushrooms that decompose wood, since a particular mushroom can function as a parasite while its host (usually a tree) is still living and switch to being a saprotroph after the host dies. Moreover, the dead wood of as still-living tree can be invaded by various wood-decomposing species, which may live in the tree for years (Stephenson 2010). Numerous mushrooms fall into this parasite/saprotroph category, including *Laetiporus sulphureus* (Figure 2.2C), *Grifola frondosa*, *Polyporus squamosus* (dryad's saddle), *Armillaria gallica* (honey mushroom), *Phellinus robiniae* (cracked-cap polypore), and *Piptoporus betulinus.*

Habitat descriptions of mushrooms often make little distinction between decorticated wood and wood with the bark still intact when referring to the usual substrate for a particular species. However, the chemical and physical properties of wood and the substrate complex consisting of wood and bark are quite different. For example, differences exist with respect to the predominant structural polysaccharides (i.e., lignin in wood and suberin in bark), and bark typically contains higher concentrations of allelopathic compounds (e.g., tannins and phenols) than wood (Hudson 1986). In addition, the very presence of bark (which usually indicates a relatively early stage of decomposition) affects the microclimate (e.g., aeration, temperature, and moisture content) of the underlying wood (Swift and Boddy 1984). These factors undoubtedly influence which mushrooms can colonize and successfully exploit a particular woody substrate. For example, various pyrenomycetes (a group of ascomycetes) almost invariably occur on bark and not wood, as was reported by Vasilyeva and Stephenson (2004, 2005) for temperate deciduous forests of the Great Smoky Mountains National Park in the United States.

In temperate deciduous forests, most of these escape notice as a result of their small size, but the number of species present is very high (Alanbagi et al. 2019). Common and relatively conspicuous examples in the forests of eastern North America are *Marasmius siccus* (orange pinwheel marasmius) and *Gymnopus dryophilus* (oak-loving Gymnopus, Figure 2.2D), both of which are found on dead leaves or are associated with the underlying humus mostly derived from dead leaves. Other species are found on small woody twigs on the forest floor. Two of the more colorful examples are *Sarcoscypha austriaca* (scarlet cup) and *Microstoma floccosum* (shaggy scarlet cup). A few mushrooms associated with forest floor litter are restricted to certain substrates, including *Mycena crocea* (walnut Mycena), which occurs only on nuts produced by members of the family Juglandaceae. Other genera represented in the assemblage of mushrooms associated with dead leaves and woody debris on the forest floor include *Coprinellus*, *Lepista*, *Marasmiellus*, *Megacollybia*, and *Pluteus*.

2.3.4 Mushrooms on Dung

Herbivore dung is undigested plant material to which animal waste products have been added. The mushrooms that decompose dung (referred to as "coprophilous") represent a specialized assemblage that includes members of the Zygomycota, Ascomycota, and Basidiomycota. Many of these are generalists and can be found on any type of herbivore dung, but others are restricted largely or completely to certain types of dung. The vast majority of fungi associated with dung is microfungi and not species that produce fruiting bodies large enough to be considered mushrooms. However, there are some true mushrooms that can be found on dung. One rather infamous group consists of members of the genus *Psilocybe* (generally referred to as the psychedelic mushrooms or the "magic mushrooms"). These mushrooms (e.g., *Psilocybe cubensis* [magic mushroom]) contain substances (psilocybin and psilocin) that, when consumed by humans, induce an altered state of consciousness during which there are false perceptions of reality. There is evidence to suggest that these psychedelic mushrooms have been used in religious ceremonies for more than 2,000 years. *Psilocybe* and other mushrooms that contain psychoactive compounds would not be considered medicinal mushrooms, although studies have shown that small doses of psilocybin can have a positive effect on the mood of individuals suffering from some mental disorders (Daniel and Haberman 2018).

Other mushrooms found on dung include *Panaeolus antillarum* (white dung panaeolus), which almost invariably occurs on horse dung, *Cyathus stercoreus* (dung-loving bird's nest mushroom), a number of other *Panaeolus* species, and species from other genera such as *Coprinopsis*.

Parasitic mushrooms produce mycelia that invade and sometimes kill living plants and, in some instances, other fungi. As noted previously, some species can function as parasites after invading a host plant but become saprophytes when the host dies. The mushrooms that decompose trees are the most conspicuous; the appearance of their fruiting indicates that they are present in the host. A number of polypores falls into this category.

The various types of mushrooms mentioned here display varying levels of specificity for the substrates and forests in which they occur. Several publications have provided fairly comprehensive information on a particular forest or region. For example, Binion et al. (2008) described mushrooms typically associated with oaks in the eastern United States, while Bossenmaier (1997) did the same for the mushrooms found in the boreal forests of Canada. Numerous publications are available on specific groups of mushrooms for more limited localities.

The spores produced by mushrooms tend to be quite small, with those of many species no more than about 10 μm in diameter. Any object this size has the potential of being carried some distance by the wind, which suggests that long-distance dispersal is possible. Indeed, some commonly encountered species of microfungi do appear to be almost ubiquitous, as they turn up virtually anywhere in the world where appropriate sampling is carried out. However, this is not usually the case for mushrooms. Some species appear to be restricted to certain geographical regions of the world. For example, the assemblage of mushrooms associated with the forests of Western Europe is different from that associated with the forests of eastern North America, although there are some species in common. For ectomycorrhizal fungi, the obvious explanation is that suitable host trees occur in one region but are absent from another.

2.4 MEDICINAL MUSHROOMS

At this point, questions might be asked as to what constitutes a medicinal mushroom. The usual definition given is that any mushroom considered useful in treating and/or preventing some disease or for promoting general health (Mwangi et al. 2022) is a medicinal mushroom. The use of mushrooms for medicinal purposes dates back to ancient Japan, China, and other countries in eastern and southeastern Asia (Stephenson 2010; Rogers 2012). Wasson and Wasson (1957) classified human societies as either mycophobic ("fungus fearing") or mycophilic ("fungus loving") depending on the presence or absence of a general prejudice against collecting and consuming fungi. Clearly, this same concept could be extended to the recognition and use of medicinal mushrooms.

The beneficial properties of medicinal mushrooms appear to be largely due to the presence of biologically active polysaccharides, although other substances such as proteins and glycoproteins are also present. The metabolic mechanism that accounts for their activity is incompletely understood and warrants additional study. The anti-inflammatory and immune-boosting properties of medicinal fungi are of considerable value, and the range of conditions they have been used to treat include diabetes, arthritis, high blood press, various respiratory diseases, gastrointestinal issues, and nervous disorders. However, their anticancer activity could be of even greater importance. Among the more common mushrooms that have demonstrated this type of activity are members of the genus *Ganoderma* (e.g., *G. applanatum*, *G. lucidum* [reishi], and *G. tsugae*), *Grifola frondosa* (hen of the woods), *Sparassis spathulata*, *Schizophyllum commune* (split-gill mushroom), and *Trametes versicolor* (turkey tail).

The use of wild mushrooms both as food and as medicine is most prominent in Asia and relatively common over much of Europe (Tripathi et al. 2017). However, wild mushrooms—medicinal or otherwise—receive very little attention in North America, especially in the United States. Nevertheless, there are some excellent books on the medicinal mushrooms found in this region of the world (e.g., Rogers 2012; Hobbs 2021). As a general rule, Australia is a mycophobic county, despite the fact that much of the population has a European origin. Interestingly, available evidence suggests that for the indigenous peoples of Australia and neighboring New Zealand, the medicinal use of mushrooms is exceedingly limited. For example, Fuller et al. (2005) attempted to document the extent of their use by the Maori, but find evidence of only six species receiving some use, and one of these was a lichenized fungus (*Usnea*). The others were one agaric (*Agrocybe parasitica*), one polypore (*Laetiporus portentosus*), one jelly fungus (*Auricularia cornera*), one puffball (*Calvatia*), and one cordyceps (*Cordyceps robertii*). These five species were used for conditions such as fever, stomach bleeding, and to treat individuals poisoned by toxic plants.

Information on the medicinal mushrooms used historically by the indigenous Aboriginal people of Australia is very limited (Kalotas 1996). Because of the harsh environment in which they lived, virtually anything edible was used as food, and this certainly include mushrooms. However, there are only a few well-established reports on medicinal mushrooms. Among the few they are known to have used are *Pycnoporus sanguineus* (orange shelf fungus), *Pisolithus* sp. (the horse dung fungus), and *Cyttaria* (beech orange) (Miller-Maiden 1889). Interestingly, one of the more widely consumed mushrooms was *Polyporus mylittae* (native bread), although it was not the mushroom itself that was the primary target but rather the large, underground sclerotium produced by this species. A similar usage is seen for a mushroom that occurs in the southeastern United States. *Wolfiporia extensa* (Indian bread) also produces a large (reported to weigh up to 18 km), usually underground sclerotium. This sclerotium, considered able to boost the immune system, was an important food item for Native Americans in the region.

Efforts to carry out surveys of the mushrooms of a particular locality or region are not uncommon (e.g., Gates et al. 2011; Andrew et al. 2013; Semwal et al. 2018; Onduso et al. 2020), but surveys of medicinal mushrooms alone appear to be relatively rare. However, there is little doubt that these mushrooms are widespread. However, knowledge of medicinal mushrooms and their subsequent uses varies considerably from place to place. The interface between traditional medicine and modern medicine remains incomplete, and the potential usefulness of medicinal mushrooms has not been exploited to its full extent.

2.5 SUMMARY

Mushrooms and other fungi are an essential but often underappreciated component of the earth's biota. They play absolutely essential roles in nature, especially with respect to decomposition and nutrient cycling. Mushrooms display a wide range of diversity with respect to their size, shape, color, biology, and ecological distribution. It has long been known that some mushrooms produce substances that can be used to treat and/or prevent certain diseases or to promote general health in humans. The other chapters in this book provide additional information relating to these medicinal mushrooms.

2.6 ACKNOWLEDGMENTS

Appreciation is extended to Mourad Ben Hassine Ben Ali, Dennis Desjardin, Todd Elliott, and Emily Johnson for providing the images used in Figures 2.1 and 2.2. Thanks are extended to Heino Lepp and Todd Elliott for reviewing an earlier version of this manuscript.

REFERENCES

Alanbagi, R.A., F.F. Ashuwzili & S.L. Stephenson. 2019. Effect of prescribed burning on the taxonomic and functional diversity of litter-associated macrofungi in the Pea Ridge National Military Park, Arkansas. *Current Research in Environmental & Applied Mycology* 5: 25–35.

Alexopoulos, C.J., C.W. Mims & M. Blackwell. 1996. *Introductory Mycology*, 4th ed. John Wiley & Sons, New York.

Andrew, E.E., T.R. Kinge, E.M. Tabi, N. Thiobal & A.M. Mih. 2013. Diversity and distribution of macrofungi (mushrooms) in the Mount Cameroon region. *Journal of Ecology and the Natural Environment* 5: 318–334.

Archer, D.B., I.F. Connerton & D.A. MacKenzie. 2008. Filamentous fungi for the production of food additives and processing aids. *Advances in Biochemical Engineering/Biotechnology* 111: 99–147.

Aydoğdu, M. & M. Gölükçü. 2017. Nutritional value of huitlacoche, maize mushroom caused by *Ustilago maydis*. *Food Science and Technology Campinas* 37: 531–535.

Binion, D.E., S.L. Stephenson, W.C. Roody, H.H. Burdsall, O.K. Miller, Jr & L.N. Vasilyeva. 2008. *Macrofungi Associated with Oaks of Eastern North America*. West Virginia University Press, Morgantown, WV.

Bossenmaier, E.F. 1997. *Mushrooms of the Boreal Forest*. University Extension Press, University of Saskatchewan, Saskatoon, Canada.

Buyck, B., D. Thoen & R. Watling. 1996. Ectomycorrhizal fungi of the Guinea-Congo region. *Proceedings of the Royal Society of Edinburgh, Section B: Biological Sciences* 104: 313–333.

Daniel, J. & M. Haberman. 2018. Clinical potential of psilocybin as a treatment for mental health conditions. *Mental Health Clinician* 7: 24–28.

Desjardin, D.E. & D.E. Hemmes. 2002. *Mushrooms of Hawai'i: An Identification Guide*. Ten Speed Press, Berkeley, CA.

Elliott, T.F. & S.L. Stephenson. 2018. *Mushrooms of the Southeastern United States*. Timber Press, Portland, OR.

Fuller, R., P. Buchanan & A. Roberts. 2005. Medicinal uses of fungi by New Zealand Maori people. *International Journal of Medicinal Mushrooms* 7: 402.

Gates, G.M., C. Mohammed, T. Wardlaw, D.A. Ratkowsky & N.J. Davidson. 2011. The ecology and diversity of wood-inhabiting macrofungi in a native Eucalyptus obliqua forest of southern Tasmania, Australia. *Fungal Ecology* 4: 56–67.

Hall, I., S.L. Stephenson, P.K. Buchanan, W. Yun & A.L.J. Cole. 2003. *Edible and Poisonous Mushrooms of the World*. Timber Press, Portland, OR.

Halling, R.E. & G.M. Mueller. 2004. *Common Mushrooms of the Talamanca Mountains, Costa Rica*. New York Botanical Garden, New York.

Henkel, T.W., J. Terborgh & R.J. Vilgalys. 2002. Ectomycorrhizal fungi and their leguminous hosts in the Pakaraima Mountains of Guyana. *Mycological Research* 106: 515–531.

Hibbett, D.S., R. Bauer, M. Binder, et al. 2014. Agaricomycetes. In McLaughlin, D.J. & J.W. Spatafora (eds.), *The Mycota. Part A. Systematics and Evolution*, 2nd ed., Vol. II, pp. 373–429. Springer-Verlag, Berlin/Heidelberg, Germany.

Hibbett, D.S., M. Binder, J.F. Bischoff, et al. 2007. A higher level phylogenetic classification of the Fungi. *Mycological Research* 111: 509–547.

Hilszczanska, D. 2012. Medicinal properties of macrofungi. *Lesne Prace Bad (Forest Research Papers)* 73: 347–353.

Hobbs, C. 2021. *Christopher Hobb's Medicinal Mushrooms: The Essential Guide: Boost Immunity, Improve Memory, Fight Cancer, Stop Infection, and Expand Your Consciousness*. Storey Publishing, North Adams, MA.

Hudson, H.J. 1986. *Fungal Biology*. Edward Arnold, Baltimore.

Justo, A., O. Miettinen, D. Floudas, et al. 2017. A revised family-level classification of the Polyporales (Basidiomycota). *Fungal Biology* 121: 798–824.

Kalotas, A.C. 1996, Aboriginal knowledge and use of fungi. In Mallett, K. & C. Grgurinovic (eds.), *Fungi of Australia Volume 1B*, pp. 268–295. ABRS/CSIRO, Canberra.

Lincoff, G.H. 1989. *The Audubon Society Field Guide to North American Mushrooms*. Knopf, New York.

Mandeel, Q.A. & A.A.A. Al-Laith. 2007. Ethnomycological aspects of the desert truffle among native Bahraini and non-Bahraini people of the Kingdom of Bahrain. *Journal of Ethnopharmacology* 110: 118–129.

Miller, O.K., Jr. & H.H. Miller. 2006. *North American Mushrooms: A Field Guide to Edible and Inedible Species*. Falcon-Guide, Helena, MT.

Miller-Maiden, J.H. 1889. *The Useful Native Plants of Australia (Including Tasmania)*. Turner and Henderson, Sydney.

Mwangi, R.W., J.M. Macharia, I.N. Wagara & R.L. Bence. 2022. The antioxidant potential of different edible and medicinal mushrooms. *Biomedicine and Pharmacotherapy* 147. https://doi.org/10.1016/j.biopha.2022,112621

Onduso, F.N., N. Alshammari & S.L. Stephenson. 2020. An assemblage of microfungi associated with a boreal forest community in northern Manitoba, Canada. *Species* 21: 330–336.

Patocka, J., E. Nepovimova, K. Kuca & W. Wu. 2021. Cyclosporine A: Chemistry and toxicity—a review. *Current Medicinal Chemistry* 28: 3925–3934.

Robinson, R. 2010. First record of *Amanita muscaria* in Western Australia. *Australasian Mycologist* 29: 4–6.

Rogers, R. 2012. *The Fungal Pharmacy: The Complete Guide to Medicinal Mushrooms and Lichens of North America*. North Atlantic Books, Berkeley, CA.

Roody, W.C. 2003. *Mushrooms of West Virginia and the Central Appalachians*. University Press of Kentucky, Lexington, KY.

Ryberg, M., E. Larsson & U. Molau. 2009. Ectomycorrhizal diversity on Dryas octopetala and Salix reticulata in an alpine cliff ecosystem. *Arctic, Antarctic, and Alpine Research* 41: 506–514.

Semwal, K.C., V.K. Bhatt & S.L. Stephenson. 2018. A survey of macrofungal diversity in the Bharsar Region, Uttarakhand Himalaya, India. *Journal of Asia-Pacific Biodiversity* 11(4): 560–565. https://doi.org/10.1016/j.japb.2018.09.006.

Shavit, E. & E. Shavit. 2014. The medicinal value of desert truffles. In Kagan-Zur, V., N. Roth-Bejerano, Y. Sitrit & A. Morte (eds.), *Desert Truffles. Soil Biology Volume 38*, pp. 323–340. Springer, Berlin, Heidelberg. https://doi.org/10.1007/978-3-642-40096-4_20.

Smith, M.E., T.W. Henkel, C. Aime, A.K. Fremier & R. Vilgalys. 2011. Ectomycorrhizal fungal diversity and community structure on three co-occurring leguminous canopy tree species in a Neotropical rainforest. *New Phytologist* 192: 699–712.

Steidinger, B.S., T.W. Crowther, J. Liang, et al. 2019. Climatic controls of decomposition drive the global biogeography of forest-tree symbioses. *Nature* 569: 404–408. https://doi.org/10.1038/s41586-019-1128-0

Stephenson, S.L. 2010. *The Kingdom Fungi: The Biology of Mushrooms, Molds, and Lichens*. Timber Press, Portland, OR.

Swift, M.J. & L. Boddy. 1984. Animal-microbial interactions in wood decomposition. In Anderson, J.M., A.D.H. Rayner & D.W.H. Walton (eds.), *Invertebrate-Microbial Interactions*, pp. 89–181. Cambridge University Press, Cambridge.

Tietel, Z. & S. Masaphy. 2018. True morels (*Morchella*)—nutritional and phytochemical composition, health benefits and flavor: a review. *Critical Reviews in Food Science and Nutrition* 58(11): 1888–1901. https://doi.org/10.1080/10408398.2017.1285269

Treindl, A.D. & A. Leuchtmann. 2019. A king amongst dwarfs: *Boletus edulis* forms ectomycorrhizae with dwarf willow in the Swiss Alps. *Alpine Botany* 129(2): 185–189. https://doi.org/10.1007/s00035-019-00218-2

Tripathi, N.N., P. Vishwakarma & P. Singh. 2017. Biodiversity of macrofungi with special reference to edible forms: a review. *Journal of the Indian Botanical Society* 96: 144–187.

Valdez-Morales, M., L.C. Carlos, M.A. Valverde, E. RamírezChávez & O. ParedesLópez. 2016. Phenolic compounds, antioxidant activity and lipid profile of huitlacoche mushroom (*Ustilago maydis*) produced in several maize genotypes at different stages of development. *Plant Foods for Human Nutrition* 71: 436–443.

Vasilyeva, L.N. & S.L. Stephenson. 2004. Pyrenomycetes of the Great Smoky Mountains National Park. I. *Diatrype* Fr. (Diatrypaceae). *Fungal Diversity* 17: 191–201.

Vasilyeva, L.N. & S.L. Stephenson. 2005. Pyrenomycetes of the Great Smoky Mountains National Park. II. *Diatrypella* (Ces et de Not.) Nitschke and *Cryptovalsa* Ces et De Not. (Diatrypaceae). *Fungal Diversity* 19: 189–200.

Veeraraghavan, V.P., S. Hussain, J. Papayya Balakrishna, L. Dhawale, M. Kullappan, J. Mallavarapu Ambrose & S. Krishna Mohan. 2021. A comprehensive and critical review on ethnopharmacological importance of desert truffles: *Terfezia claveryi*, *Terfezia boudieri*, and *Tirmania nivea*. *Food Reviews International*: 1–20.

Warcup, J.H. 1980. Ectomycorrhizal associations of Australian indigenous plants. *New Phytologist* 85: 531–535.

Wasson, V.P. & R.G. Wasson. 1957. *Mushrooms, Russia and History*. Vols. 1 and 3. Pantheon, New York.

Ye, L., H. Li, P.E. Mortimer, J. Xu, H. Gui, S.C. Karunnarathna, A. Kumar, K.D. Hyde & L. Shi. 2019. Substrate preference determines macrofungal biogeography in the Greater Mekong Sub-region. *Forests* 10: 824. http://doi.org/10.3390/f10100824.

Young, T. 2004. *A Field Guide to the Fungi of Australia*. University of New South Wales Press, Randwick, Australia.

3 Wild Medicinal Mushrooms of the Northwest Himalayas of Jammu and Kashmir, India and Their Traditional and Modern Applications

Komal Verma, Tahir Mehmood, Sanjeev Kumar, Uzma Altaf, Yash Pal Sharma

3.1 INTRODUCTION

Mushrooms have long been used as a vital source of food and medicine around the world. They have traditionally been valued for their culinary and nutritional worth, but they are now also being increasingly appreciated for their important therapeutic capabilities, not only as dietary supplements (functional foods), but also as nutritional supplements, health products, and mycotherapeutic agents (Chang and Miles 2004).

Out of the 150,000 reported fungal species, more than 27,000 mushrooms have been identified and documented globally (Antonelli et al. 2020; Cannon et al. 2018). Of these macrofungal species, 2,700 are thought to be the best for eating, while about 700 are considered medicinal mushrooms (Chang and Wasser 2017; Li et al. 2021). Approximately 444 macrofungal species from Jammu and Kashmir have been documented in various localities, with many also demonstrating ethnomedicinal qualities and therapeutic properties (Verma et al. 2019; Kumar et al. 2021).

Throughout the world, mushrooms are widely acknowledged as 'nature's little pharmaceutical factories', containing a diverse array of unique bioactive chemicals such as secondary metabolites and enzymes. In Asian regions, they have been utilized since ancient times to promote and preserve good health as well as to treat ailments, though in the Western world, this approach is relatively new. Medicinal mushrooms are thought to have a variety of pharmacological properties, such as antidiabetic, antihyperlipidemic, anti-inflammatory, immunomodulatory, cytotoxic, antimicrobial, antioxidant, anticancer, antioxidant, antiallergic, hepatoprotective, probiotic, and prebiotic properties, among others (Guggenheim et al. 2014; Spelman et al. 2017; Jeitler et al. 2020; Elhusseiny et al. 2021; Ratnaningtyas et al. 2022). These qualities are attributable to the presence of several bioactive chemicals in both the mycelium and the fruiting body, the biological effects of which vary depending on chemical composition, distribution, and macrofungal species. Because of the increased interest in the use of natural products in conventional medicine, research has increasingly focused on discovering and characterizing such micro chemicals and establishing their mechanisms of action.

Many studies have been conducted thus far on the varied activities of mushroom metabolites, highlighting their enormous potential for use in the medical field. Because cancer is still one of the most difficult medical obstacles to overcome, specific efforts have been made to investigate the anticancer and immunomodulatory characteristics of mushrooms. Clinical trials are needed to

DOI: 10.1201/b23190-3

demonstrate the efficacy and safety of medicinal mushrooms in the complicated human body system (Guggenheim et al. 2014; Frost 2016; Jeitler et al. 2020).

In recent decades, the exploration and utilization of medicinal mushrooms in India have received considerable attention. Several mushroom species collected and identified from the Northwest Himalayas of Jammu and Kashmir were discovered to have therapeutic potential. The present treatise includes a description of some of these mushroom species and their biological activity.

3.2 MATERIALS AND METHODS

Several macrofungal forays were made to different sites in the Himalayan region of Jammu and Kashmir during the years 2019–2021. During these field excursions, several medicinal mushrooms were collected and identified, and some of them are briefly described here in the present chapter. Standard procedures were used for the collection, preservation, and morphotaxonomic characterization of macrofungal specimens (Singer 1986; Largent et al. 1977) Macromorphological features were recorded from fresh specimens in the field, and the a habitat and associated host trees were noted. Photographs of the fresh basidiomata were taken using a Nikon D5300 camera. The Methuen Handbook of Color was used to specify color codes (Kornerup and Wanscher 1978). Fine, hand-cut sections of dried basidiomata were used to study micromorphological characteristics. Species were identified based on an analytical examination of the specimens based on relevant published literature (e.g., Hesler and Smith 1979; Arora 1986; Singer 1986; Kibby and Fatto 1990; Heilmann-Clausen et al. 1998; Das and Sharma 2005; Semwal et al. 2014; Bhatt et al. 2018). A thorough search using PubMed, Scopus, Web of Science, and other databases was performed for a literature review to collect knowledge of the medicinal importance of the mushrooms. Personal surveys were also conducted to gain an ethnomycological knowledge in the different areas.

3.3 RESULTS AND DISCUSSION

Twenty species of macrofungal belonging to 19 genera and 15 families were documented in the present survey. Out of the collected species, *Geopora arenicola* and *Morchella esculenta* belong to the phylum *Ascomycota*, and the remaining species are members of the phylum *Basidiomycota*. A short macromorphological description of these macrofungal species along with photographs and descriptions of habits and habitats are provided herein. Moreover, edibility status and medicinal properties are also described (Figures 3.1–3.4).

3.1.1 Characterization and Medicinal Importance of Some Wild Mushrooms of the Northwest Himalayan Region

Family *Morchellaceae*

1 ***Morchella esculenta*** (L.) Pers., Syn. Meth. Fung. (Gottingen) 2: 618 (1801) **Figure 3.1a**

Pileus 16–37 mm wide, oval to elongated, apex obtuse, yellowish brown to grayish yellow (4B4) becoming brown to darker on drying; pits angular, deep and large towards the base, shallow and small towards apex, concolorous with the ridges, secondary ridges distinct; stipe 18–50 × 8–22 mm, cylindrical, occasionally swollen at the base, hollow, yellowish white (4A2) to pale yellow, (4A3), pubescent, hollow with irregular lacunae at the base; spore deposit pale yellow (4A3); odor mild; taste pleasant.

Habit and habitat: Solitary to widely scattered to gregarious in coniferous forests of *Cedrus deodara* and *Pinus wallichiana*.

Specimen examined: India: Jammu and Kashmir, Bhaderwah, Padri, Lat. 32.980000°N, Lon. 75.720000°E, elevation 3046 m, 17 March 2019, Komal Verma and Tahir Mehmood, KVTM 012.

Edibility status: Choice edible in the study area.

Medicinal properties: *Morchella esculenta*, also known as the morel mushroom, is a very rich source of polysaccharides, which are potent sources of natural antioxidants and play a vital role in the reduction of oxidative stress (Kim et al. 2011; Cai et al. 2018). Furthermore, antimicrobial activity has been demonstrated by morel mushrooms (Ahmad et al. 2014), and methanolic extracts comprising several fatty acids and sterols exhibited anti-tumor activity against lung cancer cells in humans (Lee et al. 2018). Under both *in vitro* and *in vivo* conditions, heteropolysaccharides isolated from *M. esculenta* showed anti-melanogenic properties by reducing melanin production without showing any cytotoxicity (Cai et al. 2018). When *M. esculenta was* used as a food source for mice, it enhanced the abundance of useful gut microbiota as well as short-chain fatty acids. These properties were found to be exhibited by polysaccharides of *M. esculenta* (Huo et al. 2020).

Family *Pyronemataceae*

2 *Geopora arenicola* (Lév.) Kers, Svensk Bot. Tidskr. 68 (3): 345 (1974) **Figure 3.1 b**

Ascocarps 11–42 mm in diam., sessile, about 30 mm deep inside soil; external surface buff brown, minutely roughened with flexuous hairs which bind the substratum so that the whole exterior of apothecium is encrusted with soil; inner surface cream-colored to pure white.

Habit and habitat: Humicolous, widely scattered to gregarious in conifer forests of *Cedrus deodara* and *Pinus wallichiana.*

Specimen examined: India: Jammu and Kashmir, Kathua, Sarthal, Bani, Lat. 33.080555°N, Lon. 76.849166°E, elevation 2563 m, 9 June 2019, Komal Verma and Tahir Mehmood, KVTM-26.

Edibility status: Widely consumed in the study area.

Medicinal properties: *Geopora arenicola*, locally known as 'Kundi', is considered a potent source of many phenolic compounds. By reducing oxidative reactions and neutralizing free radicals, phenols from this mushroom are regarded as powerful antioxidants. Moreover, they prevent lipid peroxidation and exhibit cardioprotective, antimicrobial, and vasodilator effects (Keles et al. 2011; Lalotra et al. 2016).

Family *Polyporaceae*

3 *Lentinus tigrinus* (Bull.) Fr., Syst. Orb. Veg. (Lundae) 1: 78 (1825) **Figure 3.1c**

Basidiomata immature to mature, 35–60 mm long; pileus 17–72 mm wide, light yellow (4A4) covered by olive brown (4D3) squamules which are appressed and dense towards the center, dry, convex having central depression, margins entire and without striation, cuticle not peeling; Context 3–4 mm wide, fleshy; Lamellae light yellow (4A4), sub-decurrent, furcated at the center of the pileus and near margins, crowded, lamellulae present, tapering towards margin; Stipe 20–40 × 3–12 mm, light yellow (4A2), hollow, slender, apex white (4A1), centric, scales present; Context fleshy, solid; Odor mild.

Habit and habitat: Lignicolous, solitary to gregarious to caespitose, common, growing on stumps of *Salix alba.*

Specimens examined: India: Jammu and Kashmir, Kashmir, Anantnag, Kamad, Lat. 33.686692°N, Lon. 75.151327°E, elevation 1681 m, 17 June 2019, Uzma Altaf and Yash Pal Sharma, HBJU-898.

Edibility status: Widely edible in the study area.

Medicinal properties: *Lentinus tigrinus* has been reported to exhibit anticancer properties by exerting cytotoxic and pro-apoptotic effects on cancer cell lines. LTPp, a soluble protein fraction isolated from *L. tigrinus*, causes apoptosis in human breast cancer MCF-7 cells and prostate cancer PC3 cells (Mohammadnejad et al. 2019). Moreover, the hexane and acetonitrile extracts of *L. tigrinus* have been found to exhibit radical-scavenging activity. Acetonitrile extracts were found to contain considerable amounts of phenolic compounds, which indicates their potential as a source of antioxidants (Dulay et al. 2017). *Lentinus tigrinus* has been found to manage diabetes mellitus in alloxan-induced diabetic mice because of its hypoglycemic effect. Moreover, extracts of *L. tigrinus* exhibited antibacterial activity against *Pseudomonas aeruginosa* and *Escherichia coli* (Dulay et al. 2014; Sevindik 2018). Some sterols, (e.g., cerevisterol, ergosterol, and stellosterol) were extracted from the basidiomata of *Lentinus tigrinus* and might also contribute to its therapeutic properties (Ragasa et al. 2018).

4 ***Trametes versicolor*** (L.) Lloyd, Mycol. Notes (Cincinnati) 65: 1045 (1921) **Figure 3.1d**

Pileus 30–85 mm broad, hemispherical, frequently fused with another bracket, surface dry, tomentose or velvety, various color, concentric rings of various shades of brown to reddish; context pale white (2A2), leathery; yellowish white (2A2) to grey pore surface, rounded to angular, 4–6 per mm, tough, fibrous, thickness 5 mm; indistinct taste and odor; white spore print.

Habit and habitat: Growing in clusters or rosettes on the deadwood of *Quercus* spp.

Specimen examined: India, Jammu and Kashmir, Jammu, Doda, Desa, Lat. 33.173888°N, Lon. 75.517772°E, elevation 2454 m, 20 August 2019, Komal Verma and Tahir Mehmood, KV-03.

Edibility status: Not consumed but used as medicine.

Medicinal properties: *Trametes versicolor* is a medicinal mushroom known by various names, including 'Chinese medicinal mushroom', 'turkey tail mushroom', and 'Yun Zhi' in China i.e., cloud-like mushroom. These names are given probably due to its surface which is variously colored, wavy, and covered with fluff (Chu et al. 2002). This mushroom exhibits various biological activities due to the presence of several bioactive polysaccharides. It has been used as an immunopotentiator for years because of the therapeutic activities of protein-bound polysaccharides. polysaccharopeptide krestin (PSK) and polysaccharopeptide (PSP) are the two most popular commercial polysaccharopeptide preparations of its mycelia extracts. Of these, PSK has been approved for cancer treatment in Japan. Isolated in the 1960s, PSK accounted for more than 25% of Japanese anticancer drug expenditure by 1987 (Sullivan et al. 2006; Mattos-Shipley et al. 2016). PSP and PSK induce the production of immunoglobulin-G, interleukin-6, macrophages, interferons, and T-lymphocytes. Moreover, they counteract the immunosuppressive effects of radiotherapy, chemotherapy, and blood transfusion; inhibit tumor production and cancer proliferation by inducing the production of an antioxidant pool (i.e., glutathione peroxidase and superoxide dismutase [SOD]); and improve liver function. The antibacterial and antifungal activities of extracts of *T. versicolor* also been demonstrated against common pathogens such as *Pseudomonas aeruginosa*, *Escherichia coli*, *Klebsiella pneumoniae*, *Staphylococcus aureus*, *Candida albicans*, *Streptococcus pneumonia*, and *Listeria monocytogenes* due to the activation of polymorphonuclear cells and increased secretion of antimicrobial cytokines (e.g., tumor necrosis factor [TNF], interleukin-1 [IL-1]) (Chu et al. 2002; Bains and Chawla 2020). Its extracts also demonstrated effective anti-inflammatory properties and can thus be used in the development of anti-inflammatory drugs (Bains and Chawla 2020). Under *in vitro* conditions, the polysaccharide preparation of *T. versicolor* exhibited direct toxicity to cancerous cells, including breast cancer (Chauhan et al. 2019), lung cancer (Knezevic et al. 2018), and colon cancer (HT-29; SW-480, LS174-T) cells (Roca-Lema et al. 2019). In addition, the polysaccharides of *T. versicolor* have been demonstrated to ameliorate obesity in rodents (Li et al. 2019) and show quite promising results against coronavirus disease of 2019 (COVID-19); they may be used in the future against this disease (Arunachalam et al. 2022)

Family *Schizophyllaceae*

5 *Schizophyllum commune* Fr. Observ. Mycol. (Havniae) 1: 103 (1815) **Figure 3.1e**

Basidiomata immature to mature, growing in rows and shelf-like patterns; pileus 70–300 mm wide, surface dry and glabrous, brownish orange (6C5), margin wavy or lobed, crenate and longitudinally striate, cuticle not peeling, no change in color after bruising; Context very thin, grey (8E1), tough and leathery in texture; lamellae 10–20 mm wide, grey (8E1), edges concolorous, close to rather crowded, edges grooved lengthwise, radiating form point of attachment, tapering towards the margin, well-spaced and each gill composed of two adjacent plates, lamellulae present tapering towards the center; pseudostipe small; odor indistinct.

Habit and habitat: Lignicolous, uncommon, gregarious to caespitose, growing on logs of *Juglans regia.*

Specimens examined: India, Jammu, and Kashmir, Kashmir, Anantnag, Dialgam, Lat. 33.681123°N, Lon. 75.153657°E, elevation 1681 m, June 2019, Uzma Altaf and Yash Pal Sharma, HBJU-887.

Edibility: Not consumed in the study area but exhibits medicinal properties.

Medicinal properties: Schizophyllan, a β-glucan derived from *Schizophyllum commune*, has anti-tumor and immunomodulatory properties (Yao et al. 2016). This mushroom produces compounds, (e.g., iminolactons) that possess the ability to inhibit the growth of cancer cell lines (Liu et al. 2021). Moreover, polysaccharide extracts exhibited antioxidant properties and demonstrated significant acetyl cholinesterase inhibitory activity, thus emerging as a potential source for treating Alzheimer's disease (Mišković et al. 2021). Recently, peptides were also extracted based on *de novo* peptide sequencing, which demonstrated antioxidant properties but showed variations according to the concentration of peptides (Wongaem et al. 2021).

Family *Agaricaceae*

6 *Macrolepiota procera* (Scop.) Singer, Pap. Mich. Acad. Sd. 32: 141 (1948) **Figure 3.1 f**

Pileus 52–106 mm in diameter, convex then flat, umbonate, surface white, marked with prominent brown scales on the entire surface, snuff brown at the center, margins inrolled; lamellale up to 6 mm wide, free, crowded, unequal. wavy, concolorous with pileus; context white; stipe 70–90 × 6–14 mm wide, central, equal, solid, smooth, swollen at the base, concolorous with pileus; annulus present, superior, movable, evanescent, creamish; odor mild.

Habit and habitat: Humicolous, solitary to scattered in cultivated fields and grasslands.

Specimen examined: India: Jammu and Kashmir, Rajouri, Daharal, Lat. 33.476388°N, Lon. 74.417777°E elevation 2682 m, TMKV-06, July 2018, Komal Verma, Tahir Mehmood, and Yash Pal Sharma, HBJU-901.

Edibility: Consumed in some parts of the study area.

Medicinal properties:*Macrolepiota procera*, also known as parasol mushroom, is often cherished as a daily diet product. Besides its unique flavor and aroma, it is highly valued as a medicinal mushroom. It has been reported to exhibit antiproliferative, cytotoxic, anti-invasive, and apoptotic effects on A549 lung cancer cells (Seçme et al. 2018). Moreover, it exhibits immunoenhancing effects and regulates the functioning of the digestive system, specifically the pancreas (Fernandes et al. 2014). This mushroom is rich in phenolic compounds, (e.g., cinnamic acid, gallic acid, vanillic acid, β-carotene, and lycopene) and its extracts have demonstrated strong DPPH radical-scavenging activity (Erbiai et al. 2021).

7 ***Agaricus campestris*** (L.) Sp. pl. 2: 1173 (1753) **Figure 3.2a**

Basidiomata small to medium sized; pileus 35–110 mm in diam., initially globose during younger stage becoming convex to broadly convex upon maturation, white (6A1), surface dry covered with triangular squamules brownish grey (6C2) in color which are dense at disc, margins entire and appendiculate, turns slight yellowish upon reaction with 10% $FeSO_4$ and shows no reaction with 10% KOH; context white (6A1) and fleshy; lamellae free, crowded, forkation absent, intercalated with different lengths of lamellulae tapering towards the center, orange color (6B2) during younger stage becoming light brownish (6D6) and finally dark brownish (6F7) upon maturation; stipe 30–65 × 20–35 mm, concolorous with the pileus, equal (cylindrical), thick, annulus present, surface smooth above the ring and fibrillar below the ring, turns slight yellowish on reaction with 10% $FeSO_4$ and no reaction with 10% KOH; context white (6A1) and solid; annulus single, membranous, whitish, lower surface floccose and smooth upper surface, non-persistent, pendant, edge entire; odor pleasant; taste delicious.

Habit and habitat: Humicolous, gregarious to caespitose, growing in open grasslands.

Specimen examined: India: Jammu and Kashmir, Kashmir, Anantnag, Kamad, Lat. 33.686692°N, Lon. 75.151327°E, elevation 1681 m, April 2018, Uzma Altaf and Yash Pal Sharma, HBJU-869.

Edibility: Choice edible in the study area.

Medicinal properties: Antioxidant activity was exhibited by a hydroalcoholic extract of *Agaricus campestris*, showing a correlation with its total phenolic content (Bubueanu et al. 2015). Gasecka et al. (2018) also assessed the radical-scavenging activity of *A. campestris* through the DPPH method, which was also found to be directly correlated to its phenolic content. Different anti-tumor activities were demonstrated by different solvent extracts of *A. campestris*. Likewise, methanolic extracts exhibited anticancer activities (Kosanic et al. 2017).

8 ***Coprinus comatus*** (O.F. Müll.) Pers., Tent. Disp. Meth. Fung. (Lipsiae): 62 (1797) Figure 3.2 **b**

Pileus 62–85 × 41–62 mm, conical at first becoming expanded convex when mature, white when young turning blackish brown from the margins at extreme maturity due to auto-digestion; surface semi-dry with conspicuous shaggy scales on the entire surface; margins complete when young but dissolving away with maturity; lamellae 3–5 mm wide, crowded, free, whitish when young, becoming black with maturity then liquefying with age; stipe 85–115 mm long, white, smooth, central,

FIGURE 3.1 (a) *Morchella esculenta*, (b) *Geopora arnicola*, (c) *Lentinus tigrinus*, (d) *Trametes versicolor*, (e) *Schizophyllum commune*, (f) *Macrolepiota procera*.

hollow, fleshy with bulbous base; evanescent, fibrous, whitish annulus; odor pleasant when young; black spore print.

Habit and habitat: Humicolous, solitary to scattered in grasslands.

Specimens examined: India: Jammu and Kashmir, Kathua, Machedi, Lat. 32.698333° N, Lon. 75.599166°E, elevation 1710 m, July 2021, Komal Verma, Tahir Mehmood, and Yash Pal Sharma KVTM-08.

Edibility status: Young specimens are edible in the study area.

Medicinal properties: The extracts of *Coprinus comatus* exhibited antiproliferative effects in estrogen receptor-positive (MCF–7) and human estrogen receptor-negative (MDA-MB-231,

BT-20) cell lines (Gu and Leonard 2006). Moreover, selective antiproliferative effects in human T- cell leukemia Jurkat cells have been exhibited by glycan-binding protein (Y3) isolated from *Coprinus comatus* (Zhang et al. 2017). This species contains high levels of ergothioneine, which has antioxidant properties, and aminobutyric acid (GABA), which has potential relaxation and anti-anxiety effects (Cohen et al. 2014). Furthermore, research on different types of tumor cells has demonstrated its potential anti-tumor effect against breast (Asatiani et al. 2011), ovarian (Rouhana-Toubi et al. 2013), and prostate cancer by suppressing androgen receptor function (antiandrogenic activity) (Dotan et al. 2011). Additionally, *Coprinus comatus* has been demonstrated to exhibit hypoglycemic activity by reducing blood glucose levels (Han et al. 2006). Moreover, ethanolic extracts of *C. comatus* also exhibited antidiabetic properties *in vivo* as the extract reduced blood glucose levels (Ratnaningtyas et al. 2022).

Family *Physalacriaceae*

9 *Flammulina velutipes* (Curtis) Singer, Li/ba 22: 307 (1951) [1949] **Figure 3.2c**

Pileus 90–150 mm, orange-yellow at edges and darker at the center, convex when young becoming plano-convex on maturation, irregular, lobate, cuticle smooth, viscid, with striate margins; lamellae pale yellow turning light brown, slightly adnexed, unequal; stipe 65–115× 30–60 mm, cylindrical and slightly tapered towards the base, curved or twisted, yellow, velvety all over, quickly becoming blackish brown; annulus absent; context pale yellow, fine, watery and soft in the cap, leathery and fibrous in the stipe; odor mild; taste mild; spore print creamish white.
Habit and habitat: Lignicolous, caespitose on dead tree trunks of *Juglans regia* and *Salix alba.*

Specimens examined:*India:* Jammu and Kashmir, Kathua, Bani, Lat. 32.667586°N, Lon. 75.803397°E, elevation 1630 m, 17 August 2020, TM04, Komal Verma, Tahir Mehmood, and Yash Pal Sharma, HBJU-902.

Edibility: Considered edible in the study area.

Medicinal properties: *Flammulina velutipes (golden needle mushroom, or winter mushroom) is a popular edible fungus that is well known for its excellent flavor and nutritional content. Many bioactive compounds of this mushroom have been identified, including glycoproteins, phenols, polysaccharides, proteins, polyunsaturated fatty acids (PUFAs), and sesquiterpenes. These compounds have been shown to have a variety of biological activities, including anti-atherosclerotic, anti-tumor, and anticancer activity; cholesterol-lowering effects; thrombosis inhibition activity; antihypertensive, anti-aging, and antioxidant properties, memory restoration and learning deficit attenuation effects; anti-inflammatory, immunomodulatory, and antibacterial properties; ribosome inactivation effects; and melanosis inhibition activity* (Tang et al. 2016). Interestingly, the plasma-treated fruiting bodies of *F. velutipes* showed significant anticancer potential by inducing apoptotic cell death markers in breast cancer cells. The reason behind this potential is the increased expression of apoptotic genes (Mitra et al. 2020).

Family *Pleurotaceae*

10 *Pleurotus ostreatus* (Jacq.) P. Kumm., Führ. Pilzk. (Zerbst): 104 (1871) **Figure 3.2 d**

Pileus 25–65 mm wide, creamish, depressed infundibuliform, dark brown at the center and slightly brown at margins, margins inrolled; lamellae sub-decurrent, unequal, crowed at the margins, subdistant at the base, wavy, concolorous; stipe eccentric to centric, 20–36 × 4–9 mm, concolorous with the pileus, context fleshy, solid; odor pleasant; taste mild.

Habit and Habitat: Lignicolous, growing in groups on living hardwood trees or fallen stumps.

Specimen examined: India: Jammu and Kashmir, Poonch, DKG, Lat. 33.583402°N, Lon. 74.362225°E, elevation 2146 m, August 2020, Komal Verma, Tahir Mehmood, and Yash Pal Sharma TMKV08.

Edibility: Choice edible in the study area.

Medicinal properties:*Pleurotus ostreatus is an edible fungus with high protein, carbohydrate, mineral, and vitamin contents and low fat. Its key components include polysaccharides, lectin, polypeptides, amino acids, and phenol oxidase. Pleurotus ostreatus possesses anti-tumor, antioxidant, blood cholesterol-lowering, antiviral, anti-inflammatory, antibacterial, and immunoregulatory properties (*Sun and Liu 2009). *Pleurotus ostreatus* ethanolic extracts have high DPPH-scavenging activity, indicating antioxidant properties, and inhibits *Klebsiella pneumoniae, Salmonella typhi, Staphylococcus aureus, Proteus mirabilis, Escherichia coli, Pseudomonas aeruginosa*, and *Bacillus subtilis* (Oyetayo and Ariyo 2013).

The cell walls of oyster mushrooms contain β-glucan (pleuran), a type of immunostimulant known as a pathogen associated molecular pattern (PAMP). The immune system's mechanisms recognize these molecules as the structures against which the immune response has to be triggered. Direct interaction of β-glucan with macrophages and immunocompetent cells produces the desired effect. Oyster mushrooms are therefore helpful for enhancing the body's immune system. This mushroom's component of β-glucan has strong anticancer effects on tumors and sarcomas (Mowsumi and Choudhary 2010; Facchini et al. 2014).

Antiviral activity against adenovirus and herpes simplex II has been demonstrated by extracts of *P. ostreatus*. Moreover, the extracts also exhibited potent antioxidant activity (Elhusseiny et al. 2021). Proteome analysis revealed the expression of bioactive proteins such as Rab GDP dissociation inhibitors (Rab GDPI), serine proteinase, and lectin. Rab GDPIs may represent a promising future anticancer strategy, as they play a vital role in membrane trafficking in tumor cells because of their ability to regulate Rab GTPases (Stenmark 2009). Interestingly, similar results regarding antiviral activities were found by Urbancikova et al. (2020), who proposed that pleuran (β-glucan) reduced the duration of symptoms during a herpes simplex I virus outbreak. A modified version of pleuran extracted from this mushroom is a potential weapon against COVID-19 because it can reduce virus reproduction (Arunachalam et al. 2022).

Family *Cantharellaceae*

11 ***Cantharellus cibarius*** Fr., Syst. Mycol. (Lundae) 1: 318 (1821) **Figure 3.2e**

Basidiomata 35–65 mm long; pileus 18–49 × 32–90 mm, pastel yellow or light yellow (1A4) in color, dry, smooth, convex at the younger stage, and infundibuliform at maturity after developing depression in the center, margins inrolled, wavy, and smooth, cuticle peeling 50%, pileus cuticle changing color to orange upon treatment with 10% KOH; context yellowish, firm, changing color to dark yellowish upon treatment with 10% KOH; true gills absent, present in the form of veins which are decurrent, edges concolorous, crowded, forkation present, turning dark yellowish upon treatment with 10% KOH; stipe 30–53 × 5–13 mm, equal (cylindrical), pastel yellow or light yellow (1A4) in color, dry, annulus absent, central, stipe cuticle changing color to reddish black upon treatment with 10% KOH; context solid, yellowish, fibrous, changing color to a reddish yellow on treatment with 10% KOH; odor mild.

Habit and habitat: Humicolous, gregarious, growing in coniferous forests of *Abies pindrow* and *Cedrus deodara*.

FIGURE 3.2 (a) *Agaricus campestris*, (b) *Coprinus comatus*, (c) *Flammulina velutipes*, (d) *Pleurotus ostreatus*, (e) *Cantharellus cibarius*, (f) *Sparassis crispa*.

Specimen examined: India: Jammu and Kashmir, Doda, Bhaderwah, Chatergala Lat. 32.88708°N, Lon.75.791305°E, elevation 3408 m, 28 August 2020, TMKV–10, Komal Verma, Tahir Mehmood, and Yash Pal Sharma.

Edibility: Choice edible in the study area.

Medicinal properties:*Cantharellus cibarius is well known in the culinary world for its distinct aroma, flavor, and texture. This mushroom has antibacterial, anti-inflammatory, antioxidant, anti-hypoxic, antihyperglycemic, cytotoxic, wound-healing, and iron-chelation properties* (Vlasenko et al. 2019). Considerable antimicrobial activities have been demonstrated by methanolic extracts of *C. cibarius* against *Bacillus subtilis, Candida albicans, Esherichia coli,* etc. (Mendez-Espinoza et al. 2013). Caffeic acid has anti-inflammatory and antioxidant effects; it was found to be an active antioxidant in *C. cibarius* (Muszyńska et al. 2016). Polysaccharides (acidic β-glucan) extracted from fruiting bodies of *C. cibarius* using gel-permeation chromatography and anion exchange promote macrophage activation by boosting NO, TNF, and IL-6 production (Qu et al. 2021).

Family *Sparassidaceae*

12 ***Sparassis crispa*** (Wulfen) Fr., Syst. Mycol. (Lundae) 1:465(1821) **Figure 3.2f**

Basidiocarps 120–200 mm wide and 130–185 mm high, large, rounded, cauliflower-like, sessile, composed of several broad flattened crimped-like branches usually wavy, warm buff; branches fleshy, 23–40 mm wide, irregularly lobed.

Specimen examined: India, Jammu and Kashmir, Jammu, Kathua, Bani, Sarthal, Bhaderwah, Chaterkala, Lat. 32.825752°N, Lon.75.72892°E, elevation 3059 m, 12 August 2020, Komal Verma, Tahir Mehmood, and Yash Pal Sharma, KV-07.

Habit and habitat: Humicolous, scattered, growing in coniferous forests of *Abies pindrow* and *Cedrus deodara.*

Edibility: Widely edible.

Medicinal properties:*Sparassis crispa* is both an edible and medicinal mushroom with anti-tumor, anticarcinogenic, immune-enhancing, hematopoietic, anti-angiogenic, anti-inflammatory, antiviral, antiallergic, antidiabetic, wound-healing, antioxidant, anticoagulant, and antihypertensive properties, as well as cytokine induction activity. *Sparassis crispa* also contains sparassol, which gives it antimicrobial abilities (Sharma et al. 2022). It also has a high concentration of β-glucan, which contains antifungal chemicals, methyl-2,4-dihydroxy-6-methylbenzoate, methyl-dihydroxymethoxy-methylbenzoate, and benzoate derivatives that are effective in the treatment of a variety of human illnesses (Han et al. 2019; Sharma et al. 2022). Several studies have indicated that *S. crispa* has the potential for wound healing (Sharif-Rad et al. 2020). The compound sparoside A extracted from *S. crispa* shows an inhibitory effect on allergic rhinitis (Wang et al. 2019). The bactericidal activity of ethanolic *S. crispa* extract appeared to exhibit a zone of inhibition against *E. coli* (Niazi and Ijaz 2021). In addition to all these activities, crude polysaccharides obtained from this mushroom have a potential antioxidant activity (Nowacka-Jechalke et al. 2021).

Family *Laetiporaceae*

13 ***Laetiporus sulphureus*** (Bull.) Murrill, Ann. Mycol. 18 (1/3): 51 (1920) Figure **3.3a**

Basidiomata, annual, pileate, sessile substipitate stipe like base, Pileus initially knob-shaped to applanate, 80–160 × 30–40 mm, upper surface yellow to orange when young, fading to brownish yellow with age; margin often undulate, rounded, minutely tomentose to glabrous, azonate to faintly zonate; white context, tubes concolorous with the pore surface, azonate, brittle, chalky on drying; odor pleasant; taste nutty.

Specimen examined: *India: Jammu and Kashmir, Doda, Bhaderwah, Chatergala, Lat.* 32.877083°N, Lon. 75.791305°E, *elevation 3408 m, 5 September 2020, Tahir Mehmood, Komal Verma, and Yash Pal Sharma, TM-YPS-021.*

Habit and habitat:*Lignicolous,* growing on living tree trunks of *Quercus* sp.

Medicinal properties:*Laetiporus sulphureus* is one of the traditionally used medicinal mushrooms. The two main triterpenes extracted from *L. sulphureus* are known as laetiporins C and D (Sandargo et al. 2019; Hyde et al. 2019). In addition, there are four other triterpenoids: trametenolic acid, 15 hydroxytrametenolic acid, eburicoic acid, and fomefficinic acid are also present. Cancer cell lines were resistant to the proliferative effects of these triterpenes (Hassan et al. 2021). It also possesses bioactive compounds that have positive health effects, in addition to being rich in many vital macro- and micronutrients. Additionally, benzofurans, flavonoids, coumarins, and N-containing chemicals have been discovered to be present in *L. sulphureus* (Khatua et al. 2017). Moreover, the extracts obtained from the fruiting bodies of this mushroom showed a number of medicinal properties such as anti-tumor, antioxidant, antimicrobial, anti-inflammatory, immunomodulatory, and antihyperglycemic activities attributed to biologically active components such as phenolics, triterpenes, and polysaccharides (Zhang et al. 2018).

Family *Hericiaceae*

14 ***Hericium erinaceus*** (Bull.) Pers., Comm. fung. clav. (Lipsiae): 27 (1797) **Figure 3.3b**

Basidiomata 56–166 × 40–70 mm. Primary branches with a width of up to 11 mm, yellowish white (2A2), and little spines (up to 3 mm long) all over, which give rise to secondary branches. Secondary branches are somewhat thinner (up to 7 mm), without spines, and bearing tertiary branches. Spines are dense, tapered, hanging, white, and up to 20 mm long; context yellowish white (2A2), unchanged with $FeSO_4$, turning lemon yellow with KOH and dark orange with guaiacol; taste and odor excellent; spore deposit white to yellowish white (2A2).

Habit and habitat: Solitary, growing on dead wood of *Cedrus deodara* in coniferous forests.

Specimens examined: India: Jammu and Kashmir, Kathua, Bani, Sangroon, Lat. 32.683833°N, Lon. 75.803397°E, Elevation 1630 m, 16 August 2019, *Tahir Mehmood, Komal Verma, and Yash Pal Sharma,* TMKV-010.

Medicinal properties: *Hericium erinaceus* is a highly prized mushroom known for its medicinal properties. It is extensively used in traditional Chinese medicine and has long been recognized for its medicinal value due to the presence of many bioactive metabolites, therapeutic polysaccharides, and β-glucans. This mushroom is considered very beneficial for the nervous system and is thus commonly referred to as nature's nutrient for the neurons (Chong et al. 2020; Venturella et al. 2021). Diterpenoids, hericenones (C-H), and benzyl alcohol derivatives extracted from the mycelium and fruiting body of *H. erinaceus* show neurotropic and neuroprotective effects. Moreover, they can induce nerve growth factor (NGF) synthesis (Venturella et al. 2021). *H. erinaceus* also exhibits anticancer, anti-oxidative, anti-inflammatory, antidiabetic, antimicrobial, immunostimulant, hypolipidemic, and antihyperglycemic properties. The hericinones and erinacines extracted from *H. erinaceus* are most frequently used for the treatment of neurodegenerative diseases such as Alzheimer's and cognitive impairment (Spelman et al. 2017; Chong et al. 2020; Venturella et al. 2021). Erinacin A and C obtained from this mushroom have potential anti-tumor, anti-neuroinflammatory, and neuroprotective properties. Moreover, *H. erinaceus* possesses antimicrobial activity against various pathogens as well as significant antioxidant activity, and the β-glucans extracted showed significant anticancer properties (Ghosh et al. 2021). β-glucan polysaccharides from *H. erinaceus* exhibit a broad spectrum of antiviral activity (Arunachalam et al. 2022).

Family: *Meripilaceae*

15 ***Grifola frondosa*** (Dicks.) Gray, Nat. Arr. Brit. Pl. (London) 1: 643 (1821) **Figure 3.3c**

Basidiomata 185–350 × 130–215 mm, stipitate to multipileate; pileus 12–80 mm broad, imbricate, flabelliform, wavy fronds, surface dry, smooth or minutely tomentose, azonate to faintly zonate, light brown (5D5), dark brown (6D7–8) with age; context yellowish white (2A2), thin, up to 8 mm; margin thin; hymenophore decurrent, with angular pores, 2–4 per mm, white to yellowish white (2A2); stipe 60–100 × 25–30 mm, branching from a substantial base, white at the base, and context yellowish white (2A2), flexible; taste mild; odor pleasant.

Habit and habitat: Growing on the living stem of *Quercus* sp. in mixed temperate forests.

Specimens examined: India: Jammu and Kashmir, Jammu, Rajouri, Darhal, Lat. 33.483497°N, Lon. 74.442080°E, elevation 2039 m, 15 July 2019, *Tahir Mehmood, Komal Verma, and Yash Pal Sharma,* TMKV-06.

Medicinal properties: *Grifola frondosa* (Maitake) is a very edible and therapeutic fungus known as 'the hen of the woods' and 'the king of mushrooms'. This mushroom contains a number of bioactive metabolites, one of which is anticancer D-fraction or *Grifola frondosa* polysaccharide (β-glucan proteoglycan) (Alonso et al. 2013; Venturella et al. 2021). It stimulates not only macrophages, T cells, and NK cells, but also numerous genes involved in apoptotic stimulation, cell growth and proliferation inhibition, cell cycle arrest, and tumor cell migration and metastasis suppression, aiding in cancer prevention (Alonso et al. 2017, 2018; Venturella et al. 2021). It can also effectively activate the cellular immune system. The other two polysaccharide fractions derived from *G. frondosa*, F2 and F3, demonstrate hypoglycemic effects *in vitro*, aiding in diabetes management (Xiao et al. 2015). Other functions demonstrated by this mushroom include antihypertensive, anti-obesity, and anti-hepatitis activities (Venturella et al. 2021). *Grifola frondosa* contains antiviral polysaccharides (β-glucan proteoglycan) and can be utilized to treat COVID-19 infections (Arunachalam et al. 2022).

Family *Auriculariaceae*

16 ***Auricularia auricula-judae*** (Bull.) Quél., Enchir. fung. (Paris): 207 (1886) **Figure 3.3 d**

FIGURE 3.3 (a) *Laetiporus sulphureus*, (b) *Hericium erinaceus*, (c) *Grifola frondose*, (d) *Auricularia auricula-judae.*

Basidiomata 20–110 mm wide, ear-shaped to unevenly lobed, slimy, semi-transparent, sessile; exterior surface tan-brownish orange (5C6); margin entire at first, becoming somewhat lobed with age; context thin (up to 15 mm), cartilaginous, hardening on drying; hymenium pale orange (5A3), veined, irregular, and frequently irregular; taste and odor unclear; no spore deposit observed.

Habit and habitat: growing in clusters on dead or decaying hardwood and conifers.

Specimens examined: India: Jammu and Kashmir, Kashmir, Anantnag, Dialgam, Lat. 33.681123°N, Lon. 75.153657°E, elevation 1681 m, 14 June 2019, Uzma Altaf and Yash Pal Sharma, HBJU-793.

Medicinal properties:*Auricularia auricula-judae* is an important mushroom that has traditionally been used as a source of medicine in many countries due to its antidiabetic, anticancer, antihypertensive, anti-inflammatory, immunomodulatory, and antibacterial characteristics (Waktola and Temesgen 2018). In genetically diabetic mice, polysaccharides isolated from *A. auricula-judae* had a hypoglycemic effect, resulting in lower fasting plasma glucose concentrations and increased glucose tolerance (Liu et al. 2021). This mushroom contains mannose, glucose, xylose, arabinose, rhamnose, and galactose in its polysaccharides. These polysaccharides have been demonstrated to lower levels of total cholesterol, triglycerides, and low density lipoprotein (LDL) (Liu et al. 2021). The acidic heteropolysaccharides found in this mushroom, called glucuronoxylomannans (MEA and MHA), have anticancer properties (Chen et al. 2020). *Auricularia auricula-judae* polysaccharides have also demonstrated potential for treating wounds (Mapoung et al. 2021). Isolated acidic polysaccharides from *A. auricula-judae* show anticoagulant activity by enhancing the catalysis of thrombin inhibition by antithrombin and decreasing platelet aggregation (Wang and Guo 2020; Liu et al. 2021). *Auricularia auricula-judae* also exhibits antibacterial action against gram-positive and gram-negative bacteria in addition to the previously described characteristics (Islam et al. 2021).

Family *Ganodermataceae*

17 *Ganoderma lucidum* (Curtis) P. Karst., Revue Mycol., Toulouse 3(9): 17 (1881) **Figure 3.4a**

Pileus 25–120 mm broad, elongated to fan-shaped, surface smooth, glossy, shiny, concentrically sulcate, brown (5E7) to reddish brown (8E8); context white; tubes whitish to brown; pore surface white to yellowish white (3A2) turning to brown (7D6) when bruised, pore rounded to ovate; stipe 35–110 × 10–25 mm, eccentric, subcylindrical, brown (5E7) to reddish brown (8E8); taste and odor indistinct.

Habit and habitat: Solitary to dispersed, found to grow at the base of living and dead wood of *Quercus* sp.

Specimens examined: India: Jammu and Kashmir, Kathua, Bani, Sangroon, Lat. 32.683833°N, Lon. 75.803397°E, elevation 1630 m, 5 September 2019, *Tahir Mehmood, Komal Verma, and Yash Pal Sharma*, TMKV-018.

Medicinal properties:*Ganoderma lucidum*, sometimes known as the 'mushroom of immortality', has been utilized in East Asia for generations. Even now, it is one of the most widely used medicinal mushrooms. *Ganoderma lucidum* has long been used as a traditional medicine to treat a wide range of human problems, including respiratory problems, hypertension, hepatitis, hypercholesterolemia, cardiac disease, diabetes, and gastric cancer (Bulam et al. 2019). These characteristics are primarily due to the presence of two main types of metabolic compounds: triterpenes/triterpenoids (including ganoderic acids, ganodermic acid, ganodermic alcohols, lucidones, and lucinedic acids) and

FIGURE 3.4 (a) *Ganoderma lucidum*, (b) *Tremella fuciformis*, (c) *Russula cyanoxantha*, (d) *Russula brevipes*.

polysaccharides (α-1, 3, β-1, 3 and β-1,6-D-glucans and ganoderan with glucose as a major sugar component). Triterpenes derived from this mushroom have anticancer, antimetastatic, cytotoxic, and enzyme-inhibitory activities; polysaccharides have antiangiogenic and immune system-stimulating properties (Trigos and Medellín 2011; Bulam et al. 2019). Furthermore, these polysaccharides have been shown to enhance the anti-tumor immune response by increasing the activity of natural killer cells and cytotoxic T-lymphocytes (Bulam et al. 2019; El Sheikha 2022). Extract from *G. lucidum* has been shown to be carcinostatic in a range of cancer cell lines, including breast, pancreatic, lung, colon, skin, and prostate cancers (Yang et al. 2019; El Sheikha 2022). This mushroom not only has anticancer effects, but it also has cardioprotective properties (Xie et al. 2016). Ganoderic acid D is found in *G. lucidum* also exerts an anti-aging effect by suppressing the formation of reactive oxygen species (ROS) and senescence-associated indicators, as well as reducing oxidative stress-induced stem cell senescence (Xu et al. 2020). *Ganoderma lucidum* appears to be the most effective mushroom species against SARS-CoV-2, reducing the virus's impact (Bourhia et al. 2020).

Tremellaceae

18 ***Tremella fuciformis*** Berk., Hooker's J. Bot. Kew Gard. Misc. 8: 277 (1856) **Figure 3.4 b**

Basidiomata up to 120 × 80 mm, white, composed of thin, upright, branching lobes; shining surface; context slimy, whitish to transparent white.

Habit and habitat: Growing on the dead wood of *Quercus* sp. in a damp, temperate mixed forests.

Specimens examined: India: Jammu and Kashmir, Kathua, Bani, Roulka, Lat. 32.6838333°N, Lon. 75.8033972°E, elevation 1630 m, *28* July 2020, *Tahir Mehmood, Komal Verma, and Yash Pal Sharma*, TMKV-018.

Medicinal properties: *Tremella fuciformis* has long been utilized in traditional Chinese medicine as a tonic for 'weakness and aging'; it has also been widely used in Chinese cuisine because of its high levels of nutritional fibers, protein, minerals, antioxidants, and vitamin D (Shahrajabian et al. 2020). *Tremella fuciformis* polysaccharides effectively reduce oxidative stress and apoptosis in skin fibroblasts, and thus may operate as strong anti-oxidative agents for oxidative stress-related skin illnesses and aging (Shen et al. 2017). Thus, *T. fuciformis* is considered a potent agent for the treatment of cancer and Alzheimer's disease. This mushroom's polysaccharide has anti-inflammatory, wound-healing, antioxidant, anti-aging, and anti-radiation properties (Newman and Cragg 2016). *Tremella fuciformis is a safe, non-toxic, and natural active ingredient that has been employed as a strong antineoplastic agent. Numerous cosmetic companies use polysaccharides from Tremella as moisturizing ingredients* (Liu and He 2012; Wu et al. 2016). *Tremella fuciformis* polysaccharides protected mice from dextran sulfate sodium (DSS)-induced colitis. In high doses, it was suggested to prevent colonic shortening, lower the activity of colonic myeloperoxidase and serum diamine oxidase (DAO), decrease D-lactate concentrations, and prevent colonic tissue injury in colitic mice due to anti-inflammatory cytokine production. As a result, these polysaccharides have the potential to be developed as a food supplement for the prevention of a variety of intestinal illnesses (Xu et al. 2021). Furthermore, these polysaccharides contain a variety of acidic and neutral heteroglycans that have pharmacological features such as anti-tumor, hypocholesterolemic, and hypoglycemicactivity (Ma et al. 2021).

Family *Russulaceae*

19 ***Russula cyanoxantha*** (Schaeff.) Fr. **Figure 3.4 c**

Pileus 75–88 mm in diameter, convex to plano-convex with depressed center, smooth, margins cracked at maturity, pileus surface viscid to slimy when moist, grey purple to dark grey purple, greenish purple; gills 3–5 mm wide, sub-decurrent to decurrent, entire, forked near the base, even margins, regular, creamish white; stipe central, cylindrical, 45–6.3 × 16–21 mm, creamish with a light shade of purple, veined; context solid; taste mild; odor pleasant; spore print white.

Habit and habitat: Humicolous, solitary to scattered, growing in association with *Pinus wallichiana* and *Cedrus deodara*.

Specimens examined: India: Jammu and Kashmir, Kishtwar, Chingam, Sinthan top, Lat. 33.559644°N, Lon. 75.539891°E, elevation 2970 m, 17 August 2020, Komal Verma and Yash Pal Sharma, KVR18–19.

Edibility: Consumed in the area.

Medicinal properties: *Russula cyanoxantha* appears to be an effective natural antioxidant resource that could be used in human medicine, and for animal and plant ailments. Many studies suggest that this mushroom can be utilized as a nutraceutical or functional food to provide health advantages. *Russula cyanoxantha* is a potent source of antioxidants, and its extract has relatively strong antimicrobial activity against different microorganisms (Khan and Chandra 2019). Polysaccharides extracted from the mycelial culture of *R. cyanoxantha* were found to have powerful anti-tumor properties (Zhao et al. 2011). Moreover, its extract also showed a free radical-scavenging ability in the DPPH assay (Ribeiro et al. 2006). *Russula cyanoxantha* is quite rich in many phenolic compounds such as quercetin, quercetin-3-o-rutinoside, catechin, and epicatechin, which act as a potent source of antioxidants (Butkhup et al. 2018). This mushroom contains a variety of triterpene chemicals, including ergosta-4, 6,8 and 22-tetraen-3-one, as well as polysaccharides such as β-glucan (Zhao et al. 2011; Butkhup et al. 2018).

20 ***Russula brevipes*** Peck, Ann. Rep. Reg. N.Y. St. Mus. 43: 66 (1890). **Figure 3.4d**

Pileus 65–115 mm wide, originally white to creamish, soon turned brownish, widely convex, depressed from the center, sometimes develops fractures upon maturity, edges inrolled, context white, thick, and brittle; lamellae decurrent, white or creamy; stipe is 25–45 mm long and up to 25 mm wide, rigid, hard, nearly equal, dry, and creamish.

Habit and habitat: Scattered, growing in association with *Cedrus deodara*.

Specimen examined: India: Jammu and Kashmir, Jammu, Sanasar, Ramban, Lat. 33.123502°N, Lon.75.261483°E, elevation 2047 m, July–September 2018, Komal Verma and Yash Pal Sharma, KV-021.

Edibility: Consumed in the study area, commonly known as 'Sheudi'.

Medicinal properties:*Russula brevipes* is considered a potentially antioxidant-rich mushrooms which could provide considerable health benefits, particularly in the treatment of diseases related to free radical production and oxidative stress (Sharma et al. 2019). Due to its high levels of phenolic compounds and its potency to reduce oxidative stress, *Russula brevipes* is used extensively among many tribal communities (Sharma et al. 2019).

Overall, there are many helpful bioactive chemicals in these mushrooms. These include substances like polysaccharides, phenols, triterpenes, and many others that could be therapeutically beneficial against a wide range of disorders. The potential biological functions of these mushrooms have been outlined in numerous studies. A few of these are summarized in Table 3.1.

TABLE 3.1
Medicinal Properties of Some Wild Edible Mushrooms Collected from Northwest Himalayas of Jammu and Kashmir.

S. No.	Medicinal mushroom	Bioactive compound(s)	Health benefit(s)	Reference(s)
1.	*Morchella esculenta*	Polysaccharides	Anticancer activities against colon cancer cells (HT-29), antioxidant activity, anti-melanogenic, enhances gut microbiota	Liu et al. (2016); Cai et al. (2018); Lee et al. (2018); Huo et al. (2020)
2.	*Geopora arenicola*	Phenols	Antioxidant, prevention of lipid peroxidation, antimicrobial, cardioprotective, and vasodilator effects	Keles et al. (2011); Lalotra et al. (2016).

TABLE 3.1 ***Continued***
Medicinal Properties of Some Wild Edible Mushrooms Collected from Northwest Himalayas of Jammu and Kashmir.

S. No.	Medicinal mushroom	Bioactive compound(s)	Health benefit(s)	Reference(s)
3.	*Lentinus tigrinus*	LTPp (protein)	Anticancer activity against human breast cancer cell lines (MCF-7) and prostate cancer cell lines (PC3), antidiabetic, antibacterial	Dulay et al. (2014); Sevindik (2018); Mohammadnejad et al. (2019)
4.	*Trametes versicolor*	Polysaccharopeptide krestin	Antibacterial and antifungal activities	Chu et al. (2002);
5.	*Schizophyllum commune*	Schizophyllan (β-glucan), iminolactons	Anti-tumor and immunomodulating properties, anticancer	Yao et al. (2016); Mišković et al. (2021)
6.	*Macrolepiota procera*	Polysaccharides, phenolic compounds *viz.*, cinnamic acid, gallic acid, vanillic acid; β-carotene and lycopene	Anticancer effects on A549 lung cancer cells, antioxidant activity	Secme et al. (2018)
7.	*Agaricus campestris*	Extract	Anticancer and antioxidant activity	Kosanic et al. (2017); Gasecka et al. (2018)
8.	*Coprinus comatus*	Ergotheioneine, GABA, glycan-binding protein	Antioxidant and anti-anxiety effects; anticancer activity against breast, ovarian, and prostate cancer; hypoglycemic effects; anticancer activity against human T cell leukemia Jurkat cells	Dotan et al. (2011); Cohen et al. (2014); Zhang et al. (2017)
9.	*Flammulina velutipes*	Polysaccharides, proteins, PUFAs, glycoproteins, phenols, and sesquiterpenes.	Anti-tumor and anticancer activities, anti-atherosclerotic, antihypertensive, and cholesterol-lowering effects, anti-aging, antioxidant properties, anti-inflammatory, immunomodulatory, antibacterial	Tang et al. (2016); Mitra et al. (2020).
10.	*Pleurotus ostreatus*	Ethanolic extracts	Antioxidant and antimicrobial activities	Oyetayo and Ariyo (2013)
11.	*Cantharellus cibarius*	Caffeic acid, ergosterol, ergosterol peroxide, cerevisterol	Anti-inflammatory and antioxidant properties; anticancer activity	Reis et al. (2012)
12.	*Sparassis crispa*	β-glucan	Anticancer, anti-inflammatory, antioxidant and antifungal activities	Kimura et al. (2013); Yamamoto and Kimura (2014)

TABLE 3.1 ***Continued***
Medicinal Properties of Some Wild Edible Mushrooms Collected from Northwest Himalayas of Jammu and Kashmir.

S. No.	Medicinal mushroom	Bioactive compound(s)	Health benefit(s)	Reference(s)
13.	*Laetiporus sulphureus*	Polysaccharides, hydrotrametenolic acid, triterpenes	Antidiabetic, anti-tumor, antioxidant, antimicrobial, anti-inflammatory, immunomodulation and antihyperglycemic activities, antiproliferative activity	Khatua et al. (2017); Zhang et al. (2018)
14.	*Hericium erinaceus*	β-glucan, hericenones and erinacines	Anticancer, antioxidant activity; treats neurodegenerative diseases	Thongbai et al. (2015); Han et al. (2013); Lai et al. (2013)
15.	*Grifola frondosa*	Bioactive metabolites; polysaccharide (a β-glucan proteoglycan)	Anti-tumor activity; tumor suppression; hypoglycemic, antihypertensive, anti-obesity, and anti-hepatitis effects	Alonso et al. (2017, 2018); Venturella et al. (2021); Xiao et al. (2015).
16.	*Auricularia auricula-judae*	Heteropolysaccharides (glucuronoxylomannans), polysaccharides	Anti-tumor activity, anticoagulant activity, wound-healing capacity	Ukai et al. (1983); Gurusamy and Arthe (2012); Khamule et al. (2012)
17.	*Ganoderma lucidum*	Polysaccharides	Anti-tumor activity by promoting the activity of NK cells and cytotoxic T-lymphocytes	Sun et al. (2011); Bishop et al. (2015)
18.	*Tremella fuciformis*	Polysaccharides	Anti-oxidant activity, apoptosis in skin fibroblasts, skin diseases; anti-inflammatory; wound-healing, anti-aging and anti-radiation	Shen et al. (2017); Newman and Cragg (2016)
19.	*Russula cyanoxantha*	Phenolic compounds, polysaccharides	Anti-tumor, antimicrobial activity, antioxidant activity	Khan and Chandra (2019); Zhao et al. (2011); Butkhup et al. (2018)
20.	*Russula brevipes*	Phenolic compounds	Antioxidant-rich, anti-oxidative	Sharma et al. (2019)

3.1.2 Traditional Knowledge of the Important Medicinal Mushrooms in the Region

The present study was based on regular macrofungal forays and personal interviews conducted in different parts of the study area. Valuable information concerning the use of wild medicinal mushrooms by the people of various settlements was extracted. Key informants and residents, especially older ones, of the study area were interviewed using a specially designed, semi-structured questionnaire in their local languages (Kishtwari, Kashmiri, Gojri, and Boddi). Pertinent details regarding the consumption pattern, vernacular names, and market price of the species were also recorded (Table 3.2).

TABLE 3.2
Macrofungal Species along with Their Vernacular Names and Modes of Consumption.

S. No.	Species of fleshy fungus	Vernacular name	Used fresh/dried	Traditional uses: Medicinal uses	Traditional uses: Edibility
1	*Morchella esculenta*	Kich (1, 2, 3), Shamo Tetu (4)	Dried and fresh	Used for treating colds (Nazil[1]) and cough (Cxas[1]), heart ailments (Jigrukdoud[1]), and constipation (Qabzi[3]), enhances lactation in females.	Choice edible
2	*Geopora arenicola*	Kondal (1), Padur (2), Khuduz (3), Shamo (4)	Dried and fresh	Used for treating fever (Taf[1]) and cough (Cxas[2]).	Edible
3	*Coprinus comatus*	Hoon kich (1), Kalu hader (2), Kal head (3) Shamo shi (4)	Fresh	Used to eradicate nutritional disorders (Qilatae-dam[1]).	Edible
4	*Lentinus tigrinus*	Yid-haedar (1), Haedar	Fresh	Used for cough (Cxas[1]).	Edible
5	*Schizophyllum commune*	Helud (1), Lulde khas (2), dhando shak (3), Shamo (4)	Fresh and dried	-	-
6	*Macrolepiota procera*	Tope header (1), Chaltee (2, 3), Shamo (4)	Fresh	-	Edible
7	*Flammulina velutipes*	Tele header (1), Dailoo (2), Bhutoo (3), Shamo (4)	Fresh	Used for diabetes (Ziyabatees[1])	Edible
8	*Pleurotus ostreatus*	Sirjli (1), Saroori (2,3) Shairee (3), Shamo (4)	Fresh	Used for treating diabetes (Ziyabatees[1]) and bronchitis (Dam[3]).	Edible
9	*Cantharellus cibarius*	Wan haeder (1), Haldee chaltee (2,3), Shistu (4)	Fresh and dried	Used in wound healing, general body weakness (Kamzuri[5])	Edible
10	*Agaricus campestris*	Header (1), Chaltee (2, 3), Shamo (4)	Fresh	Used for treatment of cough (Cxas[1])	Edible
11	*Ramaria apiculata*	Shairee (1, 2, 3), Shamo (4)	Fresh	Used for hypertension (fishariekhoon[1]), general weakness (Kamzuri[5])	Edible
12	*Sparassis crispa*	Bhaud shairee (1), Roa Gabur (1,2,3), Shamo (4)	Fresh and dried	Used for diabetes (Ziyabatees[1]) and diarrhea (Misil[1])	Edible
13	*Laetiporus sulphureus*	Kukad shaeodi (3)	Fresh and dried	Used for heart problems (Jigrukdoud[1])	Edible
14	*Hericium erinaceus*	Shairee (1, 2, 3) Shamo (4)	Fresh	Used for treating cough (Cxas[1]), asthma (Shaah[3]), cancer (Sartan[1]), and for many diseases	Edible
15	*Grifola frondosa*	Kokad-shaeodi (3)	Fresh and dried	Used for treating fever (Taf[1]) and cough (Cxas[1]).	Edible
16	*Auricularia auricula-judae*	Kan kulis (1, 2), Kan padu (3) Shamo (4)	Fresh	Its soup is used for treating dehydration (Khoski[1])	Edible
17	*Ganoderma lucidum*	Sup-shaeodi (5)	Dried	Used for treating cold (Nazil[1]) and its soup is used as a tonic (Taqatbadawan khetri[1])	-
18	*Trametes versicolor*		Dried	-	-
19	*Russula cyanoxantha*	Merch (3), Nilli chatteri (5)	Fresh	-	Edible
20	*Russula brevipes*	Chitti shaeodi (5), chatri	Fresh	Used to eradicate nutritional disorders	Edible

Languages: 1 = Kashmiri, 2 = Gojri, 3 = Kishtwari, 4 = Bodhi, 5= Dogri, -= Not known

3.4 CONCLUSIONS AND FUTURE PERSPECTIVES

Because of their varied topography, climatic conditions, and dense forests cover, the northwestern Himalayas are considered one of the most diverse of all habitats, providing a conducive environment for the growth of wild mushrooms. Due to the richness in variety of mushrooms and their novel biochemical constituents, there is a good chance that the area contains many more interesting wild medicinal mushroom species and, hence, is wide open for exploration. Furthermore, indigenous knowledge gleaned from the ethnomycological survey can be used as input for mycomedicine and food security research. To diversify cultivation, the results of study support the domestication and cultivation of indigenous culturable edible mushrooms. Furthermore, the immunomodulatory, antiviral, and anti-inflammatory properties of mushroom extracts may be ideal for treating COVID-19 infection and severe lung inflammation. Therefore, more emphasis should be given to the bioprospection of these mushrooms for bioactive constituents. With the growing world population and dwindling natural resources, increased malnutrition, and pandemics, priority should be placed on maximizing the benefits of mushroom nutraceuticals that can potentially lead to a better knowledge of the development of functional foods and medicines.

3.5 ACKNOWLEDGMENTS

The authors are grateful to the head of the Department of Botany (SAP-DRS-II), University of Jammu, Jammu for providing laboratory facilities. Financial assistance received from the JK Science Technology & Innovation Council (vide no. JKST&IC/SRE/929–31) is duly acknowledged.

REFERENCES

Ahmad, N., F. Mahmood, S.A. Khalil, R. Zamir, et al. 2014. Antioxidant activity *via* DPPH, gram positive and gram-negative antimicrobial potential in edible mushrooms. *Toxicology and Industrial Health* 30: 826–834.

Alonso, E.N., M.J. Ferronato, M.E. Fermento, N.A. Gandini, A. López Romero, J.A Guevara, M.M. Facchinetti & A.C. Curino. 2018. Antitumoral and antimetastatic activity of Maitake D-fraction in triple-negative breast cancer cells. *Oncotarget* 9: 23396–23412. http://doi.org/10.18632/oncotarget.25174.

Alonso, E.N., M.J. Ferronato, N.A. Gandini, M.E. Fermento, D.J. Obiol, A. López Romero, J. Arévalo, M.E. Villegas, M.M. Facchinetti & A.C. Curino. 2017. Antitumoral effects of D-fraction from *Grifola frondosa* (Maitake) mushroom in breast cancer. *Nutrition and Cancer* 69: 29–43. http://doi.org/10.1080/01635581.2017.1247891.

Alonso, E.N., M. Orozco, A.E. Nieto & G.A. Balogh. 2013. Genes related to suppression of malignant phenotype induced by Maitake D-Fraction in breast cancer cells. *Journal of Medicinal Food* 16: 602–617. http://doi.org/10.1089/jmf.2012.0222

Antonelli, A., C. Fry, R.J. Smith, et al. 2020. *State of the World's Plants and Fungi 2020*. Royal Botanic Gardens, Kew.

Arora, D. 1986. *Mushroom Demystified: A Comprehensive Guide to the Fleshy Fungi*. Ten Speed Press, Berkeley, CA.

Arunachalam, K., P.S. Sasidharan & X. Yang. 2022. A concise review of mushrooms antiviral and immunomodulatory properties that may combat against COVID-19. *Food Chemistry Advances* 1: 2772–2780.

Asatiani, M.D., S.P. Wasser, E.D. Nevo, N. Ruimi, J.A. Mahajna & A. Reznick. 2011. The shaggy inkcap medicinal mushroom, *Coprinus comatus* (O.F. Mull.: Fr.) Pers. (*Agaricomycetideae*) substances interfere with H_2O_2induction of the NF-κB pathway through Inhibition of IκBα phosphorylation in MCF7 breast cancer cells. *International Journal of Medicinal Mushrooms* 13: 19–25.

Bains, A. & P. Chawla. 2020. *In vitro* bioactivity, antimicrobial and anti-inflammatory efficacy of modified solvent evaporation assisted *Trametes versicolor* extract. *Biotech* 10: 404–410.

Bhatt, R.P., U. Singh & P. Uniyal. 2018. Healing mushrooms of Uttarakhand Himalaya, India. *Current Research in Environmental & Applied Mycology (Journal of Fungal Biology)* 8: 1–23.

Bishop, K.S., C.H.J. Kao, Y. Xu, M.P. Glucina, et al. 2015. From 2000 years of *Ganoderma lucidum* to recent developments in nutraceuticals. *Phytochemistry* 114: 56–65.

Bourhia, M., F.E.Z. Amrati, R. Ullah, et al. 2020. Coronavirus treatments: What drugs might work against COVID-19? *Natural Product Communications* 15: 1–7. http://doi.org/10.1177/1934578X20945442

Bubueanu, C., G. Popa & L. Pirvu. 2015. Comparative analysis of polyphenolic profiles and antioxidant activity of *Agaricus bisporus* and *Agaricus campestris*. *Scientific Bulletin Series F. Biotechnologies* 14: 29–33.

Bulam S., N.Ş. Üstün & A. Pekşen. 2019. Health benefits of *Ganoderma lucidum* as a medicinal mushroom. *Turkish Journal of Agriculture—Food Science and Technology* 7: 84–93. http://doi.org/10.24925/turjaf.v7isp1.84-93.2728.

Butkhup, L., W. Samappito & S. Jorjong. 2018. Evaluation of bioactivities and phenolic contents of wild edible mushrooms from northeastern Thailand. *Food Science and Biotechnology* 27: 193–202. http://doi.org/10.1007/s10068-017-0237-5

Cai, Z.N., W. Li, S. Mehmood, W.J. Pan, Q.X. Wu, Y. Chen & Y.M. Lu. 2018. Effect of polysaccharide FMP-1 from *Morchella esculenta* on melanogenesis in B16F10 cells and zebrafish. *Food & Function Journal* 9: 5007–5015.

Cannon, P.F., B. Aguirre-Hudson, M.C. Aime, M.I. Bidartondo, E. Gaya, D. Hawksworth, P. Kirk, L.J. Leitch & R. Lücking. 2018. *State of the World's Fungi*. Royal Botanic Gardens, Kew, pp. 1–92.

Chang, S.T. & P.G. Miles. 2004. *Mushrooms: Cultivation, Nutritional Value, Medicinal Effect, and Environmental Impact*. 2nd ed. CRC Press, Boca Raton, FL.

Chang, S.T. & S.P. Wasser. 2017. The cultivation and environmental impact of mushrooms. In: *Oxford Research Encyclopedia. Environmental Science-Agriculture and the Environment*. Oxford University Press, Oxford, pp. 1–39.

Chauhan, P.S., M. Kumarasamy, A. Sosnik & D. Danino. 2019. Enhanced thermostability and anticancer Activity in breast cancer cells of laccase immobilized on pluronic-stabilized nanoparticles. *ACS Applied Materials & Interfaces* 11: 39436–39448.

Chen, N., H. Zhang, X. Zong, S. Li, J. Wang, Y. Wang & M. Jin. 2020. Polysaccharides from *Auricularia auricula*: Preparation, structural features and biological activities. *Carbohydrate Polymers* 247: 116750.

Chong, P.S., M. Fung, K.H. Wong & L.W. Lim. 2020. Therapeutic potential of *Hericium erinaceus* for depressive disorders. *International Journal of Molecular Sciences* 21: 163–168.

Chu, K.K., S.S. Ho & A.H. Chow. 2002. *Coriolus versicolor*: A medicinal mushroom with promising immunotherapeutic values. *The Journal of Clinical Pharmacology* 42: 976–984.

Cohen, N., J. Cohen, M.D. Asatiani, V.K. Varshney, H.T. Yu, Y.C. Yang, Y.H. Li, J.L. Mau & S.P. Wasser. 2014. Chemical composition and medicinal value of fruit bodies and submerged cultured mycelia of nutritional culinary-medicinal higher Basidiomycetes mushrooms. *International Journal of Medicinal Mushrooms* 16: 273–291.

Das, K. & J.R. Sharma. 2005. *Russulaceae of Kumaon Himalaya*. Botanical Survey of India, Ministry of Environment and Forests, Govt. of India, Kolkata.

Dotan, N., P.W. Solomon & J. Mahajna. 2011. Inhibition of the Androgen receptor activity by *Coprinus comatus* substances. *Nutrition and Cancer* 63: 1316–1327. http://doi.org/10.1080/01635581.2011.607542

Dulay, R.M.R., M.C. Arenas, S.P. Kalaw, R.G. Reyes & E.C. Cabrera. 2014. Proximate composition and functionality of the culinary-medicinal tiger saw gill mushroom, *Lentinus tigrinus* (Higher *Basidiomycetes*), from the Philippines. *International Journal of Medicinal Mushrooms* 16: 85–94.

Dulay, R.M.R., A. Luilene, J.S. Miranda, S.P. Malasaga, R. Kalaw, G. Reyes & C.T. Hou. 2017. Antioxidant and antibacterial activities of acetonitrile and hexane extracts of *Lentinus tigrinus* and *Pleurotus djamour*. *Biocatalysis and Agricultural Biotechnology* 9: 141–144.

Elhusseiny, S.M., T.S. El-Mahdy, M.F. Awad, N.S. Elleboudy, M.M.S. Farag, M.A. Yassein & K.M. Aboshanab. 2021. Proteome analysis and in vitro antiviral, anticancer and antioxidant capacities of the aqueous extracts of *Lentinula edodes* and *Pleurotus ostreatus* edible mushrooms. *Molecules* 26: 4623. http://doi.org/10.3390/molecules26154623

El Sheikha, A.F. 2022. Nutritional profile and health benefits of *Ganoderma lucidum* "Lingzhi, Reishi, or mannentake" as functional foods: Current scenario and future perspectives. *Foods* 11: 1030.

Erbiai, E.H., L.P. Da Silva, R. Saidi, Z. Lamrani, J.C.G. Esteves da Silva & A. Maouni. 2021. Chemical composition, bioactive compounds, and antioxidant activity of two wild edible mushrooms *Armillaria mellea* and *Macrolepiota procera* from two countries (Morocco and Portugal). *Biomolecules* 11: 575. http://doi.org/10.3390/biom11040575.

Facchini, J.M., E.P. Alves, C. Aguilera, R.M. Gern, et al. 2014. Antitumor activity of *Pleurotus ostreatus* polysaccharide fractions on Ehrlich tumor and Sarcoma 180. *International Journal of Biological Macromolecules* 68: 72–77.

Fernandes, Â., J.C. Barreira, A.L. Antonio, M.B.P. Oliveira, A. Martins & I.C. Ferreira. 2014. Effects of Gamma irradiation on chemical composition and antioxidant potential of processed samples of the wild mushroom *Macrolepiota procera*. *Food Chemistry* 149: 91–98.

Frost, M. 2016. Three popular medicinal mushroom supplements: A review of human clinical trials. *All Faculty Publications* 1609.

Gąsecka, M., Z. Magdziak, M. Siwulski, et al. 2018. Profile of phenolic and organic acids, antioxidant properties and ergosterol content in cultivated and wild growing species of *Agaricus*. *European Food Research and Technology* 244: 259–268. http://doi.org/10.1007/s00217-017-2952-9

Ghosh, S., S. Nandi, A. Banerjee, S. Sarkar, N. Chakraborty & K. Acharya. 2021. Prospecting medicinal properties of Lion's mane mushroom. *Journal of Food Biochemistry*: e13833. http://doi.org/10.1111/jfbc.13833

Gu, Y. & J. Leonard. 2006. In vitro effects on proliferation, apoptosis and colony inhibition in ER-dependent and ER-independent human breast cancer cells by selected mushroom species. *Oncology Reports* 15: 417–423.

Guggenheim, A.G., K.M. Wright & H. Zwickey. 2014. Immune modulation from five major mushrooms: Application to integrative oncology. *Integrative Medicine (Encinitas)* 13: 32–44.

Gurusamy, R. & R. Arthe. 2012. Effect of medicinal mushroom, *Auricularia auricula-judae* polysaccharides against EAC cell lines. *Research Journal of Biotechnology* 7: 14–17.

Han, C., J. Yuan, Y. Wang & L. Li. 2006. Hypoglycemic activity of fermented mushroom of *Coprinus comatus* rich in vanadium. *Journal of Trace Elements in Medicine and Biology* 20: 191–196. http://doi.org/10.1016/j.jtemb.2006.06.003.

Han, Z.H., J.M. Ye & G.F. Wang. 2013. Evaluation of in vivo antioxidant activity of *Hericium erinaceus* polysaccharides. *International Journal of Biological Macromolecules* 52: 66–71.

Han, J.M., E.K. Lee, S.Y. Gong, J.K. Sohng, Y.J. Kang & H.J. Jung. 2019. *Sparassis crispa* exerts anti-inflammatory activity via suppression of TLR-mediated NF-κB and MAPK signaling pathways in LPS-induced RAW264.7 macrophage cells. *Journal of Ethnopharmacology* 231: 10–18. http://doi.org/10.1016/j.jep.2018.11.003

Hassan, K., B. Matio Kemkuignou & M. Stadler. 2021. Two new triterpenes from basidiomata of the medicinal and edible mushroom, *Laetiporus sulphureus*. *Molecules* 24(26): 7090. http://doi.org/10.3390/molecules26237090.

Heilmann-Clausen, J., A. Verbeken & J. Vesterholt. 1998. The genus *Lactarius*. *Fungi of Northern Europe* 2: 1–287.

Hesler, L.R. & A.H. Smith. 1979. *North American Species of Lactarius*. The University of Michigan Press, Ann Arbor.

Huo, W., P. Qi, L. Cui, L. Zhang, L. Dai, Y. Liu, S. Hu, Z. Feng, T. Qiao & J. Li. 2020. Polysaccharide from wild morels alters the spatial structure of gut microbiota and the production of short-chain fatty acids in mice. *Bioscience of Microbiota, Food and Health* 39: 219–226.

Hyde, K.D., J. Xu, S. Rapior, R. Jeewon, S. Lumyong, A.G.T. Niego, P.D Abeywickrama, J.V.S. Aluthmuhandiram, R.S. Brahamanage, S.S. Brook, et al. 2019. The amazing potential of fungi: 50 ways we can exploit fungi industrially. *Fungal Diversity* 97: 1–136. http://doi.org/10.1007/s13225-019-00430-9.

Islam, T., K. Ganesan & B. Xu. 2021. Insights into health-promoting effects of Jew's ear (*Auricularia auricula-judae*). *Trends in Food Science & Technology* 114: 552–569. http://doi.org/10.1016/j.tifs.2021.06.017.

Jeitler, M., A. Michalsen, D. Frings, M. Hübner, M. Fischer, D.A. Koppold-Liebscher, V. Murthy, & C.S. Kessler. 2020. Significance of medicinal mushrooms in integrative oncology: A narrative review. *Frontiers in Pharmacology* 11: 580–656.

Keles, A., L. Koca & H. Genccelep. 2011. Antioxidant properties of wild edible mushrooms. *Journal of Food Processing and Technology* 2: 130–135.

Khamule, R., N. Naksupan, A. Ounaroon & N. Saelim. 2012. *Skin Wound Healing Promoting Effect of Polysaccharides Extracts from Tremella fuciformis and Auricularia auricula on the Ex Vivo Porcine Skin Wound Healing Model*. 4th International Conference on Chemical, Biological and Environmental Engineering IPCBEE Press, Singapore, 43: 20.

Khan, F. & R. Chandra. 2019. Bioprospecting of wild mushrooms from India with respect to their medicinal aspects. *International Journal of Medicinal Mushrooms* 21: 181–192. http://doi.org/10.1615/IntJMedMushrooms.2019029969.

Khatua, S., S. Ghosh & K. Acharya. 2017. *Laetiporus sulphureus* (Bull.: Fr.) Murr. as food as medicine. *Pharmacognosy Journal* 9: 1–15.

Kibby, G. & R. Fatto. 1990. *Keys to the Species of Russula in Northeastern North America*. Kibby Fatto Enterprises, Somerville, NJ.

Kim, J.A., E. Lau, D. Tay & E.J.C.D. Blanco. 2011. Antioxidant and NF-k D inhibitory constituents isolated from *Morchella esculenta*. *Natural Product Research* 25: 1412–1417.

Kimura, Y., C. Satou, S. Fujioka, W. Shoji, K. Umeda, T. Ishizuka, H. Yawo & S. Higashijima. 2013. Hindbrain V2a neurons in the excitation of spinal locomotor circuits during Zebrafish swimming. *Current Biology* 23: 843–849.

Knezevic, A., M. Stajic, I. Sofrenic, T. Stanojkovic, I. Milovanovic, V. Tesevic & J. Vukojevic 2018. Antioxidative, antifungal, cytotoxic and anti-neurodegenerative activity of selected *Trametes* species from Serbia. *PLoS ONE* 13: e0203064.

Kornerup, A. & J.H. Wanscher. 1978. *Methuen Handbook of Colour*, 3rd ed. Eyre Methuen Ltd., London, p. 252.

Kosanić, M., B. Ranković, A. Rančić & T. Stanojković. 2017. Evaluation of metal contents and bioactivity of two edible mushrooms *Agaricus campestris* and *Boletus edulis*. *Emirates Journal of Food and Agriculture*29: 98–103.

Kumar, A., T. Mehmood, N.S. Atri & Y.P. Sharma. 2021. Revised and an updated checklist of the *Amanitaceae* from India with its specific distribution in the Indian States. *Nova Hedwigia* 112: 223–240.

Lai, P.L., M. Naidu, V. Sabaratnam, K.H. Wong, R.P. David, U.R. Kuppusamy, N. Abdullah & S.N.A. Malek. 2013. Neurotrophic properties of the lion's mane medicinal mushroom, *Hericium erinaceus* (higher Basidiomycetes) from Malaysia. *International Journal of Medicinal Mushrooms* 15: 539–554.

Lalotra, P., P. Bala, S. Kumar & Y.P. Sharma. 2016. Biochemical characterization of some wild edible mushrooms from Jammu and Kashmir. *The Proceedings of the National Academy of Sciences, India, Section B: Biological Sciences* 88: 539–545. https://doi.org/10.1007/s40011-016-0783-2

Largent, D., D. Johnson & R. Watling. 1977. *How to Identify Mushrooms to Genus III: Microscopic Features*. Mad River Press Inc., Eureka, CA, p. 122.

Lee, S.R., H.-S. Roh, S. Lee, H.B. Park, T.S. Jang, Y.-J. Ko, K.-H. Baek & K.H. Kim. 2018. Bioactivity-guided isolation and chemical characterization of antiproliferative constituents from morel mushroom (*Morchella esculenta*) in human lung adenocarcinoma cells. *Journal of Functional Foods* 40: 249–260.

Lee, Y.H., H.M. Gwon, O. Gu, J.I. Choi & D.H. Jeon. 2018. Mycelial growth characteristics of *Sparassis latifolia* according to liquid media and incubation conditions. *Journal of Mushrooms* 16: 9–102.

Li, H., Y. Tian, Jr N. Menolli, L. Ye, S.C. Karunarathna, J. Perez-Moreno, M.M Rahman, M.H. Rashid, et al. 2021. Reviewing the world's edible mushroom species: A new evidence-based classification system. *Comprehensive Reviews in Food Science and Food Safety* 20: 1982–2014.

Li, X., P. Chen, P. Zhang, Y. Chang, M. Cui & J. Duan. 2019. Protein-Bound beta-glucan from *Coriolus versicolor* has potential for use against obesity. *Molecular Nutrition & Food Research* 63: e1801231.

Liu, C., Y. Sun, Q. Mao, X. Guo, P. Li, Y. Liu & N. Xu. 2016. Characteristics and antitumor activity of *Morchella esculenta* polysaccharide extracted by pulsed electric field. *International Journal of Molecular Sciences* 17: 986.

Liu, E., Y. Ji, F. Zhang, B. Liu & X. Meng. 2021. Review on *Auricularia auricula-judae* as a functional food: Growth, chemical composition, and biological activities. *Journal of Agricultural and Food Chemistry* 69: 1739–1750. http://doi.org/10.1021/acs.jafc.0c05934.

Liu, H. & L. He. 2012. Comparison of moisturizing properties of polysaccharides from *Tremella fuciformis* and hyaluronic acid. *Journal of Anhui Agricultural Sciences* 40: 13093–13094.

Ma, X., M. Yang, Y. He, C. Zhai & C. Li. 2021. A review on the production, structure, bioactivities and applications of *Tremella* polysaccharides. *International Journal of Immunopathology and Pharmacology* 35: 20–58.

Mapoung, S., S. Umsumarng, W. Semmarath, P. Arjsri, P. Thippraphan, S. Yodkeeree, Limtrakul, & P. Dejkriengkraikul. 2021. Skin wound-healing potential, of polysaccharides from medicinal mushroom *Auricularia auricula-judae* (Bull.). *Journal of Fungi* 7: 247. http://doi.org/10.3390/jof7040247.

Mattos-Shipley, K.M.J., K.L. Ford, F. Alberti, A.M. Banks, A.M. Bailey & G.D. Foster. 2016. The good, the bad and the tasty: The many roles of mushrooms. *Studies in Mycology* 85: 125–157. http://doi.org/10.1016/j.simyco.2016.11.002

Mendez-Espinoza, C., E. Garcia-Nieto, A.M. Esquivel, M.M. Gonzalez, E.V. Bautista, C.C. Ezquerro & L.J. Santacruz. 2013. Antigenotoxic potential of aqueous extracts from the chanterelle mushroom, *Cantharellus cibarius* (higher *Basidiomycetes*) on human mononuclear cell cultures. *International Journal of Medicinal Mushrooms* 15: 325–332. http://doi.org/10.1615/IntJMedMushr.v15. i3.90

Mišković, J., M. Karaman, M. Rašeta, N. Krsmanović, S. Berežni, D. Jakovljević, F. Piattoni, A. Zambonelli, M.L. Gargano & G. Venturella. 2021. Comparison of two *Schizophyllum commune* strains in production of acetylcholinesterase inhibitors and antioxidants from submerged cultivation. *Journal of Fungi* 7: 115.

Mitra, S., P. Bhartiya, N. Kaushik, L.N. Nguyen, R. Wahab, S. Bekeschus, E.H. Choi & N.K. Kaushik. 2020. Plasma-treated *flammulina velutipes* derived extract showed anticancer potential in human breast cancer cells. *Applied Sciences* 10: 8395.

Mohammadnejad, S., H.R. Pourianfar, A. Drakhshan, I. Jabaleh & M. Rezayi. 2019. Potent antiproliferative and pro-apoptotic effects of a soluble protein fraction from culinary-medicinal mushroom *Lentinus tigrinus* on cancer cells. *Journal Food Measurement and Characterization* 13: 3015–3024.

Mowsumi, F.R. & M.B.K. Choudhary. 2010. Oyster mushroom: biochemical and medicinal prospects. *Bangladesh Journal of Medical Biochemistry* 3: 23–28.

Muszyńska, B., K. Kała, A. Firlej & K. Sułkowska-Ziaja. 2016. *Cantharellus Cibarius*—Culinary-medicinal mushroom content and biological Activity. *Acta Poloniae Pharmaceutica* 73: 589–98.

Newman, D.J. & G.M. Cragg. 2016. Natural products as sources of new drugs from 1981 to 2014. *Journal of Natural Products* 79: 629–661.

Niazi, A.R. & H. Ijaz. 2021. Proximate analysis and in vitro biological activities of caulifower mushroom, *Sparassis crispa* (*Agaricomycetes*), from Pakistan. *International Journal of Medicinal Mushrooms* 23: 79–84.

Nowacka-Jechalke, N., R. Nowak, M.K. Lemieszek, W. Rzeski, U. Gawlik-Dziki, N. Szpakowska & Z. Kaczýnski. 2021. Promising potential of crude polysaccharides from *Sparassiscrispa* against colon cancer: an in vitro study. *Nutrients* 13: 161.

Oyetayo, V.O. & O.O. Ariyo. 2013. Antimicrobial and antioxidant properties of *Pleurotus ostreatus*, cultivated on different tropical woody substrates. *Journal of Waste Conversion, Bioproducts and Biotechnology* 6: 223–226.

Qu, Y., X. Zhao, H. Guo, Y. Meng, Y. Wang, Y. Zhou & L. Sun. 2021. Structural analysis and macrophage activation of a novel β-glucan isolated from *Cantharellus cibarius*. *International Journal of Molecular Medicine* 47: 50. https://doi.org/10.3892/ijmm.2021.4883

Ragasa, C.Y., M.C.S. Tan, M.E. De Castro, MM. De Los Reyes, G.G. Oyong & C.C. Shen. 2018. Sterols from *Lentinus tigrinus*. *Pharmacognosy Journal* 10: 1079–1081.

Ratnaningtyas, N.I., H. Hernayanti, N. Ekowati & F. Husen. 2022. Ethanol extract of the mushroom *Coprinus comatus* exhibits antidiabetic and antioxidant activities in streptozotocin-induced diabetic rats. *Pharmaceutical Biology* 60: 1126–1136.

Reis, F., A. Martins, L. Barros, et al. 2012. Antioxidant properties and phenolic profile of the most widely appreciated cultivated mushrooms: a comparative study between in vivo and in vitro samples. *Food and Chemical Toxicology* 50: 1201–1207. https://doi.org/10.1016/j.fct.2012.02.013

Ribeiro, B., J. Rangel, O.P. Valenta, P. Baptista, R.M. Seabra & P.B. Andrade. 2006. Contents of carboxylic acids and two phenolics and antioxidant activity of dried Portuguese wild edible mushrooms. *Journal of Agricultural and Food Chemistry* 54: 8530–8537.

Roca-Lema, D., O. Martinez-Iglesias, C. Fernandez de Ana Portela, A. Rodriguez-Blanco, M. Valladares Ayerbes, A. Diaz-Diaz, A. Casas-Pais, C. Prego & A. Figueroa. 2019. In vitro anti-proliferative and anti-invasive effect of polysaccharide-rich extracts from *Trametes versicolor* and *Grifola frondosa* in colon cancer cells. *International Journal of Medical Sciences* 16: 231–240.

Rouhana-Toubi, A., S.P. Wasser, A. Agbarya & F. Fares. 2013. Inhibitory Effect of Ethyl AcetatecExtract of the shaggy inc cap medicinal mushroom, *Coprinus comatus* (higher basidiomycetes) fruit bodies on cell growth of human ovarian cancer. *International Journal of Medicinal Mushrooms* 15: 457–470.

Sandargo, B., C. Chepkirui, T. Cheng, L. Chaverra-Muñoz, B. Thongbai, M. Stadler & S. Hüttel. 2019. Biological and chemical diversity go hand in hand: Basidiomycota as source of new pharmaceuticals and agrochemicals. *Biotechnology Advances* 37: 107344.

Seçme, M., O. Kaygusuz, C. Eroglu, Y. Dodurga, O.F. Colak & P. Atmaca. 2018. Potential anticancer activity of the parasol mushroom, *Macrolepiota procera* (*Agaricomycetes*), against the A549 human lung cancer cell line. *International Journal of Medicinal Mushrooms* 20: 1–10.

Semwal, K.C., S.L. Stephenson, V.K. Bhatt & R.P. Bhatt. 2014. Edible mushrooms of the Northwestern Himalaya, India: a study of indigenous knowledge, distribution and diversity. *Mycosphere* 5: 440–461.

Sevindik, M. 2018. Investigation of antioxidant/oxidant status and antimicrobial activities of *Lentinus tigrinus*. *Advances in Pharmacological and Pharmaceutical Sciences* 1718025: 1–4.

Shahrajabian, M.H., S. Wenli, S. Hong & Q. Cheng. 2020. Chemical compounds and health benefits of *Tremella*, a valued mushroom as both cuisine and medicine in ancient China and modern era. *Amazonian Journal of Plant Research* 4: 692–697.

Sharif-Rad, J., M. Butnariu, S.M. Ezzat, C.O. Adetunji, M. Imran, S.R. Sobhani, T. Tufail, T. Hosseinabadi, K. Ramírez-Alarcón, M. Martorell, A. Maroyi & N. Martins. 2020. Mushrooms-rich preparations on wound healing: from nutritional to medicinal attributes. *Frontiers in Pharmacology* 11: 567518.

Sharma, N., A. Tapwal, R. Verma, D. Kumar, E. Nepovimova & K. Kuca. 2022. Medicinal, nutritional, and nutraceutical potential of *Sparassis crispa* s. lat.: A review. *IMA Fungus* 13: 8.

Sharma, Y.P., R. Sharma, S. Khatua & K. Acharya. 2019. Morphotaxonomy and comparative mycochemical study and antioxidant activity of hydromethanol, infusion and decoction extracts from *Russula brevipes*. *Indian Phytopathology* 72: 445–452.

Shen, T., C. Duan, B. Chen, M. Li, Y. Ruan, D. Xu, D. Shi, D. Yu, J. Li & C. Wang. 2017.*Tremella fuciformis* polysaccharide suppresses hydrogen peroxide-triggered injury of human skin fibroblasts via upregulation of SIRT1. *Molecular Medicine Reports* 16: 1340–1346.

Singer, R. 1986. *The Agaricales in Modern Taxonomy*, 4th ed. Bishan Singh Mahendra Pal Singh, Dehradun.

Spelman, K., E. Sutherland & A. Bagade. 2017. Neurological activity of Lion's mane (*Hericium erinaceus*). *Journal of Restorative Medicine* 6: 16–26.

Stenmark, H. 2009. Rab GTPases as coordinators of vesicle traffic. *Nature Reviews Molecular Cell Biology* 10: 513–525.

Sullivan, R., J.E. Smith & N.J. Rowan. 2006. Medicinal mushrooms and cancer therapy: translating a traditional practice into western medicine. *Perspectives in Biology and Medicine* 49: 159–170.

Sun, L.X., Z.B. Lin, X.J. Li, M. Li, et al. 2011. Promoting effects of *Ganoderma lucidum* polysaccharides on B16F10 cells to activate lymphocytes. *Basic and Clinical Pharmacology and Toxicology* 108: 149–154.

Sun, Y. & J. Liu. 2009. Purification, structure and immunobiological activity of a water-soluble polysaccharide from the fruiting body of *Pleurotus ostreatus*. *Bioresource Technology* 100: 983–986.

Tang, C., P.C. Hoo, L.T. Tan, P. Pusparajah, T.M. Khan, L.H. Lee, B.H. Goh & K.G. Chan. 2016. Golden Needle Mushroom: A culinary medicine with evidenced-based biological activities and health promoting properties. *Frontiers in Pharmacology* 7: 474.

Thongbai, B., S. Rapior, K.D. Hyde, K. Wittstein & M. Stadler. 2015. *Hericium erinaceus*, an amazing medicinal mushroom. *Mycological Progress* 14: 1–23.

Trigos, Á., & J.S. Medellín. 2011. Biologically active metabolites of the genus *Ganoderma*: Three decades of myco-chemistry research. *Re Vista Mexicana D E Micología* 34: 63–83.

Ukai, S., T. Kiho, C. Hara, M. Morita, A. Goto, N. Imaizumi & Y. Hasegawa. 1983. Polysaccharides in fungi. XIII. Antitumor activity of various polysaccharides isolated from *Dictyophora indusiata*, *Ganoderma japonicum*, *Cordyceps cicadae*, *Auricularia auricula-judae*, and *Auricularia* species. *Chemical and Pharmaceutical Bulletin* 31: 741–744.

Urbancikova, I., D. Hudackova, J. Majtan, Z. Rennerova, P. Banovcin & M. Jesenak. 2020. Efficacy of pleuran (β-Glucan from *Pleurotus ostreatus*) in the management of herpes simplex virus type 1 infection. *Evidence-Based Complementary and Alternative Medicine*: 8562309. http://doi.org/10.1155/2020/8562309

Venturella, G., V. Ferraro, F. Cirlincione & M.L. Gargano. 2021. Medicinal mushrooms: bioactive compounds, use, and clinical trials. *International Journal of Molecular Sciences* 22: 634.

Verma, K., S.A.J. Hashmi, N.S Atri & Y.P. Sharma. 2019. A checklist of agaricoid russulaceous mushrooms from Jammu and Kashmir, India. *Kavaka* 53: 42–47.

Vlasenko, V., D. Turmunkh, E. Ochirbat, D. Budsuren, K. Nyamsuren & J. Samiya. 2019. Medicinal potential of extracts from the chanterelle mushroom, *Cantharellus cibarius* (Review) and prospects for studying its strains from differs plant communities of ultra-continental regions of the Asia. *BIO Web of Conferences* 16: 39.

Waktola, G. & T. Temesgen. 2018. Application of mushroom as food and medicine. *Advances in Biotechnology and Microbiology* 11: 97–101. http://doi.org/10.19080/AIBM.2018.11.555817

Wang, Y. & M. Guo. 2020. Purification and structural characterization of polysaccharides isolated from *Auricularia cornea*. *Carbohydrate Polymers* 230: 115680.

Wang, Z., J. Liu, X. Zhong, J. Li, X. Wang, L. Ji & X. Shang. 2019. Rapid characterization of chemical components in edible mushroom *Sparassis crispa* by UPLC-orbitrap MS analysis and potential inhibitory effects on allergic rhinitis. *Molecules* 24: 3014. http://doi.org/10.3390/molecules24163014

Wongaem, A., O. Reamtong, P. Srimongkol, P. Sangtanoo, T. Saisavoey & A. Karnechanatat. 2021. Antioxidant properties of peptides obtained from the split gill mushroom (*Schizophyllum commune*). *Journal of Food Science of Technology* 58: 680–691. http://doi.org/10.1007/s13197-020-04582-4

Wu, Y., M.H. Choi, J. Li, H. Yang & H.J. Shin. 2016. Mushroom cosmetics: the present and future. *Cosmetics* 3: 22. http://doi.org/10.3390/cosmetics3030022

Xiao, C., Q. Wu, Y. Xie, J. Zhang & J. Tan. 2015. Hypoglicemic effects of *Grifola frondosa* (M) polysaccharides F2 and F3 through improvement of insulin in diabetic rats. *Food &Function* 6: 3567–3575.

Xie, Y., F. Yang, W. Tan, X. Li, C. Jiao, R. Huang & B.B. Yang. 2016. The anti-cancer components of *Ganoderma lucidum* possesses cardiovascular protective effect by regulating circular RNA expression. *Oncoscience* 3: 203–207.

Xu, Y., L. Xie, Z. Zhang, W. Zhang, J. Tang, X. He, J. Zhou & W. Peng. 2021.*Tremella fuciformis* polysaccharides inhibited colonic inflammation in dextran sulfate sodium-treated mice via foxp3+ t cells, gut microbiota, and bacterial metabolites. *Frontiers in Immunology* 12: 648162. http://doi.org/10.3389/fimmu.2021.648162

Xu, Y., H. Yuan, Y. Luo, Y. Zhao & J. Xiao. 2020. Ganoderic acid D protects human amniotic mesenchymal stem cells against oxidative stress-induced senescence through the PERK/NRF2 signaling pathway. *Oxidative Medicine and Cellular Longevity* 2020: 18.

Yamamoto, K. & T. Kimura. 2014. Orally and topically administered *Sparassis crispa* (Hanabiratake) improved healing of skin wounds in mice with streptozotocin-induced diabetes. *Bioscience, Biotechnology and Biochemistry* 77: 1303–1305.

Yang, M., T. Belwal, H.P. Devkota, et al. 2019. Trends of utilizing mushroom polysaccharides (MPs) as potent nutraceutical components in food and medicine: a comprehensive review. *Trends in Food Science & Technology* 92: 94–110.

Yao, H.M., G. Wang, Y.P. Liu, M.Q. Rong, C.B. Shen, X.W. Yan, X.D. Luo & R. Lai. 2016. Phenolic acids isolated from the fungus *Schizophyllum commune* exert analgesic activity by inhibiting voltage-gated sodium channels. *Chinese Journal of Natural Medicines* 14: 661–670.

Zhang, J., J. Lv, L. Zhao, et al. 2018. Antioxidant and antimicrobial activities and chemical composition of submerged cultivated mycelia of *Laetiporus sulphureus*. *Chemistry of Natural Compounds* 54: 1187–1188. http://doi.org/10.1007/s10600-018-2591-9

Zhang, P., K. Li, G. Yang, C. Xia, J.E. Polston, G. Li, S. Li, et al. 2017. Cytotoxic protein from the mushroom *Coprinus comatus* possesses a unique mode for glycan binding and specificity. *The Proceedings of the National Academy of Sciences* 114: 8980–8985.

Zhao, Y.Y., X. Shen, X. Chao, et al. 2011. Ergosta-4, 6, 8(14), 22-tetraen-3-one induces G2/M cell cycle arrest and apoptosis in human hepatocellular carcinoma HepG2 cells. *Biochim Biophys Acta* 1810: 384–390.

4 Ethnomycology, Bioprospection, and Uses of Mushrooms in Costa Rica

Carlos Rojas, Ignacio Arroyo, Robin G. Doss

4.1 INTRODUCTION

4.1.1 The General Context of Mushroom Utilization in Costa Rica

Mushrooms, as a natural resource, have most likely been used by the various inhabitants of the Costa Rican territory since human migrations have taken place in this part of the world. This use, however, does not necessarily imply a specific utilization of mycological resources and has rather been related to the intrinsic exploration of food sources that accompanies the geographical movements of human groups (McNeill 1984; Guerra-Doce 2015). Thus, the original colonization of the Americas, the successive migrations established in this region, and the ethnic mix resulting from European Caribbean and Asian settlement have determined the current behavior of mushroom utilization in Costa Rica. This contextual framework can currently be extended to all ethnic inhabitant groups, indigenous or not, of the Costa Rican territory.

4.1.1.1 Pre-Hispanic and Current Indigenous Groups

Most of the pre-Hispanic groups in Costa Rica were of Chibcha affinity and belonged to the pre-Hispanic indigenous zone known as the Isthmo-Colombian area (Hoopes and Fonseca 2003). Thus far, there is no documentary evidence or direct proof of the consumption of mushrooms as food in pre-Hispanic Costa Rica (Jones 1991). This is interesting given the contrasting evidence found in other regions of Middle America (i.e., Mesoamerica, just north of Costa Rica), where the use of mycological resources as food has been widely documented.

The exception to this lack of pre-Hispanic mushroom use in Costa Rica comes from evidence represented in engravings made in a "metate" (a flat stone traditionally used to grind corn and other grains) showing the mushroom *Amanita muscaria*, and from a mushroom-shaped sculpture with images of humans and animals upside down, presumably suggesting the hallucinogenic use of fungi (Jones 1991). These artifacts were found in the northwestern (Guanacaste) zone of the country and have been associated with Mesoamerican influence, possibly commercial, in that region. In the southern Pacific zone of Costa Rica, there is suggestive documentation of the use of *Psilocybe* sp. for recreational practices in the Diquís area from a series of inhalers found during archaeological excavations (Arce-Cerdas and Rodríguez-Arce 2019).

Although multiple efforts have been made to establish the traditional use of biological resources in contemporary indigenous groups of Costa Rica, no significant use of mushrooms has been documented in such populations. This suggests limited current utilization of mycological resources by these groups. The scarce mycological knowledge currently transmitted among them is well summarized in the phrase of the Bribri leader Justo Torres Layan, that "young people do not know this form of food" (Vicerrectoria de Acci6n Social 2020).

DOI: 10.1201/b23190-4

Within the Bribri however, the use of boiled and roasted *Pleurotus* sp. (**skì'kuö**) in banana leaves is known (Vicerrectoría de Acción Social 2017), and the fourth-grade public curriculum adapted for this indigenous group includes an activity of culinary preparation and consumption of mushrooms (**skì'**), in which teachers must also talk about the scarcity of fungal reproductive structures in the mountains (MEP 2019). In the Cabécar, the use of *Pleurotus* sp. has been documented (**shötoro**, Madriz-Masis 1999). However, according to Chalampuente Flores (2012), the practice of consuming wild oyster mushrooms has been lost due to a lack of public interest, even though people throughout the Alto Chirripó area still extract the same fungus from the forest during the months associated with the first rains.

Among the Boruca, there is better documentation of mushroom use, but this does not imply a high degree of utilization or cultural importance of mycological resources. This indigenous group is perhaps the one that shows the most evidence of edible mushroom utilization in Costa Rica. According to Quesada and Rojas (1999), there are four distinct terms for fungi in Boruca. **Bricuá** is used for "a specific—yet unidentified—type of red mushroom", **brítcua** for "a specific—yet unidentified—type of edible mushroom", **rótcua** for "cup", and **tótcua** for "edible white mushroom". According to Quesada Pacheco (1996), the term **óscua** is also used to refer to an edible, white, ear-shaped mushroom. The Boruca have various ways of preparing mushrooms, but as in the case of the Cabecar, it is documented that the use of mycological resources has been lost over time (Garro-Valverde 2010).

For all indigenous groups in Costa Rica, field collection is primarily unplanned and depends on the production of reproductive structures associated with rainfall at different times of the year. Since ethnomycological documentation in Costa Rican indigenous groups is scarce, the available information is still very scattered. As such, effort is required to identify the precise species consumed by indigenous groups. For example, among the Boruca, the consumption of *Hygrocybe conica* (a toxic species according to Mata [2003]), has been suggested (Garro-Valverde 2010), but this observation has not been scientifically validated.

4.1.1.2 General Population

More than 98% of the Costa Rican population does not belong to any indigenous group, even though 33% of the average genetic markers are of Native American origin (Campos-Sánchez et al. 2014). According to Zumbado and Raabe (1976), upon the arrival of the first Europeans, half of the population of Costa Rica lived in the northwestern part of the country, while the so-called "Central Valley" (the most developed area in Costa Rica) was barely populated by 15% of the indigenous people. However, colonial development focused primarily on the central parts of the country (Garita-Hernández 2000), and the importance of this area grew to the point of becoming the most densely populated area in the country. Thus, the Costa Rican Central Valley, which is home to 70% of the current population, primarily has a post-colonial origin and exhibits European (mainly Spanish) influence.

Interestingly, recent ethnomycological research has determined that knowledge of mushrooms is greater in various cities of the Central Valley than in other areas of the country, and that from such knowledge, the Costa Rican population visualizes mushrooms primarily as a food resource (Rojas and Molina-Murillo 2018). Based on the available information, however, it is not possible to determine whether this recognition of mycological resources in the population of the Central Valley is associated with European cultural influence or with other confounding variables such as the higher rate of formal education in the area (see Rojas et al. 2019). Either way, this awareness may constitute a socio-cultural entry point to strengthen the appreciation of this group towards mycological resources and has been previously identified as an argument to increase environmental conservation efforts sensitive to the needs of the population (Molina-Murillo et al. 2015).

Although previous research has shown low levels of general mycological knowledge in Costa Rica and has noted that wild mushrooms are absent from Costa Rican markets (Rojas et al.

2017), the work of Molina-Murillo et al. (2015) yielded the first quantifications of mushroom consumption frequency in this country. Fifty-one percent of the population interviewed in an urban area, with high education and purchasing power, said they consume mushrooms monthly, and 24% responded that they consume them weekly. These values contrast with nationwide quantifications, where 38% of Costa Ricans reported consuming mushrooms monthly and 13.5% reported consuming mushrooms weekly (unpublished data, N=1369, 90% confidence). Although it can be said that there is a market corresponding to approximately 50% of the population of Costa Rica, 41% of the same population responded that they never consume mushrooms. These data indicate that the culinary use of mushrooms is not generalized in Costa Rica and that it is rather concentrated in specific regions of the Central Valley, particularly in districts with a high development index.

Such a medium-scale appreciation of mycological resources [as food] by the Costa Rican population does not support the dialectical positioning of mycophilia-mycophobia that is frequently found in the ethnomycological literature. In fact, other recent data focused on young populations in Central America have indicated that, in this region, cultural background does not play as important a role in the perception towards mushrooms as social factors associated with living in cities do (e.g., internet-mediated globalization) (Rojas et al. 2019). In this manner, it can be said that the Costa Rican society is at best "mycoindifferent", and that such a perception of mycological resources is of recent construction (in contrast to a "traditional" preference or rejection, for which there is simply no evidence). Thus, given the dynamic and influential position of the younger generations of Costa Ricans in urban situations, it is possible that more avenues of appreciation for mushrooms will continue opening in the future and that the national market for products of mycological origin will increase as well.

This is supported by the fact that cultivated mushrooms have gained space in supermarkets and in traditional fruit and vegetable markets (farmers' markets, which in Costa Rica exist in many municipalities, see Figure 4.1) in the last decade. In the 1980s, such activities were unheard of (Corella-Vargas 1985). In these spaces, champignons and oyster mushrooms have been positioned relatively well. Furthermore, recent data on the general appreciation of mycological resources among the Costa Rican population indicate that 44% of the people value mushrooms primarily for food, while 26% believe that their highest value could be found in the medical industry (Rojas and Molina-Murillo 2018). These data suggest that the Costa Rican population shows good levels of social receptivity for the development of mycological products in the food or medical industries.

4.2 EDIBLE SPECIES OF MUSHROOMS KNOWN IN COSTA RICA

Based on preliminary estimations, there are about 152 species of edible mushrooms known to occur in Costa Rica (Table 4.1). From these, 11 (about 9%) are Ascomycetes and the rest are Basidiomycetes. Among the former are species in the genera *Aleuria*, *Cookeina*, *Elaphomyces*, *Hydnotrya*, *Leotia*, and *Morchella*. Most of these edible species occur in specific habitats, such as the high-elevation oak forests of Costa Rica. Therefore, the functional conservation of the forests where these species occur is vital to ensuring the healthy population dynamics of these fungi.

Based on the literature (Wright and Albertó 2002; Mata 2003; Mata et al. 2003; Boa 2004; Jordan 2004; Halling and Mueller 2004; Dulger 2005; Landeros et al. 2006; Miller Jr. and Miller 2006; Bessette et al. 2007; Garibay-Orijel et al. 2010; Phillips 2010; Roberts and Evans 2011; Davis et al. 2012; Guzmán and Piepenbring 2012; Gamboa-Trujillo et al. 2014; Carranza-Velázquez et al. 2017; Sharma and Gautam 2017; Bessette et al. 2018; Castro-Sánchez et al. 2019; Flores Cavada et al 2018; Carranza-Velázquez and Mata 2019; Uitzil-Colli and Guzmán-Dávalos 2019; Ganatra et al. 2021; Li et al. 2021; Sofian et al. 2021; Hemmes and Desjardin 2022; Zhou et al. 2022), there are 10 excellent choices of edible mushroom in Costa Rica:

FIGURE 4.1 Shiitake production on oak stumps in the high elevations of Costa Rica. The experience of Finca Boryana, an initiative started by the second author, is a successful example of the technical skills and interest in mycology acquired by locals during the last two decades.

Cantharellus cibarius, *C. lateritius*, *Hericium coralloides*, *H. erinaceus*, *Hydnum repandum*, *Morchella elata*, *M. esculenta*, *M. guatemalensis*, *M. rufobrunnea*, and *Pleurotus pulmonarius*. Similarly, about 27 species, or 21.6%, are edible and considered a good choice, including some species of *Auricularia*, *Craterellus*, *Gyroporus*, *Leccinum*, *Pleurotus*, and *Russula*. Fifty six edible species of mushroom in Costa Rica can be consumed but do not necessarily represent good culinary choices (37%). Information regarding the quality of 55 species (36%) is lacking from the reviewed literature.

Among the edible mushrooms in Costa Rica, 16 species have been previously reported to have medicinal properties or medical applications. These correspond to two species each of *Hericium* and *Tremella*, as well as *Aleuria aurianta*, *Auricularia auricula-judae*, *Boletopsis grisea*, *B. leucomelaena*, *Byssomerulius incarnatus*, *Dacryopinax spathularia*, *Fistulina hepatica*, *Lepista nuda*, *Lycoperdon perlatum*, *Pisolithus arhizus*, *Russula adusta*, and *Schizophyllum commune*. Recent efforts to inform the general population of the medicinal properties of some fungi have been implemented by the National Museum of Costa Rica (Museo Nacional de Costa Rica 2022), but the leaders of these initiatives are still hesitant to target specific demographic groups. The International Union for Conservation of Nature currently monitors the conservation status of a handful of edible mushrooms species present in Costa Rica, but all of them are in the "least concern" category. As such, no threatened edible mushrooms have been found yet in this country.

Given the increase in mushroom consumption due to the rising awareness of associated health benefits and the recognition of mushrooms as functional foods, is necessary to standardize analysis methods and address edibility using an evidence-based classification system. Subjective judgments about taste, texture, or smell may differ and must be kept separate from an objective assessment of safe consumption. Therefore, categorizing mushroom species and assigning a consensual edibility status may help to reduce conflicting reports, prevent mushroom-related poisoning, and protect public health. Further studies must clearly indicate sources of evidence to support their assessments (Li et al. 2021).

Following the classification system of Li et al. (2021), 122 species (80.3%) on the list of edible mushrooms in Costa Rica presented herein are supported by clear evidence of consumption without adverse or harmful effects and are considered safe and suitable to eat. Furthermore, there is clear evidence that 11 species, or 8.8%, can be consumed once cooked or prepared in such a way as to make them safe and suitable to eat (including species that may result in allergic or adverse reactions when consumed with alcohol, for example). On the other hand, evidence of safe consumption is uncertain or incomplete in four taxa (3.2%), and six species (4.8%) have an unconfirmed status of edibility due to mixed reports that include adverse or harmful reactions when consumed.

TABLE 4.1
List of 152 Edible Mushrooms of Costa Rica Based on Published Information in Recent Years. Both Final Edibility Status (FES) and Quality as Edible Choice were Determined Based on the Literature Review.

Species	FES	Quality
Aleuria aurantia (Pers.) Fuckel	E2	1
Amanita fulva Fr.	E2	2
Amanita garabitoana Tulloss, Halling & G.M. Muell.	E1	2
Apioperdon pyriforme (Schaeff.) Vizzini	E1	1
Armillaria puiggarii Speg.	E1	2
Artomyces pyxidatus (Pers.) Jülich	E1	1
Aseroe rubra Labill.	E1	1
Aureoboletus auriporus (Peck) Pouzar	E1	2
Auricularia auricula-judae (Bull.) Quél.	E1	2
Auricularia delicata (Mont. ex Fr.) Henn.	E1	2
Auricularia fuscosuccinea (Mont.) Henn. [as '*fusco-succinea*']	E1	2
Auricularia mesenterica (Dicks.) Pers.	E1	2
Auricularia nigricans (Sw.) Birkebak, Looney & Sánchez-García	E1	1
Austroboletus gracilis (Peck) Wolfe	E1	0
Boletellus ananas (M.A. Curtis) Murrill	U	1
Boletopsis grisea (Peck) Bondartsev & Singer	E1	1
Boletopsis leucomelaena (Pers.) Fayod [as '*melaleuca*']	E2	1
Boletus quercophilus Halling & G.M. Muell.	E1	3
Boletus variipes Peck	E1	0
Butyriboletus frostii (J.L. Russell) G. Wu, Kuan Zhao & Zhu L. Yang	E1	2
Byssomerulius incarnatus (Schwein.) Gilb.	E1	0
Calvatia cyathiformis (Bosc) Morgan	E2	1
Calvatia gigantea (Batsch) Lloyd	E2	1
Calvatia rugosa (Berk. & M.A. Curtis) D.A. Reid	E3	0
Campanophyllum proboscideum (Fr.) Cifuentes & R.H. Petersen	E1	0
Cantharellus cibarius Fr.	E1	3
Cantharellus lateritius (Berk.) Singer	E1	3
Clathrus ruber P. Micheli ex Pers.	E3	1
Clavaria fragilis Holmsk.	E1	1
Clavaria zollingeri Lév.	E1	0

TABLE 4.1 ***Continued***
List of 152 Edible Mushrooms of Costa Rica Based on Published Information in Recent Years. Both Final Edibility Status (FES) and Quality as Edible Choice were Determined Based on the Literature Review.

Species	FES	Quality
Clavulina amethystina (Bull.) Donk	E1	0
Clavulina coralloides (L.) J. Schröt.	E1	0
Clavulinopsis fusiformis (Sowerby) Corner	E1	0
Clavulinopsis laeticolor (Berk. & M.A. Curtis) R.H. Petersen	E1	1
Clitopilus prunulus (Scop.) P. Kumm.	E1	0
Cookeina speciosa (Fr.) Dennis	E1	0
Cookeina sulcipes (Berk.) Kuntze	E1	0
Cookeina tricholoma (Mont.) Kuntze	E1	0
Cookeina venezuelae (Berk. & M.A. Curtis) Le Gal	E1	0
Coprinellus disseminatus (Pers.) J.E. Lange [as '*disseminata*']	E2	1
Coprinellus domesticus (Bolton) Vilgalys, Hopple & Jacq. Johnson	E1	0
Coprinellus micaceus (Bull.) Vilgalys, Hopple & Jacq. Johnson	E2	1
Coprinus comatus (O.F. Müll.) Pers.	E1	1
Cordyceps militaris (L.) Fr.	E1	0
Cortinarius iodes Berk. & M.A. Curtis	E1	1
Craterellus boyacensis Singer	E1	2
Craterellus ignicolor (R.H. Petersen) Dahlman, Danell & Spatafora	E1	2
Cuphophyllus pratensis (Pers.) Bon	E1	2
Dacryopinax spathularia (Schwein.) G.W. Martin	E1	1
Favolus tenuiculus P. Beauv.	E1	2
Fistulina hepatica (Schaeff.) With.	E2	1
Flammulina velutipes (Curtis) Singer	E1	2
Gliophorus psittacinus (Schaeff.) Herink	E2	1
Gymnopus dryophilus (Bull.) Murrill	E2	1
Gymnopus nubicola Halling	E1	2
Gyroporus castaneus (Bull.) Quél.	E1	2
Gyroporus subalbellus Murrill	E1	0
Harrya chromipes (Frost) Halling, Nuhn, Osmundson & Manfr. Binder [as '*chromapes*']	E1	2
Hemileccinum subglabripes (Peck) Halling	E1	0
Hericium americanum Ginns	E1	0
Hericium coralloides (Scop.) Pers.	E2	2
Hericium erinaceus (Bull.) Pers.	E1	3
Hohenbuehelia petaloides (Bull.) Schulzer	E1	1
Humidicutis marginata (Peck) Singer	E1	1
Hydnopolyporus fimbriatus (Cooke) D.A. Reid	E1	0
Hydnotrya tulasnei (Berk.) Berk. & Broome	E1	1
Hydnum repandum L.	E1	3
Hydnum umbilicatum Peck	E1	0
Hygrocybe cantharellus (Schwein.) Murrill [as 'Hydrocybe']	E1	0
Hygrocybe miniata (Fr.) P. Kumm.	E1	0
Hygrophoropsis aurantiaca (Wulfen) Maire	E2	1
Hygrophorus russula (Schaeff. ex Fr.) Kauffman	E1	2
Hygrophorus sordidus Peck	E1	1
Hypholoma capnoides (Fr.) P. Kumm.	E2	1
Kuehneromyces mutabilis (Schaeff.) Singer & A.H. Sm.	E1	1
Laccaria amethystina Cooke	E1	1
Laccaria gomezii Singer & G.M. Muell.	E1	0

TABLE 4.1 ***Continued***
List of 152 Edible Mushrooms of Costa Rica Based on Published Information in Recent Years. Both Final Edibility Status (FES) and Quality as Edible Choice were Determined Based on the Literature Review.

Species	FES	Quality
Laccaria laccata (Scop.) Cooke	E1	1
Laccaria ohiensis (Mont.) Singer	E1	0
Lactarius atroviridis Peck	E1	0
Lactarius indigo (Schwein.) Fr.	E1	2
Lactarius psammicola A.H. Sm.	E1	0
Lactarius rimosellus Peck	E1	0
Lactifluus deceptivus (Peck) Kuntze	E1	1
Laetiporus persicinus (Berk. & M.A. Curtis) Gilb.	E1	0
Laetiporus sulphureus (Bull.) Murrill	E2	2
Leccinellum rugosiceps (Peck) C. Hahn	E1	0
Leccinum monticola Halling & G.M. Muell.	E2	2
Lentinellus ursinus (Fr.) Kühner	E1	1
Lentinula aciculospora J.L. Mata & R.H. Petersen	E3	2
Lentinula boryana (Berk. & Mont.) Pegler	E1	2
Lentinula raphanica (Murrill) Mata & R.H. Petersen	E1	0
Lentinus berteroi (Fr.) Fr. [as '*bertierii*']	E1	0
Lentinus concavus (Berk.) Corner	E2	0
Lentinus crinitus (L.) Fr.	E1	1
Lentinus scleropus (Pers.) Fr.	E1	0
Lentinus velutinus Fr.	E1	0
Leotia lubrica (Scop.) Pers.	E3	1
Lepista nuda (Bull.) Cooke	E2	2
Lepista sordida (Schumach.) Singer	E1	2
Lepista tarda (Peck) Murrill	E1	2
Lycoperdon perlatum Pers.	E2	1
Lyophyllum loricatum (Fr.) Kühner	E1	1
Macrocybe titans (H.E. Bigelow & Kimbr.) Pegler, Lodge & Nakasone	E1	1
Macrolepiota colombiana Franco-Mol.	E1	0
Marasmius cladophyllus Berk.	E1	0
Marasmius cohaerens (Pers.) Cooke & Quél.	E1	0
Marasmius haematocephalus (Mont.) Fr.	E1	0
Morchella herediana L.D. Gómez	E2	0
Oudemansiella canarii (Jungh.) Höhn.	E1	2
Panaeolus antillarum (Fr.) Dennis	E1	0
Panus neostrigosus Drechsler-Santos & Wartchow	E1	0
Panus strigellus (Berk.) Overh.	E1	0
Phaeotremella foliacea (Pers.) Wedin, J.C. Zamora & Millanes	E1	0
Phallus impudicus L.	E2	0
Phallus indusiatus Vent.	E1	1
Pholiota spumosa (Fr.) Singer	E1	0
Phyllotopsis nidulans (Pers.) Singer	E1	0
Pisolithus arhizus (Scop.) Rauschert	E2	1
Pleurotus albidus (Berk.) Pegler	E1	2
Pleurotus djamor (Rumph. ex Fr.) Boedijn	E1	2
Pleurotus dryinus (Pers.) P. Kumm.	E1	0
Pleurotus pulmonarius (Fr.) Quél.	E1	3
Pluteus cervinus (Schaeff.) P. Kumm.	E2	2

TABLE 4.1 ***Continued***
List of 152 Edible Mushrooms of Costa Rica Based on Published Information in Recent Years. Both Final Edibility Status (FES) and Quality as Edible Choice were Determined Based on the Literature Review.

Species	FES	Quality
Pseudohydnum gelatinosum (Scop.) P. Karst.	E1	1
Pulveroboletus ravenelii (Berk. & M.A. Curtis) Murrill	E1	1
Ramaria araiospora Marr & D.E. Stuntz	E1	1
Ramaria stricta (Pers.) Quél.	E1	1
Ramariopsis kunzei (Fr.) Corner	E1	1
Rhodocollybia maculata (Alb. & Schwein.) Singer	E1	1
Rhodocollybia prolixa (Fr.) Antonín & Noordel.	E1	0
Russula adusta (Pers.) Fr.	E2	0
Russula brevipes Peck	E1	1
Russula compacta Frost	E1	1
Russula flavida Frost ex Peck	E3	0
Russula foetens Pers.	E2	1
Russula xerampelina (Schaeff.) Fr.	E1	2
Schizophyllum commune Fr. [as '*Schizophyllus communis*']	E1	1
Sebacina schweinitzii (Peck) Oberw.	E1	1
Squamanita umbonata (Sumst.) Bas	E1	0
Strobilomyces confusus Singer	E1	0
Strobilomyces strobilaceus (Scop.) Berk.	E1	1
Suillus brevipes (Peck) Kuntze	E1	1
Tremella fuciformis Berk.	E1	2
Tremella mesenterica (Schaeff.) Pers.	E2	1
Tricholoma caligatum (Viv.) Ricken	E1	1
Tricholoma portentosum (Fr.) Quél.	E1	3
Tricholomopsis aurea (Beeli) Desjardin & B.A. Perry	E1	0
Volvariella bombycina (Schaeff.) Singer	E1	1
Volvariella diplasia (Berk. & Broome) Singer	E1	0
Volvariella volvacea (Bull.) Singer	E1	2
Xerocomellus chrysenteron (Bull.) Šutara	E1	1

Final Edibility Status (FES): E1 = completely edible, E2 = edible under certain conditions, E3 = edibility is uncertain, U = nor confirmed.

Quality is stated as reported in the literature and corresponds to: 0 = no information, 1 = not of choice, 2 = good choice, 3 = excellent choice.

4.2.1 Academic Growth, Bioprospection, and Legislation

The development of formal mycology in Costa Rica has been strongly linked to local biological academia. This development began with the contribution of Dr. José Alberto Sáenz Renauld, who was among the first professors in the School of Biology of the University of Costa Rica (García 2009). Dr. Saénz Renauld, a pharmacist by training, founded the sections of Mycology and Mycorrhizae towards the end of the 1950s, setting into motion the training of the first generation of Costa Rican mycologists (Rojas and Valverde 2019).

In recent decades, valuable contributions have been published that further elucidate the mycological diversity of Costa Rica, and efforts have been made to train a new generation of local professionals in mycology. Among the earlier works are the field guides to Macrofungi of Costa Rica (Mata 2003; Mata et al. 2003). More recent publications include the mushroom guides of San Gerardo de Dota (Carranza and Mata 2019) and Corcovado National Park (Carranza-Velázquez et al. 2017). In 2010, two practical guides for the cultivation of *Pleurotus ostreatus* (Mata et al. 2010) and *Lentinula edodes* (Mata and Navarro 2010) were published with the aim of raising awareness of different techniques for producing edible mushrooms.

Mycology in Costa Rica is currently active and undergoing a process of diversification of approaches. For example, bioprospective studies (Rojas-Jiménez et al. 2016) and those with a focus on sustainability (Rojas and Calvo 2014), agricultural integration (Álvarez Montero 2013), and engineering (Jimenez Rey 2018) have gained space in the work agenda of this country. Alternative approaches also include analysis of the economic perspectives of mushroom production (Díaz-Gutiérrez 2004) and assessments of the quality of materials used in mushroom production (WingChing-Jones and Alvarado 2009).

So far, and based on what we know, there are no large-scale biotechnological initiatives involving edible fungi in the country. The active selection of edible fungal strains is perhaps the only form of biotechnological manipulation currently being carried out, mainly for the purpose of improving productivity and minimizing losses. The biotechnological approach, however, has been very commonly applied to recent activities regarding non-edible fungi in Costa Rica. In this field, researcher Giselle Tamayo has conducted multiple bioprospective trials with Costa Rican species (e.g., Cao et al. 2012; Lee et al. 2016). Similarly, the study of pathogenic fungi has benefited from biotechnological approaches, and the Centro de Investigación en Biotecnología from the Instituto Tecnológico de Costa Rica (www.tec.ac.cr) has several research projects intended to maintain fungal strains and develop strategies for the application of fungi as biological pest control.

More than 20 years ago, the Costa Rican law number 7788 of 30 April 1998 (Biodiversity Law) established guidelines for the conservation of biodiversity and the sustainable use of resources. This law developed parameters for the fair distribution of benefits and costs derived from biodiversity. As such, this innovative and inclusive document for its time (the issue of co-benefits was formulated in many countries only after the Nagoya Protocol in 2010) has been highly valued at the international level and was called in 2006 "the most comprehensive legislation for the implementation of the Convention on Biological Diversity" (Miller 2006).

Article 7 of this law recognized fungi as "unicellular and multicellular organisms, lacking chlorophyll and belonging to the Phylum Fungi" and established that these organisms are biological components of ecosystems. In this manner, Costa Rica has legally recognized fungi and their role in nature since 1998 and has established guidelines for their conservation based on national development criteria. For example, it is established in this law that fungi contain biochemical and genetic elements of potential interest to society and that the commercial use of such elements must be regulated by national entities and entail a socioeconomic benefit for the Costa Rican society.

In a complementary manner, Law 7317 from 30 October 1992 (Wildlife Conservation Law) and its regulation number 40548-MINAE determined that fungi of the groups Chytridiomycota, Zygomycota, Ascomycota, and Basidiomycota are part of the national flora and are subject to regulations on the use of national soil. These regulations control the extraction, exploitation, and conservation of fungi by non-indigenous groups (any use made by indigenous populations is culturally respected in the legislation). Thus, in Costa Rica, there are clear legal guidelines for extracting, even for research purposes, fungi—mainly mushrooms—in their wild state from natural populations, and there are also parameters that establish the potential control of fungal species that can be considered invasive to the territory.

From the point of view of mycological production, the national regulatory framework falls within guidelines of the Ministry of Agriculture and Livestock (www.mag.go.cr), none of which is specific to mushrooms. In terms of labeling, competence lies with the Ministry of Industry and Commerce (www.meic.go.cr), which establishes a series of minimum information criteria for the labeling of prepackaged food products. Fresh mushrooms, however, do not fall into this group, and therefore labelling only applies to imported canned and/or dried mushrooms. Thus, in addition to the fact that the label must be written in Spanish, indications of the country of origin, product ingredients, expiration date, and registration number with the Ministry of Health, among others, are required (see Technical Regulation RTCR100:1997, labeling of prepackaged foods). Additives used in mushroom products must follow Central American regulations such as the Technical Regulation RTCA 67.04.54:10. In the case of organic products, including mushrooms, the 2002 "Guide for the Labelling of Organic Products" (www.mag.go.cr) is followed.

4.2.2 Diversity and Trends of Ecological Study

In terms of fungal diversity, Ruiz-Boyer and Rodríguez-González (2020) mentioned that there are about 3000 known species of fungi in the country, but Obando (2013), several years earlier, had already communicated about 3800 species. In a more recent estimation of fungal diversity, Bermúdez Rojas and Obando Acuña (2021) estimated that about 3000 species are present. A quick search on the Global Biodiversity Information Facility (GBIF.org 2022) shows about 4100 accepted names of fungi present in the Costa Rican territory. Such an estimate does not consider another 4000 names that might be inconsistent with current taxonomic treatments or classified as doubtful identifications, from which perhaps a small percentage could be added to the accepted list. In this manner, it seems conservative enough to say that currently there are between 3000–4000 fungal species reported in Costa Rica. Such value represents only a small percentage of the expected fungal diversity in the country, currently calculated at about 65000 species (Bermúdez Rojas and Obando Acuña 2021).

A large percentage of such diversity has been recorded in the Costa Rican highlands, where several research projects took place in recent decades (i.e., Halling and Mueller 2004; Carranza-Velázquez and Mata 2019). In these studies, most of the recorded fungi were macrofungal ectomycorrhizal species associated with *Quercus* forests. However, lower elevations, mainly in the wet areas of the country, such as the Caribbean and the central and southern Pacific regions, have also been studied in recent years (i.e., Ruiz-Boyer 2015). In other areas, lowland and premontane moist or wet forests have been studied. Some documentation of ecological patterns in the Costa Rica dry forests, the only seasonal forest ecosystems in this territory, also took place in the last decade (Rojas et al. 2017).

Recent ecological studies have encompassed leaf litter dynamics (Bills and Polishook 1994), host specificity of wood-inhabiting fungi (Lindblad 2000), the relationship between chytrids and amphibian decline (Puschendorf et al. 2006), and the documentation of endophytic fungal dynamics associated with ferns (Del Olmo-Ruiz and Arnold 2013). Investigations have also included the study of soil fungi as part of the microbiome associated with tree roots (McGee et al. 2019), the resilience of fungal populations to climatic effects such as El Niño-Southern Oscillation (Rojas et al. 2017), fungal communities in the canopy of wet tropical forests (Looby et al. 2020), and the impact of human disturbance over fungal populations in soils (Eaton et al. 2021).

4.2.3 Perspectives on the Integration of Fungi in Initiatives with Human Impact

Fungi in general, and mushrooms in particular, are very beautiful elements of natural systems. Their role is fundamental for the functionality of the biosphere, and given the extreme diversity of species on the planet, there are certainly many potential uses that are yet unknown to human populations. Even though some work has been done to increase mycoliteracy in Costan Rican populations and integrate fungi with the discussion of national problems and solutions, more work can be carried out (i.e., in public schools) to continue promoting or sustaining the general appreciation of these resources.

As mentioned by Elizondo-Cruz and Blanco-Segura (2010), bioliteracy activities are crucial for the development of a balanced mindset regarding the conservation/utilization pattern of natural resources in Costa Rica. As mentioned before, the work of Rojas and Molina-Murillo (2018) on fungi clearly demonstrated that an effective mycoliteracy model could be implemented in areas of Costa Rica with poor knowledge of this group of organisms. Rojas and Valverde (2019) called for such implementation with internet-based tools such as websites, apps, and social media to encourage a parallel growth in public involvement in science (i.e., an increase in citizen science initiatives). With the first quarter of the current century nearly gone, the power of such tools to mobilize initiatives, including conservation ideas, worldwide is higher than that of other strategies previously used and promoted.

Costa Rican society is highly technological (García-Martínez et al. 2021), and the power of the internet to standardize fungal information has already been documented (Molina-Murillo et al. 2015). The citizen science approach could be very reliable for generating information on fungi (see www.inaturalist.org/), and there is no clear reason not to integrate it with formal scientific projects. In the end, the more laypeople get to know about a natural resource, the higher the probability of those people integrating such knowledge into their lifestyle, potentially leading to conservation and sustainable utilization of those resources. Such an approach will likely be the next step in the promotion of mycology in Costa Rica.

At the same time, the sustained effort to integrate mycology with developmental agendas in STEAM disciplines is likely to continue in the years to come. Costa Rica has never in its academic history had as many people officially involved in research on fungi in the different colleges and universities as it has today (Rojas and Valverde 2019). This is reflected in the more than 70,000 collections of fungi in local herbaria. Even two decades ago, it was not very common to find projects involving fungi in disciplines other than those in the natural sciences, but now fungi are present in most academic disciplines in one way or another. As we have seen before, the development of applications involving fungi is more common today.

Finally, the increasing production of mushrooms for consumption in Costa Rica is likely to continue as well. Conceptual contributions and economic support, albeit still at low levels, are present in Costa Rica. As an example, the Costa Rican Government very recently decided to build, with public funds, an oyster mushroom production plant in the south of the country (Presidencia de la República 2022). Despite this example, most mushroom production in Costa Rica is still privately funded (i.e., the shiitake production by Finca Boryana from the second author), and most mushroom producers have expressed inefficient commercial links between producers, distributors, and retailers. These are obstacles to overcome, and the effect that their resolution can have on the local production of mushrooms is likely positive and therefore aligned with the promotion of mycophagy.

4.3 ACKNOWLEDGMENTS

The research for this chapter was supported by the University of Costa Rica (Vicerrectoría de Investigación 731-B8–900) and by Finca Boryana for the first and second authors, respectively. We would like to acknowledge the support of Maialu S.A., a private initiative for supporting bioliteracy in Latin America, on behalf of the third author.

REFERENCES

Álvarez Montero, K.J. 2013. *Formulación de un compost a partir de excretas de animales y diferentes subproductos agroindustriales para su utilización como substrato para la producción del hongo champiñón (Agaricus bisporus)* [Unpublished thesis] School of Zootechnics, University of Costa Rica.

Arce-Cerdas, M.A. & J.M. Rodríguez-Arce. 2019. Ritual consumption of psychoactive fungi and plants in ancestral Costa Rica. *Journal of Psychedelic Studies* 3(2): 179–197.

Bermúdez Rojas, T. & V. Obando Acuña. 2021. Biodiversidad en cifras: avances en el conocimiento de especies en Costa Rica. *Biocenosis* 32(2): https://doi.org/10.22458/rb.v32i2.3899

Bessette, A.E., A.R. Bessette & M.W. Hopping. 2018. *A Field Guide to Mushrooms of the Carolinas*. Chapel Hill, NC: The University of North Carolina Press.

Bessette, A.E., W.C. Roody & A.R. Bessette. 2007. *Mushrooms of the Southeastern United States*. Syracuse, NY: Syracuse University Press.

Bills, G.F. & J.D. Polishook. 1994. Abundance and diversity of microfungi in leaf litter of a lowland rain forest in Costa Rica. *Mycologia* 86(2): 187–198.

Boa, E. 2004. *Wild Edible Fungi: A Global Overview of Their Use and Importance to People (Non-Wood Forest Products)*. Rome: Food & Agriculture Organization (FAO).

Campos-Sánchez, R., H. Raventós & R. Barrantes. 2014. Ancestry informative markers clarify the regional admixture variation in the Costa Rican population. *Human Biology* 85(5): 721–740.

Cao, S., L. Cryan, K.A. Habeshian, C. Murillo, G. Tamayo-Castillo, M.S. Rogers & J. Clardy. 2012. Phenolic compounds as antiangiogenic CMG2 inhibitors from Costa Rican endophytic fungi. *Bioorganic & Medicinal Chemistry Letters* 22(18): 5885–5888.

Carranza-Velázquez, J., W. Marín Mendez, A. Ruiz Boyer & J.F. DiStefano Gandolfi. 2017. *Guía de los macrohongos más comunes del Parque Nacional Corcovado*. Estación La Leona. Editorial UCR, Costa Rica.

Carranza-Velázquez, J. & M. Mata. 2019. *Guía de los macrohongos en San Gerardo de Dota*. Editorial UCR, Costa Rica.

Castro-Sánchez, E. I., A.I. Moreno-Calles, S. Meneses-Eternod, B. Farfán-Heredia, J. Blancas & A. Casas. 2019. Management of wild edible fungi in the Meseta Purépecha, Mexico. *Sustainability* 11: 3779.

Chalampuente Flores, D.S. 2012. *Seguridad alimentaria en comunidades indígenas de Costa Rica: el caso de comunidades Cabécar de Alto Chirripó* [Unpublished Master's thesis]. Centro Agronómico Tropical de Investigación y Enseñanza, Costa Rica.

Corella-Vargas, R. 1985. Potencial del cultivo de hongos comestibles en Costa Rica. *Uniciencia* 2(1): 81–82.

Davis, R.M., R. Sommer & J.A. Menge. 2012. *Field Guide to Mushrooms of Western North America*. Berkeley: University of California Press.

Del Olmo-Ruiz, M.D. & A.E. Arnold. 2013. Interannual variation and host affiliations of endophytic fungi associated with ferns at La Selva, Costa Rica. *Mycologia* 106(1): 8–21.

Díaz-Gutiérrez C. 2004. Análisis económico para el establecimiento de un proyecto de producción de hongos comestibles (*Pleurotus ostreatus*), en la zona de Atenas, Alajuela. *Tecnología en Marcha* 17(4): 3–7.

Dulger, B. 2005. Antimicrobial activity of ten Lycoperdaceae. *Fitoterapia* 76(3–4): 352–354.

Eaton, W.D., K.M. McGee, R. Donnelly, A. Lemenze, M. Larimer & M. Hajibabaei. 2021. The impacts of a logging road on the soil microbial communities, and carbon and nitrogen components in a Northern Zone Costa Rican forest. *Applied Soil Ecology* 164: 103937.

Elizondo-Cruz, R. & R. Blanco-Segura. 2010. Developing the bioliteracy of school children for 24 years: A fundamental tool for ecological restoration and conservation in perpetuity of the Área de Conservación Guanacaste, Costa Rica. *Ecological Restoration* 28(2): 193–198.

Flores Cavada, E., A. Carrillo Parra, C.A. Wehenkel, F. Garza Ocañas & J.C. Hernández Díaz. 2018. Diversity of macromycetes in pine forests at the municipality of Madera, Chihuahua. *Revista mexicana de ciencias forestales* 9(50): 342–360.

Gamboa Trujillo, J.P., F. Wartchow, C. Cerón, E. Aules, C. Aigaje, L.H. Calvalcanti & T.B. Gibertoni. 2014. Traditional use of *Gymnopus nubicola* as a food resource in a kichwa community, Pichincha, Ecuador. *Mycosphere* 5(1): 180–186.

Ganatra, M. B., V. Potapov, S. Vainauskas, A.Z. Francis, C.M. McClung, C.I. Ruse, C. I., . . . & C.H. Taron. 2021. A bi-specific lectin from the mushroom *Boletopsis grisea* and its application in glycoanalytical workflows. *Scientific Reports* 11(1): 1–16.

García, J.E. 2009. Brief history of the school of biology of the University of Costa Rica (1957–2009). *Journal of Tropical Biology* 57(Suppl. 1): 1–14.

García-Martinez, J.A., E.J. Fuentes-Abeledo & E.R. Rodríguez-Machado. 2021. Attitudes towards the Use of ICT in Costa Rican University Students: The influence of sex, academic performance, and training in technology. *Sustainability* 13(1): 282.

Garibay-Orijel, R., F. Ruan-Soto & E. Estrada-Martínez. 2010. El conocimiento micológico tradicional, motor para el desarrollo del aprovechamiento de los hongos comestibles y medicinales. In: Martínez-Carrera, D. (ed.), *Hacia un desarrollo sostenible del sistema de producción-consumo de los hongos comestibles y medicinales en Latinoamérica: avances y perspectivas en el siglo XXI*, Red Latinoamericana de Hongos Comestibles y Medicinales, México, D.F., pp. 243–270.

Garita-Hernández, F. 2000. La influencia extranjera en la tiponimia costarricense. *Filología y lingüística* 26(2): 139–160.

Garro-Valverde, L. 2010. *Saberes y sabores de los boruca*. Leila Garro-Valverde, Costa Rica.

GBIF.org. 2022. *GBIF Occurrence Download*. Accessed on 14 June 2022. Available at: https://doi.org/10.15468/dl.man3p8

Guerra-Doce, E. 2015. Psychoactive substances in prehistoric times: Examining the archaeological evidence. *Time & Mind* 8(1): 91–112.

Guzmán G. & M. Piepenbring. 2012. *Los hongos de Panamá*. Instituto de Ecología A.C. Xalapa, Mexico.

Halling R.E. & G.M. Mueller. 2004. *Common Mushrooms of the Talamanca Mountains, Costa Rica*. New York Botanical Garden, New York.

Hemmes, D. E. & D.E. Desjardin. 2022. *Mushrooms of Hawai'i: An Identification Guide*. Echo Point Books & Media, Vermont.

Hoopes J.W. & O.M. Fonseca. 2003. Goldwork and chibchan identity: endogenous change and diffuse unity in the isthmo-colombian area. In: Quilter, J. and Hoopes, J.M (Eds.), *Gold and Power in Ancient Costa Rica, Panama and Colombia*. Dumbarton Oaks Research Library and Collection, Washington, DC, pp. 49–90.

Jiménez Rey, F. 2018. *Elaboración de un material biológico a partir del hongo Pleurotus ostreatus para su utilización en aplicaciones de ingeniería* [Unpublished undergraduate thesis]. School of Agricultural and Biosystems Engineering, University of Costa Rica.

Jones, U. 1991 Metates and Hallucinogens in Costa Rica. *Papers from the Institute of Archaeology* 2(0): 29–34.

Jordan, M. 2004. *The Encyclopedia of Fungi of Britain and Europe*. Frances Lincoln, London, p. 81.

Landeros, F., J. Castillo, G. Guzmán & J. Cifuentes. 2006. Los hongos (macromicetos) conocidos en el Cerro el Zamorano (Querétaro-Guanajuato), México. *Scientia Fungorum* 22: 25–31.

Lee, S., G. Tamayo-Castillo, C. Pang, J. Clardy, S. Cao & K.H. Kim. 2016. Diketopiperazines from Costa Rican endolichenic fungus Colpoma sp. CR1465A. *Bioorganic & Medicinal Chemistry Letters* 26(10): 2438–2441.

Li, H., Y. Tian, N. Menolli Jr, L. Ye, S.C. Karunarathna, J. Perez-Moreno, . . . & P.E. Mortimer. 2021. Reviewing the world's edible mushroom species: A new evidence-based classification system. *Comprehensive Reviews in Food Science and Food Safety* 20(2): 1982–2014.

Lindblad, I. 2000. Host specificity of some wood-inhabiting fungi in a tropical forest. *Mycologia* 92(3): 399.

Looby, C.I., E.C. Hollenbeck & K.K. Treseder. 2020. Fungi in the canopy: How Soil Fungi and extracellular enzymes differ between canopy and ground soils. *Ecosystems* 23: 768–782.

Madriz-Masis J.P. 1999. Ethnobotanical exploration of edible wild flora in the tropical rainforests of the Tayní aboriginal reserve, Limón, Costa Rica. *Tecnología en Marcha* 13(3): 188–192.

Mata, M. 2003. *Macrofungi of Costa Rica*. Editorial INBio, Costa Rica.

Mata, M., R. Halling & G.M. Mueller. 2003. *Macrofungi of Costa Rica Vol. 2*. Editorial INBio, Costa Rica.

Mata, M. & E. Navarro. 2010. *Guía práctica para el cultivo del hongo comestible, Lentinula edodes (Shiitake) sobre troncos*. Editorial INBio, Costa Rica.

Mata, M., S. Soto & E. Navarro. 2010. *Guía práctica para el cultivo de hongos ostra* (*Pleurotus ostreatus*). Editorial INBio, Costa Rica.

McGee, K.M., W.D. Eaton, T.M. Porter, S. Shokralla & M. Hajibabaeia. 2019. Soil microbiomes associated with two dominant Costa Rican tree species, and implications for remediation: A case study from a Costa Rican conservation area. *Applied Soil Ecology* 137: 139–153.

McNeill, W. 1984. Human migration in historical perspective. *Population and Development Review* 10(1): 1–18.

MEP. 2019. *Guía del programa cultura bribri-sulá*. Ministerio de Educación Pública, Costa Rica.

Miller, M.J. 2006. Biodiversity policy making in Costa Rica: Pursuing indigenous and peasant right. *The Journal of Environment and Development* 15(4): 359–381.

Miller Jr., O.K. & H.H. Miller. 2006. *North American Mushrooms: A Field Guide to Edible and Inedible Fungi*. Falcon Guide, Guilford, CN.

Molina-Murillo, S., C. Rojas, S. Somerville, R.G. Doss & B. Lemus. 2015. What do you think of fungi? Perceptions by "educated" groups in Honduras and Costa Rica. *Brenesia* 83–84: 30–36.

Museo Nacional de Costa Rica. 2022. *Hongos "orejas de palo" en los bosques de Costa Rica*. Accessed on 20 June 2022. Available at: www.museocostarica.go.cr/divulgacion/articulos-educativos/hongos-orejas-de-palo/

Obando, V. 2013. *Vigésimo Informe del Estado de la Nación en Desarrollo Humano Sostenible (2013): Conservación y Biodiversidad*. Programa Estado de la Nación, Costa Rica.

Phillips, R. 2010. *Mushrooms and Other Fungi of North America*. Firefly Books, Buffalo, NY.

Presidencia de la República. 2022. *Invierten más de ¢87 millones en planta procesadora de hongos ostra en San Vito de Coto Brus*. Accessed on 20 June 2022. Available at: www.presidencia.go.cr/comunicados/2022/03/invierten-mas-de-%C2%A287-millones-en-planta-procesadora-de-hongos-ostra-en-san-vito-de-coto-brus/

Puschendorf, R., F. Bolaños & G. Chaves. 2006. The amphibian chytrid fungus along an altitudinal transect before the first reported declines in Costa Rica. *Biological Conservation* 132(1): 136–142.

Quesada, M.A. & C. Rojas. 1999. *Boruca-Spanish, Spanish-Boruca Dictionary*. Editorial UCR, Costa Rica.

Quesada Pacheco, M.A. 1996. *Narraciones Borucas*. Editorial UCR, Costa Rica.

Roberts, P. & S. Evans. 2011. *The Book of Fungi: A Life-Size Guide to Six Hundred Species from Around the World*. University of Chicago Press, Chicago.

Rojas, C. & E. Calvo. 2014. Forest biomass, carbon stocks and macrofungal dynamics: A study case in Costa Rica. *International Journal of Forestry Research* 2014: 607372.

Rojas, C. & S. Molina-Murillo. 2018. Citizen perception of fungal resources and potential for myco-literacy in Costa Rica. *Scientia Fungorum* 48: 23–31.

Rojas, C., S. Molina-Murillo & R.G. Doss. 2019. Perceptions of the use and value of fungi by two groups of young adults from Central America. *UNED Research Journal* 11(3): 345–351.

Rojas, C. & R. Valverde. 2019. Towards an integrated approach of conducting fungal research in Costa Rica. *Uniciencia* 33(1): 101–117.

Rojas, C., R. Valverde & R. Morales. 2017. Functional variability of macrofungal populations in four different forest types of Costa Rica. *Mycosphere* 8(9): 1288–1296.

Rojas-Jimenez, K., M. Hernandez, J. Blanco, L.D. Vargas, L.G. Acosta-Vargas & G. Tamayo. 2016. Richness of cultivable endophytic fungi along an altitudinal gradient in wet forests of Costa Rica. *Fungal Ecology* 20: 124–131.

Ruiz-Boyer, A. 2015. Hongos. In: Sánchez, J. (Ed.), *Guía de Biodiversidad del Caribe Sur de Costa Rica*. Museo Nacional de Costa Rica, San José, Costa Rica, pp. 199–244.

Ruiz-Boyer, A. & A. Rodríguez-González. 2020. Lista preliminar de hongos (Ascomycota y Basidiomycota) y mixomicetos (Myxomycota) de la Isla del Coco, Puntarenas, Costa Rica. *Revista de Biología Tropical* 68(1): 33–56.

Sharma, S.K. & N. Gautam. 2017. Chemical and bioactive profiling, and biological activities of coral fungi from Northwestern Himalayas. *Scientific Reports* 7: 46570.

Sofian, F., N. Kikuchi, T. Koseki, Y. Kanno, S. Uesugi & Y. Shiono. 2021. Antioxidant p -terphenyl compound, isolated from edible mushroom, *Boletopsis leucomelas*. *Bioscience, Biotechnology, and Biochemistry* 86(3): 300–304.

Uitzil-Colli, M. O. & L. Guzmán-Dávalos. 2019. El género *Morchella* (Pezizales, Ascomycota) en Jalisco, México. *Scientia Fungorum* 49: e1209.

Vicerrectoría de Acción Social. 2017. *Proyecto EC-408 Diversidad lingüística de Costa Rica: Se' má Diccionario-recetario de la alimentación tradicional bribri*. Universidad de Costa Rica, Costa Rica.

Vicerrectoría de Acción Social. 2020. *Proyecto IE-118 Tradición oral sobre los macrohongos en las comunidades indígenas de Bajo Coén y Coroma, en Bribri, Talamanca*. Universidad de Costa Rica, Costa Rica.

WingChing-Jones, R. & G. Alvarado. 2009. Valor nutricional del heno de translava inoculado con el hongo *Pleurotus ostreatus*. *Agronomía Costarricense* 33(1): 147–153.

Wright, J.E. & E. Albertó. 2002. *Guía de hongos de la región pampeana. I. Hongos con laminillas*. LOLA, Buenos Aires, Argentina.

Zhou, H.M., Q. Zhao, Q. Wang, F. Wu & Y.C. Dai. 2022. Two new species of *Boletopsis* (Bankeraceae: Thelephorales) from Southwest China. *MycoKeys* 89: 155–169.

Zumbado, F. & C. Raabe. 1976. *Evaluación de la distribución geográfica de la población de Costa Rica*. Instituto Geográfico Nacional, Costa Rica.

5 A Checklist of Medicinal Mushrooms in Kerala State, India

C. K. Pradeep, A. Manoj Kumar

5.1 INTRODUCTION

Mushrooms have long been relished as a delicacy for their nutritional aspects. Organoleptic and medicinal properties, along with their economic significance, make them perfect candidates of interest as neutraceuticals. Nutritional analyses have revealed that edible mushrooms possess all the three credentials of an ideal food—taste, nutritive value, and physiological benefits (Chang and Buswell 2003).

Regarding their nutritional value, mushrooms are abundant in protein. The actual nutritional value of mushrooms is recognized not just by farmers and researchers, but also by consumers in general (Chang and Wasser 2012). Apart from their wonderful flavour, mushrooms contain a favourable blend of nutrients, functional proteins, and polyunsaturated fatty acids, placing them one of the most desirable positions in low-calorie diets. Edible mushrooms are rich in vitamins B (B1, B2, B12), C, D, and E (Finkel and Holbrook 2000; Heleno et al. 2010; Mattila et al. 2001). Moreover, the presence of trace elements like selenium (which is considered an excellent antioxidant) and the lack of cholesterol has inspired their use as an ideal food.

Other than their value from a nutritional perspective, edible mushrooms are also unique in terms of aroma, colour, texture, and taste, rendering them attractive for consumption by humans. They are storehouses of physiologically active substances, including high molecular weight polysaccharides, dietary fibre, RNA components, lectins, proteoglucans, peptidoglucans, chitinous substances, and heteroglucans, along with low molecular weight organic substances such as novel phenols, steroids, and terpenoids. Many mushroom varieties, in addition to being of importance as nutritious sources of food, are also ingredients in health products and are used medicinally as sources of pharmaceutically active substances like antiviral and anti-tumour molecules. Proprietary products including beverages, cosmetics, dietary supplements, etc. are currently available in the market and are in great demand. The term 'mushroom nutraceuticals' encompasses both medicinal and nutritional aspects of biomolecules extractable from the entire fungus, fruiting bodies, or mycelium culture fluids (Chang and Buswell 1996). Mushroom extract nutraceuticals, even in high doses, show very low toxicity compared with other drugs. Hence, mushrooms can certainly be considered dietary supplements and health foods that mankind and modern society must value in the present and future.

5.2 MEDICINAL PROPERTIES OF MUSHROOMS

Historically, mushrooms have long been thought to have medicinal value. In the Orient, several thousand years ago, many edible and certain non-edible mushrooms were recognized as having valuable health benefits (Bensky and Gamble 1993; Hobbs 1995).

Traditionally, several cultures have utilized the antineoplastic and immunomodulatory potential of many mushrooms for the prevention and treatment of diseases. There has been an overwhelming

DOI: 10.1201/b23190-5

upsurge in the interest in exploring and exploiting the pharmaceutical potential of mushrooms, highlighting them as mini pharmaceutical factories which produce molecules with splendid biological properties (Ferreira et al. 2010; Patel and Goyal 2012).

In spite of the short shelf life of mushrooms, properties such as the chemical composition of interest, useful secondary metabolites, and healthy macro- and micronutrients make them useful from the medicinal and nutraceutical perspectives.

Owing to excellent medicinal properties along with nutraceutical capacity, edible mushrooms can be directly included in the human diet. A resulting overall health improvement may occur due to the synergistic effects of all bioactives present (Barros et al. 2007a; Vaz et al. 2010).

In recent years, edible mushrooms have invited attention as commercial sources of antioxidants (Ferreira et al. 2009; Kozarski et al. 2014; Khatua et al. 2013). Micronutrients like Cu, Zn, Fe, and Mn present in mushrooms form cofactors of enzymes possessing antioxidant properties and are regarded as antioxidant micronutrients (Bhattacharyya et al. 2014). Normal cellular metabolic functions within the gastrointestinal (GI) tract generate by-products called reactive oxygen species (ROS). Even though the intestinal mucosa provides a primary protective barrier, ingested materials and pathogenic microbes can cause oxidative injury and GI inflammatory responses involving immune cells. This may lead to GI cancers, ulcers, and inflammatory bowel disease. Oxidative stress may aggravate and play a part in the aforementioned disorders (Bhattacharyya et al. 2014). Selenium, one of the key nutrients for humans, is present as selenoproteins in many mushrooms and can minimize the cytotoxic effects of ROS. Several selenium compounds such as selenocysteine, selenomethionine, methyl selenocysteine, and seleno-polysaccharides have been identified in mushrooms (Milovanovic et al. 2015). Direct consumption of edible mushrooms might help in boosting antioxidant defence to mitigate oxidative stress. The health benefits and antioxidant effects of edible mushrooms might have prompted their traditional use as a popular delicacy (Kozarski et al. 2015a).

5.3 EDIBLE MUSHROOMS AND MEDICINAL VALUES

As many common edible species of mushrooms also have therapeutic properties, one may not find it easy to draw a clear distinction between those which are edible and those having medicinal properties (Valverde et al. 2015; Chen et al. 2012). Besides antioxidant properties, mushrooms have invited attention for their biological activities, such as hepatoprotective, hypolipidemic, antidiabetic, anticoagulant, anticomplementary, antiviral, anti-tumour, immunostimulant and immunological activities, which render them suitable for use in food, biomedicine, cosmetics, environmental protection, agriculture, and wastewater management (Kozarski et al. 2014; Van Griensven 2009; Loria-Kohen et al. 2014).

It is generally accepted that extracts of fungi contain many components, each of which has its own specific biological effects (Wasser and Weis 1999; Wasser 2010). Antioxidant compounds found in fruiting bodies, mycelia, and broth have been identified as flavonoids, phenolics, polysaccharides, glycosides, ergothioneine, tocopherols, ascorbic acid, and carotenoids (Yang et al. 2002; Acharya et al. 2004; Puttaraju et al. 2006; Ajith and Janardhanan 2007; Klaus et al. 2011; Kozarski et al. 2011; Chen et al. 2012; Kozarski et al. 2012; Weigand-Heller et al. 2012; Muszynska et al. 2013; Stojkovic et al. 2014; Kozarski et al. 2015b). Moreover, high mannitol content and a low glycaemic index make edible mushrooms beneficial for diabetics. Low sodium levels are beneficial for patients with hypertension, whereas high potassium and phosphorus are important for their orthomolecular aspects (Chang and Wasser 2012). In Asian home remedies, mushrooms find an important place in the treatment of several diseases and disorders caused by improper metabolism and reactive oxygen species (ROS) (Khatua et al. 2013). Oxidative stress caused by an imbalanced metabolism and an excess of ROS leads to a range of health disorders in humans. Our oxidative homeostasis is regulated by the dietary intake of antioxidants in addition to the endogenous antioxidant defence system. A number of synthetic antioxidants can make a significant improvement in defence mechanisms; however, the associated risk of adverse toxic effects under some conditions gives natural antioxidants an advantage. This advantage has led to a rapid upsurge in research to consider edible mushrooms as potent sources of antioxidants and other beneficial chemicals. Mushrooms are generally capable of accumulating trace elements, allowing for their integration in the food chain (Radulescu et al. 2010; Falandysz and Borovicka 2013).

Approximately 700 species of higher basidiomycetes are of significance pharmacologically. For thousands of years, various cultures like the Chinese have made use of mushrooms in traditional medicine. Apart from the general medicinal benefit of mushrooms for boosting immunity, many mushrooms offer other specific benefits, such as the presence of polysaccharides (like β-glucans). These polysaccharides help to 'awaken' the immune system. It was also recorded that a formula containing a combination of several mushrooms could trigger a synergistic immune response, thus making the combined formula more beneficial than a single mushroom.

Several bioactive molecules, including those having anti-tumour properties, have been identified in numerous mushroom species. Terpenoids represent the primary class of bioactive compounds isolated from mushrooms. These polysaccharides are the best known and most potent mushroom-derived substances, with immune modulating and anti-tumour properties. Mizuno et al. (1995a), in their work, mentioned research on the anti-tumour activities of mushrooms. These authors found potent anti-tumour properties in aqueous extracts of some mushrooms like *Ganoderma lucidum* (Polyporaceae). Polysaccharides with high anti-tumour activity were reported from shiitake mushrooms (Chihara et al. 1970). A Japanese company commercialized one of these polysaccharides, lentinan, which could be used for the treatment of stomach cancer (Mizuno et al. 1995a). Thereafter, numerous reports were published describing the isolation and study of substances with medicinal properties from many mushroom species. Apart from lentinan, two other cytostatic drugs, krestin and schizophyllan, have been isolated in Japan. Krestin, from the mycelium of *Coriolus versicolor*, is used in the treatment of breast, lung, and GI cancers. Schizophyllan (Sonifilan) extracted from *Schizophyllum commune* culture is used in the treatment of cervical cancer (Mizuno et al. 1995b).

Mushroom polysaccharides have traditionally been used for the prevention and treatment of many cancers, infectious illnesses, and autoimmune diseases. Edible and medicinal mushrooms thus can be considered widely used dietary supplements and nutraceuticals.

Active metabolites of mushroom origin can be sourced from pure mycelial culture, culture filtrates, or whole fruiting bodies. Efforts are under way to obtain active biomolecules and metabolites from mycelia through submerged fermentation cultures. *Lentinus edodes* (shiitake), *Ganoderma lucidum* (reishi), *Inonotus obliquus* (chaga), and many other species have been collected and traditionally used by the people of China, Korea, Eastern Russia, and Japan (Wasser 2002). High fibre and protein and low fat contents in the extracts of edible mushrooms make them ideal for the prevention of dietetic hyperglycaemia (Gunde-Cimerman 1999). Examples of medicinal mushrooms whose extracts have the potential to lower blood glucose are *Cordyceps sinensis*, *Tremella aurantia*, *Auricularia auricula-judae*, and *Ganoderma lucidum* (Kiho et al. 1995; Hikimo et al. 1989). Recent epidemiological studies from Asia have demonstrated that mushroom intake protects against cancer, specifically breast and GI cancers. The anticancer action of mushrooms is connected to immunomodulation by glucans (branched polysaccharides) and glycoproteins. Aberrantly activated signalling pathways in oncocytes are inhibited by some natural compounds of mushroom origin which modulate specific target molecules related to cell functions such as angiogenesis, cell proliferation, and cell survival (Jiang and Sliva 2010).

5.4 MEDICINAL MUSHROOMS AND PLANT PATHOGENS

Studies on the effect of metabolites from basidiomycete fungi on viruses pathogenic to plants are meagre (Di Piero et al. 2010). Strains of *Ganoderma applanatum* and *G. lucidum* were found to inhibit the tobacco mosaic virus (TMV). Cultured biomass filtrate of the polypore *Fomes fomentarius* was also effective against TMV transmission (Zjawiony 2004). A novel lectin, *Agrocybe aegerita* lectin, was purified from the basidiocarps of the edible mushroom *Agrocybe aegerita* (Sun et al. 2003). Aqueous extracts from basidiocarps of *Lentinula edodes* and *Agaricus brasiliensis* exhibited antiviral activity against the aphid-borne mosaic virus (Di Piero et al. 2010). A pink quinone, β-L-glutaminyl-3,4-benzoquinone, in basidiocarps of *Agaricus bisporus* is a potent inhibitor of plant viral infections and has been found to inhibit TMV on *Nicotiana glutinosa* (Tavantzis and Smith 1982). Recently, basidiomycete fungi have received much attention due to their ready availability, medicinal value, low toxicity, and limited side effects on non-target organisms. The discovery of more novel natural products from basidiomycetes for the control of vector mosquitoes, plant diseases, and nematodes will lead to eco-friendly crop protection strategies.

5.5 CHECKLIST OF MEDICINAL MUSHROOMS OF KERALA STATE

5.5.1 *Agaricus campestris* L.

Pileus 4.5–9 cm diam., at first hemispherical, convex to plano-convex with a flattened centre or becoming entirely flat in old specimens; surface pure white, discolouring dull ochraceous or buff in dry conditions, particularly at the disk, smooth and glabrous, at times cracking to appear subsquamulose. Lamellae free, pinkish grey to pinkish brown finally chocolate brown, crowded, with lamellulae. Stipe 4.5–6 × 0.5–1.5 cm, central, cylindrical, often tapering towards base, stuffed; surface white, above annulus smooth, below annulus smooth to fibrillose. Annulus superior, descending, evanescent. Context white, firm, and discolouring when cut. Smell pleasant. Spores 6.5–8.5 × 4.5–6 µm, ellipsoid, smooth, with a brown wall. Basidia 21–24 × 7–8.5 µm, clavate, 4-spored. Cheilocystidia none. Pileipellis a repent epicutis.

Locality: India, Kerala state, Idukki district: Munnar
Season: May

5.5.1.1 Medicinal Properties

Agaricus campestris has been recorded as a traditional remedy for diabetes mellitus (Gray and Flatt 1998). The lectins from *A. campestris* stimulated insulin release in isolated islets of Langerhans from rats (Ewart et al. 1975; Ahmad et al. 1984a, b). *Agaricus campestris* is mentioned in Ayurveda as 'Chatraka' for the treatment of fatigue and is boiled and consumed with ghee (clarified butter) to improve sperm count. It is a good source of proteins and vitamins B, K, C, and D. Chatraka mushroom is cut into small pieces and boiled in milk. This boiled milk is consumed to treat bodily emaciation and to improve physical strength. In conditions of hyperacidity, the mushroom is taken along with ghee or sugar candy. This mushroom is fried with ghee and consumed to improve sexual vigour.

5.5.2 *Agaricus sylvaticus* Shaeff

Pileus 6–9.5 cm diam., convex, often subumbonate, finally applanate; surface cinnamon brown, conspicuously squamulose with appressed dark brown fibrillose squamules on a whitish background; margin appendiculate. Lamellae free, pale greyish brown becoming chocolate brown, crowded with lamellulae. Stipe 5.5–11 × 6–1.2 cm, cylindrical with a small basal bulb, solid or narrowly hollow; surface greyish white, with concolourous fibrills. Annulus superior, white, fragile, pendant, and evanescent. Context up to 5 mm thick at the disk, white, blushing reddish when cut, colour changing slowly on exposure to pinkish vinaceous or dark purple. Spores 5.5–6 × 3.5–4 µm, ovoid, fuscous, with a thickened wall and guttulate contents. Basidia 15–22 × 6–7 µm, clavate, 4-spored. Lamella edge sterile, with crowded cheilocystidia, 10–18 × 8–12 µm, inflated-clavate, piriform, subglobose or globose, thin-walled, hyaline. Pileipellis is a disrupted epicutis of repent radially arranged hyphae, interrupted by thin-walled hyphae made up of short, broad elements, 14–45 × 5–15 µm, rarely branching, with dark fuscous vacuolar contents. All hyphae lack clamp connections.

Locality: India, Kerala state, Thiruvananthapuram district: JNTBGRI campus; Wayanad district: Ponkuzhi.
Season: April-December.

5.5.2.1 Medicinal Properties

Agaricus sylvaticus comprises approximately 41% protein, 36% carbohydrates, and 7% lipids, plus water, ash, and fibre. Extract of *A. sylvaticus* was known to have biological activities, including antiviral, antibacterial, antioxidant, anticancer, immunostimulatory, and anticomplementary effects (Munkhgerel et al. 2011). Four compounds, namely mannitol (Costa et al. 2011), urea (Al-Momany and Saleh 2009), ergosterol (Barros et al. 2008), and ergosterol peroxide (Fortes et al. 2007), were isolated from fruiting bodies of *A. sylvaticus*. Owing to its anti-tumour, antimicrobial, and anticomplementary activities, ergosterol and its peroxide derivatives have been researched frequently recently (Fan et al. 2006; Subbiah and Abplanalp 2003). Ergosterol peroxide (EPO) is a major anti-tumour sterol produced by medicinal or edible mushrooms. *Agaricus sylvaticus*-derived pharmaceuticals can lower the ill effects of hepatitis B and C virus infections. Dietary supplementation with *A. sylvaticus* reduced glycaemia levels in post-surgery cancer patients (Fortes et al. 2008) and provided blood pressure benefits for post-surgery colorectal cancer patients (Fortes and Novaes 2011). Dietary administration of *A. sylvaticus* improved overall nutritional status, reduced abnormal bowel functions, and minimized nausea in patients undergoing chemotherapy for breast cancer (Valadares et al. 2013).

5.5.3 *Auricularia nigricans* (Sw.) Birkebak, Looney, and Sánchez-García

Fruiting body substipitate to sessile; broadly to deeply concave, sometimes with an incurved margin, convoluted with age; up to 6 cm × 1–4 mm, thick, brittle when dry and cartillaginous when fresh; colour variable from hairs reflecting ash grey to moderately brown pigmented to darkly pigmented yellowish brown and sometimes to oxide red with age; rubbery; hymenium smooth, soft, light pinkish dull red when young to dark grey; hymenium densely tomentose to hispid, tufted around margins; medulla present; hyphae branched, hyaline with a slightly thick wall (1.5–2.5 µm), embedded in a copious amount of mucilage. Spores allantoid, often with one large oblong guttule, hyaline, 14.5–17.0 × 5.0–7.0 µm. Basidia 52–73 × 3.2–6.5 µm, 3-septate, cylindrical, sterigmata occasionally observed. Clamp connections present.

Locality: India, Kerala state: throughout
Season: June-December

5.5.3.1 Medicinal Properties

Wood ears are abundantly rich in protein and iron content when compared to regular mushrooms or even vegetables. They are also low in fat, calories, and carbohydrates. A cup of wood ear mushrooms can meet about 20% of daily vitamin B2 (riboflavin) requirements and is rich in vitamin K. Recent reports suggest hypoglycemic, anticoagulant, and cholesterol-lowering effects. *Auricularia* species have formed part of folk medicine for many years in China and have been used traditionally as a stomach tonic and for the treatment of haemorrhoids. Chinese people regularly included *Auricularia* species in their dishes, as these were believed to improve breathing, circulation, and overall health. Recent studies on *A. nigricans* have identified an anticoagulant chemical, and consuming the fungus may help in evading blood clot-related cardiovascular diseases.

In traditional medicine, *A. nigricans* has been used to increase physical and mental energy, control excessive uterine bleeding and haemorrhoids, and treat abdominal pain and toothache. It is also used for the treatment of lumbago, muscle spasm, debility after childbirth, pain, poor blood circulation, dysentery, nausea, slow wound healing in aged persons, and as an antidote for mushroom poisoning (Liu and Bau 1980). It is also said to strengthen the lungs. Avci et al. (2016) tested the antimicrobial properties of *A. polytricha* against *Candida albicans*, *Escherichia coli*, *Enterococcus faecalis*, *Pseudomonas aeroginosa*, and *Staphylococcus aureus* and demonstrated that *A. polytricha* has both antioxidant and antimicrobial properties.

5.5.4 *Calocybe indica* Purkay. and A. Chandra

Pileus 4–14 cm diam., at first convex, expanding applanate; surface chalky white, smooth, and shiny, sometimes with appressed scales at or around the disc. Lamellae adnexed, greyish white, crowded. Context 1.5 cm thick at the disc, white. Stipe 9–10 × 1.7–2 cm, central, sometimes eccentric, cylindrical, stout, solid, broadening towards the base, fibrous; surface chalky white. Spore print white. Taste mild, slightly radish-like. Spore print white. Spores 5.5–6.5 × 4–5 µm, broadly ellipsoid, thin-walled, inamyloid. Basidia 25.5–30.5 × 6.5–8.5 µm, clavate, thin-walled, with carminophilous granulation, 4-spored. Cystidia are not differentiated, often elongated cylindrical basidioles with contents present. Context consisting of thin-walled hyphae with clamp connections. Pileipellis an epicutis.

Locality: India, Kerala state, Thiruvananthapuram district: JNTBGRI campus
Season: November

5.5.4.1 Medicinal Properties

Besides their rich nutritional composition that includes carbohydrates, crude fat, proteins, and fibre, *Calocybe indica* has medicinal properties of much significance (Prabu and Kumuthakalavalli 2014). Ethanolic extracts from *C. indica* prevent hepatotoxicity (Chatterjee et al. 2011). Anticancer activity of this mushroom was also reported against Dalton's lymphoma ascites (Prabu and Kumuthakalavalli 2014), human Ewing's sarcoma MHH-ES-1, and breast cancer MCF-7 (Ghosh 2015). The mushroom also has antidiabetic properties (Krishnamoorthy 2003). It is reported to contain different types of minerals, vitamins, and flavonoid compounds and has immunoenhancing abilities (Mandal et al. 2011). It possesses all the essential amino acids, especially threonine, levels of which were exceptionally high when compared to those in a reference protein. Likewise, methionine, isoleucine, and valine content were also found to be higher than that in the reference protein. Milky mushroom (*C. indica*) was found to be of superior protein quality regarding essential amino acid content and composition. *Calocybe indica* contains a significant degree of antioxidant activity with high levels of phenols, protein, carbohydrates, lipids, fibre, moisture, and ash. *Calocybe indica* is used by the Kewat tribes of Uttar Pradesh, India, as a medicine for stomach pain (Vishwakarma et al. 2016). Mushroom extracts were found to have an anti-hyperglycaemic and anti-lipid peroxidation effect (Subbiah and Balan 2015). Nisha and Kumuthakalavalli (2016) evaluated *C. indica* for its anticancer activity against the lung carcinoma (A549) cell line. A methanolic extract from *C. indica* showed significant inhibitory activity against *Bacillus lentus* and *Bacillus cereus*, as well as a moderate effect against other microorganisms. The aqueous extract was found to show significant activity against *Bacillus cereus* and *Pseudomonas azotogensis* (Alfala et al. 2017).

5.5.5 *Cantharellus cibarius* Fr.

Pileus 3–10 cm diam., convex at first becoming nearly plane, disc shallowly depressed, finally broadly infundibuliform; surface deep yellow to butter yellow fading in age and when exposed to weather, smooth and glabrous. Lamellae decurrent, narrow, cream to light orange, forked, not intervenose. Stipe 2.5–6.5 × 0.7–0.15 cm, central, cylindrical to compressed, solid; surface light yellow to yellowish white, smooth, and glabrous. Smell pleasant, fruity, like apricots when dried. Context yellowish white. Spores 7.5–12 × 5–6.5 µm, ellipsoid, hyaline, inamyloid, smooth. Basidia elongated-clavate, 64–72 × 8–9.6 µm, 4–6 spored. Lamella-edge fertile, cystidia absent. Pileipellis is a repent epicutis of radially arranged, thin-walled hyphae.

Locality: India, Kerala state, Thiruvananthapuram district: JNTBGRI campus; Kallar; Wayanad district: Muthanga, Ponkuzhi
Season: January, May-September, November-December

5.5.5.1 Medicinal Properties

Cantharellus cibaruius is rich in vitamin A, B-complex, D, and carotene pigment. It possesses all the essential amino acids, along with glutamic acid, which boost immunity and may help in fighting reheumatoid arthritis, infections, and cancer. There is considerable evidence that it inhibits the clotting of blood and may fight cardiovascular diseases. Ethanolic and methanolic extracts of *C. cibarius* exhibited antioxidant, antimicrobial and phytochemical potential (Aina et al. 2012). Terpenoids, phenols, flavonoids, anthraquinones, alkaloids, and saponins in both extracts had antioxidant properties. *Candida albicans* and *Escherichia coli* were inhibited by both methanolic and ethanolic extracts, while methanolic extracts also inhibited *Salmonella typhi*. This suggested that *C. cibarius* metabolites and phytochemicals possess potent bioactivity capable of free radical-scavenging and inhibition of harmful microorganism growth.

Kozarski et al. (2015a) demonstrated that the methanolic extract of *C. cibarius* had anti-oxidative, cytotoxic, antihypertensive, and antibacterial properties. The antioxidant activity of methanol extract from *C. cibarius* exhibited good selectivity in its cytotoxic effect on breast carcinoma, human cervix adenocarcinoma, HeLa, and human myelogenous leukemia cells, with limited effects on control human lung bronchial epithelial cells and human foetal lung fibroblasts. Extracts also showed selective antimicrobial activity against Gram-positive bacteria, especially *Enterococcus faecalis*. *Cantharellus cibarius* has been a part of traditional cuisine, especially in Western countries, owing to its medicinal and health benefits. Its wound-healing and anti-inflammatory medicinal properties have led to its use in treating wounds.

5.5.6 *Coprinus comatus* (O. F. Mull.) Pers.

Pileus 5–6 cm diam., elongated campanulate or conical, becoming torn or revolute at the margin; surface white then blackish, with shaggy white or pale brownish scales, often with darker tips. Lamellae free, white finally black. Stipe 10–15 × 1–2 cm, central, cylindrical, attenuating upwards from a thickened bulbous, sometimes rooting base, hollow; surface white or clay buff, smooth and shiny. Odour none or pleasant. Spore print fuscous black. Spores 11.5–18.5 × 7–10 µm, ovoid to short ellipsoid, thick-walled, smooth, blackish brown, truncated by a broad apical germ pore. Basidia abundant, trimorphic, all 4-spored. Cheilocystidia dimorphic, globose and clavate, globose cystidia, 15.6–35.5 × 8.5–24.2 μm, clavate cystidia 34–51 × 11.4–17 μm in size, thin-walled. Pileus cuticle hyphoid with projecting veil remnants. Clamp connections are present.

Locality: India, Kerala state, Wayanad district: Muthanga
Season: February, March, August

5.5.6.1 Medicinal Properties

Coprinus comatus has been used to treat piles, digestive disorders, diabetes, and circulatory diseases in traditional Chinese medicine. It is claimed to control diabetes mellitus, regulate blood glucose levels, and exert anti-inflammatory and preventive effects on patients with cardiovascular diseases and diabetes. Its protective functions on islets of Langerhans have also been demonstrated. After oral intake of *Coprinus* powder, blood sugar levels reduced by 41%, 31%, and 20% after 1 ½, 3, and 6 h respectively. Besides its tremendous property to control diabetes, *C. comatus* is also useful in the treatment of haemorrhoids, speeding up metabolism and relieving constipation. Furthermore, *C. comatus* has been scientifically demonstrated to reduce the risk of cancer and sarcoma. Many physiological functions of *C. comatus* have been reported, such as anticancer properties (Cui et al. 2002), immunomodulatory effects (Xing et al. 2003), and hypoglycemic activity (Gu and Ju 1996). *Coprinus comatus* contains the antioxidant ergothionein, which shows anti-tumour activity against hormone dependent tumours and has a very high lectin content. The naturally occurring antioxidant arsenal in *C. comatus* includes tocopherols, phenols, polysaccharides, and flavonoids (Li et al. 2010). In their studies, Dotan et al. (2011) demonstrated that *C. comatus* can act as a natural antiandrogenic modulator and may serve in the treatment of prostate diseases. The antinociceptive activity of *C. comatus* was studied by Ren and colleagues (Ren et al. 2014), who found that *C. comatus* showed antioxidant, anti-inflammatory, peripheral antinociceptive, and antihyperalgesic activity in various models of inflammatory pain.

5.5.7 *Ganoderma lucidum* (Curtis) Karst.

Basidiocarps annual, centrally or laterally stipitate, rarely substipitate, solitary or imbricate, orbicular to semi-circular, sometimes variously lobate, corky in consistency, woody when dry. Pileus surface oxblood red, reddish brown to terra cotta towards the margin, colour slightly varying in different sporophores; glabrous, smooth, shiny, laccate, and concentrically sulcate. Stipe fox brown or black, glabrous, cylindrical to flattened, reduced, fibrous, solid. Hymenophore pale grey becoming brownish on bruising, even and smooth; pores almost visible to the naked eye, extending to the margin, slightly decurrent, round, 4–6 per mm. Context corky, zonate, cream, darker when old, shiny. Hyphal system trimitc. Pileal cuticle a typical hymeniderm consisting of end cells of hyphae anticlinal to the pileal surface, clavate, thick-walled, about 45–50 × 12.5–15 μm, compactly arranged, with encrustation. Basidia broadly ellipsoid, 10–20 × 8–11 μm with basal clamp connections. Basidiospores 8–11 × 6–8 μm, yellowish brown, thick-walled, ovoid, truncate, outer wall hyaline, inner wall thick, finely verrucose and coloured, inamyloid. Cuticular layer composed of a palisade of brown, cylindrical-clavate hyphal ends.

Locality: India, Kerala state: throughout
Season: January-December

5.5.7.1 Medicinal Properties

Ganoderma, known as 'reishi mushroom' is also termed the 'elixir of life', 'king of herbal medicines, and 'mushroom of immortality', as it is considered a high-class supplement to promote longevity.

Several records and scripts in traditional Chinese medicine certify this. It is well established that the *Ganoderma* species stimulate the regeneration of a damaged liver and reduce renal fat deposition. This mushroom also has benefits against immune system disorders and cancer. It can help to regulate blood pressure and normalize cholesterol levels. Due to its immunomodulating effects and ability to work against viral infections and autoimmune disorders, reishi is a highly researched basidiomycete. Its ability to oxygenate blood makes it an ideal material to treat altitude sickness. Hobbs (1995) outlined several pharmacological effects of extracts from whole reishi mushrooms in his book, 'Medicinal Mushrooms'. These effects have been demonstrated to include antiallergic, analgesic, antibacterial, anti-inflammatory, anti-tumour, antioxidant, hypotensive, antiviral, relaxing nervine, cardiotonic, expectorant and antitussive, adaptogen, anti-retroviral, immunomodulatory, and hepatoprotective properties. Strong antihepatotoxic activity against galactosamine-induced cytotoxicity in primary cultured rat liver cells was shown by ganodermic acids R and S isolated from mycelial cultures of *Ganoderma*. Ganosporeric acid A is yet another liver protective compound isolated from the ether-soluble fraction of *G. lucidum* spores (Chen and Yu 1991). *Ganoderma lucidum* extracts scavenge the hyperoxide radicals responsible for aggravating aging in humans (Liu et al. 1997).

5.5.8 *Hygrocybe coccinea* (Schaeff.) P. Kumm.

Pileus 4 cm in diam., convex; surface chrome orange with reddish tints, smooth, viscid, translucently striate, non-hygrophanous; margin wavy. Lamellae adnexed, white. Stipe 8 × 0.6 cm, central, cylindrical, stuffed, saffron yellow, whitish at the extreme base, slimy. Odour and taste none. Spore print white. Spores 7–10 × 4.5–6 µm, oblong, ellipsoid to rarely subamygdaliform, sometimes constricted, smooth, thin-walled, hyaline. Cheilocystidia and pleurocystidia absent. Hymenophoral trama regular. Pileipellis a slightly gelatinized cutis. Clamp connections present.

Locality: India, Kerala state, Kollam district: Achenkovil
Season: June

5.5.8.1 Medicinal Properties

Phenols are the major antioxidant components detected in *Hygrocybe coccinea*, followed by anthocyanidins, flavonoids, ascorbic acid (found in small amounts), β-carotene, and lycopene (Sharma and Gautam 2015). Phenolic compounds were detected in higher amounts than other bioactive compounds. High levels of α- and β-tocopherol correspond with high oxidative activity and are thus regarded to play a role in cardiovascular protection (Reis et al. 2011). The abundance of phenolic compounds accounts for the high antioxidant properties of the species (Barros 2007b). *Hygrocybe coccinea* mushroom extracts are inhibitory against sarcoma (Zusman et al. 1997).

5.5.9 *Hygrocybe parvula* (Peck) Murrill

Pileus 2–4 cm in diam., convex becoming applanate with a small depression at the centre, often perforate and continuous with the stipe; surface light yellow or pale yellow, sometimes paler in older specimens, pellucid striate, dry. Lamellae sub-decurrent to decurrent, milky-white to yellowish white with lamellulae. Stipe 3–5 × 3–7 mm, central, cylindrical, hollow, milk white or yellow, smooth, dry. Context brittle, waxy, thin. Odour and taste mild or none. Spore print white. Spores 6.5–7.5 × 3.5–4.5 µm, ellipsoid or ellipsoid-elongated to cylindrical, constricted, thin-walled, smooth. Basidia 49–60 × 5.5–6.5 µm, clavate to cylindro-clavate, 4-spored. Cheilocystidia and pleurocystidia absent. Hymenophoral trama subregular. Pileipellis a cutis. Clamp connections present.

Locality: India, Kerala state, Thiruvananthapuram district: JNTBGRI campus; Wayanad district: Muthanga; Kollam district: Kulathupuzha, Thenmala, Aryankavu.
Season: June-November

5.5.9.1 Medicinal Properties

Various therapeutically important secondary metabolites have been discovered in the phytochemical screening of *Hygrocybe parvula*, including saponins, alkaloids, glycosides, phenols, flavonoids, and terpenoids (Chittaragi et al. 2013). These phytoconstituents, along with sesquiterpine, alkaloid, glycosides, and phenolic compounds, exhibit inhibitory action against bacterial growth. They protect plants against fungal and bacterial infections (Mather and Gonzalel 1982). Chloroform, petroleum ether, and methanol extracts of *H. parvula* were found to possess significant inhibitory effects against nine plants and ten human pathogenic fungi. High concentrations of microelements such as zinc, copper, manganese, and iron were found to occur in the fruiting bodies, whereas other elements such as lead and cadmium were found in moderate to slight concentrations (Chittaragi and Naika 2014). Crude extracts of *H. parvula* were evaluated for anthelmintic activity against *Pheretima posthuma* (Chittaragi and Kodiyalmath 2014). In the study, the effectiveness of the drug was judged on the basis of the loss of spontaneous movement or death of trematodes. All extracts showed effective anthelmintic activity at a 100 mg/ml concentration.

5.5.10 *Hymenopellis colensoi* (Dörfelt) R. H. Petersen

Pileus 5–6 cm diam., convex to applanate, often with a slight depression at the centre; surface yellowish brown or sometimes darker, viscid when wet, rugose at the centre. Lamellae adnate, white. Stipe 14–15 × 0.6–0.8 cm, central, cylindrical, fistulose, tapering to a pseudorrhiza of 5–6 cm in length below the soil surface; surface greyish brown, pruniose. Spores ovoid to ellipsoid, 12–20 × 9–12 µm, hyaline, smooth with a slightly thickened wall. Basidia 60–68 × 10–15 µm, clavate, 4-spored. Cheilocystidia abundant, 50–60 × 12–20 µm, clavate or constricted, with a pale brown vacuolar pigment. Pleurocystidia 50–130 × 6–30 µm, cylindrical, clavate or lageniform, with a pedicellate base, thin- or sometimes with a slightly thickened wall. Hymenophoral trama regular. Pileipellis hymeniodermic forms an epithelium embedded in a gelatinous matrix. Clamp connections present.

Locality: India, Kerala state, Thiruvananthapuram district: JNTBGRI campus; Kollam district: Cheenikkla, Thenmala; Palakkad district: Nellyampathy; Idukki district: Munar, Rajamala; Wayanad district: Meppady

Season: June-November.

5.5.10.1 Medicinal Properties

Hymenopellis colensoi (i.e., *Oudemansiella radicata*) is an edible wild mushroom with high economic value. Its antifungal and antibiotic activity (due to the metabolite oudenone), as well as its responses to cadmium and lead, are well-established. It has been reported that *H. colensoi* shows promising therapeutic and inhibitory effects on Erhrlich carcinoma and sarcoma 180 a due to the mycochemical compound oudenone (Anke et al. 1979). Liu and colleagues isolated lectin and examined its characteristics (Liu et al. 2013). *In vivo* hepatoprotective and *in vitro* antioxidant activities of alkali-soluble polysaccharides (ORAP) and water-soluble polysaccharides (ORWP) were investigated by Liu et al. (2017). They reported the potential of this mushroom as an excellent source of hepatoprotective polysaccharides and antioxidants, suggesting that it could be promising for the development of a nutraceutical or functional foods to improve health.

5.5.11 *Lentinus sajor-caju* (Fr.) Fr.

Pileus 5–10 cm diam., convex with a deeply umbilicate centre then deeply cyathiform to infundibuliform, soft coriaceous drying hard and rigid; surface pale grey, cream or brownish grey, dry, glabrous and smooth or with minute appressed squamules towards the centre. Lamellae deeply decurrent, extending down to the annulus, very crowded, and whitish. Stipe 2–4 × 0.5–1.5 cm, short, central to excentric, cylindrical, equal, solid; surface concolourous to the pileus or paler. Annulus superior, tough, and fleshy when fresh, often disappearing with age. Context white, thin, fibrous, coriaceous, composed of a dimitic hyphal system. Spores 4–8 × 1.5–2.5 µm, cylindrical, thin-walled, hyaline. Basidia 15–20 × 4–5 µm, narrowly clavate, 4-spored. Lamella-edge sterile, cheilocystidia tufted, 30–55 × 2–5 µm, with obtuse or nodulose apices. Hyphal pegs abundant, consisting of fascicles of narrow hyphae, originating from the trama. Hymenophoral trama irregular. Pileipellis an epicutis.

Locality: India, Kerala state, Thiruvananthapura district: JNTBGRI campus; Kollam district: Aryankavu, Cheenikkala
Season: April, June, August-December

5.5.11.1 Medicinal Properties

Lentinus sajor-caju is highly valued as nutritious food in many Asian and European countries. Currently, *Lentinus sajor-caju* is cultivated all over the world. It has low lipid and starch levels and contains quality vitamins (B1, B2, and C) and proteins. It can reduce bad cholesterol levels in the blood (Pramanik et al. 2007). High soluble fibre and low fat contents help in preventing hyperlipidaemia (Schneider et al. 2011). Several extracts from this mushroom showed remarkable therapeutic potential, including antioxidant (Kanagasabapathy et al. 2013; Boonsong et al. 2016), anticancer (Finimundy et al. 2013), anti-inflammatory (Silveira et al. 2015) properties. *Lentinus sajor-caju* contains zyloglucan and xylanproteins and has exhibited anti-tumour activity against sarcoma 180 tumour cells *in vivo* (Zhuang et al. 1993).

5.5.12 *Lentinus squarrosulus* Mont.

Pileus 2.5–7 cm diam., convex soon depressed, umbilicate to deeply infundibuliform, sometimes perforate, fleshy when fresh, drying hard and rigid; surface white, becoming cream or pale brown with age, covered with appressed to recurved, concentrically arranged squamules. Lamellae deeply decurrent, white to pale buff, crowded. Stipe 1.2–3.5 × 0.2–0.7 cm, typically central often excentric, cylindrical, rarely lateral, attenuated towards the base, firm, solid; surface concolourous with the pileus, covered with irregular floccose squamules. Context up to 2 mm thick, fleshy, white, coriaceous, with a dimitic hyphal system. Spore print white. Spores 5–7.5 × 1.5–2.5 µm, cylindrical, hyaline, thin-walled, smooth. Basidia 17–18 × 4.5–5.5 µm, 4-spored. Lamella-edge sterile with abundant cystidiform elements, 20–55 × 2–5 µm. Hyphal pegs numerous, extending beyond the basidial layer, arising directly from the trama. Hymenophoral trama irregular and intermixed. Pileipellis a loose epicutis.

Locality: India, Kerala state: throughout.
Season: January-December

5.5.12.1 Medicinal Properties

A study by Ahmad et al. (2014) revealed that extracellular crude extract from *Lentinus squarrosulus* culture broths exhibited excellent antioxidant activity (in DPPH and β-carotene systems). Mycelium extracts exhibited high β-glucan in intracellular polysaccharides. Extracellular crude extracts could be used as anti-oxidative agents in the food industry, and intracellular polysaccharides may be used as tumour-preventing agents in the treatment of cancer. High content of β-carotene (provitamin-A), an excellent antioxidant (Ross et al. 2011), and flavonoids (Hussein et al. 2015) have been identified in *L. squarrosulus*.

5.5.13 *Lepista sordida* (Schumach.) Singer

Pileus 1.5–6 cm diam., convex, subumbonate, finally depressed; surface pale lilaceous pink to lilaceous brown, hygrophanous. Lamellae sinuate adnate, pale lilaceous, up to 8 mm broad with lamellulae. Stipe 3–7 × 0. 4–1 cm, equal, soon hollow; surface concolourous with the pileus, fibrous, fleshy. Context thin, up to 5 mm wide, soft, consisting of thin-walled hyphae 3–5 µm wide with clamp connections. Spore print pink. Spores 5.5–7 × 3–4.5 µm, ellipsoid, thin-walled, hyaline, with fine, distinct, rugulose ornamentation, inamyloid. Basidia 20–25 × 5–7 µm, clavate, 4-spored. Cystidia absent. Hymenophoral trama regular. Pileal surface a repent epicutis.

Locality: India, Kerala state, Thiruvananthapuram district: JNTBGRI campus; CPCRI campus, Palode, Varkala; Kollam district: Thenmala, Kulathupuzha
Season: January-December

5.5.13.1 Medicinal Properties

Four water-soluble polysaccharides extracted from the fruiting bodies of *Lepista sordida* were found to suppress the growth of tumour cells. Miao et al. (2013) successfully isolated and purified four polysaccharides (LSPa1, LSPb1, LSPb2, and LSPc1) from the fruiting bodies of *L. sordida* and characterized their physicochemical properties. They analysed and evaluated the cytotoxicity and tumour inhibition effects of these on Hep-2 human laryngocarcinoma cells *in vitro*. Results clearly indicated that three acidic polysaccharides exhibited stronger lactate dehydrogenase release-promoting and antiproliferative effects on Hep-2 cells than neutral polysaccharides. Further investigations of the anti-tumour activity of LSPc1 were conducted in mice engrafted with Hep-2 cancer cells. Hence, both *in vivo* and *in vitro* studies demonstrate the potential use of LSPc1 as an anticancer agent. Mazur et al. (1996) isolated two new active diterpenoids from *Lepista sordida* in screening for inducers of the differentiation of human leukaemia cells.

5.5.14 *Macrocybe gigantea* (Massee) Pegler and Lodge

Pileus 20–35 cm diam., fleshy, conico-convex then expanding convex to flat; surface white to greyish white, paler towards the margin, glabrous. Lamellae sinuate, straw yellow, densely crowded. Stipe 15–18 × 6 cm, central, cylindrical, solid finally fistulose; surface concolourous with the pileus, fibrillose-striate. Odour mushroomy. Context up to 3 cm thick at the disc, white, firm. Spore deposit white. Spores 5–6.5 × 3.5–4.5 µm, ovoid to short ellipsoid, inamyloid. Basidia 25–37 × 5–8 µm, narrowly clavate to subcylindrical. Cystidia absent. Hymenophoral trama regular. Pileipellis a compact, repent cutis. Clamp connections present.

Locality: India, Kerala state, Thiruvananthapuram district: JNTBGRI campus
Season: January-March; May-November.

5.5.14.1 Medicinal Properties

Studies by Gaur and Rao (2016) suggested that *M. gigantea* contains high antioxidant potential and thus may represent a good nutraceutical source. Recently, Roy et al. (2017) evaluated for proximate composition and mineral nutrients, and the efficacy of the mycelial extract was evaluated for antimicrobial activity against Gram-positive and Gram-negative bacteria, with results suggesting a broad-spectrum antimicrobial potential. Acharya et al. (2012) suggested that *M. gigiantea* is one of the mushrooms most frequently used in hepatoprotection studies, and this property is due to the presence of large amounts of antioxidant phenolics and flavonoids. Medicinal and nutraceutical research on the giant mushroom *M. gigantea* was conducted at the Ethnomedicine and

Ethnopharmacology Division and Microbiology Division of Jawaharlal Nehru Tropical Botanic Garden and Research Institute (JNTBGRI), Thiruvananthapuram, with an objective of assessing the medicinal and nutraceutical potential of *M. gigantea* with special emphasis on immunomodulation and hepatoprotection. The hepatoprotective potential of *M. gigantea* was determined using an acetaminophen-induced hepatotoxicity study and carbon tetrachloride-induced liver damage. Hepatoprotection and immunomodulation studies revealed that *M. gigantea* possesses significant hepatoprotective and immunomodulatory activity.

5.5.15 *Macrolepiota dolichaula* (Berk. and Broome) Pegler and Rayner

Pileus 6–20 cm diam., fleshy, globose in the bud, expanding to convex, then plano-convex to finally applanate with a broad obtuse umbo; surface ivory to brown in the bud, soon disrupting to form minute concolourous granular, punctate to verrucose squamules on a whitish to pale pinkish background; surface discolouring reddish and then brownish on bruising. Lamellae free, remote, collariate, creamy white, turning reddish on bruising, up to 2 cm wide, densely crowded. Stipe 11–28 × 1.5–2.5 cm, central, cylindrical or tapering upwards from a clavate to bulbous base, stuffed becoming narrowly hollow, brittle; surface white, turning reddish brown on handling, with appressed fine squamules. Annulus superior, white, membranous, becoming movable with age, double-edged, flaring, with white cottony patches below. Odour pleasant. Context up to 2 cm wide at the disc, soft, white, unchanging. Spore print white. Spores 9–13.5 × 6–9.75 µm, ovo-ellipsoid to ellipsoid with an apical germ pore, with a thick, complex wall, dextrinoid, metachromatic. Basidia 31.5–43.5 × 10.5–15 µm, clavate, 4- spored. Lamella-edge sterile. Cheilocystidia 18–38.5 × 7.5–12 µm, clavate to elongated-clavate, thin-walled. Hymenophoral trama subregular. Pileal cuticle, a trichoderm. Clamp connections present.

Locality: India, Kerala state, Thiruvananthapuram district: JNTBGRI campus, Bharathanoor, Peringamala, Ex-service colony, Palode, Kulathupuzha; Kollam district: Madathara, Kadakkal.

Season: January-March; May-November

5.5.15.1 Medicinal Properties

Macrolepiota dolichaula was found to be a good source of bioactive compounds like flavonoids, phenolics, alkaloids, β-carotene, and lycopene. Antioxidant properties of flavonoids function against free radical scavengers, thus preventing oxidative cell damage and cancer. This mushroom is a good source of digestible carbohydrates, proteins, vitamins, and fibre, as well as flavonoids, phenolic compounds, lycopene, β-carotene, and alkaloids (Kumari and Atri 2014). *Macrolepiota dolichaula* also contains significant amounts of vitamins A, B1, B2, B3, C, and D and exhibits laccase activity (Kumari et al. 2012). Two water-soluble polysaccharides (PS-I and PS-II) were isolated from its fruiting body extracts (Samantha et al. 2013), and a fucogalactan isolated from this mushroom exhibited immunostimulatory and antioxidant properties *in vitro* (Samantha et al. 2015).

5.5.16 *Macrolepiota mastoidea* (Fr.) Singer

Pileus 8.5–10.5 cm diam., convex, sometimes with a broad obtuse umbo; surface white with a brown or reddish brown centre, smooth at the disc or sometimes cracked forming large patches of scales, disrupting to concolourous granular scales near the margin; margin exceeding the lamellae, appendiculate. Lamellae free, remote, creamy white, up to 1.3 cm, crowded. Stipe 19–26.5 × 1–1.3 cm, central, cylindrical, narrowly tapering upwards from a clavate-bulbous base, stuffed; surface white turning dull brown on handling, covered with concolourous fine squamules. Annulus superior, double-edged, with a flaring pendant region, becoming movable with age, white with a brown rim. Context soft, white, unchanging. Spore print white. Spores 11–16.5 (18) × 7.7–10.5 (11.5) µm, broadly ellipsoid to ellipsoid with a broad germ pore, smooth with a thick complex wall, dextrinoid, metachromatic. Basidia 21–38.5 × 11–13.2 µm, 2–4 spored. Lamella-edge sterile with cheilocystidia 25.3–46.6 × 12–19.5 µm, clavate to inflated-clavate with a broad obtuse apex, often with a pedicel, thin-walled, hyaline. Hymenophoral trama regular. Pileipellis a trichoderm at the disc and a disrupted epicutis towards the margin.

Locality: India, Kerala state, Kollam district: Kulathupuzha, Thenmala
Season: September

5.5.16.1 Medicinal Properties

Macrolepiota mastoidea has various pharmacological potential uses due to its antiviral, antimicrobial, antiallergic, anti-tumour, anti-inflammatory, immunomodulating, hypoglycaemic, antiatherogenic, and hepatoprotective properties (Lindequist et al. 2005). It also exhibits potent antioxidant activities (Yang et al. 2002; Mau et al. 2004; Cheung et al. 2003; Ribeiro et al. 2006; Ferreira et al. 2007). *Macrolepiota mastoidea* contains a significant amount of antioxidant activity with high amounts of phenol (Vishwakarma et al. 2016).

5.5.17 *Macrolepiota procera* (Scop.) Sing.

Pileus 11 cm diam., convex; surface white, covered with chocolate brown scales when young, later expanding into a central calotte, which is entire and smooth, around which there are plate-like scales, recurved and easily removable, the remaining surface is covered by dirty white furfuraceous scales which vanish on handling; margin exceeding the lamellae, appendiculate. Lamellae free, remote from the stipe, collariate, developing some brown spots on aging, crowded. Stipe 18 × 1 cm, central, cylindrical, with a basal bulb, hollow; surface creamy white with brown or chocolate brown squamules arranged in zig-zag bands, girdling over the whole length. Annulus superior, becoming movable with age, complex, double-edged, upper part white with floccose scales and the underside with chocolate brown scales. Context up to 1 cm thick, soft, white, unchanging. Spore print white. Spores 10–13.3 × 7.7–10 µm, broadly ellipsoid to ellipsoid with a conspicuous germ pore, hyaline with a thick complex wall, dextrinoid, metachromatic. Basidia 25.3–29.6 × 12–17 µm, clavate 2–4-spored. Lamella-edge sterile with crowded cheilocystidia. Cheilocystidia 20–40.4 × 10–19 µm clavate, elongated-clavate to pyriform. Hymenophoral trama regular. Pileipellis a disrupted epicutis with clamp connections. Caulocystidia abundant, and cylindrical.

Locality: India, Kerala state, Palakkad district: Chembotty
Season: December

5.5.17.1 Medicinal Properties

Apart from its excellent nutritional value, *Macrolepiota procera* is a valuable source of natural antioxidants (Popescu et al. 2016). The fruiting body is used for the treatment of diabetes and high blood pressure by the Bhuj and Bhar tribes of Uttar Pradesh (Vishwakarma et al. 2016). *Macrolepiota procera* contains 18 amino acids, of which eight are essential, and the fruiting body extracts possess anti-tumour and antibacterial activity (Didukh et al. 2003). Lukanc et al. (2015) isolated and characterized trypsin inhibitors from *M. procera*, the exceptional characteristics of which are advantageous for their possible use in medicine for drug development, in biotechnology for purification processes, in agriculture for pest control and crop protection, and for studies on trypsin regulation in various diseases.

5.5.18 *Oudemansiella canarii* (Jungh.) Hohn.

Pileus 2.5–11 cm diam., convex then applanate; surface white to deep sepia, often darker at the centre, soon fading to greyish cream or white, viscid to glutinous when moist translucent striate, smooth.

Lamellae adnexed to adnate, white to pale greyish, close with lamellulae. Stipe 1.5–6 × 0.3–1 cm, cylindrical, equal or with a clavate or bulbous base, solid; surface whitish with concolourous floccose squamules. Context fleshy, white. Spores 14–24 × 11–23 µm, globose to subglobose, inamyloid. Basidia 65–80 × 16–24 µm, clavate. Lamella-edge sterile; cheilocystidia and pleurocystidia similar, 80–150 × 10–40 µm, ventricose fusoid with a broadly rounded apex, hyaline, thin-walled. Hymenophoral trama regular. Pileal surface epithelium of globose to pyriform or lageniform elements, 25–55 × 6–15 µm, thin-walled, with or without brownish contents. Clamp connections present.

Locality: India, Kerala state: Thiruvananthapuram district: JNTBGRI campus, Kallar, Pazhakutty; Kollam district: Thenmala, Achankovil: Ernakulam district: Bhuthathankettu, Edayar; Idukki district: Munnar; Wayanad district: Nadavayal
Season: January, April-November

5.5.18.1 Medicinal Properties

Oudemansiella canarii is an edible mushroom with good potential as a food source (Ruegger et al. 2001). Rosa et al. (2005) found that the crude extract of *O. canarii* has significant antimicrobial activity against pathogenic yeasts *Candida krusei*, *Candida albicans*, *Candida tropicalis*, and *Candida glabrata*. The presence of oudemansin A in *O. canarii* extracts shows strong activity against several pathogenic *Candida* species. Oudemansins exhibited antifungal effects at very low concentrations against *Candida albicans* and dermatophytes belonging to the genera *Epidermophyton*, *Trichophyton*, and *Microsporum* (Anke 1997). All oudemansins exhibit strong but reversible cytostatic activities against both animal and human cell lines Anke (1997).

5.5.19 *Pleurotus cystidiosus* O. K. Miller

Pileus 4.5–11 cm diam., flabelliform; surface brown at the centre, elsewhere brownish orange, smooth and glabrous. Lamellae decurrent, white when fresh, becoming yellowish white, edge darker, and crowded. Stipe 2.5–3.5 × 0.6–0.8 cm, almost lateral, cylindrical, solid, equal or tapered towards the base; surface greyish brown, whitish towards the base. Context white, up to 1 cm wide, turning yellow, fleshy when fresh. Spore print white to cream-coloured. Spores 12–16 × 4.0–6.0 µm, cylindrical-oblong, thin-walled. Basidia 37–44 × 5.2–9 µm, 2- to 4-spored. Gill edge fertile, cheilocystidia absent. Pleurocystidia with granular contents, clavate, 50 × 7–7.8 µm. Pileipellis a repent epicutis. Context monomitic, formed of thin-walled, hyphae, 3.8–9.4 µm, with clamp connections.

Locality: India, Kerala state, Thiruvananthapuram district: JNTBGRI campus
Season: August

5.5.19.1 Medicinal Properties

Pleurotus cystidiosus is reported to contain two potential angiotensin converting enzyme (ACE) inhibitory peptides which were isolated and characterized by Lau et al. (2013). Compared with chemosynthetic drugs, naturally derived ACE inhibitors are safer for consumption and have fewer negative effects. Consumption of *P. cystidiosus* may be beneficial as a remedy for bodily pain arising from daily activities (Kudahewa et al. 2008). Extracts of *P. cystidiosus* have anticancer and antioxidant properties and exhibit cytotoxic activity against Hep-2 carcinoma cells. *Pleurotus cystidiosus* has disease curative and preventative properties apart from the basic function of providing nutrients (Menikpurage et al. 2012).

5.5.20 *Pleurotus flabellatus* Sacc.

Pileus 3–8 cm diam., orbicular, reniform to irregularly flabelliform with an attenuated base, more rarely with an excentric or central stipe, convex or depressed when young; surface white glabrous or slightly tomentose towards the base; margin at first incurved, then undulate and becoming upturned with age, sometimes incised. Lamellae decurrent, moderately crowded, with lamellulae. Stipe presents or absent, if present then typically lateral, sometimes excentric or even central, short and thick, solid. Context thin, white, fleshy coriaceous, with a monomytic hyphal system of generative hyphae, with clamp connections. Spore print white. Spores 6–8.5 × 2.5–4 µm, oblong cylindrical, hyaline, thin-walled. Basidia 15–20 × 3.5–4.5 µm, clavate, 4-spored. Lamella edge sterile; cheilocystidia 18–26 × 5–10 µm, clavate, cylindrical, at times constricted, with a broad obtuse apex, sometimes with an elongated, mucronate outgrowth. Hymenophoral trama subregular to irregular. Pileipellis a repent epicutis, with scattered dermatocystidia.

Locality: India, Kerala state, Thiruvananthapuram district: JNTBGRI campus; Kollam district: Cheenikkala, Thenmala, Rosemala; Ernakulam district: Iringole Sacred Grove; Palakkad district: Nenmara.
Season: February, June-December

5.5.20.1 Medicinal Properties

Ethanolic extract of *P. flabellatus* was demonstrated to be an effective antioxidant in different *in vitro* assays, including ferric iron-reducing, ferrous iron-chelating, DPPH free radical-scavenging, and total antioxidant analyses (Dasgupta et al. 2013).

5.5.21 *Pleurotus giganteus* (Berk.) Karunarathna and K. D. Hyde

Pileus 5–60 cm diam., convex becoming applanate, finally infundibuliform; surface initially dark fuscous brown then fading with age, remaining darker at the centre, disrupted into small, fuscous, appressed, concentrically arranged brownish squamules, denser at the centre, scattered elsewhere; margin thin, finely serrate, strongly involute then straight, upturned in older specimens. Lamellae decurrent, creamy white, often interveined, crowded. Stipe 5–61 × 0.7–6 cm, central, solid, thick and fleshy, tapering into a radicant base below ground; surface concolourous with the pileus, brownish squamulose towards the apex, soon becoming glabrous. Context white, thick and fleshy. Spore print white. Spores 6–8.5 × 4.5–5.5 µm, broadly ovoid to oblong ellipsoid, smooth, inamyloid. Basidia 27–32.5 × 6–7.5 µm, elongated-clavate, bearing four sterigmata. Lamella-edge sterile; cheilocystidia 15–25.5 × 6–7.5 µm, more or less lecythiform with a ventricose base and small capitellum (2–3 µm). Pleurocystidia scattered, similar to cheilocystidia. Hymenophoral trama irregular. Pileipellis with radially appressed generative hyphae, 3–5 µm wide, with semi-erect, terminal elements forming discrete fascicles. Stipitipellis with tufts of caulocystidia.

Locality: India, Kerala state, Thiruvananthapuram district: JNTBGRI campus, Jawahar colony, Kallar, Nedumangad, Varkala; Kollam district: Amayambalm Pacha
Season: April-November

5.5.21.1 Medicinal Properties

Pleurotus giganteus is a mushroom with outstanding culinary properties and has been consumed by the indigenous people of Peninsular Malaysia for many years. It has been recently demonstrated to contain bioactive compounds that mimic nerve growth factor and are responsible for neurite stimulation (Phan et al. 2012). The presence of bioactive compounds like triterpenoids and high potassium levels in the basidiocarps of *P. giganteus* could be the reason for neuroactivity. Ethanolic and aqueous extracts of *P. giganteus* possess liver protection functions and antioxidant properties (Phan et al. 2012). The mushroom extracts may contain bioactive compounds that either mimic nerve growth factor or trigger its production, resulting in neurite outgrowth. *Pleurotus giganteus* thus can be developed as a potent nutraceutical for the mitigation of neurogenerative diseases like Parkinson's and Alzheimer's. Further ongoing studies have revealed the presence of phenolics like cinnamic acid and caffeic acid, along with organic acids like succinic acid and triterpenoids (unpublished data). These compounds have a synergistic effect on neuritogenesis. There is a good scope for developing *Pleurotus giganteus* as a healthy and safe dietary supplement for brain and cognitive health (Phan et al. 2012).

5.5.22 *Pleurotus ostreatus* (Jacq.) P. Kumm.

Pileus 2.5–12 cm diam., orbicular, reniform to irregularly flabelliform, rarely with an excentric stipe; the margin is inrolled when young, smooth and often somewhat lobed or wavy; surface convex or depressed, pale brown to dark brown, glabrous. Lamellae decurrent. Stipe absent or reduced, lateral. Context thick, soft, white. Spores 8–10.5 × 3–3.5 µm, oblong cylindrical, hyaline, thin-walled, smooth. Basidia 15–21 × 3.5–4.5 µm, narrowly clavate. Gill edge sterile; cheilocystidia 19–35 × 5–9.5 µm, clavate to cylindric, sometimes the apex is attenuated into an elongated mucronate outgrowth. Pileal surface an undifferentiated epicutis. Clamp-connexions present.

Locality: India, Kerala state, Thiruvananthapuram district: JNTBGRI campus, Kallar
Season: August-October

5.5.22.1 Medicinal Properties

In addition to antiviral and antimicrobial activities, *P. ostreatus* has anti-aging properties and is believed to help with the prevention of age-associated diseases. Clinical studies demonstrate that *P. ostreatus* is capable of reducing triglyceride levels without side effects on the kidney and liver. The combined action of statins (lovastatin or mevinolin) and dietary fibres in *P. ostreatus* inhibit cholesterol synthesis. Lovastatin reduces total cholesterol, triglycerides, and LDL while increasing good cholesterol (HDL). Gunde-Cimerman and Plemenitaš (2001) proposed that adding 5% dried *P. ostreatus* to a cholesterol-rich diet could help to maintain a good blood lipid profile. It is also believed to reduce blood glucose levels and hypertension in type 2 diabetes patients (Khatun et al. 2007; Choudhury et al. 2013). Additionally, β-glucans, including pleuran and related polysaccharides from *P. ostreatus*, possess antiviral and anti-tumour properties are due to the activation of the cellular immune response against viral infections and tumour cells. The protective role of pleuran against respiratory tract infections was demonstrated in a clinical trial on athletes (Majtan 2012), suggesting potential immune-strengthening benefits from natural compounds such as β-glucan in *P. ostreatus. Pleurotus ostreatus* also contains lectin proteins, terpenes, and other compounds with fungicidal, antibiotic, and anti-tumour activities (Gunde-Cimerman 1999). Deepalakshmi and Mirunalini (2014) reported anticholestrolic, antidiabetic, antiarthritic, antibacterial, antiviral, and anticancer activities of *P. ostreatus*, along with the capacity to improve eye health. Broad-spectrum antibacterial properties of *P. ostreatus* have also been reported (Iwalokun et al. 2007). Isolation of a novel ubiquitin-like protein from *P. ostreatus* manifested inhibitory activity towards HIV-1 reverse transcriptase (Wang and Ng 2000). Ravi et al. (2013) demonstrated that *Pleurotus ostreatus* mushrooms are promising as antidiabetic nutraceuticals as well and demonstrated the promising effect of a novel low molecular weight α-glucan isolated from *P. ostreatus* on colon cancer cell proliferation by inducing programmed cell death. Suppression of human breast cancer cell proliferation by hot-water extracts of *P. ostreatus* was also reported (Martin and Brophy 2010). Jedinak and Sliva (2008) reported that methanolic extract of *P. ostreatus* suppressed the proliferation of colon and breast cancer cells. The therapeutic efficacy of protein extract from *P. ostreatus* through glutathione depletion, ROS production, and mitochondrial dysfunction against human monocytic leukaemia and colorectal adeno carcinoma cell lines have been successfully demonstrated (Wu et al. 2011), highlighting this species as a promising new source of anticancer drugs.

In China, *P. ostreatus* is indicted for muscle and joint relaxation (Yang and Jong 1989). When the sporophores are made into a powder, they are effective against numbed limbs, lumbago, and blood vessel discomfort. In the Czech Republic, it is reported that the extracts from the fruiting bodies are used as the main ingredient in dietary preparations for preventing high cholesterol (Opletal 1993).

5.5.23 *Psilocybe subcubensis* Guzman

Pileus 1.5–13 cm diam., convex to plano-convex with or without a broad obtuse umbo, then applanate to slightly uplifted; surface light yellow to greyish orange at the disc, yellowish white towards the margin, slightly slimy, smooth. Lamellae adnate, mouse grey to dark brown, crowded with lamellulae. Stipe 3–13 × 0.3–1.8 cm, central, cylindrical, tapering up, from a subbulbous base, narrowly hollow; surface white to off-white on extreme apex and base, elsewhere yellowish white, becoming bluish on bruising or handling, slightly slimy when wet. Context white, becoming bluish on handling. Annulus well developed, superior, descending, pendant, skirt-like, membranous. Spore print dark brown. Basidiospores 10–14 × 6.8–9.2(–9.6), ellipsoid to broadly ellipsoid, with a broad germ pore, thick-walled. Basidia 23–29 × 10.5–12 µm, clavate to cylindro-clavate, 4-spored. Lamella-edge sterile, cheilocystidia 26–33.5 × 6–8 µm, narrowly utriform to utriform, sublageniform with or without a capitate apex. Pleurocystidia scattered, 21–24 × 8–11.5 µm, broadly clavate to ventricose. Hymenophoral trama regular. Pileipellis an ixocutis, embedded in a gelatinous matrix. Caulocystidia 24–61 × 7.5–11 µm, clavate to cylindrical, narrowly utriform to sublageniform with a capitate apex. Clamp connections present.

Locality: India, Kerala state, Wayanad district: Nadavayal; Thiruvananthapuram district: JNTBGRI campus, Kollam district: Sankili, Ammayambalam Pacha.
Season: May-July, December.

5.5.23.1 Medicinal Properties

Nearly all ethnobotanical use of *Psilocybe subcubensis* is attributed to its hallucinogenic compounds psilocin and psilocybin (Badham 1983). Psilocybin-containing mushrooms represent potent hallucinogenic drugs that may be abused and by psychedelic drug users (Pollock 1975). At the same time, the use of psilocybin for the treatment of compulsive disorders in humans has also been reported (Delgado and Moreno 1998). Psilocybin is distributed under the brand name Indocybin® (Sandoz) and is used to support psychotherapeutic procedures.

5.5.24 *Russula cyanoxantha* (Schaeff.) Fr.

Pileus 10–10.5 cm in diam., applanate to slightly uplifted with a shallow depression at the centre; surface olive brown throughout with brown tinges or more often reddish lilac throughout, weakly tuberculate-striate towards the margin, dry, slimy when wet; margin straight, uplifted on aging, cuticle easily peeling off to expose the underlying white context. Lamellae adnate to adnexed, white to off-white, up to 6 mm wide, close to crowded, mostly without lamellulae, brittle, interveined. Stipe 5–6 × 1.5–2 cm, central, cylindrical, equal, rarely with an attenuated narrowly curved base, stuffed, brittle; surface white to off-white. Context white, up to 7 mm wide, soft. Odour none. Taste acrid. Spores 7.5–9 × 6–7.8 µm, globose to subglobose, rarely broadly ellipsoid to ellipsoid, with long amyloid isolated obtuse ridges and warts, rarely with 2–3 fine connections, not forming a complete reticulum, plage inamyloid. Basidia 25.5–37 × 7–11.5 µm clavate, cylindrical to cylindro-clavate, 4-spored, rarely 1–2-spored. Lamella-edge heteromorphic, cheilocystidia macrocystidiod, 44–58 × 5–6.7 µm, cylindrical with a mucronate apex. Pluerocystidia 57.2–66 × 6.6–8.36 µm, similar to the macrocheilocystidia. Hymenophoral trama heteromorphic with thin-walled, hyaline sphaerocytes, and interwoven thin-walled, septate, branched, hyaline hyphae. Pileipellis an interrupted ixocutis. Pileocystidia 45–90 × 3.5–4.5 µm cylindrical with a capitate apex, thin-walled, hyaline. Caulocystidia with or without a mucronate apex, 15–42.6 × 3.6–4.2 µm, filled with refractive granular contents. Clamp connections absent.

Locality: India, Kerala state, Thiruvananthapuram district: JNTBGRI Campus; Kollam district: Sankili
Season: June-July, November

5.5.24.1 Medicinal Properties

Ergone obtained from the sporocarp of *R. cyanoxantha* was found to be a potent anti-tumour compound that caused apoptosis and induced G2/M phase arrest in hepatocellular carcinoma HepG2 cells, thereby demonstrating efficacy against tumours. The ability of ergone to induce apoptosis was attributed to the activation of capsases (via both extrinsic and intrinsic pathways). The Bax/Bcl-2 ratio was increased profoundly in ergone-treated cells (Zhao et al. 2011). Gao et al. (2001) extracted a ceramide potentially involved in the apoptotic pathway from *R. cyanoxantha*. These results give important information regarding the molecular mechanism of ergone and its potential use as a chemotherapeutic agent for the treatment of liver cancer (Pérez-Moreno and Martínez-Reyes 2013). Ergones also exhibit immunosuppressive (Fujimoto et al. 2004), diuretic (Zhao et al. 2009a, b), and cytotoxic activities (Lee et al. 2009). Acetone and chloroform extracts of *R. cyanoxantha* showed antifungal activities against *Fusarium moniliforme* and *F. culmorum*. Ohuchi and Aoyagi (2010) reported that ethanol extracts from sporophores of *R. cyanoxantha* have some α-amylase inhibitory effects.

5.5.25 *Russula nigricans* Fr.

Pileus 7.7–12.4 cm in diam., plano-convex to slightly uplifted with a shallow depression at the centre; surface whitish to cream when young, becoming dark grey-brown to blackish brown from the centre outwards, slimy when wet soon dry, smooth, and glabrous. Lamellae adnate, white, up to 2 mm wide, crowded. Stipe 4.5–6.2 × 1.5–2 cm, central, cylindrical, equal, solid, brittle; surface white to off-white, smooth. Context white, up to 2 cm wide, soft. Taste acrid. Spores 6–7.5 × 6–7.2 µm, globose to subglobose, rarely broadly ellipsoid, with amyloid small warts and obtuse ridges, connected by small connectives, forming a reticulum, plage inamyloid. Basidia 28.8–33.6 × 6–10 µm, clavate to cylindro-clavate, 4-, 2-, or rarely 1-spored. Lamella-edge heteromorphic, deeply stained pink with Sulfovanillin, orthochromatic in cresyl blue. Cheilocystidia macrocystidiod, 30–55.6 × 6.8–17.2 µm, cylindrical to cylindro-clavate, with a mucronate to pointed apex or club-shaped to ventricose with an obtuse broad apex, with brownish refractive granular. Pluerocystidia 30–68 × 6–10 µm, cylindrical with a pointed or mucronate apex, frequently with 1–2 constructions, filled with brownish contents. Hymenophoral trama homomorphic. Pileipellis an interrupted epicutis. Pileal trama heteromorphic with sphaerocytes and hyphae. Stipetipellis intermixed with thin-walled cylindrical to narrowly clavate caulocystidia, 154–64 × 4–6 µm. Clamp connections absent.

Locality: India, Kerala state, Thiruvananthapuram district: JNTBGRI campus
Season: June, November

5.5.25.1 Medicinal Properties

Fruiting bodies of *R. nigricans* have been found to show anti-tumour activity (Mao 1998). Tan et al. (2004) revealed the existence of some novel triterpenoids, sesquiterpenoids, ceramides, and lepidamine (an N-containing aristolane derivative) in *R. nigricans*. They also isolated nigricanin, a phenolic compounds with an ellagic acid skeleton that was the first ellagic acid-like compound discovered in higher fungi. Ohuchi and Aoyagi (2010) found that *R. nigricans* have a conspicuous α-amylase inhibitory effect which can be utilized to treat diabetes mellitus, a chronic metabolic disease.

5.5.26 *Schizophyllum commune* Fr.

Pileus 1–5 cm diam; thin, flabelliform, laterally attached by a small base; fan-shaped, irregular to shell-shaped when attached above or below; surface pale to dark greyish brown, upper surface covered with small hairs, dry; margin lobed, often deeply incised. Hymenophore falsely lamellate. Lamellae gill like folds that split down the middle, white to greyish. Stipe absent. Context tough, leathery, thin. Spore print white. Spores 3.5–6.5 × 2–2.5 µm, ellipsoid, thin-walled, hyaline. Basidia 15–22 × 3–6.5 µm, narrowly clavate, 4-spored. Cystidia absent. Pileipellis a trichodermium.

Locality: India, Kerala state: throughout
Season: In all seasons

5.5.26.1 Medicinal Properties

In Southeast Asia, *S. commune* has been consumed as a nutritional food and is cultivated in Thailand and Malaysia. Pharmacologically, *S. commune* is particularly important as it produces the polysaccharide schizophyllan, which exhibits many medicinal properties (Ooi and Liu 1999; Wasser 2002). Schizophyllan (SPG, sonifilan, sizofiran) has been researched clinically for its anticancer activities (Mantovani et al. 1997). Recent studies by Mirfat et al. (2014) have shown that some extracts of *S. commune* could cure bacterial and fungal infections and thus can be used as a potential antimicrobial agent. Patients suffering from chronic hepatitis B and liver cirrhosis showed good improvement when administered extracts from *S. commune* (Kakuma 1991). Schizophyllan has also be demonstrated to increase protection against infection by *Staphylococcus* sp. (Matsuyama et al. 1992).

5.5.27 *Termitomyces clypeatus* R. Heim

Pileus up to 2–7 cm diam., at first cylindrico-conical then convex with central, strongly spiniform perforatorium, expanding to campanulate with a spiniform perforatorium; surface silky, pale greyish brown, dry, smooth. Lamellae free, pinkish, 6–8 mm broad, crowded. Stipe 6–15 × 0.7–1 cm, central, solid, cylindrical, swollen slightly at the base before developing a long tapering blackish pseudorrhiza; upper surface creamy white to pale greyish brown, smooth or often longitudinally striate, tough, fibrous, solid. Spore print pink. Spores 5–5–8.5 × 4–5.5 µm, ellipsoid, smooth, inamyloid, thin-walled. Basidia 18–30 × 8.5–15 µm, clavate, 4-spored. Lamella-edge sterile or heteromorphous. Cheilocystidia 15–32 × 10–20 µm, broadly clavate to pyriform. Pleurocystidia scattered, similar. Hymenophoral trama indistinctly bilateral, with diverging hyphe. Pileipellis a repent epicutis. All hyphae lacking clamp connections.

Locality: India, Kerala state, Thiruvananthapuram district: JNTBGRI campus, Kallar; Kollam district: Thenmala, Achenkovil; Wayanad district: Muthanga
Season: May-October

5.5.27.1 Medicinal Properties

Aqueous extracts of *T. clypeatus* exhibited potential antioxidant activity (Mondal et al. 2016), and *T. clypeatus* is considered an excellent and viable source of natural antioxidants. *Termitomyces clypeatus* extracts were toxic against several cancer cell lines. *Termitomyces clypeatus* can also be used as a nutraceutical in preparing functional food (Ghorai et al. 2009). *Termitomyces clypeatus* extracts are natural source of antioxidants and are used for general health preservation and cancer prevention. Antibacterial activity against *Pseudomonas aeruginosa* has also been attributed to extracts from *T. clypeatus*.

5.5.28 *Termitomyces heimii* Natarajan

Pileus 6–12 cm diam., at first subglobose to subumbonate becoming convex to plano-convex, more often applanate, with a broad umbo; surface silky white, grey at the centre, smooth. Lamellae free, white then pinkish, 4–8 mm broad, crowded. Stipe 5–8 × 0.6–1.2 cm, white, smooth, cylindrical, stuffed; surface white, smooth-striate. Annulus superior, white, persistent. Pseudorrhiza 11–35 cm × 3–6 mm, white, smooth, hollow. Context 3–6 mm thick, fleshy, white. Spore print pink. Spores 5–7 (8) × 3.5–4.5 µm, ovoid to ellipsoid, smooth, inamyloid. Basidia 15–22 × 6–7 µm, clavatel, 4-spored. Lamella-edge sterile, cheilocystidia 20–52 × 10–25 µm, ovoid to pyriform. Pleurocystidia scattered, 20–65 × 10–26 µm, inflated-clavate. Pileipellis a repent epicutis of radially arranged, up to 5 µm wide hyphae.

Locality: India, Kerala state: throughout
Season: March-December

5.5.28.1 Medicinal Properties

Malek et al. (2012) reported the presence of 28 compounds (mainly fatty acids, esters of fatty acids, and ergosterol) in the extracts of *T. heimii*. Ergosterol and linoleic acid are abundant in *T. heimii.* Ergosterol is the precursor of ergocalciferol, which is administered as a dietary calcium supplement to patients suffering from osteoporosis and hypocalcinaemia. Linoleic acid, on the other hand, is a polyunsaturated fatty acid and belongs to a class of essential fatty acids called omega-6 fatty acids. It is a precursor tp several compounds vital for health, like arachidonic acid, which is involved in reducing inflammation. Egmond et al. (1996) suggested that linoleic acid is important for proper growth and development in infants. Linoleic acid produces prostaglandins required for general health maintenance. *Termitomyces heimii* also contains nicotinamide, which has demonstrated anti-inflammatory and anti-anxiety properties (Jaconello 1992; Niren 2006)

5.5.29 *Termitomyces microcarpus* (Berk. and Broome) R. Heim

Pileus 0.5–2.5 cm diam., convex to campanulate, expanding with or without a papillate umbo; surface white to cream sometimes streaked with grey or brown, silky-striate, smooth. Lamellae free, white then pinkish, 1–2 mm wide, moderately crowded. Stipe 2–4 (–10) × 0.2–0.4 cm, slender, cylindrical, solid; surface white or paler, fibrillose-striate, usually with a very short rooting base but lacking a true pseudorriza. Spore print pinkish cream. Spores 5–8 × 3.5–4.5 µm, ovoid to ellipsoid, smooth, inamyloid. Basidia 20–25 × 6–7 µm, clavate, 4-spored. Lamella-edge with scattered cheilocystidia. Cheilocystidia 15–40 × 8–15 µm, pyriform, clavatel, utriform, fusoid or vesiculose. Pleurocystidia scattered, similar to cheilocystidia. Pileipellis a repent epicutis of radially parallel narrow, thin-walled, 2–7.5 µm wide hyphae. All hyphae lack clamp connections.

Locality: India, Kerala state: throughout
Season: January-April, June-November

5.5.29.1 Medicinal Properties

Termitomyces microcarpus is a good source of nutrients including dietary fibre, minerals, and protein (Degreef et al. 1997; Agrahar-Murugkar and Subbulakshmi 2005). Consumption of this mushroom was associated with reduced serum levels of total and LDL cholesterol. Since high serum levels of total and LDL cholesterol are responsible for coronary heart diseases, it seems that the dietary incorporation of *T. microcarpus* could help to address this increasing global health problem. Rajoriya and Gupta (2015) found high radical-scavenging effects in *T. microcarpus.* A polyphenol-rich fraction from *T. microcarpus* was tested for its antioxidant capacity and evaluated for the presence of bioactive components (Mitra et al. 2016). The extract was found to contain two important phenolic compounds (cinnamic acid and pyrogallol) along with other bioactive compounds such as lycopene, β-carotene, and ascorbic acid. Antioxidant and antiradical activities of methanolic extracts of *Termitomyces microcarpus* were noticed by Tibuhwa (2012). *Termitomyces microcarpus* was also found to have high levels of β-carotene, flavonoids and phenols. *Termitomyces microcarpus* showed activity against the Gram-negative bacteria *Proteus vulgaris* and Gram-positive bacteria *Bacillus cereus* (Giri et al. 2012).

5.5.30 *Volvariella volvacea* (Bull.) Singer

Pileus 5–10 cm in diameter, ovoid, companulate to convex with a broad umbo; surface brown, paler towards the margin, radially fibrillose-squamulose, moist to subviscid. Lamellae free, up to 7 mm wide, white, then pinkish brown, crowded. Stipe 5–8 × 0.8–1.5 cm, central, curved, cylindrical or attenuated above, solid; surface whitish, smooth, and glabrous. Volva saccate, membranous-fleshy, irregularly lobed, greyish brown. Context up to 5 mm thick at the disc. Spore print pinkish buff. Spores 7–9 × 4.5–5.4 μm, ovoid to ellipsoid, smooth. Basidia 21–23 × 7–10 μm, clavate, 4-spored. Lamella edge sterile or heteromorphous; cheilocystidia 33–55 × 10–15 μm, lageniform, fusoid ventricose, often with an elongated, mucronate outgrowth, thin-walled, hyaline. Pleurocystidia similar, Hymenophoral trama inversely bilateral. Pileipellis a repent epicutis.

Locality: India, Kerala state, Thiruvananthapuram district: JNTBGRI campus, CPCRI campus; Kollam district: Cheenikkala, Sankili; Wayanad district: Muthanga.
Season: January-April, June-July, September–October

5.5.30.1 Medicinal Properties

Volvariella volvacea contains a high amount of fibre (chitin), protein, fats, vitamins (vitamin C, biotin, thiamine, riboflavin), amino acids (all essential amino acids like arginine, alanine, serine, glycine, etc.), carbohydrates, unsaturated fatty acids, and essential minerals (sodium, potassium, and phosphorus) with low caloric value (Chang and Buswell 1996). Secondary metabolites from *V. volvacea* represent a good source of terpenes, polypeptides, steroids, and phenolic compounds such as phenolic acids, flavonoids, and tannins, which contribute to high antioxidant capacity. Methanol and water extracts of *V. volvacea* were found to have rich anti-oxidative activities which may aid in the prevention of cardiovascular diseases, cancer, (Cheung et al. 2003) neuro-degenerative diseases (Joseph et al. 1999), inflammation, and problems caused by cell and cutaneous aging.

5.6 CONCLUSIONS

Throughout the world, mushrooms have found a place as nutritional and tasty foods in several societies. They are widely appreciated for their nutritional and chemical properties, in addition to their unique flavours. Owing to low fat levels and high moisture, carbohydrate, and protein contents, they are suitable for low-calorie diets. Edible and medicinal mushrooms may even act as saviours of humankind, preventing diseases like HIV/AIDS, cancer, and diabetes that could lead to kidney diseases, heart attack, blindness, stroke, and limb amputation. As dietary supplements, they are now widely accepted. Mushrooms represent an important, understudied source of novel bioactive compounds useful for improving human health, such as nutraceuticals or medicine. Antiallergic, analgesic, antibacterial, anticarcinogenic, antifungal, anticoagulant, anti-inflammatory, antihypertensive, antioxidant, antinociceptive, antivenom, antipyretic, antiviral (including anti-HIV), cholesterol-lowering, immune-enhancing, and hepatoprotective properties have been found in several species of wild mushrooms. In addition to this importance as medicinal and functional foods, mushrooms represent an attractive source of bioactive compounds useful for sustainable agriculture, since antagonistic effects against pests, plant pathogenic fungi, bacteria, viruses, and nematodes have been detected in many mushroom genera. It is quite clear that the potential characterization and purification of bioactive compounds from wild mushrooms might lead to the formulation of novel therapeutic compounds.

Mushrooms still need to be thoroughly studied for humankind to exploit their full potential. Kerala, floristically one of the richest Indian states with its diverse habitats, may hold many medicinal mushrooms. This immensely diverse and important group of organisms is gradually diminishing due to several natural and manmade factors. This chapter was aimed to provide a preliminary account of the wide variety of medicinal mushrooms available in Kerala, along with their medicinal values. Knowledge of mushroom varieties, and their seasonal occurrence patterns will enable researchers to execute advanced investigations in this field.

In conclusion, mushrooms indigenous to Kerala have enormous potential as sources of bioactive agents for biopharmaceutical exploitation. Ecological surveys, molecular identification, and assessment of medicinal potential and nutritional profiles are excellent research domains. These efforts will ensure the complete and sustainable exploitation of the understudied value of mushrooms.

REFERENCES

Acharya, K., S. Chatterjee, G. Biswas, A. Chatterjee & G.K. Saha. 2012. Hepatoprotective effect of a wild edible mushroom on carbon tetrachloride-induce hepatoprotective in mice. *International Journal of Pharmacy and Pharmaceutical Sciences* 4: 285–288.

Acharya, K., K. Samui, M. Rai, B.B. Dutta & R. Acharya. 2004. Antioxidant and nitric oxide synthase activation properties of *Auricularia auricula*. *Indian Journal of Experimental Biology* 42: 538–540.

Agrahar-Murugkar, D. & G. Subbulakshmi. 2005. Nutritional value of edible wild mushrooms collected from the Khasi hills of Meghalaya. *Food Chemistry* 89: 599–603.

Ahmad, N., A.K. Bansal & J.R. Kidwai. 1984a. Effect of PHA-B fraction of *Agaricus bisporus* lectin on insulin release and 45Ca2+ uptake by islets of Langerhans in vitro. *Acta Diabetologica* 21: 63–70.

Ahmad, N., M.M. Khan, A.K. Rastogi & J.R. Kidwai. 1984b. Effect of age on *Agaricus bisporus* PHA-B stimulated insulin release and 45Ca2+ uptake in vitro by islets of Langerhans. *Acta Diabetologica* 21: 349–355.

Ahmad, R., S. Muniandy, N.I.A. Shukri, S.M.U. Alias, A.A. Hamid, W.M. Wan Yusoff, S. Senafi & F. Daud. 2014. Antioxidant properties and Glucan compositions of various crude extracts from *Lentinus squarrosulus* mycelial culture. *Advances in Bioscience and Biotechnology* 5: 805–814.

Aina, D.A., S.G. Jonathan, O.J. Olawuyi, D.O. Ojelab & B.M. Durowoju. 2012. Antioxidant, antimicrobial and phytochemical properties of alcoholic extracts of *Cantharellus cibarius*—a Nigerian musroom. *New York Science Journal* 5(10): 114–120.

Ajith, T.A. & K.K. Janardhanan. 2007. Indian medicinal mushrooms as a source of antioxidant and antitumor agents. *Journal of Clinical Biochemistry and Nutrition* 40: 157–162.

Alfala, S.A. & B. Jaykar. 2017. Isolation and characterization and antimicrobial activity of *Calocybe Indica* APK-2. *Scholars Academic Journal of Pharmacy* 6(4): 120–125.

Al-Momany, A.M. & G. Saleh. 2009. A comprehensive study on *Agaricus* species of North Cyprus. *World Journal of Agriculture Science* 5(2): 195–200.

Anke, T. 1997. Strobilurins. In: *Fungal Biotechnology*. Ed. Anke, T. Chapman & Hall Weinhein, London.

Anke, T., H.J. Hecht, G. Schramm & W. Steglich. 1979. Antibiotics from basidiomycetes. IX. Oudemansin an antifungal antibiotic from *Oudemansiella mucida* (Schrader ex Fr.) Hoehnel (Agaricales). *Journal of Antibiotics* 32: 1112–1117.

Avci, E., G. Çağatay, G.A. Avcı, M. Suiçmez & S.C. Cevher. 2016. An edible mushroom with medicinal significance; *Auricularia polytricha*. *Hittite Journal of Science and Engineering* 3(2): 111–116.

Badham, E.R. 1983. Ethnobotany of psilocybin mushrooms, especially *Psilocybe cubensis*. *Journal of Ethnopharmacology* 10: 254.

Barros, L., P. Baptista, D.M. Correia, S. Casal, B. Oliveira & I.C.F.R. Ferreira. 2007a. Fatty acid and sugar compositions and nutritional value of five wild edible mushrooms from Northeast Portugal. *Food Chemistry* 105(1): 140–145.

Barros, L., P. Baptista, D.M. Correia, J. Sa Morais & I.C.F.R. Ferreira. 2007b. Effects of conservtion treatment and cooking on the chemical composition and antioxidant activity of Portuguese wild edible mushrooms. *Journal of Agriculture and Food Chemistry* 55(12): 4781–4788.

Barros, L., S. Falcão, P. Baptista, C. Freire & M. Vilas-Boas. 2008. Antioxidant activity of *Agaricus* sp. mushrooms by chemical, biochemical and electrochemical assays. *Food Chemistry* 3: 61–66.

Bensky, D. & A. Gamble. 1993. *Chinese Herbal Medicine: Materia Medica, Chinese Herbal Medicine Materia Medica*. Eastland Press Inc, Seattle.

Bhattacharyya, A., R. Chattopadhyay, S. Mitra & S.E. Crowe. 2014. Oxidative stress: an essential factor in the pathogenesis of gastrointestinal mucosal diseases. *Physiological Reviews* 94: 329–354.

Boonsong, S., W. Klaypradit & P. Wilaipun. 2016. Antioxidant activities of extracts from five edible mushrooms using different extractants. *Agricultre and Natural Resources* 50: 89–97.

Chang, S.T. & J.A. Buswell. 1996. Mushroom nutriceuticals. *World Journal of Microbial Biotechnology* 12: 473–476.

Chang, S.T. & J.A. Buswell. 2003. Medicinal mushrooms-a prominent source of nutriceuticals for the 21st century. *Current Topics in Nutraceutical Research* 1(3): 257–283.

Chang, S.T. & S.P. Wasser. 2012. The role of culinary-Medicinal Mushrooms on human welfare with a Pyramid Model for Human Health. *International Journal of Medicinal Mushrooms* 14: 95–134.

Chatterjee, S., A. Dey, R. Dutta, S. Dey & K. Acharya. 2011. Hepatoprotective effect of the ethanolic extract of *Calocybe indica* on mice with CCl4 hepatic intoxication. *International Journal of PharmTech Research* 3(4): 2162–2168.

Chen, R.Y. & D.Q. Yu. 1991. Studies on the triterpenoid constituents of the spores from Ganoderma lucidum. *Acta Pharmaceut Simica* 26: 267–273.

Chen, S.Y., K.J. Ho, Y.J. Hsieh, L.T. Wang & J.L. Mau. 2012. Contents of lovastatin, γ-aminobutyric acid and ergothioneine in mushroom fruiting bodies and mycelia. *LWT Food Science Technology* 47: 274–278.

Cheung, L.M., P.C.K. Cheung & V.E.C. Ooi. 2003. Antioxidant activity and total phenolics of edible mushroom extracts. *Food Chemistry* 81: 249–255.

Chihara, G., J. Hamuro & Y.Y. Maeda. 1970. Fractionation and purification of the polysaccharides with marked antitumor activity, especially Lentinan, from *Lentinus edodes*. *Cancer Research* 30: 2776–2781.

Chittaragi, A. & J. Kodiyalmath. 2014. Anthelmintic screening of crude extracts of *Hygrocybe parvula*. *Journal Club for Applied Science* 1(I): 21–27.

Chittaragi, A. & R. Naika. 2014. Determination of trace elements on some wild mushroom samples encountered from Western Ghats of Karnataka. *Journal of Chemical and Pharmaceutical Research* 6(7): 2124–2135.

Chittaragi, A., R. Naika, K.B. Aruna & K.K. Jayashree. 2013. *In- vitro* antibacterial activity of *Hygrocybe parvula* (Peck) Pegler. *International Journal of Pharmacy & Life Sciences* 4(11): 3080–3086.

Choudhury, M.B.K., T. Rahman, A.J. Kakon, N. Hoque, M. Akhtaruzzaman, M.M. Begum, M.S.K. Choudhuri & M.S. Hossain. 2013. Effects of *Pleurotus ostreatus* on blood pressure and glycemic status of hypertensive diabetic male volunteers. *Bangladesh Journal of Medical Biochemistry* 6(1): 5–10.

Costa, J.V., M.R.C.G. Novaes & E.R. Asquieri. 2011. Chemical and antioxidant potential of *Agaricus sylvaticus* mushroom grown in Brazil. *Journal of Bioanalysis & Biomedicine* 3: 49–54.

Cui, M., H. Zhang & L. An. 2002. Tumor growth Inhibition by polysaccharide from *Coprinus comatus*. *World Chinese Journal of Digestology* 10: 287–290.

Dasgupta, A., M. Rai & A. Krishnendu. 2013. Chemical composition and antioxidant activity of a Wild Edible Mushroom *Pleurotus flabellatus*. *International Journal of PharmTech Research* 5(4): 1655–1663.

Deepalakshmi, K. & S. Mirunalini. 2014. *Pleurotus ostreatus*: an oyster mushroom with nutritional and medicinal properties. *Journal of Biochemical Technology* 5(2): 718–726.

Degreef, J., F. Malaisse, J. Rammeloo & E. Baudart. 1997. Edible mushrooms of the Zambezian woodland area: a nutritional and ecological approach. *Biotechnology, Agronomy, Society and Environment* 1: 221–231.

Delgado, P.L. & F.A. Moreno. 1998. Hallucinogens, serotonin, and obsessive-compulsive disorder. *Journal of Psychoactive Drugs* 30: 359–366.

Didukh, M.Y., S.P. Wasser & E. Nevo. 2003. Medicinal value of species of the family Agaricaceae Cohn (higher Basidiomycetes): current stage of knowledge and future perspectives. *International Journal of Medicinal Mushrooms* 5: 133–152.

Di Piero, R.M., Q.S. Novaes & S.F. Pascholati. 2010. Effect of *Agaricus brasiliensis* and Lentinula edodes mushrooms on the infection of passionflower with Cowpea aphid-borne mosaic virus. *Brazilian Archives of Biology and Technology* 53: 269–278.

Dotan, N., S.P. Wasser & J. Mahajna. 2011. The culinary—medicinal mushroom *Coprinus comatus* as a natural antiandrogenic modulator. *Integrative Cancer Therapies* 10(2): 148–159.

Egmond, V., W.A. Andreas, R. Michael, R.K. Kosorok, L. Anita & M.F. Philip. 1996. Effect of linoleic acid intake on growth of infants with cystic fibrosis. *American Journal of Clinical Nutrition* 63: 746–752.

Ewart, R.B.L., S. Kornfeld & D.M. Kipnis. 1975. Effect of Lectins on hormone release from isolated rat islets of Langerhans. *Diabetes* 24: 705–714.

Falandysz, J. & B. Borovicka. 2013. Macro and trace mineral constituents and radionuclides in mushrooms: health benefits and risks. *Applied Microbiology and Biotechnology* 97: 477–501.

Fan, L., H. Pan, A.T. Soccol, A. Pandey & C.R. Soccol. 2006. Advances in mushroom research in the last decade. *Food Technology and Biotechnology* 44: 303–311.

Ferreira, I.C.F.R., P. Baptista, M. Vilas-Boas & L. Barros. 2007. Free-radical scavenging capacity and reducing power of wild edible mushrooms from northeast Portugal. *Food Chemistry* 100: 1511–1516.

Ferreira, I.C.F.R., L. Barros & R.M.V. Abreu. 2009. Antioxidants in wild mushrooms. *Current Medicinal Chemistry* 16: 1543–1560.

Ferreira, I.C.F.R., J.A. Vaz, M.H. Vasconcelos & A. Martins. 2010. Compounds from wild mushrooms with antitumor potential. *Anti-Cancer Agents in Medicinal Chemistry* 10(5): 424–436.

Finimundy, T.C., G. Gambato, R. Fontana, M. Camassola, M. Salvador, S. Moura, J. Hess, J.A.P. Henriques, A.J.P. Dillon & M. Roesch-Ely. 2013. Aqueous extracts of *Lentinula edodes* and *Pleurotus sajor-caju* exhibit high antioxidant capability and promising in vitro antitumor activity. *Nutrition Research* 33: 76–84.

Finkel, T. & N.K. Holbrook. 2000. Oxidants, oxidative stress and the biology of ageing. *Nature* 408: 239–247.

Fortes, R.C. & M.R.C.G. Novaes. 2011. The effects of *Agaricus sylvaticus* fungi dietary supplementation on the metabolism and blood pressure of patients with colorectal cancer during post-surgical phase. *Nutrición Hospitalaria* 26: 176–186.

Fortes, R.C., V.C. Recova, A.L. Melo & M.R.C.G. Novaes. 2007. Quality of life of patients with colorectal cancer on dietary supplementation with *Agaricus sylvaticus* fungus: after six months of segment: randomized and placebocontrolled clinical trial. *Revista Brasileira de Coloproctologia* 27(2): 130–138.

Fortes, R.C., V.L. Recôva, A.L. Melo & M.R.C.G. Novaes. 2008. Effects of dietary supplementation with medicinal fungus in fasting glycemia levels of patients with colorectal cancer: a randomized, double-blind, placebo-controlled clinical study. *Nutrición Hospitalaria* 23: 591–598.

Fujimoto, H., E. Nakamura, E. Okuyama & M. Ishibashi. 2004. Six immune suppressive features from an Ascomycete, *Zopfiella longicaudata*, found in a screening study monitored by immune modulatory activity. *Chemical and Pharmaceutical Bulletin* 52: 1005–1008.

Gao, J.M., Z.L. Dong & J.K. Liu. 2001. A new ceramide from the basidiomycete *Russula cyanoxantha*. *Lipids* 36(2): 175–180.

Gaur, T. & P.B. Rao. 2016. Antioxidant potential of the Giant Mushroom, *Macrocybe gigantea* (Agaricomycetes), from India in different drying methods. *International Journal of Medicinal Mushrooms* 18(2): 133–140.

Ghorai, S., S.P. Banik, D. Verma, S. Chowdhury, S. Mukherjee & S. Khowala. 2009. Fungal biotechnology in food and feed processing. *Food Research International* 42: 577–587.

Ghosh, S.K. 2015. Study of anticancer effect of *Calocybe indica* mushroom on breast cancer cell line and human Ewings sarcoma cancer cell lines. *New York Science Journal* 8(5): 10–15.

Giri, S., G. Biswas, P. Pradhan, S.C. Manda & K. Acharya. 2012. Antimicrobial activities of basidiocarps of wild edible mushrooms of West Bengal, India. *International Journal of PharmTech Research* 4(4): 1554–1560.

Gray, A.M. & P.R. Flatt. 1998. Insulin-releasing and insulin-like activity of *Agaricus campestris* (mushroom). *Journal of Endocrinology* 157: 259–266.

Gu, Y. & Y. Ju. 1996. Food and officinal mushroom-Coprinus comatus. *Vegetable* 13: 10–12.

Gunde-Cimerman, N. 1999. Medicinal value of the genus *Pleurotus* (Fr.) P. Karst. (Agaricales s.l., Basidiomycetes). *International Journal of Medicinal Mushrooms* 1(1): 69–80.

Gunde-Cimerman, N. & A. Plemenitaš. 2001. Hypocholesterolemic activity of the genus *Pleurotus* (Jacq.: Fr.) P. Kumm. (Agaricales s. I., Basidiomycetes). *International Journal of Medicinal Mushrooms* 3(4): 395–397.

Heleno, S.A., L. Barros, M.J. Sousa, A. Martins & I.C.F.R. Ferreira. 2010. Tocopherols composition of Portuguese wild mushrooms with antioxidant capacity. *Food Chemistry* 119(4): 1443–1450.

Hikimo, H., M. Ishiyama, Y. Suzuki & C. Konno. 1989. Mechanisms of hypoglycaemic activity of ganoderan B: a glycan of *Ganoderma lucidum* fruitbodies. *Planta Medica* 55: 423–428.

Hobbs, C.H. 1995. *Medicinal Mushrooms: An Exploration of Tradition, Healing and Culture*. Botanica Press, Summertown, TN.

Hussein, J.M., D.D. Tibuhwa, A.M. Mshandete & A.K. Kivaisi. 2015. Antioxidant properties of seven wild edible mushrooms from Tanzania. *African Journal of Food Science* 9(9): 471–479.

Iwalokun, B.A., U.A. Usen, A.A. Otunba & D.K. Olukoya. 2007. Comparative phytochemical evaluation, antimicrobial and antioxidant properties of *Pleurotus ostreatus*. *African Journal of Biotechnology* 6(15): 1732–1739.

Jaconello, P. 1992. Niacin versus niacinamide. *Canadian Medical Association Journal* 147(7): 990.

Jedinak, A. & D. Sliva. 2008. *Pleurotus ostreatus* inhibits proliferation of human breast and colon cancer cells through p53—independent pathway. *International Journal of Oncology* 33: 1307–1313.

Jiang, J. & D. Sliva. 2010. Novel medicinal *mushroom* blend suppresses growth and invasiveness of human breast cancer cells. *International Journal of Oncology* 37: 1529–1536.

Joseph, J.A., B. Shukit-Hale & N.A. Denisova. 1999. Reversal of age related declines in neuronal signal transduction, cognitive and motor behavioral deficits with blue berry, spinach or strawberry dietary supplementation. *Journal of Neuroscience* 19: 8114–8812.

Kakuma, S. 1991. Effect of schizophyllan, a polysaccharide, on interferon gamma antibody production and lymphocyte proliferation specific for hepatitis B virus antigen in patients with chronic hepatitis B. *International Journal Immunopharmacology* 13: 969–975.

Kanagasabapathy, G., S.N.A. Malek, A.A. Mahmood, K.H. Chua, S. Vikineswary & U.R. Kuppusamy. 2013. Beta-glucan-rich extract from *Pleurotus sajor-caju* (Fr.) Singer prevents obesity and oxidative stress in C57BL/6J mice fed on a high-fat diet. *Evidence Based Complementary and Alternative Medicine*: 1–10.

Khatua, S., S. Paul & K. Acharya. 2013. Mushroom as the potential source of new generation of antioxidant: a review. *Research Journal of Pharmacy and Technology* 6: 496–505.

Khatun, K., H. Mahtab, P.A. Khanam, M.A. Sayeed & K.A. Khan. 2007. Oyster mushroom reduced blood glucose and cholesterol in diabetic subjects. *Mymensingh Medical Journal* 16(1): 94–99.

Kiho, T., H. Morimoto, M. Sakushima, S. Usui & S. Ukai. 1995. Polysaccharides in fungi. XXXV. Antidiabetic activity of an acidic polysaccharide from the fruiting bodies of *Tremella aurantia*. *Biological and Pharmaceutical Bulletin* 18: 1627–1629.

Klaus, A., M. Kozarski, M. Niksic, D. Jakovljevic, N. Todorovic & L.J.L.D. van Griensven. 2011. Antioxidative activities and chemical characterization of polysaccharides extracted from the basidiomycete *Schizophyllum commune*. *LWT-Food Science and Technology* 44: 2005–2011.

Kozarski, M., A. Klaus, D. Jakovljevic, N. Todorovic, J. Vunduk, P. Petrovic, M. Niksic, M.M. Vrvic & L. van Griensven. 2015b. Antioxidants of edible mushrooms. *Molecules* 20: 19489–19525.

Kozarski, M., A. Klaus, M. Niksic, D. Jakovljevic, J.P.F.G. Helsper & L.J.L.D. van Griensven. 2011. Antioxidative and immunomodulating activities of polysaccharide extract of the medicinal mushrooms *Agaricus bisporus*, *Agaricus brasiliensis*, *Ganoderma lucidum* and *Phellinus linteus*. *Food Chemistry* 129: 1667–1675.

Kozarski, M., A. Klaus, M. Niksic, L.J.L.D. van Griensven, M.M. Vrvic & D.M. Jakovljevic. 2014. Polysaccharides of higher fungi: biological role, structure and antioxidative activity. *Chemical Index* 68: 305–320.

Kozarski, M., A. Klaus, M. Niksic, M.M. Vrvic, N. Todorovic, L.J.L.D. van Griensven & D. Jakovljevic. 2012. Antioxidative activities and chemical characterization of polysaccharide extracts from the widely used mushrooms *Ganoderma applanatum*, *Ganoderma lucidum*, *Lentinus edodes* and *Trametes versicolor*. *Journal of Food Composition and Analysis* 26: 144–153.

Kozarski, M., A. Klaus, J. Vunduk, Z. Zizak, M. Niksic, D. Jakovljevic, M.M. Vrvic & L.J.L.D. van Griensven. 2015a. Nutraceutical properties of the methanolic extract of edible mushroom *Cantharellus cibarius* (Fries): primary mechanisms. *Food Function* 6: 1875–1886.

Krishnamoorthy, A.S. 2003. *Current Vistas in Mushroom Biology and Production*. Eds. Upadhay, R.C., Sing, S.K. & Rai, R.D. Mushroom Society of India, Solan, Himachal Pradesh, India, 131–135.

Kudahewa, D.D., D.T.U. Abeytunga & W.D. Ratnasooriya. 2008. Antinociceptive activity of *Pleurotus cystidiosus*, an edible mushroom, in rats. *Pharmacognosy Magazine* 4: 37–41.

Kumari, B. & N.S. Atri. 2014. Nutritional and nutraceutical potential of wild edible macrolepiotoid mushrooms of North India. *International Journal of Pharmacy and Pharmaceutical Sciences* 6(2): 200–204.

Kumari, B., N.S. Atri & R.C. Upadhyay. 2012. Screening and evaluation of extra-cellular oxidases in some Termitophilous and Lepiotoid mushrooms. *World Journal of Agricultural Sciences* 8: 409–414.

Lau, C.C., N. Abdullah & A.S. Shui. 2013. Novel angiotensin I-converting enzyme inhibitory peptides derived from an edible mushroom, *Pleurotus cystidiosus* O.K. *Miller BMC Complementary and Alternative Medicine* 13: 313–323.

Lee, S.S., Y.S. Chang & M.N.R. Noraswati. 2009. Utilization of macrofungi by some indigenous communities for food and medicine in Peninsular Malaysia. *Forest Ecology and Management* 257(10): 2062–2065.

Li, B., F. Lu, X. Suo & H. Nan. 2010. Antioxidant properties of cap and stipe from *Coprinus comatus*. *Molecules* 15: 1473–1486.

Lindequist, U., T.H.J. Niedermeyer & W.G. Julich. 2005. The pharmacological potential of mushrooms. *Annals of Oncology* 2: 285–299.

Liu, B. & Y.S. Bau. 1980. *Fungi Pharmacopoeia (Sinicas)*. Kinoko Co., Oakland, CA.

Liu, F., V.E. Ooi & S.T. Chang. 1997. Free radical scavenging activities of mushroom polysaccharide extracts. *Life Science* 60: 763–771.

Liu, Q., T. Ng & H. Wang. 2013. Isolation and characterization of a novel Lectin from the Wild Mushroom *Oudemansiella radicata* (Relhan.: Fr.) Sing. *Biotechnology and Bioprocess Engineering* 18: 465–471.

Liu, Q., M. Zhu, X. Geng, H. Wang & T.B. Ng. 2017. Characterization of polysaccharides with antioxidant and hepatoprotective activities from the edible mushroom *Oudemansiella radicata*. *Molecules* 22(2): 234.

Loria-Kohen, V., T. Lourenco-Nogueira, I. Espinosa-Salinas, F.R. Marin, C. Soler-Rivas & A. Ramirez de Molina. 2014. Nutritional and functional properties of edible mushrooms: a food with promising health claims. *Journal of Pharmacy and Nutrition Sciences* 4: 187–198.

Lukanc, T., J. Brzin, J. Kos & J. Sabotič. 2015. Trypsin-specific inhibitors from the *Macrolepiota procera*, *Armillaria mellea* and *Amanita phalloides* wild mushrooms. *Acta Biochemica Polonica* 64(1): 21–24.

Majtan, J. 2012. Pleuran (β-glucan from *Pleurotus ostreatus*): an effective nutritional supplement against upper respiratory tract infections? *Medicine and Sport Science* 59: 57–61.

Malek, S.N.A., G. Kanagasabapathy, V. Sabaratnam, N. Abdullah & H. Yaacob. 2012. Lipid components of a Malaysian edible mushroom, *Termitomyces heimii* Natarajan. *International Journal of Food Properties* 15: 809–814.

Mandal, E.K., K. Maity, S. Maity, S.K. Gantait, S. Maiti, T.K. Maiti, S.R. Sikdar & S.S. Islam. 2011. Chemical analysis of new water-soluble (1→6) -, (1→4) -α, β-glucan and water-insoluble (1→3) -, (1→4) -β-glucan (Calocyban) from alkaline extract of an edible mushroom, Calocybe indica (Dudh Chattu). *Carbohydrate Research* 346: 2237–2243.

Mantovani, G., A. Bianchi, L. Curreli, M. Ghiani, G. Astara, B. Lampis, et al. 1997. Clinical and immunological evaluation of schizophyllan (SPG) in combination with standard chemotherapy in patients with head and neck squamous cell carcinoma. *International Journal of Oncology* 10(1): 213–221.

Mao, X.L. 1998. *Economic Fungi of China*, 1st edn. Sciences Press, Beijing, p. 376.

Martin, K.R. & S.K. Brophy. 2010. Commonly consumed and specialty dietary mushrooms reduce cellular proliferation in MCF-7 human breast cancer cells. *Experimental Biology and Medicine* 235: 1306–1314.

Mather, S.B. & L. Gonzalel. 1982. Identification of terpenoids from leaves of *Piptocarpha peritora* and their biological activities. *Journal of Natural Products* 45: 495–496.

Matsuyama, H., R.E.P. Mangindaan & T. Yano. 1992. Protective effect of schizophyllan and scleroglucan against *Streptococcus* sp. infection in yellow tail (Seriola quinqueradiata). *Aquaculture* 101: 97–203.

Mattila, P., K. Könkö, M. Eurola, J.M. Pihlava, J. Astola, L. Vahteristo, V. Hietaniemi, J. Kumpulainen, M. Valtonen & V. Piironen. 2001. Contents of vitamins, mineral elements, and some phenolic compounds in cultivated mushrooms. *Journal of Agricultural and Food Chemistry* 49(5): 2343–2348.

Mau, J.L., C.N. Chang, S.J. Huang & C.C. Chen. 2004. Antioxidant properties of methanolic extract from *Grifola frondosa*, *Morchella esculenta* and *Termitomyces albuminosus* mycelia. *Food Chemistry* 87: 111–118.

Mazur, X., U. Becker, T. Anke & O. Sterner. 1996. Two new bioactive diterpenes from *Lepista sordida*. *Phytochemistry* 43(2): 405–407.

Menikpurage, I.P., S.S.S.B.D.P. Soysa & D.T.U. Abeytunga. 2012. Antioxidant activity and cytotoxicity of the edible mushroom, *Pleurotus cystidiosus* against Hep-2 carcinoma cells. *Journal of National Science Foundation Sri Lanka* 40(2): 107–114.

Miao, S., X. Mao, R. Pei, S. Miao, C. Xiang, Y. Lv, X. Yang, J. Sun, S. Jia & Y. Liu. 2013. Antitumor activity of polysaccharides from *Lepista sordida* against laryngocarcinoma *in- vitro* and *in-vivo*. *International Journal of Biological Macromolecules* 60: 235–240.

Milovanovic, I., M. Staji, T. Stanojkovic, A. Knezevic & J. Vukojevic. 2015. Effects of selenium presence in mycelia of *Ganoderma* species (higher basidiomycetes) on their medicinal properties. *International Journal of Medicinal Mushrooms* 17: 11–20.

Mirfat, A.H.S., A. Noorlidah & S. Vikineswary. 2014. Antimicrobial activities of split gill mushroom *Schizophyllum commune* Fr. *American Journal of Research Communication* 2(7): 113–124.

Mitra, P., N.C. Mandal & K. Acharya. 2016. Mycochemicals and antioxidant activity of polyphenol-rich fraction of *Termitomyces microcarpus*. *International Food Research Journal* 23(6): 2384–2389.

Mizuno, T., H. Saito, T. Nishitoba & H. Kawagishi. 1995a. Antitumor active substances from mushrooms. *Food Reviews International* 11: 23–26.

Mizuno, T., T. Sakai & G. Chihara. 1995b. Health foods and medicinal usages of mushrooms. *Food Reviews International* 11: 69–81.

Mondal, A., D. Banerjee, R. Majumder, T.K. Maity & S. Khowala. 2016. Evaluation of in vitro antioxidant, anticancer and in vivo antitumour activity of *Termitomyces clypeatus* MTCC 5091. *Pharmaceutical Biology* 54(11): 2536–2546.

Munkhgerel, L., N. Erdenechimeg, M. Dumaa, G. Zhang, P. Odonmajig & D. Regdel. 2011. Chemical and biological investigation of the *Agaricus silvaticus* Schaeff ex. Seer. *Mongolian Journal of Chemistry* 12(38): 92–97.

Muszynska, B., K. Sulkowska-Ziaja & H. Ekiert. 2013. Phenolic acids in selected edible basidiomycota species: *Armillaria mellea, Boletus badius, Boletus edulis, Cantharellus cibarius, Lactarius deliciosus* and *Pleurotus ostreatus. Acta Scientiarum Polonorum Hortorum Cultus* 12: 107–116.

Niren, N.M. 2006. Pharmacologic doses of nicotinamide in the treatment of inflammatory skin conditions: a review. *Cutis* 77: 11–16.

Nisha, A.P. & R. Kumuthakalavalli. 2016. Anticancer activity of milky mushroom (*Calocybe indica* var. APK2) against A549 human lung cancer cell line studies. *International Journal of Pharma and Bio Sciences* 7(3): 73–80.

Ohuchi, K. & Y. Aoyagi. 2010. Inhibitory effect of mushroom extracts on α-amylase and α-aglucosidase. *Journal of the Japanese Society for Food Science and Technology* 57(12): 532–538.

Ooi, V.E.C. & F. Liu. 1999. A review of pharmacological activities of mushroom polysaccharides. *International Journal of Medicinal Mushrooms* 1: 195–206.

Opletal, L. 1993. Phytotherapeutic aspects of diseases of the circulatory systems 2. The oyster mushroom and its potential use. *Cesk Farm* 42: 160–166.

Patel, S. & A. Goyal. 2012. Recent developments in mushrooms as anticancer therapeutics: a review. *3 Biotech* 2: 1–15.

Pérez-Moreno, J. & M. Martínez-Reyes. 2013. Edible ectomycorrhizal mushrooms: biofactories for sustainable development. In: *Biosystems Engineering: Biofactories for Food Production in the Century XXI.* Eds. Ramón Guevara-Gonzalez & Irineo Torres Pacheco. Springer, London.

Phan, C., W. Wong, P. David, M. Naidu & V. Sabaratnam. 2012. *Pleurotus giganteus* (Berk.) Karunarathna & K.D. Hyde: nutritional value and in vitro neurite outgrowth activity in rat pheochromocytoma cells. *BMC Complementary and Alternative Medicine* 12: 102.

Pollock, S.H. 1975. The psilocybin mushroom pandemic. *Journal of Psychedelic Drugs* 7: 73–84.

Popescu, M., T. Costea, I. Nencu, L.E. Duţu & C.E. Gîrd. 2016. Polyphenols contents and antioxidant activity of some Romanian wild edible mushrooms. *Farmacia* 64(2): 231–236.

Prabu, M. & R. Kumuthakalavalli. 2014. Nutritional and phytochemical studies on *Pleurotus florida* (Mont.) Singer and *Calocybe indica* P&C. *World Journal of Pharmaceutical Research* 3(3): 4907–4913.

Pramanik, M., I. Chakraborthy, S. Mondal & S.S. Islam. 2007. Structural analysis of a water-soluable glucan (Fr I) of an edible mushroom, *Pleurotus sajor-caju. Carbohydrate Research* 342: 2670–2675.

Puttaraju, N.G., S.U. Venkateshaiah, S.M. Dharmesh, S.M.N. Urs & R. Somasundaram. 2006. Antioxidant activity of indigenous edible mushrooms. *Journal of Agriculture and Food Chemistry* 54: 9764–9772.

Radulescu, C., C. Stihi, G. Busuioc, A.I. Gheboianu & I.V. Popescu. 2010. Studies concerning heavy metals bioaccumulation of wild edible mushrooms from industrial area by using spectrometric techniques. *Bullettin of Environmental Contamination and Toxicology* 84: 641–646.

Rajoriya, A. & N. Gupta. 2015. Proximate and antioxidant activity of mycelia of *Termitomyces microcarpus* and *Amanita loosii. Agriculture Research & Technology* 1(1): 13–16.

Ravi, B., R.E. Renitta, M. Lakshmi Prabha, R. Issac & S. Naidu (2013). Evaluation of antidiabetic potential of oyster mushroom (*Pleurotus ostreatus*) in alloxan-induced diabetic mice. *Journal Immunopharmacology and Immunotoxicology* 35(1): 101–109.

Reis, F.S., E. Pereira, L. Barros, M.J. Sousa, A. Martins & I.C.F.R. Ferreira. 2011. Biomolecule profiles in inedible wild mushrooms with antioxidant value. *Molecules* 16(6): 4328–4338.

Ren, L., Y. Hemar, C.O. Perera, G. Lewis, G.W. Krissansen & P.K. Buchanan. 2014. Antibacterial and antioxidant activities of aqueous extracts of eight edible mushrooms. *Bioactive Carbohydrates and Dietary Fibre* 3: 41–51.

Ribeiro, B.V., J. Rangel, P. Valentao, P. Baptista, R.M. Seabra & P.B. Andrade. 2006. Contents of carboxylic acids and two phenolic and antioxidant activity of dried Portuguese wild edible mushrooms. *Journal of Agricultural and Food Chemistry* 54: 8530–8537.

Rosa, L., B.B. Cota, K.M.G. Machado, C.A. Rosa & C.L. Zani. 2005. Antifungal and other biological activities from *Oudemansiella canarii* (Basidiomycota). *World Journal of Microbiology & Biotechnology* 21: 983–987

Ross, A.C., Q. Chen & Y. Ma. 2011. Vitamin A and retinoic acid in the regulation of β-cell development and antibody production. *Vitamins & Hormones* 86: 103–126.

Roy, D.A., A.K. Saha & P. Das. 2017. Proximate composition and antimicrobial activity of three wild edible mushrooms consumed by ethnic inhabitants of Tripura in northeast India. *Studies in Fungi* 2(1): 17–25.

Ruegger, M.J.S., S.M.T. Tornisielo, V.L.R. Bononi & M. Capelari. 2001. Cultivation of the edible mushroom *Oudemansiella canarii* (Jungh.) Hohn., in lignocellulosic substrates. *Brazilian Journal of Microbiology* 32: 211–214.

Samantha, S., A.K. Nandi, I.K. Sen, P. Maity, M. Pattanayak, K.S.P. Devi, S. Khatua, T.K. Maiti, K. Acharya & S.S. Islam. 2015. Studies on antioxidative and immunostimulating fucogalactan of the edible mushroom *Macrolepiota dolichaula*. *Carbohydrate Research* 413: 22–29.

Samantha, S., A.K. Nandi, I.K. Sen, P.K. Maji, K.S.P. Devi, T.K. Maiti & S.S. Islam. 2013. Structural characterization of an immunoenhancing glucan isolated from a mushroom Macrolepiota dolichaula. *International Journal Biological Macromolecules* 61: 89–96.

Schneider, I., G. Kressel, A. Meyer, U. Krings & R.G. Berger. 2011. Lipid lowering effects of oyster mushroom (*Pleurotus ostreatus*) in humans. *Journal of Functional Foods* 3: 17–24.

Sharma, S.K. & N. Gautam. 2015. Chemical, bioactive, and antioxidant potential of twenty wild culinary mushroom species. *Biomed Research International* 346508: 1–12.

Silveira, M.L.L., F.R. Smiderle, F. Agostinia, E.M. Pereiraa, M. Bonatti-Chavesa, E. Wisbecka, A.C. Ruthes, G.L. Sassaki, T.R. Cipriani, S.A. Furlan & M. Iacomini. 2015. Exopolysaccharide produced by *Pleurotus sajor-caju*: its chemical structure and anti-inflammatory activity. *International Journal of Biological Macromolecules* 75: 90–96.

Stojkovic, D., F.S. Reis, J. Glamoclija, A. Ciric, L. Barros, L.J.L.D. van Griensven, M. Sokovic & I.C.F.R. Ferreira. 2014. Cultivated strains of *Agaricus bisporus* and *A. brasiliensis*: chemical characterization and evaluation of antioxidant and antimicrobial properties for final healthy product-Natural preservatives in yoghurt. *Food Function* 5: 1602–1612.

Subbiah, K.A. & V. Balan. 2015. A comprehensive review of tropical milky white mushroom (*Calocybe indica* P&C). *Mycobiology* 43(3): 184–194.

Subbiah, R.M.T. & W. Abplanalp. 2003. Ergosterol (major sterol of baker's and brewer's yeast extracts) inhibits the growth of human breast cancer cells in vitro and the potential role of its oxidation products. *International Journal for Vitamin and Nutrition Research* 73: 19–23.

Sun, H., C.G. Zhao, X. Tong & Y.P. Qi. 2003. A lectin with mycelia differentiation and antiphytovirus activities from the edible mushroom *Agrocybe aegerita*. *Journal of Biochemistry and Molecular Biology* 36: 214–222.

Tan, J.W., J.B. Xu, Z.J. Dong, D.Q. Luo & J.K. Liu. 2004. Nigricanin, the first ellagic acid derived metabolite from the Basidiomycete Russula nigricans. *Helvetica Acta Sinica* 87: 1025–1028.

Tavantzis, S. & S. Smith. 1982. Isolation and evaluation of a plant-virus-inhibiting quinone from sporophores of *Agaricus bisporus*. *Phytopathology* 72: 619–621.

Tibuhwa, D.D. 2012. Antiradical and antioxidant activities of methanolic extracts of indigenous termitarian mushroom from Tanzania. *Food Science and Quality Management* 7: 13–23.

Valadares, F., M.R.C.G. Novaes & R. Cañete. 2013. Effect of *Agaricus sylvaticus* supplementation on nutritional status and adverse events of chemotherapy of breast cancer: a randomized, placebo-controlled, double-blind clinical trial. *Indian Journal of Pharmacology* 45: 217–222.

Valverde, M.E., T. Hernandez-Perez & O. Paredes-Lopez. 2015. Edible mushrooms: improving human health and promoting quality life. *International Journal of Microbiology* 2015: 1–14.

Van Griensven, L.J.L.D. 2009. Culinary-medicinal mushrooms: must action be taken? *International Journal of Medicinal Mushrooms* 11: 281–286.

Vaz, J.A., S.A. Heleno, A. Martins, G.M. Almeida, M.H. Vasconcelos & I.C.F.R. Ferreira. 2010. Wild mushrooms *Clitocybe alexandri* and *Lepista inversa*: in vitro antioxidant activity and growth inhibition of human tumour cell lines. *Food and Chemical Toxicology* 48: 2881–2884.

Vishwakarma, P., P. Singh & N.N. Tripathi. 2016. Nutritional and antioxidant properties of wild edible macrofungi from North-Eastern Uttar Pradesh, India. *Indian Journal of Traditional Knowledge* 15(1): 143–148.

Wang, H.X. & T.B. Ng. 2000. Isolation of a novel ubiquitin-like protein from *Pleurotus ostreatus* mushroom with anti-human immune deficiency virus, translation-inhibitory and ribonuclease activities. *Biochemical and Biophysical Research Communications* 276: 587–593.

Wasser, S.P. 2002. Review of medicinal mushrooms advances: good news from good allies. *HerbalGram* 56: 28–33.

Wasser, S.P. 2010. Medicinal mushroom science: history, current status, future trends, and unsolved problems. *International Journal of Medicinal Mushrooms* 12: 1–16.

Wasser, S.P. & A.L. Weis. 1999. Therapeutic effects of substances occurring in higher Basidiomycetes mushrooms: a modern perspective. *Critical Reviews in Immunology* 19: 65–96.

Weigand-Heller, J., P.M. Kris-Etherton & R.B. Beelman. 2012. The bioavailability of ergothioneine from mushrooms (*Agaricus bisporus*) and the acute effects on antioxidant capacity and biomarkers of inflammation. *Preventive Medicine* 54: S75–S78.

Wu, J.Y., C.H. Chen, W.H. Chang, K.T. Chung, Y.W. Liu, F.J. Lu & C.H. Chen. 2011. Anti-cancer effects of protein extracts from *Calvatia lilacina, Pleurotous ostreatus* and *Volvariella volvacea. Evidence-Based Complementary and Alternative Medicine* 2011, Article ID 982368. https://doi.org/10.1093/ecam/neq057

Xing, F., H.X. Wang, C. Han & Y. Wang. 2003. Study on the immunocompetence of polysaccharides from the *Coprinus comatus. Journal of Food Science* 24: 139–141.

Yang, J.H., H.C. Lin & J.L. Mau. 2002. Antioxidant properties of several commercial mushrooms. *Food Chemistry* 77: 229–235.

Yang, Q.Y. & S.C. Jong. 1989. *Medicinal Mushrooms in China. Mushroom Science. XII (Part I): 631–634.* Proceedings of the 12th Internationak Congress on the Sceince and Cultiation of Edible Fungi. From K. Grabbe and O. Hilber (eds.) Braunschweig-Germany: Institut fur Bodenbiologie, Bundesforschungsanstolt fur Londwirtschoft.

Zhao, J.K., H.X. Wang & T.B. Ng. 2009a. Purification and characterization of a novel lectin from the toxic wild mushroom *Inocybe umbrinella. Toxicon* 53: 360–366.

Zhao, Y.Y., X. Shen, X. Chao, C.C. Ho, X.L. Cheng, Y. Zhang, R.C. Lin, K.J. Du, W.J. Luo, J.Y. Chen & W.J. Sun. 2011. Ergosta-4, 6, 8(14), 22-tetraen-3-one induces G2/M cell cycle arrest and apoptosis in human hepatocellular carcinoma HepG2 cells. *Biochimica et Biophysica Acta* 1810(4): 384–390.

Zhao, Y.Y., R.M. Xie, X. Chao, Y.M. Zhang, R.C. Lin & W.J. Sun. 2009b. Bioactivity-directedisolation, identification of diuretic compounds from *Polyporus umbellatus. Journal of Ethnopharmacology* 126: 184–187.

Zhuang, C., T. Mizuno, A. Shimada, H. Ito, C. Suzuki, Y. Mayuzumi, H. Okamoto, Y. Ma & J. Li. 1993. Antitumor protein-containing polysaccharides from a Chinese mushroom Fengweigu or Houbitake, *Pleurotus sajor-caju* (Fr.) Sing. *Bioscience, Biotechnology, and Biochemistry* 57(6): 901–906.

Zjawiony, J.K. 2004. Biologically active compounds from Aphyllophorales (Polypore) Fungi. *Journal of Natural Products* 67: 300–310.

Zusman, I., R. Reifen, O. Livni, P. Smirnoff, P. Gurevich, B. Sandler, A. Nyska & R. Gal. (1997). Role of apoptosis, proliferating cell nuclear antigen and p53 protein in chemically induced colon cancer in rats fed corncob fibre treated with the fungus *Pleurotus ostreatus. Anticancer Research* 17(3): 2105–2113.

6 Cultivation of Medicinal Mushrooms for a World in Need

Steven L. Miller, Adam Sayner

6.1 INTRODUCTION

For thousands of years, indigenous people have made extensive use of the materials occurring in forests to meet their health needs. Forest dwellers in Southeast Asia, for example, use around 6,500 different plants and fungi to treat illnesses (Mongabay 2012). Shaman were the first medical specialists in indigenous communities, and their traditional methods are known to be effective in treating both physical and psychological ailments. The World Health Organization (WHO) estimates that 80 percent of the people in developing countries still rely on traditional medicine for their primary healthcare needs (Jackson 1989). In short, without money, access, or faith in modern healthcare, many people continue to depend on forests for their medical requirements.

Systems of traditional medicine are the result of the longstanding use of natural medicines. By virtue of its claim of being the cradle of mankind, Africa is likely home to the oldest practitioners of traditional medicine and the most assorted of all therapeutic systems. The continent's biological and cultural diversity affords it a wide variety of healing practices (Zibiah 2021). India is similarly important site of traditional medicine, with millions of daily practitioners (Adhikari and Paul 2018; Syamili 2020).

Perhaps best-known system, however, is traditional Chinese medicine, which is well documented to have begun thousands of years ago (Bensky et al. 2015). This practice, largely rejected as superstitious after the fall of the Qing dynasty in 1911, has enjoyed a stunning resurgence in the 21st century (Xue et al. 2010; Ulloa 2012). This resurgence has developed partly because the number of western-trained doctors is dwarfed by the size of the Chinese population. There is also a huge demand for preventive medicines that people believe will help avoid the need for more expensive treatment in hospitals. Interestingly, the inverse is also true. There has been wider recognition of the preventative strengths of traditional Chinese medicine because people have more money in their pockets and care more about their well-being (Economist 2017). Traditional Chinese medicine makes common use of over 300 plants and fungi (Bensky et al. 2015). Of these, a number of medicinal mushrooms and other fungi hold a special place among the most valued category of traditional medicines.

The top ten most common health issues worldwide (WHO 2021; Health Benefits Times 2022) include cancer, heart disease, digestive health issues, arthritis and osteoporosis, heartburn (esophageal ulcers or esophageal cancer), hypertension, diabetes, headache, and high blood pressure. A search for the most common diseases around the world (Johansen 2019) reveals such ailments as lower respiratory infections (bronchitis, pneumonia, influenza, tuberculosis), diarrheal diseases and dehydration (bacteria or intestinal viruses), HIV/AIDS, coronary heart disease, stroke, and cancers (respiratory, stomach and colon, prostate).

Approximately 130 medicinal functions have been identified in the fruiting bodies, extracts of fruiting bodies, cultured mycelia, and cultured broth of medicinal mushrooms and fungi (reviewed by Anke (1989); Didukh et al. (2003); Zaidman et al. (2005); Dai et al. (2009); Wasser (2010); Lo and Wasser (2011); Chang and Wasser (2012); De Silva et al. (2013); and Wasser (2014), including anti-tumor, immunomodulatory, antioxidant, radical-scavenging, cardiovascular, antihypercholesterolemic, antiviral, antibacterial, antiparasitic, antifungal, detoxicating, hepatoprotective, and

DOI: 10.1201/b23190-6

antidiabetic effects. The best usage of medicinal mushroom drugs has been in preventing and treating immune disorders, especially in immunodeficient and immuno-suppressed patients; they are also used for patients undergoing chemotherapy or radiotherapy and those suffering from different types of cancer; chronic blood-borne viral infections such as hepatitis B, C, and D; different types of anemia; human immunodeficiency virus/acquired immunodeficiency syndrome (HIV/AIDS); herpes simplex virus (HSV); chronic fatigue syndrome; Epstein-Barr virus; chronic gastritis and gastric ulcers caused by *Helicobacter pylori*; and dementia (especially Alzheimer's disease) (Wasser 2014). In addition, the recent integration of traditional Chinese and modern medicine practices has begun to solve multiple worldwide health problems, and modern clinical practice in Japan, Korea, Russia, and several other countries relies on mushroom-derived preparations (Reshetnikov et al. 2001; Chang and Wasser 2012). Therefore, it is easy to see that medicinal fungi offer effective treatment options for many or most of the conditions and diseases that currently affect human populations.

Unfortunately, several factors have coalesced resulting in issues that undermine the traditional use of fungi for medicinal needs. Not only has massive deforestation across the globe negatively impacted the ability of many people to dependably harvest and use certain plants and fungi for medicine but it has also dramatically influenced the discovery of potential new medicines (Jackson 1989). In addition, the popularity of some high-potency fungal species (e.g., *Ophicordyceps sinensis* and *Fometopsis officinalis*) has led to overharvesting, placing these fungi on endangered species lists (Wang and Yao 2011; Pietka and Szczepkowski 2011).

Fortunately, the fruiting bodies of a large number of fungi that produce highly effective, medically active compounds can be cultivated under artificial conditions. Many of these fungi are wood decomposers in nature and can be cultivated not just on woody substrates but more importantly on many regionally common agricultural waste substrates available throughout the world. Other medicinal compounds can be harvested from the mycelia alone, broadening the number of fungi that can be grown to produce effective, medically active compounds.

Regardless of background, race, and religion, human beings around the world are susceptible to a range of common diseases and illnesses. As a basic human right, everyone should have access to safe, effective treatments, although that has not been the case due to economic, political, and environmental issues. The science of cultivating mushrooms of many varieties, both culinary and medicinal, has now reached a level of maturity where people around the world and in a variety of circumstances are capable of successful cultivation. The development of a network of medicinal mushroom farms could provide safe, effective, inexpensive, and widely-accessible healthcare for the most common health issues.

It should be pointed out that there might not be a universal belief in the power of medicinal mushrooms. While there is a firm and longstanding tradition of the use of fungi for many maladies, at least some clinical trials have failed to show efficacy (Wasser 2017; Rogers 2020). In addition, some studies have found differences in the medicinal compounds recovered from wild harvested versus cultivated mushrooms (Guo et al. 2012; Zhang et al. 2015) and even in the same medicinal mushroom cultivated on different substrates and casing soil in different regions (Chen et al. 2008; Jing et al. 2014; Phan et al. 2014; Li et al. 2016). Admittedly, a great deal is yet to be learned regarding medicinal fungi and whether specific strains, wild *vs.* artificial cultivation, and particular preparation, use, or extraction protocols affect the benefits of medicinal fungi. However, it should also be said that there is scientific evidence that certain fungi provide an advantage in the treatment of many diseases and conditions (Rogers 2020). Also, there is little evidence that any of these fungi cause harm when taken as directed. Our stance is that many people around the world have no access to any type of medical treatment and are suffering from a range of medical issues known or suspected to be treatable by use of medicinal fungi. What is the harm in pursuing this option?

Numerous books have been written outlining the detailed steps required to accomplish mushroom cultivation (e.g., Cotter 2014; Stamets 2000; McCoy 2016; Bray 2020, 2021; Press 2021; Rowsan 2021a; Heferle 2022; Wallin 2020 to name a few), and hundreds of high-quality tutorials are available for viewing online (e.g., GroCycle, NorthSpore, Oak and Spore, FreshCap, Paul Stamets, Southwest Mushrooms, Du Toit-Henke, Chido Govera, UFA Agro Research Center, Mushroom Kenya, Mushroom Growers Handbook 1 2016a, and Mushroom Growers Handbook 2 2016b, among others).

This single chapter does not attempt to restate this information. Rather, the goals of this chapter are to provide a pragmatic summary of important considerations and resources for low-tech approaches that may be feasible for the widespread cultivation of medicinal mushrooms in a variety of areas and circumstances, and to explore possible models that could work to help develop a network of medicinal mushroom cultivation farms so that mycological products can be made available to those who have no access to healthcare or medicine. It is hoped that this chapter will encourage governmental agencies, NGOs, and individuals to consider supporting a network of cultivation farms focused on medicinal fungi that can benefit a large number of people almost anywhere medical treatment is unavailable.

6.2 COMMON, POPULAR, AND EFFECTIVE MEDICINAL FUNGI

Table 6.1 is a listing of the most common, popular, and effective medicinal fungi, as well as their common names, approximate distributions, and ability to be cultured and artificially cultivated.

TABLE 6.1
Alphabetical (mostly) List of Common, Popular, and Effective Medicinal Fungi with Their Common Names, Distributions, and Cultivation Status.

Medicinal fungus	Common name	Distribution	Cultivation status
Agaricus subrufescens Peck (=*A. brasiliensis* S. Wasser et al.; =*A blazei* sensu Heinem.)	Almond mushroom, mushroom of the sun, God's mushroom, mushroom of life, jisongrong, himematsutake	Temperate and warm temperate, to tropical areas, probably worldwide	+ compost blocks, outdoor compost beds, interplanted in vegetable beds
Auricularia auricula-judae (Bull.) J. Schröt. and closely related species	Jelly ear, Judas's ear, Jew's ear	Worldwide *in totus*, but some species are East Asian, European, or North American	+ sawdust blocks, logs
Coprinus comatus (O.F. Müll.) Pers.	Shaggy mane, inky cap, Shaggy ink cap, lawyer's wig	Native to Europe and North America, but now widely introduced around the world	+ compost blocks, straw blocks, outdoor mushroom beds
Ophicordyceps (=*Cordyceps*) *sinensis* (Berk.) G. H. Sung et al.	Caterpillar fungus, winter worm/summer herb, yartsa gunbu, yarcha-gumba	Native to the high meadows of the Tibetan Plateau of Southwest China, Himalayan regions of Bhutan and Nepal	– entomopathogenic on the larvae of moths in the family Hepialidae. The mycelia can be cultured, but so far, the fruiting bodies cannot be cultivated and must be collected from the wild
Cordyceps militaris (L) Fr.	Cordyceps, scarlet caterpillar club	Throughout forested parts of Northern Hemisphere, including Mexico	+ entomopathogenic on insect larvae or pupae, especially moths, cultivated on silkworm pupae, mealworms, brown rice, liquid fruiting media
Flammulina velutipes (W.Curt.:Fr.) Singer and related species	Velvet shank, velvet foot, enoki, enokitake, winter mushroom	Circumboreal	+ sawdust and plant waste blocks and enoki jars
Fomes fomentarius (L.) Fr.	Tinder fungus, false tinder fungus, tinder conk, Amadou, Birch polypore	Circumboreal, northern and southern Africa	+ logs, sawdust blocks, soil covering, at least some bioactive compounds are also produced by the mycelium in culture
Ganoderma lucidum Karst, including *G. resinaceum*, *G. sessile*, *G. sinense*, *Ganoderma* spp. *sensu latu*	Reishi, lingzhi, ling chih, divine mushroom of immortality	Forested areas worldwide	+ logs, buried logs, sawdust blocks
Grifola frondosa (Dicks.:Fr.) Gray	Maitake, hen of the woods, ram's head	Native to China, Europe and North America	+ logs, sawdust blocks, straw blocks, seed husks

TABLE 6.1 ***Continued***

Alphabetical (mostly) List of Common, Popular, and Effective Medicinal Fungi with Their Common Names, Distributions, and Cultivation Status.

Medicinal fungus	Common name	Distribution	Cultivation status
Hericium erinaceus (Bull.) Persoon	Lions' mane, mountain-priest mushroom, bearded tooth fungus	Native to North America, Europe, Asia, Australia	+ sawdust blocks, logs
Hypsizygus tessulatus (Bull.) Singer (= *H. mamoreus* (Peck) Bigel.)	Beech mushroom, brown beech mushroom, white beech mushroom, brown clamshell mushroom, white beech mushroom, buna-shimeji, bunapi-shimeji	Throughout forested parts of the Northern Hemisphere	+ sawdust blocks, logs
Inonotus obliquus (Ach. Ex Pers.) Pilát	Chaga, clinker polypore, cinder conk, birch canker polypore, sterile conk trunk rot of birch	Circumboreal, grows especially on living birch, but also rarely on other trees such as alder, beech, and poplar	– mycelia can be cultured, but medicinal compounds are only produced when the fungus interacts with living birch trees during the formation of the sclerotium
Laricifomes officinalis (Vill.) Kotl. and Pouzar (= *Fometopsis officinalis* (Vill.) Bondartsev & Singer	Agarikon, eburiko, quinine conk	Native to old-growth forests of Europe, Asia, North America, North Africa; largely associated with larch, but also occasionally cedar and pine trees	– some attempt to inoculate standing trees and branches of larch, some bioactive compounds are also produced by the mycelium in culture
Lentinula edodes (Berk.) Pené	Shiitake, Chinese black mushroom, Black Forest mushroom	Native to East Asia, but has become naturalized in many temperate areas	+ logs, sawdust blocks, straw blocks
Phellinus linteus (Berk. et M.A. Curt.) Teng	Mesima, meshimaboku, song gen, sanghwang, black hoof mushroom	Native to East Asia	+ especially on mulberry logs
Schizophyllum commune Fr.:Fr	Split-gill	worldwide	+ sawdust blocks
Sparassis crispa (Wulfen) Fr., and related *S. spathulata* (Schwein.) Fr., *S. brevipes* Krombh., *S. latifolia* Y.C. Dai and Z Wang, *S. americana* R.H. Petersen	Cauliflower mushroom	*In totus* temperate forested areas worldwide with some species only in Europe, some only in North America, others in Asia	+ sawdust blocks, fruiting bottles
Trametes versicolor (L.:Fr.)	Turkey tail	Forested areas worldwide	+ logs, sawdust blocks
Tremella fuciformis Retz.:Fr.	Snow fungus, snow ear, silver ear fungus, white jelly mushroom, white cloud ear	Worldwide in forested tropical and subtropical ecosystems, also warm temperate East Asia and North America	+ parasite of wood-loving ascomycetes, must be cultivated in dual-culture
Tremella mesenterica Retz.	Yellow brain, golden jelly fungus, yellow trembler, witches' butter	Widely distributed in temperate and tropical regions of Africa, Asia, Australia, Europe, and North and South America	+ parasite of wood-decay fungi in the genus *Peniophora*, must be cultivated in dual-culture, at least some bioactive compounds are also produced by the mycelium
Wolfiporia extensa (Peck) Gins (= *Wolfiporia cocos* F.A. Wolf)	Chinese Tuckahoe, fu-ling, hoelen, poria, China root, matsuhodo	Widely distributed in East Asia	+ culture bottles with pine logs and root pieces

Note: a brief summary of cultivation approaches is provided, but the list of ingredients is neither exhaustive nor complete. This list does not include psychotropic fungi, which exhibit incredible promise for treating mental disorders. "+" under "Cultivation status" indicates that fruiting bodies of that fungus have been reliably produced under artificial conditions; "–" indicates that fruiting bodies of that fungus have not been successfully produced under artificial conditions on a commercial scale.

Specifics on the compounds produced by these fungi and the medical use of these mushrooms are provided elsewhere in this book. It should be noted, however, that the natural distribution of these fungi should not be an impediment to their cultivation and use. Spores and cultured mycelia in liquid slurries or in substrate blocks are commonly and widely shared across the globe, and a particular fungus does not necessarily need to be endemic to a specific region to be considered for cultivation there (this is brought up again herein).

6.3 COMPARISON OF HIGH-, LOW-, OR HYBRID-TECH CULTIVATION APPROACHES

6.3.1 Critical Decision—Sterilization or Pasteurization of the Fruiting Substrate?

A critical decision must be made early on regarding the initial overall approach to the cultivation of medicinal mushrooms that will affect all downstream requirements and processes. This decision might be dictated by circumstance, depending on the amount of funding available for cultivation facility set up and operation, and centers on how the fruiting substrate will be handled and prepared. The basic options include pressure sterilization or pasteurization. In pressure sterilization, culture units (bags, bottles, etc.) containing the fruiting substrates are placed in a pressure vessel such as a pressure cooker or autoclave and heated to at least 121°C under a pressure of at least 15 lbs/in^2 (103.4 kPa). The duration of pressurization is at least 15 min but could be longer depending on the substrate and the size of the culture unit. The temperature and pressure are released gradually until reaching ambient conditions. The pressure allows the temperature to rise above the boiling point of water without allowing the liquid to boil. The combination of heat and pressure for an appropriate time kills all potential contaminating microbes and fungi, including actively growing hyphae and dormant spores of molds and other fungi, as well as bacteria and their resistant propagules.

In pasteurization, the substrate is subjected to some form of treatment that will knock back contaminating organisms to the point that the fungi to be cultivated have an advantage in utilizing and colonizing the substrate. Pasteurization does not kill all microbes, just reduces their number and activity so they can be outcompeted by the desired fungus. There are several types of pasteurization commonly used in fungal cultivation, such as steam (including tyndallization or fractional sterilization), hot water, and chemical methods.

The steam pasteurization process utilizes flowing steam from a boiler of some type that passes through the bulk substrate or around individual culture units containing the substrate. This is usually accomplished in a large metal box. Figures 1a and b show homemade steam pasteurizing chambers. Figures 1 c and d show wood-fired steam generators made from 55 gallon drums. Figure 1e shows a mass fired steam generator made from a bee keg. Ideally, the temperature of the substrate should reach 150–180°F (62–82°C) for up to five days. Tyndallization uses repetitive heating and cooling cycles over several days. This allows the resistant propagules to germinate and begin to grow, while subsequent steam cycles to kill them. No pressurization is required, so the container does not have to be constructed to withstand pressure, only steam for up to several days. Steam pasteurization is better than dry heat pasteurization because dry heat desiccates the substrates, which then must be rehydrated, potentially adding back contaminating microbes.

In hot-water (scalding) pasteurization, the fruiting substrate is immersed in hot water 150–180°F (62–82°C) for an appropriate length of time (usually 1–5 h), and then the substrate is drained before use. Alternatively, the hot water can be added directly to the substrate in the culture units, which are generally covered with insulating material so the heat level stays high for several hours or overnight, and then allowed to cool before inoculation. The soaking substrate in a hot-water bath will kill off most of the organisms on and in the substrate, but it will leave behind other potentially competitive contaminating microbes that can survive that temperature range, especially those that produce thick-walled dormant propagules. Chopped straw or large wood chips contained in wire mesh baskets can easily be dunked into a large metal container of hot water and hot-water pasteurized. Other substrates with smaller particle sizes (e.g., sawdust, pelleted substrates) usually must be contained in a finer mesh bag of some type or pasteurized *in situ* in the fruiting units.

Several types of chemical pasteurization can be used. Perhaps the most famous is the hydrogen peroxide method developed by R. Rush Wayne (Wayne 1996), which is similar to the hot-water addition method. In this method, the substrate is immersed in hot sterile water to which a small amount of 3% hydrogen peroxide is added. The substrate is allowed to cool and drain to field capacity before being inoculated.

Another popular method is cold-water liming. A lime bath is prepared with 2 g hydrated lime powder per l cold water and the substrate is added and mixed thoroughly. This is allowed to sit for

FIGURE 6.1 Equipment for steam pasteurization of fruiting substrates. a). Metal steam pasteurizing cabinet, courtesy Carie Cave. (b) Steam pasteurization cabinet, courtesy of Unclesam Dalaay. (c-d) Wood-fired steam generators made from 55 gal drums, (c) courtesy of (in Thai 2248), (d) courtesy of Aod Hang Over (in Thai 2496). (e) Gas-fired steam generator made from beer-keg and 55 gal drum, courtesy of Samuel Touchon.

12–24 h. The substrate is then allowed to cool and drain to field capacity before being inoculated. The pH of the bath should approach 12.0 to be effective.

Formaldehyde (500 ppm) has also been used to pasteurize fruiting substrates for mushroom cultivation (Nanje Gowda and Manvi 2019). Because many wood-decomposing fungi produce enzymes that can detoxify large, ring-shaped molecules such as formaldehyde, phenol, and creosote, they can tolerate residual formaldehyde in the substrate. It should be noted that all chemical methods of mushroom substrate pasteurization carry safety risks. Hydrogen peroxide, hydrated lime, and formaldehyde are extremely caustic and are harmful upon skin contact or inhalation, so appropriate safety equipment and apparel must be worn. A great deal of care must be taken when handling these chemicals before, during, and after use.

6.3.2 High-Tech Mushroom Cultivation

Once the decision to sterilize or pasteurize has been made, downstream requirements and other considerations will fall in line regarding the cultivation approach. General categories of cultivation approaches occur along a spectrum and include high-tech, low-tech, or hybrid-tech cultivation techniques. Hi-tech is likely the approach that most people know about and is typical of most big, commercial mushroom production facilities. High-tech approaches include complete pressure sterilization of the substrate, often in large batches, 100% sterile transfer protocols, and perhaps a broader range of fungal species that can be cultivated. This approach typically uses a nutritionally supplemented fruiting substrate that results in abundant, highly repeatable, and dependable production. To achieve this, many of the processing steps are mechanized with specialized machinery for bulk substrate handling, processing, and inoculation, thereby reducing the necessary labor force, and precise environmental controls are required for incubation, fruiting, and post-processing. The start-up costs associated with high-tech cultivation facilities are often quite high and therefore out of the reach of most people.

6.3.3 Low-Tech Mushroom Cultivation

On the other end of the spectrum is the low-tech approach, which typically utilizes pasteurization of the substrate, little or no nutritional supplementation, clean but not necessarily 100% sterile transfer and inoculation protocols, and possibly a reduced set of fungal species that can be cultivated because environmental controls necessary for certain species might be lacking. Often the specialized machinery characteristic of the high-tech approach is replaced with innovation and labor input, along with less precise environmental controls for incubation, fruiting, and post-processing. It must be noted that there are many low-tech mushroom cultivation facilities scattered around the world capable of producing large quantities of mushrooms, but they do not focus on medicinal mushrooms (GroCycle 2020a, b, c).

6.3.4 Hybrid-Tech Mushroom Cultivation

The hybrid-tech approach lies somewhere between the high-tech and low-tech approaches. The hybrid-tech approach might still rely on pasteurization of the substrate but may also include some machinery for handling bulk substrates, be able to handle more cultivation units, and provide more precise environmental controls for fruiting.

6.4 NECESSITIES AND NICETIES OF LOW-TECH MUSHROOM CULTIVATION FACILITIES

Regardless of the approach, some requirements are common to almost all intended mushroom grow facilities. These include a mushroom-oriented lab with an incubation area, fruiting area, and pre- and postproduction areas. These commonalities are scalable and can be accomplished for any need/capability from hobby to small-scale, semi-commercial, or commercial endeavors depending on the goal (personal use, economic/income, or providing critical products for those who need them). There may be tradeoffs between automated equipment that saves labor costs and a readily available labor force, the ability to successfully and reliably produce fungi *vs.* the maximization of efficiency through N supplementation to increase yield and meet regular quotas, and time input *vs.* financial input.

6.4.1 Mushroom Lab

A mushroom lab can be critical depending on the overall approach (see discussion on spawn availability in section 6.8), and a highly useable lab with all necessary functions can be developed under surprisingly simple conditions. At a minimum, the lab area must be separated in space and/or function from all other activities, especially preproduction and fruiting areas, which often generate dust and a large spore load. The area does not need to be overly large or permanent, however, the walls and floor should ideally be constructed of smooth, washable materials that can easily be thoroughly cleaned and frequently disinfected. This lab space could even be a "tent" with plastic fabric sides and a roof on a concrete floor.

Similarly, the lab should be centered around a sturdy, level, cleanable bench or table with a surface made of smooth, hard, and non-porous material (stainless steel, laminate, and even glass are possible alternatives). Miscellaneous lab equipment such as scalpels, flame burners (alcohol burners, gas burners, propane torches, or even disposable multi-purpose lighters), hot-plates, tape, marking pens, isopropyl or ethyl alcohol, glass jars of various sizes, plastic bags, spray bottles, and clean rags or paper towels round out the list of necessities. Glass Petri dishes, or bottles, if they can be secured, are nice because they can be reused.

The category of niceties in the lab includes a still air box or laminar flow hood, a pressure cooker or autoclave, refrigeration, and an impulse sealer. Depending on the budget of the cultivation facility, it is often economical to produce agar slants and plates for maintaining the cultures of fungi to be cultivated, and possibly liquid cultures and spawn, in-house. To do so, in addition to the lab itself, a still air box is recommended for working with viable cultures in an easily cleanable, draft-free environment that protects the cultures from moving air, hair, and the breath of the workers, which often introduces contaminating microbes and other fungi. These can easily be manufactured, even from recycled materials or plastic panels. A step up would be a forced air laminar flow hood that uses a fan to move air through a series of filters across the bench top. The idea is to protect the culture media and transfer process from contaminants derived from the workspace and workers' hair or breath by forcing ultrafiltered air across the workspace. Large, commercial-grade laminar flow hoods can be quite expensive, but other less expensive models are available and there are readily available plans for more than adequate hoods online for those who can build one themselves.

Niceties also include a steam pressure cooker or autoclave for pressure sterilization (see the previous discussion of sterilization *vs.* pasteurization). Household pressure cookers of various sizes have been used with good results, with the only reservations being the volume of available interior space and the heat source. Autoclaves accomplish the same function, but generally are larger with built-in and controllable timers and feature thermometers attached to an integral boiler or steam generator. Of course, it is possible to cultivate medicinal fungi without a pressure cooker or autoclave. Instead of producing agar and liquid media for culture and preserving the strains under cultivation and for inoculating spawn, mycelial and spore syringes as well as commercially prepared spawn inocula, and even presterilized or prepasteurized fruiting substrate is available for purchase from commercial sources if the budget allows it. Likewise, purchasing premade agar plates, agar slants, liquid media, and inocula of various kinds could go a long way towards negating the need for making and sterilizing one's own media and inocula but can be expensive.

The capability to store media, cultured mycelia, spores, and reagents under cool conditions is often critical, and some would argue that refrigeration should be listed under necessities instead of niceties. The size of the refrigeration unit necessary will be determined by the uses and space needs. A small "dorm-sized" refrigerator or one to several household refrigerators might suffice.

If pasteurized or sterilized fruiting or spawning substrate will be produced in 5–10 lb (2.5–4.5 kg) single-use, filter-patch plastic bags, an impulse sealer is a must. These contain an electrical thermal strip that melts 2–4 layers of plastic together for a complete seal.

Another handy off-the-shelf implement is a moisture meter. This is useful both in the lab and in the preproduction area to optimize the hydration of the substrate.

Many years of fieldwork all over the world has taught us that the lab space is critical, but that preparation, knowledge, and practice of sterile techniques are the biggest keys to successful culture work, not necessarily an ultra-fancy high-tech lab. Even the equivalent of a small hotel bathroom or large shower are perfectly acceptable alternatives that can be found or constructed just about anywhere.

6.4.2 Incubation Room/Area

An area where inoculated fruiting or spawning substrate units can be maintained at an acceptable temperature until fully colonized and/or developed to the point of pinning/fruiting is critical. Typically, these do not require tight environmental controls beyond temperature (i.e., they do not require control of light, humidity, or CO_2 levels). The temperature should be maintained loosely around 65–70°C, which may require a fan for air movement and cooling and possibly even heating and/or air conditioning depending on the climate of the facility and the fungi under cultivation. This area should be easy to clean and disinfect and could be another "tent" similar to that proposed as an alternative for the lab space, with sufficient shelf space to house successive iterations of inoculated cultivation units for up to several months.

6.4.3 Fruiting Chamber/Room

Potentially a bit more challenging is the space specifically designed for fruiting. Each species and even different isolates of the same species under cultivation have their own requirements for temperature, humidity, CO_2 levels, and light for optimum fruitbody production. This can often vary depending on the substrate used for fruiting and whether nutritional supplementation is used. Ideally, construction materials for the fruiting chamber should be water- and humidity- resistant and easy to clean and disinfect. Additional tents such as the ones described for the lab and incubation chambers might be satisfactory. We note here that most mushroom cultivation facilities we have visited in the tropics are simply enclosed spaces constructed of native building materials (e.g., bamboo or thatch) or tarps which provide shade for the cultivation units and feature shelving or an innovative way to organize and optimize the space. This is acceptable because once the substrate, whether pasteurized or sterilized, is fully colonized by the mycelium of the intended fungus, the risk of contamination is greatly reduced. Size of the fruiting chamber will be determined by the number and size of cultivation units to be placed in the fruiting chamber and the length of time required before harvesting the fruiting bodies.

The fruiting chamber is where environmental controls can be most critical. Too high a temperature can often lead to inadequate humidity and poor fresh air exchange. Humidity control is also critical, and this can be more or less important depending on the ambient humidity found in different regions of the globe. If the humidity is too high, fruiting bodies may never expand fully or may show unusual growth forms such as secondary pinning on the mushrooms; excessive humidity may also lead to the collapse of some fungi and undesirable colonization by bacteria that effectively destroy the useability of the mushrooms. If the humidity is too low, the pins and developing fruiting bodies will never expand and may simply dry up. High-tech facility operators spend a great deal of money on sophisticated environmental control systems with automated monitoring. In dryer areas of the world, which require substantial humidification, homebuilt humidifiers or misting systems can be constructed cheaply and controlled with relatively inexpensive off-the-shelf control systems.

Fresh air exchange is also critical for the successful growth of medicinal fungi. CO_2 concentrations are typically high in the incubation stage because the substrate is totally or partially enclosed by the container or vessel they are growing. It is thought that in nature, most fungi use sense levels of CO_2 as a sign that they are in an acceptable environment to produce the most important part of the fruiting structure—namely, the spore-bearing tissue. Accordingly, if CO_2 is too high during the fruiting stage, the spore-bearing parts the fruiting bodies of most fungi will be stunted and poorly shaped and often will develop stipe tissue only. If they sense the CO_2 is too high, they will likely keep producing additional stipe tissue to try to reach an area away from the substrate with lower CO_2 to release their spores. Although off-the-shelf CO_2 sensors are available, typically in low-tech approaches, actual CO_2 levels do not need to be measured. Rather, lower levels of CO_2 are achieved through the use of fans and ducts that bring in fresh air and exhaust CO_2 outside the fruiting chamber. These fresh air exchanges can be controlled by off-the-shelf timers set to accomplish a certain duration of fresh air exchange per hour (e.g., three, one-minute fresh air exchanges every hour). With experience, this level of control is more than adequate and can be changed quickly and easily with the maintenance of humidity also in mind. As explained earlier, in some drier areas of the world with low relative humidity, it may be difficult to maintain proper humidity while at the same time reducing CO_2 levels. Certain strains of medicinal fungi are more forgiving, and successful fruiting can be achieved through adequate, although not precise, balance of temperature, moisture, and fresh air exchange based on experience.

Another environmental factor that must be considered for the fruiting chamber is light. Although fungi do not require light to produce their carbon and energy like plants do, fungi use light in the blue wavelengths to sense that they have reached an area outside their substrate in which to develop spore-bearing tissue (Kamada et al. 2010). Therefore, some light is required so that the typical morphology and orientation of the fruiting bodies can be achieved. Often, adequate light is provided by diffuse sunlight that filters into the fruiting chambers, especially in tropical areas where the ambient humidity is sufficient. In drier areas where construction of the fruiting chamber must be tighter to maintain temperature and humidity and opaque to block damaging levels of direct sunlight, artificial lighting throughout the fruiting chamber might be necessary. This lighting can comprise relatively inexpensive and low temperature LED lights. All these factors can greatly affect the development, expansion, and final form of the fungal fruiting structures and therefore overall productivity.

Another consideration for the fruiting chamber/room is the ability to efficiently organize the fruiting units to maximize fruiting body development and production. Fruiting units are often placed on shelves with adequate space between them so light, humidity, and fresh air can reach all parts of the developing mycelia and pins, allowing the fruiting bodies to develop all around the unit. Figure 2a and b show shelves with one-layer and four-layers of fruiting units supported by shelves. These requirements might be met with a shelf space from floor to ceiling. In other systems such as grow bags with a single opening for both inoculation and fruiting (e.g., *Ganoderma* spp., and *Pleurotus* spp.), the culture units can be stacked in many layers on top of each other with the openings facing outward, requiring relatively few shelves. Other systems utilize hanging columns or bags, requiring structures on the ceiling to support the hanging weight but no shelving *per se*. Figures 2c-f are innovative designed for hanging fruiting bags made from inexpensive twine. Innovative hanging methods have been developed where individual bags are supported in vertical chains using only polymer twine. A number of different shelving or hanging types may be required if a broad diversity of fungal species with different fruiting characteristics are to be cultivated as opposed to a single species.

There are several options for the successful use of a fruiting chamber. Obviously, depending on the fungal species and strains to be cultivated and the region of the world, the best situation would be to have the capability of controlling temperature, humidity, CO_2 levels, and light, at least to some

FIGURE 6.2 Methods for organizing fruiting units in the fruiting chamber. (a) Floor to ceiling shelves with one layer of fruiting bags, courtesy of Fun&Go. (b) Shelves with multiple layers of fruiting bags, courtesy of Dr. Mushroom. (c-f) Rope, twine, and strap bag hanging devices, (c) courtesy of Kern Nakin Mushroom Farm in Cambodia, (d) courtesy of Rhona P. Nieto Mushroom Grower in the Philippines, (e) courtesy of Pradip Das Mushroom Farming Complete Guide, (f) courtesy of Ravi Saliya Pramukh Mushroom Farm.

extent. Failing that, the selection of fungal species could be made to correlate with the conditions present in the fruiting chamber. This may drastically reduce the kinds and diversity of fungi that can be cultivated. However, there are many fungal species that have not been documented to exist and that have not been tested for medicinal properties. The list provided in Table 6.1 is simply a summary of those fungi that have been used, maybe even for millennia, and for which adequate information exists to elevate them to the top of the list of medicinal fungi. It is possible, likely even, that in certain areas, other fungi with superior medicinal value are known locally and could be brought into cultivation. This would be advantageous because environmental conditions for that fungus already exist, and therefore tight controls might not be necessary.

It should also be pointed out that artificial environmental control in even modestly sized fruiting rooms in certain regions of the world often become a significant part of the operating budget. Good examples would be humidification and heating of a fruiting room during the winter months in a temperate region or cooling of a fruiting room in hot desert regions. This is very difficult without purpose-built insulated buildings and dedicated heating and cooling systems. In these cases, seasonal cultivation along with an appropriate choice of species to cultivate may be required.

6.4.4 Preproduction Area

The preparation of even a moderate number of fruiting units requires the use and handling of a large amount of fruiting substrate and supplements. The preproduction area can often be quite dusty and messy as different substrates and supplements are mixed and hydrated. Therefore, an area dedicated to delivery, storage, measuring, preparation and mixing, and filling the culture units is necessary. High-tech or hybrid-tech facilities often have intricate machines and dedicated equipment for such processes, and operators of these facilities can buy necessary substrates and supplements in large truckloads or prepackaged in uniform bags. Figures 3a-c show the preproduction area of a low-tech mushroom facility where the fruiting substrate and amendments are being mixed by hand. Low-tech facilities typically rely on labor to accomplish these tasks and often depend on the delivery of substrates in bulk directly from the farmer or distribution center. Figure 3d shows a truckload of fruiting substrate being delivered on-site. If substrates and amendments are readily available locally, many low-tech facilities can accept delivery of the materials and do the preprocessing (e.g., drying or grinding) and preproduction immediately for each cultivation run, negating the need for proper storage. If substrates are purchased in bulk shipments that cannot be used immediately, the preproduction area must have an adequate means of sheltering the substrates until they can be processed to minimize uncontrolled wetting and contamination by bacteria, molds, and other fungi before processing. Many facilities utilize a simple concrete pad on which they measure, dump, mix, and hydrate the fruiting substrate manually before the fruiting units are filled. Others utilize simple but innovative purpose-built machines for mixing and dispensing the substrates into the fruiting units. These are often constructed after it is known at which exact steps processing bottlenecks occur.

As discussed previously, another major consideration is whether the substrates are going to be pasteurized or sterilized and how that will be accomplished. For small-scale operations, household pressure cookers, such as those often used in the low-tech lab, can be used to sterilize a few culture units. However, the repetitive processing of even a moderate number of culture units requires either a shift in approach (e.g., moving from pressure sterilization to hot water, unpressurized steam, or chemical pasteurization) or an investment in machinery with a larger capacity (e.g., commercial-sized autoclave). Again, there is ample room for innovation, with purpose-built, relatively large-capacity steaming set-ups made from excess metal drums. If the budget or availability allows, substrates that are essentially prepasteurized during their manufacturing (e.g., pelleted sawdust, straw, or seed hulls) and require little, if any, additional treatment before use can be purchased and used. Unfortunately, these substrates are not readily available outside of developed areas and may be costly.

6.4.5 Postproduction Area

The type of postproduction area and equipment required will be determined by the product and intended distribution. Continuous production of gourmet mushrooms typically requires

FIGURE 6.3 Preproduction area. (a) Preproduction area—mixing substrate and supplements by hand, Unclesam Dalaay Mushroom Growing and Marketing in the Philippines. (b) Preproduction area—filling bags by hand, Mr. RAD's Mushroom House courtesy of ReyAnn Lovelie Detanov in the Philippines. (c) Preprodution area—filling bags by hand, courtesy of Unclesam Dalaay Mushroom Growing and Marketing in the Philippines. (d) Receiving a truckload of mushroom substrate, courtesy of Aod Hang Over (in Thai).

harvesting, packaging, and delivery within a small window of time, usually no more than a few days. Often there is an absolute requirement for large-scale refrigeration and storage and use of engineered storage and transport containers, however, even under the best circumstances it is difficult to stretch the shelf life of these mushrooms beyond 3–4 days. Some medicinal mushrooms may require similar accommodations; however, others have different requirements entirely. Because medically active compounds are largely contained in the cell walls of fruiting structures and spores, a means of drying the mushrooms to around 6% moisture may replace the need for refrigeration. These driers and fans can be solar driven. Slicing the fruiting structures into smaller and thinner pieces and/or grinding or milling the pieces or accumulations of spores into fine powder may be required, but much of this can be done by hand with smaller equipment. For many producers, this could be the terminal step in production, and the final product is intended to be used in this form through the preparation of teas and tinctures by the consumer. The powders could also be placed into gelatin capsules and the small pieces of fruiting bodies made into ready-to-use tea bags.

Further postprocessing steps may include the preparation of hot-water extracts of the medically active compounds by steeping the pieces of fungal fruiting structures in hot water, essentially making large batches of tea. Similarly, tinctures or alcohol extracts may be produced by soaking the pieces in alcohol. Hot-water and alcohol extraction methods free different medically active compounds from the fungal tissue. Prepared tinctures could be bottled and made available directly from the cultivation facility. Another alternative would be to prepare the hot-water and alcohol extracts and then freeze dry them in large batches to produce extract powders similar to freeze-dried coffee or tea. These extract powders could be placed into gelatin capsules or left in bulk form for the production of mixtures of extracts from different fungi. Alternatively, the simple dried fruiting structures could be sent to a centralized postproduction facility that can produce the sliced pieces, powdered fruiting structures, and freeze-dried extract powder products. Obviously, preparing and shipping dried materials would require less expense and fewer logistics than storing and shipping refrigerated products.

6.4.6 Cost of Setting up a Low-Tech Cultivation Facility

Detailed estimates of costs for setting up a low-tech mushroom cultivation facility of the type that has been described in this chapter have been made. Oak and Spore (2020) estimated approximately $14,800 USD to set up a facility from scratch in New Zealand. GroCycle (2020b) estimated it a cost of $20,000 USD to set up a working low-tech mushroom cultivation facility in England. While this may sound like an acceptable investment for a small business in developed countries, it might be wildly out of the realm of possibility for most people in undeveloped countries (see the proposed model later in this chapter).

6.5 FRUITING UNITS AND CONTAINERS

The types of containers used for spawn production and fruiting have seen a great deal of innovation in the past decade as mushroom cultivation has moved from a relatively obscure hobby to the basis of small and large business, and then into the worldwide mainstream.

6.5.1 Gusseted Filter-Patch Bags

For many years, the standard inoculation and fruiting container has been the filter-patch mushroom grow bag (see the guide produced by Fresh Cap 2022). This is typically a gusseted, autoclavable polypropylene bag with a built-in breathable filter patch that allows fresh air exchange while keeping contamination out. Figure 1a shows a typical gusseted filter-patch bag. These bags have been (and for many continue to be) indispensable for producing substrate fruiting blocks and spawn. Size, wall thickness, filter rating, and overall cost are all important considerations when using these bags. A bag size of 20 × 13 × 46 cm has become almost standard for most applications because it will comfortably hold 5 lbs (2.5 kg) of fruiting or spawning substrate and is relatively easy to handle. Larger bags can be found that will hold up to 10 lbs (4.5 kg) of substrate. The size of the fruiting or spawn blocks is often a limiting factor due to space available in the pressure sterilizer or steam pasteurizer.

FIGURE 6.4 Gusseted filter-patch bags—a standard fruiting and spawn container. (a) Gusseted filter-patch spawn bags of rye seeds colonized by *Stropharia rugosoannulata*, courtesy of S. Miller. (b) Top fruiting of *Ganoderma* sp. in a gusseted bag, courtesy of Jacob Shanbrom. (c) Side fruiting of *Hericium erinaceus* from a gusseted bag, courtesy of Angele Poudrette Chambers. (d) Top fruiting of *Coprinus comatus* in a gusseted bag, courtesy of S. Miller. I Top and side fruiting of *Pholiota nameko* from a gusseted bag, courtesy of S. Miller.

Filter patches built into the sidewall of the bags can come with a pore size rating ranging from 0.2 to 0.5 μm. Production and quality control of the material for a 0.2 μm rating is expensive, and therefore bags with that filter-patch rating are more expensive. Patches with a 0.5 μm rating allow for greater gas exchange but also allow more contaminants to enter the bag.

Bag wall thickness can usually range from 2.2 to 4.0 mil. Thicker is typically better, with the thinner plastic often tearing during handling or rupturing under pressure sterilization or steam pasteurization. Thicker bags are typically more expensive.

It should also be noted that a modest investment in a small, specialized heat-sealer would allow anyone to make their own gusseted filter-patch mushroom bags (Du Toit-Henke 2022). The cost of these homemade bags might be substantially less, especially if the materials were purchased in bulk; however, the cost must still be within the budget.

Gusseted filter-patch bags do have one tremendous benefit, that being that most medicinal fungi can be cultivated with this culture unit. Several species fruit best from the sides of their woody substrate in nature (e.g., *Trametes*, *Grifola*, and many species of *Pleurotus*), so the three-dimensional aspect of the fruiting bags can provide this condition artificially. The bags can also be used for those medicinal fungi that are "top-fruiters" and only fruit from the upper surface of the substrate.

These bags are by and large single-use bags. This has both budgetary as well as environmental costs. For a commercial operation standpoint where the cost of production can be passed on to the consumer, the cost of each bag is largely inconsequential. However, for the low-tech cultivation facility that we are focusing on here, the cost of continually purchasing bags might be prohibitive. The single-use nature of these bags is also problematic because the plastic is produced from petroleum and must be disposed of in landfills. For the hobby farmer, this might not be that much of an issue; however, a large commercial fungal cultivation facility may use thousands of these bags each week.

6.5.2 Alternative Types of Bags

Several alternatives to the gusseted filter-patch bag are also in common use. Many cultivation facilities use tubular, thin-walled, clear plastic bags that come with one end sealed and lack filter patches, especially for the cultivation of *Pleurotus* spp. and *Ganoderma* spp., among others. Figure 6.5 a-d show single opening commercially available plastic bags that can be steam-pasteurized and used for fruiting many species of mushrooms. Similarly, clear or black plastic continuous plastic tubes come in many plastic types and diameters and can be sealed with an impulse sealer or tied at one end for filling. Depending on the plastic type and wall thickness, these can be filled with bulk hot-water-, steam-, or chemically pasteurized substrates, as the bags themselves are nearly sterile from the manufacturing process. Other types of plastic bags can be pressure sterilized. Air exchange can be achieved with a reusable plastic ring that can be plugged with newsprint and a rubber band or with an accessory cap (e.g., Rattanapron Plastics, Thailand www.facebook.com/people/Rattanapron-Plastics-Mushroom/100002155864921/) that can be steam pasteurized or pressure sterilized. Figure 6.6a-c show the use of reusable rings in single opening plastic bags. If the column approach is used, air exchange can be achieved simply by cutting small slits in the column after the fruiting substrate is completely colonized. While these bags are cheaper than the gusseted filter-patch bags described earlier, especially if purchased in large quantities, they are not free, and the cost of repeated purchase must still be considered. There also remains the environmental concern of single-use plastic and the use of petroleum in their manufacture. Environmental concerns may be lessened if the bags could be recycled, or if single-use compostable plastics could be developed for the fungal cultivation industry. Waste bags could be used as part of the fuel to heat the water during low-tech steam pasteurization. Burning the bags as fuel may slightly avoid the single-use issue; however, this simply exchanges one environmental concern for another.

6.5.3 Large Boxes, Tubs, and Trays

Probably the greatest innovations leading to decreased dependency on plastic bags have been derived from hobby mushroom cultivators wishing to avoid the continual purchase of increasingly expensive filter-patch bags and switching instead to using reusable fruiting containers. Many are still

FIGURE 6.5 Alternative types of bags used for fruiting. (a) Large, one-opening poly-bags for fruiting *Pleurotus* spp., courtesy of Ayush Raj. (b) Various colored and clear "garbage" bags for fruiting *Pleurotus* spp., courtesy of MuBindu. (c) Large, one-opening poly-bags for fruiting *Agaricus* spp., courtesy of Wirimai Desmond Chipwanya. (d) One-opening ring-bags for fruiting *Pleurotus* spp., courtesy of (in Thai 2300).

made of plastic, but the plastic is stronger and more rigid than that in the bags and can be washed and reused almost indefinitely. While some plastic formulations used to make larger boxes and tubs can withstand pressure sterilization, the size of these containers often prohibits their use in smaller pressure sterilizers or autoclaves. Consequently, their very nature has promoted the development of cultivation techniques using steam, hot-water, or chemical pasteurization of bulk substrates, or substrates such as pelleted seed hulls or some types of wood shavings that come essentially prepasteurized from their manufacturing process. Figure 6.7a and b show the use of "monotubs" where

FIGURE 6.6 Single-opening ring-bags. (a) Hanging ring-bags bags with fruiting *Pleurotus* spp., courtesy of Kern Nakin Mushroom Farm, Cambodia. (b) Two ring-bags with fruiting *Pleurotus djamor*, courtesy of Albert F. Estabillo in Mushroom Industry, Philippines. (c) Stacked ring-bags with fruiting *Hericium erinaceus*, courtesy of Juan Lopez, Lasdoó Embey, Oaxaca, Mexico.

the fruiting substrate, humidification and fresh air exchange are accomplished in a single modified but enclosed plastic tub. Figure 6.7 c-e are plastic trays that are in use to top-fruit *Psilocybe* sp, and *Cordyceps* sp.

The choice of large plastic boxes, tubs, or trays frequently limits their use to fungi that can be top-fruited (e.g., *Pleurotus eryngii*, *P. tuber-regium*, *Cordyceps militaris*, and *Agaricus subrufescens*) on pasteurized and only lightly supplemented substrates. Figure 6.7 c-e are plastic trays that

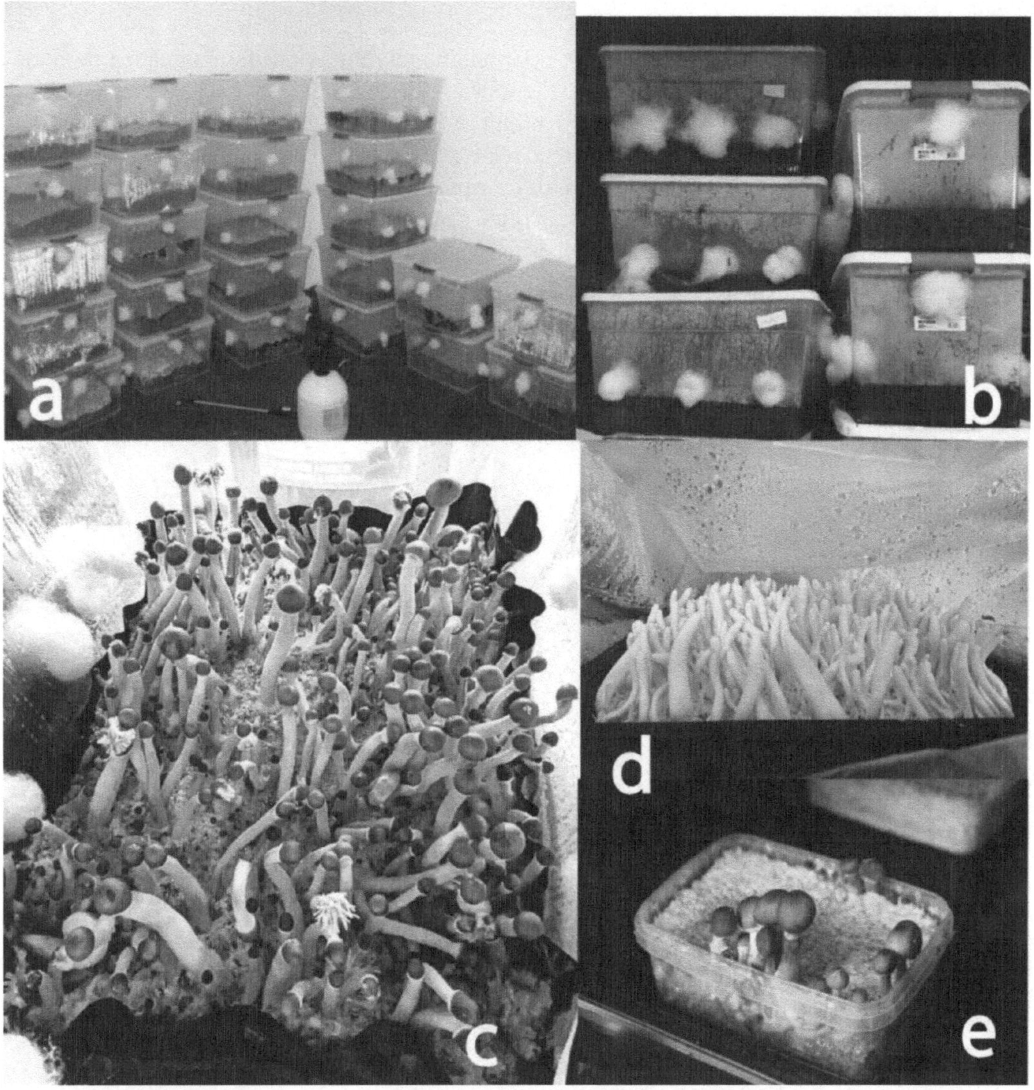

FIGURE 6.7 Monotubs and trays as fruiting containers. (a) Stacked monotubs showing substrate in the bottom, condensed moisture on the tub walls, and modifications for air exchange, courtesy of Shroomery.org. (b) Monotubs showing modifications for air exchange, courtesy of Matt Pinquoch. (c) Monotub with fruiting *Psilocybe* sp., courtesy of Rick Zuniga (d) Small tray with albino *Cordyceps militaris*, courtesy of Ryan Paul Gates. (e) Small tray with fruiting *Psilocybe cubensis*, courtesy of Shroomery.org.

are in use to top-fruit *Psilocybe* sp, and *Cordyceps* sp. Large plastic storage boxes or "totes" with lids, referred to in fungal cultivation circles as "monotubs," have become popular because a single container, once properly modified with holes for drainage, filters, and ports for the input of humidity and fresh air, can stand alone without the need for a larger fruiting chamber with environmental controls. The one possible disadvantage to large tubs is that they can become quite heavy when filled with hydrated substrate. Smaller tubs and trays can ameliorate this problem, and several can be organized in a single layer inside a larger tub or box.

6.5.4 Buckets

Buckets are similar to monotubs or trays but offer additional dimensions for cultivating both top- and side-fruiting species of medicinal fungi. Figure 6.8 a-c show stacks of buckets that have been modified with holes drilled into the sides for fruiting of *Pleurotus* app. Like the monotubs, buckets with lids can be large (e.g., 5 gal, 18.9 l), sourced relatively inexpensively, and reused almost indefinitely. This type of bucket is commonly used for transporting animal feed, large-scale food service ingredients, and construction materials, and can often be obtained at very low or even no cost under ideal conditions. Caution must be taken when using repurposed buckets if toxic or harmful materials were once stored in them.

The formulation of the plastic used in most buckets can withstand repeated *in situ* hot-water or chemical pasteurization, but usually not steam pasteurization or pressure sterilization. However,

FIGURE 6.8 Buckets as fruiting containers. (a) Stacked buckets with fruiting *Pleurotus* spp. in a fruiting chamber, courtesy of Brian Rathgaber (b) Stacked buckets with fruiting *Pleurotus* spp., courtesy of Robbie Pearce. (c) Stacked buckets with fruiting *Pleurotus* spp., courtesy of Ikeh Joy, Mushroom Cultivation Network Nigeria. Note the holes in the sides of the buckets for primordial formation and fruiting.

they can be filled with substrate that has been pasteurized elsewhere in bulk, similar to monotubs. Therefore, their use is especially good for low-tech approaches.

Like the monotubs, buckets must be modified by drilling or melting holes in the sides, top and bottom for the fungal fruiting bodies to exit the substrate. These holes are typically plugged with polyfill material or closed with tape while the fungi colonize the substrate, then opened for pinning and fruiting body development. The buckets have an added advantage in that they are engineered to be strong enough to be stacked one on top of another in a warehouse during storage and transport, and thus can be self-supporting without the need for shelving in the fruiting chamber.

Numerous other bucket-like containers could also be sourced for free or very inexpensively. Because they are not meant to be subjected to pressure sterilization, just about any container of at least 3–4 l volume could be used, as long as there was nothing toxic contained in them. These could include many types of plastic containers previously used in the food service industry (e.g., plastic coffee, milk, and oatmeal containers, inexpensive pots from nursery out-planting, and short lengths of discarded, large-diameter water pipes). Imagination and availability are the only limitations.

6.5.5 Baskets

Baskets with mesh or open holes in their design, constructed of plastic (e.g., laundry baskets) or woven from synthetic fiber, can be filled with prepasteurized fruiting substrate. Figure 6.9 shows the use of laundry baskets that are filled with substrate and are in use to fruit *Volvariella volvacea* (a) and *Pleurotus* spp. (b-c). Often the entire basket or group of baskets must be covered by a large plastic bag, sheet plastic or tarp during the incubation phase. For most fungi, this covering is totally or partially removed at the time of pinning to encourage the development of fruiting bodies. Baskets have the advantage of being inexpensive and reusable, and like buckets and bags, can promote the three-dimensional fruiting of many species of fungi.

6.5.6 Jars

Bottles composed of glass, hard plastic, or possibly even ceramic or glazed fired clay can be effectively used for, fruiting, spawning, and even liquid cultures. A mainstay of hobby mushroom cultivation is spawn production in Ball brand canning jars. Both wide-mouth and narrow mouth canning jars have their supporters. These canning jars are typically sealed with a plastic or metal screw-on lid that has been modified with a filter patch and an injection port. The filter patch allows for atmospheric equilibration of the jar during pressure sterilization and for gas exchange after inoculation, and the injection port allows for inoculation of the spawn substrate with a syringe containing a liquid culture of the fungus. These jars are made of a borosilicate glass that can withstand pressure sterilization, but these may eventually crack or break during the sterilization process. This can be ameliorated to some degree by putting the jars in a tub (if the size of the autoclave is large enough) with bottoms of the jars covered by 2–4 cm of water and allowing them to cool naturally in the sterilizer, rather than bringing the pressure down quickly. In smaller pressure cookers, the bottoms of the jars on the lowest layer should be immersed in water and the jars allowed to cool slowly.

Some jars used for fruiting are specialized and constructed of rigid plastic with a shape that has wider shoulders and a slightly narrower opening; these are typically used for top-fruiting species (e.g., enoki). In some large commercial operations, washing, filling, sterilization, inoculation, and transport of these jars around the facility are highly mechanized. While this type of mechanized processing is likely out of the possibility for low-tech cultivation facilities, used jars often appear on the secondhand market for relatively low cost.

FIGURE 6.9 Baskets as fruiting containers. (a) Laundry-type basket with fruiting *Volvariella volvacea*, courtesy of (in Thai 2233). (b) Laundry-type basket with fruiting *Pleurotus djamor*, courtesy of Chelsea Dub. (c) Laundry-type basket with fruiting *Pleurotus* spp., courtesy of Daniel Griffin.

6.5.7 Bottles

We have witnessed the repurposing of random bottles from the food service and spirits industries for the production of spawn. These bottles (e.g., ketchup, wine, and soft drink bottles) can be steam sterilized or even pressure sterilized with the grain-spawn substrate inside and plugged by polyfill or paper. I applaud the repurposing of these bottles, which can be collected cheaply from restaurants and bars. While the narrow opening of the bottles can be advantageous to reduce contamination and provide a grip when shaking the spawn, the downside is that the openings often make it difficult to get the prepared fully colonized spawn out during inoculation, and additional implements must be used to pull the spawn out. This can be somewhat ameliorated by using millet or some other small grain as spawning substrate, following a strict timeline during production, and by not allowing the mycelium to become completely, solidly overgrown within the bottle.

6.5.8 No-Container Cultivation

Cultivation of some fungi requires no specialized containers at all. The best-known example would be traditional outdoor cultivation of shiitake on especially oak logs ~0 cm diam and 1.2 m long. Figure 6.10 a-b shows the use of log cultivation for growing Shiitake. Holes are drilled in the logs and filled with sawdust spawn or colonized wooden dowls, and the inoculum is covered with wax. This process typically requires longer for the fungus to colonize the substrate and fruiting to commence (e.g., 1–2 years), but these logs can frequently produce mushrooms for 4–7 years with proper management. A number of farms have also used the log inoculation technique to grow other species of medicinal fungi (e.g., *Pleurotus* spp. and *Ganoderma* spp.), and many species could be grown with this approach. Figure 6.10 c shows partially buried logs used in the cultivation of *Pholiota nameko*. Figure 6.10 d shows stacked logs inoculated with *Hericium erinaceus*. Arguably, the simplest way to cultivate many medicinal mushrooms would the log technique, assuming that logs are abundant and available.

Some modification of the traditional shiitake log cultivation process has been used for *Ganoderma* spp. and other fungi. Rather than using relatively long logs, cultivators cut hardwood logs into uniform lengths of approximately 60 cm. The logs are then plug inoculated, as in the shiitake process, but then are buried in soil and kept moist until fruiting commences. Another modified process uses short logs and variously sized branches that are packed into bags in uniformly shaped boxes, steam pasteurized, inoculated with spores, and closed up until the logs are fully colonized (Noal Farm 2021). The colonized block of logs and branches is then buried in soil and allowed to fruit. Use of any of these approaches obviously requires that appropriate logs (they do not have to be oak but should ideally be hardwood of some type) be readily available at a relatively low cost, which may be a problem in certain deforested areas. However, other loose substrates can be packed into long, thin plastic bags forming a cultivation unit known as a "stick."

Another container-free method that often gives good results is the outdoor mushroom bed. Figure 6.11 a-f shows the use of outdoor mushroom beds for the fruiting of various fungi. Essentially, a growing area is delimited in some way, and layers of substrate material such as unpasteurized wood chips or straw are layered with colonized spawn and hydrated well. The substrate in this case is ideally sawdust to decrease contamination and rodent activity. The bed can be covered initially during the colonization of the substrate, but then it is opened to the environment and kept moist. Like log cultivation, there is often a lag time until fruiting begins, but afterwards, the beds will produce continually for several years with only periodic renewal and addition of new substrate. Modifications of this technique have been used for intercropping fungi between rows of vegetables. Figure 6.11 a, e and f show mushrooms being intercropped with vegetables.

Similar to mushroom beds in many respects is the cultivation of fungi on other substrates, (e.g., rows and mounds of rice straw or directly on bales of wheat and barley straw). Raised beds with wire mesh sides can also be used to encourage fruiting in all three dimensions of the bed.

6.6 SUBSTRATES FOR FRUITING

As mentioned earlier in this chapter, many but not all medicinal mushrooms specialize in wood decomposition in nature. Their preferred substrates are composed of lignin, cellulose, and hemicellulose. Some fungi specialize in growing on hardwood (i.e., from angiosperms) while a much

FIGURE 6.10 Using logs for no-container fruiting of fungi. (a) Newly inoculated logs for *Lentinellus edodes* fruiting—note holes drilled, filled with inoculum and covered with wax, courtesy of Jared Hill. (b) Logs with *Lentinellus edodes* fruiting, courtesy of Quentin Kresser. (c) Partially buried logs with fruiting *Pholiota nameko*, courtesy of Robert Boice. (d) Logs with fruiting *Hericium erinaceus*, courtesy of Larry Poole.

smaller group of mushrooms grow on softwood (i.e., from conifers). A relatively small percentage of basidiomycete fungi can grow on both, while others are found on hardwoods in nature but can be made to grow on softwood that has been amended with additional substrate materials. Table 6.2 is a partial listing of fruiting substrates that are or could be used to cultivate medicinal mushrooms—many more potential substrates are available in specific regions.

FIGURE 6.11 Various substrates for fruiting. (a) Spent waste from palm oil extraction for fruiting of *Volvaria volvacea*, courtesy of (in Thai 2406). (b) Corn cob waste for mushroom fruiting, courtesy of Cece Glow, Indonesia. (c) Banana waste intended for the fruiting of *Volvaria volvacea*, courtesy of Queency Laranio Apas, Philippines. (d) Waste cardboard colonized by *Pleurotus colombinus* mycelia, Cody Evans.

TABLE 6.2
A Partial Listing of Plant-Based Materials that Are or Could Be Used to Cultivate Medicinal Mushrooms.

Bamboo
Banana leaves from banana farming
Coal
Coco coir from coconut processing
Corncobs and corn stalks
Hardwood logs
Kiln-dried, pelleted substrates such as hardwood, straw, seed hull pellets
Leaves from urban landscaping
Peanut, lentil, soybean, and other legume processing by-products
Recycled bulk xerox and newsprint paper from recycling and commercial shredding efforts
Recycled cardboard
Sawdust from furniture and wood product manufacturing
Spent coffee and tea grounds from urban coffee and tea shops
Spent grains from brewing
Straw from grain harvesting—wheat, barley, rye, millet, sorghum
Sugarcane bagasse
Sunflower, safflower, mustard seed hulls
Waste from cocoa processing
Waste from coffee processing
Waste from cotton harvesting and manufacturing—cotton seed hulls, plant debris
Waste from cassava harvesting
Waste from fruit and vegetable canning—rambutan, lychee, lotus, bamboo shoots
Waste from hemp and cannabis processing
Waste from nut processing—walnuts, pecans, filberts, pistachio, cashews, almonds
Waste from palm oil manufacturing
Waste from leaf-wax processing

Note: this is by no means an exhaustive list. Many more substrates are in use and still many other untested but potential substrates are available in specific regions. These materials could be used alone or combined to achieve proper nutrition and texture.

The particle size of substrates should neither be too small or too large. Substrate texture is critical if using non-sterile approaches. An overly small particle size can reduce or eliminate airspace for gas exchange, resulting in anaerobic growth of certain bacteria. This can sometimes be ameliorated if complete pressure sterilization of the fruiting substrate is possible but can be more problematic if only pasteurization is possible. Overly large substrate particles often lead to slow colonization, increased production times, and possible contamination issues. An ideal texture is one with the ability to hold moisture at field capacity of 50–70%, depending on the substrate and the fungi to be cultivated, but containing a variety of pore spaces between individual elements of the substrate. A critical characteristic of all potential substrates is the absence of competitor organisms (e.g., straw that has been baled wet in the field or which has gotten moldy in storage should not be used). For one thing, the presence of competing organisms present in the substrate before preparation can result in their usurping of critical nutrients intended for the fungi under cultivation. It can also result in production of millions of propagules that can become activated during processing of the substrate and that can rapidly colonize the substrate before the intended fungus can become established. Some fungi and bacteria actually produce allelopathic compounds used in nature to retard the mycelium growth of the intended cultivated mushrooms.

FIGURE 6.12 Outdoor mushroom beds. (a) Polyculture raised bed with *Hypsizygus ulmarius* mushrooms and eggplant, courtesy of Everything Mushrooms. (b) Raised bed with *Pleurotus* sp., courtesy of www.helseini.fi (c) Mushroom bed with fruiting *Stropharia rugosoannulata*, courtesy of Jean-François Savoie. (d) Raised bed with fruiting *Hericium erinaceus*, courtesy of www.helseini.fi (e) Mushroom polyculture bed with *Stropharia rugosoannulata* fruiting among vegetables, courtesy Rafter Sass Ferguson. (f) Polyculture raised bed with *Pleurotus* sp. mushrooms and green beans, courtesy of Popsagone.

6.6.1 Wood (Logs, Chips, and Sawdust)

Wood is primarily composed of lignin, cellulose, and hemicellulose and has a C/N ratio ranging from 220–550 in most hardwoods and 490–1250 in most softwoods (Fengel and Wegener 1989). Some substrates contain the 1–2% nitrogen necessary for growth naturally, while others such as certain kinds of wood, straw, and agricultural wastes (e.g., nut shells) must be supplemented with nitrogen for the production of fruiting bodies. However, substrates with naturally high levels of nitrogen or over-supplemented, nitrogen-poor substrates with too much nitrogen can lead to high levels of growth of competing microbes (bacteria and other fungi) as contaminants.

Most substrates will contain sufficient amounts of macro- and microelements such as K, P, Mg, Ca, and S (Stamets 2000), but this will vary from substrate to substrate, region to region, and even season to season. A pH generally around 5–6.5 is good for most mushrooms, and this pH is commonly arrived at in single substrates or supplemented substrates, so pH is rarely monitored unless problems are encountered when testing a new substrate material.

6.6.2 Other Substrates

Particularly germane to the perspective of this chapter is that waste biomass from a large variety of agricultural crops, often with little or no value itself, can be used in the cultivation of medicinal mushrooms. Figure 6.12 a-d shows waste from palm oil, corn, and banana processing, and waste cardboard that are being used as substrates for mushroom production. Accumulations of many of these waste products can, in fact, be a problem for farmers and many are burned *in situ* as part of the regular yearly cycle. Therefore, sources of waste biomass for mushroom cultivation in some places can be found almost free for the taking. It is critical to source substrates locally and to locate cultivation farms nearby because transportation costs to move quantities of these substrates even modest distances can be prohibitive.

Most medicinal fungi can be cultivated on nearly any substrate that can provide nutrition naturally or with supplementation in sufficient quantities and densities and under appropriate hydration and environmental controls. In addition, there is considerable genetic variation in different strains of the same fungus, so it is likely that a strain can be found or recovered that will be able to use just about any woody substrate or substrate mix available in a given region.

6.7 SUBSTRATES FOR SPAWNING

Spawn is a plant-based material loaded with fungal mycelia and used to inoculate the fruiting substrate. In many instances, the spawn stage is intermediate between mycelium growing on agar or in liquid culture and the fungi growing in the fruiting substrate. As described previously, many medicinal mushrooms are wood decomposers in nature. That means they are capable of producing the enzymes necessary to utilize wood for their carbon and energy requirements. These enzymes are not constitutive, meaning that the fungus produces them only when confronted with the lignin, cellulose, and hemicellulose found in wood and other complex materials. Nutrient agar or liquid culture broth generally contain easily assimilable sugars, and when growing on these sugars, fungi use only constitutive enzymes and do not produce enzyme systems for the complex materials found in wood. Therefore, the spawn, being an intermediate in processing steps between agar and liquid culture and the wood or other complex material found in the fruiting substrate, provides enough easily assimilable carbohydrates for the fungus to start growing, but also enough lignin, cellulose, and hemicellulose to initiate enzyme production for transition to growth on complex substrates. Many types of spawn substrate can fulfill these nutritional requirements. In addition, spawn substrates ideally should comprise numerous individual units that, when fully colonized by the fungus, can be broken off and provide numerous points of inoculum in the fruiting substrate. Table 6.3 is a partial listing of commonly used plant-based materials for spawn production.

TABLE 6.3
A Partial Listing of Plant-Based Materials that Are or Could Be Used to Create Spawn for Inoculating Medicinal Mushrooms and Other Fungi on Fruiting Substrate.

Grain
Rye
Wheat
Oats
Barley
Brown rice
Millet
Popcorn
Sorghum
Non-grain
Sawdust
Coconut coir (alone or + vermiculite)
Wild birdseed mixes
Hardwood dowls (for plug spawning of logs)

6.7.1 Grain

Many grains provide the necessary nutrition to transition the actively growing fungus from agar or liquid culture to the complex materials found in the fruiting substrate and comprised individual units. Rye seeds (aka "berries") have been the standard but are increasingly difficult to find in grocery stores. However, rye and other grains to be used for spawn production can be ordered from companies that provide bulk ingredients for the food service or animal feed industries. Large filter-patch bags (40 lb, 18 kg) of millet have become the standard for many commercial spawn producers.

6.7.2 Other Substrates

Other substrates with the characteristics described earlier can be used to create spawn. Small wood chips, sawdust, and coco coir—alone or sometimes mixed with vermiculite to allow it to become friable at the time of use—are examples. Occasionally, these alternative spawning substrates are superior to standard grain because some spawn substrates are contraindicated for certain applications. A good example would be that grain spawn should not be used when inoculating outdoor mushroom beds because the seeds are a large draw for insects and rodents. Sawdust or coco coir spawn with or without vermiculite is preferred for these applications, and spent fruiting units can often be repurposed into spawn for outdoor mushroom beds.

6.8 MODELS FOR DEVELOPING A NETWORK OF MEDICINAL MUSHROOM CULTIVATION FACILITIES

The final goal of this chapter is to explore models that could work to help develop a network of medicinal mushroom cultivation facilities so that the products can be directed to those most in need at very low or no cost. There are four phases to this model: (1) identification of needs by country and region, (2) securing of funding, (3) education, and (4) implementation. The old adage "give a person a fish and he eats for a day, but teach a person to fish and he can eat for a lifetime" is wholly appropriate here. For far less money than it would take to provide full medical coverage for even a small percentage of people in need, the argument can be made that teaching people how to cultivate medicinal mushrooms and providing the basic setup costs that will enable them to do so will result in effective treatments for common medicinal problems for a large number of people.

6.8.1 Identification of Needs

The scale of identifying specific needs for medical treatment in any given area is beyond any individual, so governments, organizations sponsored by governments, and large-scale philanthropic agencies must be involved. The WHO is charged with working with countries to improve health issues, and several recent initiatives by the WHO would dovetail nicely with developing a network of medicinal mushroom cultivation farms. In 2018, the WHO released a standard series for developing traditional medicines, including medicinal mushrooms, across the world (WHO 2014-2023). The World Health Organisation (WHO) has also recently developed the Universal Health Coverage (UHC) compendium, which is a tool to help countries identify the essential health services they need and find ways to treat diseases (WHO 2022). It will be critical for the WHO to recognize potential that medicinal mushrooms offer towards the amelioration of health problems for much of the world.

6.8.2 Sources of Funding

While many mushroom farms exist with the laudable goal of making money for poverty-stricken people and providing a boost to the local or regional economy, most are aimed at the commercial culinary market and not at providing medicinal mushrooms at low or no cost. The authors of this chapter are not against capitalism and there is certainly more than a little crossover between culinary and medicinal species of fungi (e.g., shiitake, *Agaricus subrufescens* enoki, maitake, *Pleurotus* spp., etc.), and gourmet mushrooms can indeed provide critical nutrition, a lack of which, along with poverty, is also often a problem underlying health issues in many areas (GroCycle 2020a). However, if the resulting product is too expensive to be easily afforded by those in need, then it does nothing to further the goal of getting medicinal mushrooms or their derived products into the hands of people who need them but cannot afford to buy them.

Few people would likely disagree with the premise that all people should have equal access to medical treatment, but actually committing to pay for this treatment is another issue entirely. Lending platforms such as Kiva, Lend with Care, and Zidisha have been used successfully (GroCycle 2020a, c) to provide a means of allowing individuals and groups to begin growing mushrooms for market and could provide microloans directed at establishing medicinal mushroom farms. However, even though these platforms provide benefits to those who can make money from selling mushrooms to pay back the loans, for obvious reasons they may not be suitable for the purpose stated here. Funds could also be crowd-sourced. However, major funding for this model must likely come from governments, agencies, NGOs, and major philanthropic organizations, and we hope that this message will be received and that they will become enthusiastic supporters of this project.

Another source of funding could come from bioprospecting in countries and regions where a natural drug products are found; such products may become commercialized, often making a great deal of money. Many examples of successful bioprospecting are available; however there are also many famous stories of biopiracy (Newman and Cragg 2016; Neimark 2017). This route of funding must include stricter enforcement of biopiracy laws (Reid 2009).

A simple funding and business model similar to that developed by D. Heath and R. Goldberg for Bombas—"For every pair of socks that you buy, we donate a pair to somebody who's at risk, in need, or experiencing homelessness" (Perry 2019)—could also work admirably. Similar altruistic business models include TOMS shoes and Warby Parker eyeglasses (Kinne 2020). Mushroom farms producing gourmet mushrooms could sell their products to whoever can afford to buy them, but then include the cultivation of medicinal mushrooms, a portion of which would go to those in greatest need. Similarly, farmers specializing in the cultivation of medicinal mushrooms could sell a portion of their products but provide products at no cost to those who cannot afford to purchase them. Approaching medicinal mushroom cultivation with this model could attract willing venture capitalists and business entrepreneurs, provide a means of boosting local economies, and lead to sustainable and expandable cultivation of medicinal mushrooms for the production of safe and effective medicines. The initial funding, from whatever source, should be used to generate momentum towards the cultivation of medicinal mushrooms, and any model developed should work towards sustaining this momentum.

6.8.3 Education

The need for education on a number of fronts cannot be understated. The first front may, in fact, be the hardest—getting people who have never consumed mushrooms for food or used them medicinally to buy in to the concept. It has taken a concerted effort to create markets for culinary mushrooms in certain areas of the world (GroCycle, 2020a,b,c; Kigali Farms and Mbeng Adio Mushroom Farm, 2022). Similarly, while it is true that medicinal mushrooms have been part of traditional medicines in some areas for centuries, not all societies have a cultural memory of such uses of medicinal mushrooms. A similar concern could be that some cultures have indeed depended on local mushrooms for certain ailments, but they have no knowledge of other medicinal fungi native to other parts of the world that could provide better treatment for specific ailments. Thus, part of the network development must include individuals capable of convincing people that medicinal mushrooms are safe and effective for certain health issues afflicting their populations. This may best be done by training local people in the use of mushrooms for specific medicinal needs and allowing them to develop inroads of acceptance into each community or region.

The second front for education concerns teaching people how to cultivate medicinal mushrooms. There are thousands of knowledgeable people who have developed successful culinary mushroom farms, even under adverse conditions. While specifics and steps required to cultivate medicinal mushrooms may differ a little or a lot from those used for growing culinary mushrooms, there is substantial commonality between the two, especially in getting started with the basics of growing fungi. Thus, an invaluable educational resource will come from individuals in the cultivation networks who already have the knowledge and experience with growing mushrooms. However, it is also likely that more people will be necessary to provide guidance, specifically on cultivating medicinal mushrooms.

A third front for education that will likely be necessary is the instruction of individuals in the process of creating spawn. As discussed in Section 6.8.4, spawn availability is among the top critical factors for success of this project.

6.8.4 Focus on Spawn—Implementation and Making It Work

One factor hindering expansion of mushroom growers in underdeveloped areas is the availability of spawn (GroCycle 2020a, c). While it would undoubtedly be beneficial for each grower to be able to produce their own spawn for use in their facility, that idea may not be tenable, at least in the short term. For one thing, many farms in the current network of culinary mushroom growers are at the very margin of being able to cultivate mushrooms at all. They have no lab facilities or other resources; therefore, their operation depends on obtaining spawn from laboratories that specialize in spawn production. That in itself is not a bad thing. If spawn is readily available through some reliable channel, each farm need not spend money developing a laboratory to produce their own spawn or to maintain their culture collection. However, as described by GroCycle (2020a), many spawn production facilities are commercial operations or are associated with universities, and the spawn they produce is for sale. That fact is generally not compatible with the goal of providing medicinal mushroom products at low or no cost. For one thing, there may be no one to pass the cost of spawn on to. In addition, even if the medical mushroom cultivation farm could absorb the cost of purchasing spawn, there might not be a spawn producing facility in the region or even in the country.

A substantial portion of initial funding should go towards developing labs for those who may be capable of producing their own spawn and setting up regional spawn production facilities to serve farmers who cannot afford to set up or maintain their own labs. Receiving funds for setting up these spawn production facilities would be contingent upon agreement to provide spawn of medicinal fungi to growers in the region for free while at the same time selling spawn to culinary mushroom producers who can afford to purchase it.

6.8.5 Sustainable Cultivation

The crux of the models suggested here must be on how to sustain the production of medicinal mushrooms beyond initial funding. Many lessons can be learned from the New Theory Farming practices proposed by His Majesty King Bhumiphol Adulyadej for Thailand (Borgen Project 2018). One tenet of New Theory Farming is an integrated farming system where two or more different agriculture activities are practiced in the same field. This lowers costs for farmers and utilizes limited resources to their maximum potential. Following this, farmers with a wealth of unwanted agricultural waste could also develop a medicinal mushroom cultivation facility near their fields. Similarly, farmers already producing culinary mushrooms could branch out to include the cultivation of medicinal mushrooms. Farmers wishing to start mushroom farms could receive funding with the stipulation that they produce significant crops of medicinal mushrooms that will be distributed to people who have no access to healthcare.

Another tenet of New Theory Farming supports agroforestry. This approach enables cash crops to coexist with trees and forestry land. As such, it amplifies the ability to utilize land for sustainable agriculture while also promoting reforestation. In our case, sustainable forests could also provide an excellent source of plant biomass for the cultivation of medicinal mushrooms.

The concept of New Theory Farming suggests dividing available land into four parts and utilizes each for various needs. In Thailand, for example, it was suggested that the four parts include fish ponds/aquaculture, fields for crops and vegetables, a rice paddy for self-consumption, and space for living and livestock raising. This concept could easily be adapted to other regions and areas by including culinary and medicinal mushroom cultivation and forests as one of the four parts of sustainable farming.

Finally, deeply ingrained in the concept of New Theory Farming is sharing. Every area is different, with some having the ability to grow certain crops, while others might be able to produce those crops but lack the water to make a pond and grow fish. We envision the mushroom cultivation network as a community, area, or regional co-op where some can provide plant biomass as substrate for medicinal mushroom cultivation, others can grow spawn, and others still can be part of the post-processing and distribution system. Kigali Farms in Rwanda and Mbeng Adio Mushroom Farm in Cameroon (see websites in the references) are two community mushroom projects that have been very successful and embody the approach we support.

REFERENCES

Adhikari, P.P. & S.B. Paul. 2018. History of Indian traditional medicine: A medical inheritance. *Asian Journal of Pharmaceutical and Clinical Research* 11(1): 421–426.

Anke, T. 1989. Basidiomycetes: A source for new bioactive secondary metabolites. *Progress in Industrial Microbiology* 27: 51–66.

Bensky, D., S. Clavey, E. Stoger & L.L. Bensky. 2015. *Chinese Herbal Medicine: Materia Medica* (Portable 3rd ed.). Seattle, WA: Eastland Press, 1325 pp.

Borgen Project. 2018. *Five Key Factors Improving Sustainable Agriculture in Thailand*, Jan 16, 2018. https://borgenproject.org/sustainable-agriculture-in-thailand/. Accessed Sep 9, 2022.

Bray, R. 2020. *Healing Mushrooms—A Practical Guide to Medicinal Mushrooms (Urban Homesteading).* Monkey Publishing, Hamburg, Germany, 228 pp.

Bray, R. 2021. *Mushroom Cultivation: Become the MacGyver of Mushrooms—Easy Step-by-Step Instructions to Grow Any Mushroom at Home (Urban Homesteading Book 4).* Monkey Publishing, Hamburg, Germany, 149 pp.

Chang, S.T. & S.P. Wasser. 2012. The role of culinary-medicinal mushrooms on human welfare with a pyramid model for human health. *International Journal of Medicinal Mushrooms* 1: 95–134.

Chen, Y., S.B. Zhu, M.Y. Xie, S.P. Nie, W. Liu, C. Li, et al. 2008. Quality control and original discrimination of *Ganoderma lucidum* based on high-performance liquid chromatographic fingerprints and combined chemometrics methods. *Analytica Chimica Acta* 623: 146–156. http://doi.org/10.1016/j.aca.2008.06.018

Dai, Y.C., Z.L. Yang, B.K. Ui, C.H. Yu, & L.W. Zhou. 2009. Species diversity and utilization of medicinal mushrooms and fungi in China (review). *International Journal of Medicinal Mushrooms* 11: 287–302.

De Silva, D.D., S. Rapior, E. Sudarman, M. Stadler, J. Xu, S.A. Alias, et al. 2013. Bioactive metabolites from macrofungi: Ethnopharmacology, biological activities and chemistry. *Fungal Divers* 62: 1–40.

Didukh, M.Y., S.P. Wasser & E. Nevo. 2003. Medicinal value of species of the family Agaricaceae Cohn (higher Basidiomycetes) current stage of knowledge and future perspectives. *International Journal of Medicinal Mushrooms* 5: 133–152.

Du Toit-Henke, J. 2022. *For Those Who Asked about How We Make Our Own Substrate Bags.* www.facebook.com/jennifer.d.toit/videos/616807423376898/?d=n. Accessed Aug 19, 2022.

Economist. 2017. *Why Chinas' Traditional Medicine Boom is Dangerous*, Sep 1, 2017. www.economist.com/china/2017/09/01/why-chinas-traditional-medicine-boom-is-dangerous. Accessed Aug 23, 2022.

Fengel, D. & G. Wegener. 1989. *Wood: Chemistry, Ultrastructure, Reactions.* W. de Gruyter, Berlin, 613 pp.

Fresh Cap. 2022. *Mushroom Grow Bags: The Ultimate Guide.* https://learn.freshcap.com/growing/mushroom-grow-bags-the-ultimate-guide/#:~:text=Filter%20patch%20grow%20bags%20are%20an%20indispensible%20tool,to%20breath%20while%20it%20grows%20throughout%20the%20sawdust. Accessed Aug 16, 2022.

GroCycle. 2020a. *How Mushrooms Can Reduce Poverty & Malnutrition (Inc. Examples)*, May 25, 2020. www.youtube.com/watch?v=4p4hbW3r3MM. Accessed Aug 19, 2022.

GroCycle. 2020b. *How Much Does a Low Tech Mushroom Farm Cost*? Apr 20, 2020. www.youtube.com/watch?v=KoR4QVTyE1w. Accessed Aug 19, 2022.

GroCycle. 2020c. *Economics of Mushroom Production*, Jul 21, 2020. www.youtube.com/watch?v=18oNt5j5z3s. Accessed Aug 19, 2022.

Guo, L.-X., X.-M. Xu, C.F. Wu, L. Lin, S.C. Zou, T.G. Luan, et al. 2012. Fatty acid composition of lipids in wild *Cordyceps sinensis* from major habitats in China. *Biomedicine & Preventive Nutrition* 2: 42–50.

Health Benefits Times. 2022. *Top 10 Most Common Health Issues Worldwide.* www.healthbenefitstimes.com/top-10-most-common-health-issues-worldwide/. Accessed Aug 14, 2022.

Heferle, G. 2022. *Growing Gourmet Mushrooms for Market—A Complete Guide to Making a Living Growing Gourmet Mushrooms at Home.* Ebook from Fresh from the Farm, Denver, CO. ETSY 1189456792, listed on 22 June, 2022.

Jackson, D.J. 1989. Searching for medicinal wealth in Amazonia. *Smithsonian* 19(11): 94–99.

Jing, P., S.J. Zhao, M.M. Lu, A. Cai, J. Pang & L.H. Song. 2014. Multiple-fingerprint analysis for investigating quality control of *Flammulina velutipes* fruiting body polysaccharides. *Journal of Agricultural and Food Chemistry* 62: 12128–12133.

Johansen, A. 2019. Common diseases around the world. *Foreign Policy News*, Nov 22, 2019. Accessed Aug 14, 2022. https://foreignpolicynews.org/2019/11/22/common-diseases-around-the-world/

Kamada, T., H. Sano, T. Nakazawa & K. Nakahori. 2010. Regulation of fruiting body photomorphogenesis in *Coprinopsis cinerea. Fungal Genetics and Biology* 47(11): 917–921.

Kinne, A. 2020. The bombas story: Changing the world, one sock at a time. *Workhuman Blog*, June 30, 2020. www.workhuman.com/resources/globoforce-blog/the-bombas-story-changing-the-world-one-sock-at-a-time. Accessed Sep 2, 2022.

Li, Y., J. Zhang, T. Li, H. Liu & Y. Wang. 2016. A comprehensive and comparative study of *Wolfiporia extensa* cultivation regions by fourier transform infrared spectroscopy and ultra-fast liquid chromatography. *PLoS ONE* 11(12): e0168998. http://doi.org/10.1371/journal.pone.0168998

Lo, H.-C. & S.P. Wasser. 2011. Medicinal mushrooms for glycemic control in diabetes mellitus: History, current status, future perspectives, and unsolved problems (review). *International Journal of Medicinal Mushrooms* 13: 401–426.

McCoy, P. 2016. *Radical Mycology: A Treatise on Seeing & Working with Fungi*. Chthaeus Press, Portland, OR, 672 pp.

Mongabay. 2012. *Medicinal Plants*, July 22, 2012. https://rainforests.mongabay.com/1007.htm. Accessed July 22, 2022.

Mushroom Growers Handbook 1. 2016a. *Oyster Mushroom Cultivation*. https://archive.org/details/MushroomGrowersHandbook1OysterCultivation. Accessed Sep 22, 2022.

Mushroom Growers Handbook 2. 2016b. *Shiitake Cultivation*. https://archive.org/details/MushroomGrowersHandbook2ShiitakeCultivation/mode/2up. Accessed Sep 22, 2022.

Nanje Gowda, N.A. & D. Manvi. 2019. *Agro-residues Disinfection Methods for Mushroom Cultivation*. Sep 2019. www.researchgate.net/publication/344187502_Agro-residues_Disinfection_Methods_for_Mushroom_Cultivation. Accessed Aug 16, 2022.

Neimark, B.D. 2017. Bioprospecting and biopiracy. In: *The International Encyclopedia of Geography*, D. Richardson, N. Castree, M.M. Goodchild, A. Kobayashi, W. Liu, and R.A Marston, eds. New York, NY: John Wiley & Sons, pp. 1–7.

Newman, D.J. & G.M. Cragg. 2016. Natural products as sources of new drugs from 1981 to 2014. *Journal of Natural Products* 79(3): 629–661. http://doi.org/10.1021/acs.jnatprod.5b01055

Noal Farm. 2021. *How it's Grow Reishi—Red Reishi Mushroom Farm—Reishi Mushroom Harvest and Processing*, Jan 31, 2021. www.youtube.com/watch?v=j5RzRnsdJUM. Accessed Aug 29, 2022.

Oak and Spore. 2020. *Small Scale Mushroom Farm Setup Costs*, July 21, 2020. www.youtube.com/watch?v=AcBpVAOVeGo. Accessed Aug 19, 2022.

Perry, C. 2019. Socks are donated to homeless by Bombas for every pair it sells—Bombas wants to help decrease homeless shelter demand for socks. *Fox Business*, Dec 6, 2019. www.foxbusiness.com/money/socks-donation-homeless-bombas-randy-goldberg. Accessed Sep 2, 2022.

Phan, C.W., P. David, M. Naidu, K.H. Wong & V. Sabaratnam. 2015. Therapeutic potential of culinary-medicinal mushrooms for the management of neurodegenerative diseases: Diversity, metabolite, and mechanism. *Critical Reviews in Biotechnology* 35: 355–368. http://doi.org/10.3109/07388551.2014.887649

Pietka, J. & A. Szczepkowski. 2011. Assessment of the present of occurrence, threats and protection of *Fomitopsis officinalis* (Vill.: Fr.) Bond. & Sing in Poland. *Sylwan* 155(1): 31–45.

Press, B. 2021. *Mushroom Cultivation: A Practical Guide to Growing Mushrooms at Home*. Brandon, SD: Barton Press, 174 pp.

Reid, J. 2009. Biopiracy: The struggle for traditional knowledge rights. *American Indian Law Review* 34(1). https://digitalcommons.law.ou.edu/ailr/vol34/iss1/2. Accessed Aug 23, 2022.

Reshetnikov, S.V., S.P. Wasser & K.K. Tan. 2001. Higher Basidiomycota as source of antitumor and immunostimulating polysaccharides. *International Journal of Medicinal Mushrooms* 3: 361–394.

Rogers, R.D. 2020. *Medicinal Mushrooms—The Human Clinical Trials*. Alberta, Canada: Prairie Deva Press, 156 pp.

Rowsan, K. 2021a. *Mushroom Cultivation for Beginners: 21 Step-by-Step Methods to Master Your Mushroom Growing Skills in as Little as 4 Weeks*. London, England: Kris Rowsan, 190 pp.

Rowsan, K. 2021b. *The Medicinal Mushroom Book: A Practical Guide to Medicinal Mushrooms and Their Healing Powers*. London, England: Kris Rowsan, 154 pp.

Rowsan, K. 2022. *The Medicinal Mushroom Book—A Practical Guide to the Most Powerful Medicinal Mushrooms*. London, England: Kris Rowsan, 206 pp.

Stamets, P. 2000. *Growing Gourmet and Medicinal Mushrooms*, 3rd ed. Ten Speed Press, Berkeley, CA, 592 pp.

Syamili, S. 2020. A study of traditional medicines in India. *Scholars Impact* 7(4): 435–440.

Ulloa, S. 2012. Traditional Chinese medicine sees global resurgence. *GlobalSurance*, Sep 13, 2012. www.globalsurance.com/news/2012/09/13/traditional-chinese-medicine-sees-global-resurgenc/. Accessed Aug 23, 2022.

Wallin, C. 2020. *Growing Gourmet Mushrooms for Profit*. Anacortes, WA: Headstart Publishing, 120 pp.

Wang, X.-L. & Y.-J. Yao. 2011. Host insect species of *Ophiocordyceps sinensis*: A review. *ZooKeys* 127: 43–59.

Wasser, S.P. 2002. Medicinal mushrooms as a source of antitumor and immunomodulating polysaccharides. *Applied Microbiology and Biotechnology* 60(3): 258–274. http://doi.org/10.1007/s00253-002-1076-7. Epub Sep 10, 2002.

Wasser, S.P. 2010. Medicinal mushroom science: History, current status, future trends, and unsolved problems. *International Journal of Medicinal Mushrooms* 12: 1–16.

Wasser, S.P. 2011. Current findings, future trends, and unsolved problems in studies of medicinal mushrooms. *Applied Microbiology and Biotechnology* 89(5): 1323–1332. http://doi.org/10.1007/s00253-010-3067-4. Epub Dec 29, 2010.

Wasser, S.P. 2014. Medicinal mushroom science: Current perspectives, advances, evidences, and challenges. *Journal of Biomedical Science* 37(6): 345–356. http://doi.org/10.4103/2319-4170.138318.

Wasser, S.P. 2017. Medicinal mushrooms in human clinical studies. Part I. Anticancer, oncoimmunological, and immunomodulatory activities: A review. *International Journal of Medicinal Mushrooms* 19(4): 279–317.

Wayne, R.R. 1996. *Growing Mushrooms with Hydrogen Peroxide*, Dec 1996. https://idoc.pub/documents/growing-mushrooms-the-easy-way-pqn8okqm8kl1. Accessed Aug 29, 2022.

WHO. 2014-2023. The WHO traditional medicine strategy 2014–2023 http://apps.who.int/iris/bitstream/10665/92455/1/9789241506090_eng.pdf, Accessed Aug 5, 2022

WHO. 2021. *10 Global Health Issues to Track in 2021*. www.who.int/news-room/spotlight/10-global-health-issues-to-track-in-2021. Accessed Aug 14, 2022.

WHO. 2022. UHC Compendium: Health interventions for Universal Health Coverage. https://www.who.int/universal-health-coverage/compendium. Accessed Aug 14, 2022.

Xue, C.C., A.L. Zhang, K.M. Greenwood, et al. 2010. Traditional Chinese medicine: An update on clinical evidence. *Journal of Alternative and Complementary Medicine* 16(3): 301–312.

Zaidman, B.Z., M. Yassin, J. Mahajna & S.P. Wasser. 2005. Medicinal mushroom modulators of molecular targets as cancer therapeutics. *Applied Microbiology and Biotechnology* 67: 453–468.

Zhang, J., X. Zhong, S. Li, G. Zhang & X. Liu. 2015. Metabolic characterization of natural and cultured *Ophicordyceps sinensis* from different origins by 1H NMR spectroscopy. *Journal of Pharmaceutical and Biomedical Analysis* 115: 395–401.

Zibiah, A. 2021. Traditional medicine: Affirmation or resurgence. *Islands Business*, Dec 5, 2021. https://islandsbusiness.com/news-break/traditional-medicine/ Accessed Aug 5, 2022.

7 Ethnomycology

Diversity and Utilization of Medicinal Mushrooms in the Philippines

Angeles M. De Leon, Jaica Jane C. Balcruz, Justine Reine T. Matillano, Charlyn M. Miguel, John Paul R. Orevillo, Rich Milton R. Dulay

7.1 INTRODUCTION

Mushrooms are the spore-bearing structures (also known as fruiting bodies) produced by members of the phyla Basidiomycota and Ascomycota of the kingdom Fungi (Kinge et al. 2017). While mushrooms can be classified as toxic or non-toxic, they can also be classified as edible or inedible and wild or cultivated (Figure 7.1). Wild edible mushrooms are mainly cultivated for human consumption, medicinal use, and pharmacological production (Jacob et al. 2017) because of their rich bioactive compounds and nutritional compositions. Due to its favorable climatic conditions, the Philippines supports exceedingly diverse flora and fauna. Many of these organisms, including mushrooms, still exist in the wild, awaiting discovery and use for human benefit. They are typically seen growing on lignocellulosic substrates as saprophytes.

At present, macrofungi are valued greatly for their extensive economical, nutritional, and medical potential. Macrofungi can be potentially important sources of functional food and other nutraceuticals, as many species contain a variety of bioactive components including polysaccharides, dietary fiber, proteins, phenolic compounds, and vitamins (Lu et al. 2020). Some are known to be potential sources of antibacterial, antifungal, antiviral, antiparasitic, antioxidant, anti-inflammatory, and anticoagulant materials (Lindequist et al. 2005). Moreover, they are also of great use to environmental monitoring as bioindicators. Some macrofungi are known to have the ability to digest lignocellulosic and other compounds in the environment by secreting enzymes (Sumic et al. 2017). More research relating to macrofungi may further enhance and expand their value and applications.

In the past three to four decades, scientific research has demonstrated that mushrooms are significant sources of pharmacologically beneficial chemicals that can improve human health. Determining the bioactive ingredients and toxicological effects of mushrooms is therefore essential for the development of nutraceuticals.

7.2 MUSHROOMS FOUND IN THE PHILIPPINES

According to Sridhar and Deshmukh (2019), as cited by Torres et al. (2020), the estimated total number of species of macrofungi in Asia is between 53,000 to 110,000, including both the Ascomycota and Basidiomycota. With this estimated number, many studies suggest that mushrooms are among the most diverse organisms not only in Asia but also in the world. One Asian country, the Philippines, is considered to possess one of the most diverse ecosystems in the world. According to Liwanag et al. (2017), many studies have been conducted in the Philippines

DOI: 10.1201/b23190-7

FIGURE 7.1 Types of mushrooms: (a) edible, (b) inedible, (c) wild, and (d) cultivated.

regarding the different species of plants, animals, and insects that show how diverse the country is. However, the study of mushrooms is very limited. This suggests that there are numerous mushroom species that are still unidentified and may be endemic to the Philippines; these can potentially be used to discover new drugs for certain diseases. As stated by Torres et al. (2020), the number of mushrooms in the Philippines is relatively high, but there are still undiscovered and undocumented species. According to Dulay et al. (2020), it is important to elucidate more about mushrooms, especially those belonging to unrecorded species and found in places that have never been studied for mycological diversity. If we explore further, we can guarantee the identification and possible cultivation of new mushroom species with great nutraceutical potential.

One of the studies that have been conducted on the assessment of mushroom diversity is that of De Leon et al. (2021). In their study, they documented a total of 37 mushrooms belonging to 29 different species based on morphological identification, and four of these species were edible. The species documented are *Auricularia auricula-judae* (edible), *Conocybe arrhenii*, *Coprinellus disseminates* (edible), *Coprinus* sp., *Crepidotus mollis*, *Daldinia concentrica*, *Earliella scabrosa*, *Favolus acervatus*, *Fomes fomentarium*, *Ganoderma applanatum*, *G. fornicatum*, *G. lucidum*, *Hygrocybe* sp., *Irpex lacteus*, *Lentinus strigosus* (edible), *Lenzites elegans*, *Lepiota lilacea*, *Lepiota* sp., *Marasmiellus ramealis*, *Microporus xanthopus*, *Mycena* sp., *Panellus mitis*, *P. cyanescens*, *Parasola plicalitis*, *Psathyrella candolleana*, *Psathyrella* sp., *Russula* sp. *Schizophyllum commune* (edible), *Trametes elegans*, *T. gibbosa*, *T. hirsuta*, *T. versicolor*, *Trametes* sp. 1, *Trametes* sp. 2, and *Xylaria papulis*. Additionally, the study of Paguirigan et al. (2020) shows the abundance of mushrooms in the Philippines. In their study, they documented a total of 36 taxa wherein only 21 specimens were identified up to species level and the remaining 15 species were identified only at the genus level. The species they identified are *Agaricus moelleri*, *Coprinopsis* sp., *Leucocoprinus* sp., *Marasmiellus* sp., *Marasmius* sp., *Mycena* sp., *Schizophyllum commune*, *Tricholoma lascivum*, *Auricularia auricular*, *A. delicate*, *A. polytricha*, *Phellinus* sp., *Amauroderma* sp., *Ischnoderma resinosum*, *Lentinus* sp., *Microporus* sp., *M. xanthopus*, *Podoscypha* sp., *Polyporus* sp., *P. varius*, *Pycnoporus sanguineus*, *Trametes* sp., *T. hirsute*, *T. ochracea*, *Tyromyces chioneus*, *Stereum* sp., *S. rugosum*, *S. subtomentosum*, *Tremella fuciformis*, *Peziza* sp., *Plicariella scabrosa*, *Daldinia concentrica*, *Hypoxylon fragiforme*, *Xylaria* sp., *X. longiana*, and *X. polymorpha*. Furthermore, the study of Nacua et al. (2018) recorded and documented a total of 21 mushrooms, of which 14 were identified to the species level and the remaining seven were identified only at the genus level. The mushrooms they documented are *Cantharellus* sp., *Cookeina speciosa*, *Coprinus* sp., *Dacryopinax spathularia*, *Daedalea* sp., *Hexagonia* sp., *Fomitopsis* sp., *Ganoderma lucidum*, *G. sinense*, *Heterobasidion annosum*, *Lactifluus piperatus*, *Polystictus* sp., *Lentinus tigrinus*, *Lepiota cristata*, *Panaeolus papilionaceus*, *Russula virescens*, *Stereum ostrea*, *S. hirsutum*, *Stereum* sp., *Trametes pubescens*, and *T. versicolor*. Moreover, the study of Jacob et al. (2017) also shows the abundance of mushrooms in the Philippines. In their study, they listed a total of 31 species of mushrooms, of which 27 were identified up to the species level and the remaining four were identified up to the genus level. The mushrooms are *Auricularia polytricha*, *A. auricular-judae*, *Termitomyces striatus*, *Calvatia cythiformis*, *Ganoderma lucidum*, *G. applanatum*, *G. adspersum*, *G. japonicum*, *Daedalea dickinsii*, *Phellinus linteus*, *Pleurocybella porrigens*, *Polyporus sanguineus*, *Pleurotus porrigens*, *Agaricus arvensis*, *Podoscypha* sp., *Trametes pubescens*, *T. elegans*, *T. hirsute*, *T. versicolor*, *Geastrum triplex*, *Lentinus sajor-caju*, *Psathyrella candolleana*, *Clavulina cristata*, *Phellinus ignarius*, *Schizophyllum commune*, *Phellinus gilvus*, *Lentinus tigrinus*, *Marasmius* sp., *Coprinus* sp., *Cookenia* sp., and *Geastrum fimbriatum*. Lastly, the study of Dulay et al. (2020) also shows the abundance of mushrooms in the Philippines with a recorded total of 72 specimens, 39 of which were identified down to the species level while the remaining 33 species were identified down to the genus level. The mushrooms they recorded are *Agaricus comtulus*, *Agaricus* sp. 1, *Agaricus* sp. 2, *Agaricus* sp. 3, *Agaricus* sp. 4, *Agaricus xanthodermus*, *Armillaria* sp., *Auricularia auricula-judae*, *A. polytricha*, *Auricularia* sp. 1, *Auricularia* sp. 2, *Cantharocybe* sp., *Chaetocalathus* sp., *Chlorophyllum molybdites*, *Clitopilus prunulus*, *Conocybe* sp., *Coprinellus disseminates*, *Coprinellus* sp., *Cortinarius pakistanicus*, *Coprinopsis clastophylla*, *Colletotrichum musae*, *Cyathus* sp., *Cymatoderma elegans*, *Dacryopinax spathularia*, *Daedalea flavida*, *Deconica coprophila*, *Earliella scabrosa*, *Favolus emerici*, *Favolus* sp. 1, *Favolus* sp. 2, *Ganoderma australe*, *G. gibbosum*, *G. lucidum*, *G. neo-japonicum*, *Geastrum schmidelii*, *Gerronema keralense*, *Gymnopilus lepidotus*, *Gymnopilus* sp., *Gymnopus* sp., *Haageocereus tenuis*, *Hohenbuehelia* sp., *Hydropus marginellus*, *Hymenagaricus* sp., *Lentinus* sp., *Leucocoprinus* sp., *Lycoperdon* sp., *Maramiellus palmivorus*, *M. candidus*, *Marasmius* sp.1, *Marasmius* sp.2, *Oudemansiella canarii*, *Phallus indusiatus*, *Phallus* sp., *Pleurotus djamor*, *Pluteus multiformis*, *Polyporus* sp. 1, *Polyporus* sp. 2, *Psathyrella* sp., *Pterula* sp., *Schizophyllum commune*, *Serpula similis*, *Termitomyces bulborhizus*, *T. microcarpus*, *Trametes hirsuta*, *Trametes* sp., *Tremella* sp., *Volvariella dunensis*, *Xanthagaricus flavosquamosus*, *Xanthagaricus* sp., *Xylaria polymorpha*, *Xylaria* sp. 1, and *Xylaria* sp. 2. These five different studies show the abundance of mushrooms in the Philippines, including numerous species that are edible and not edible, with or without medicinal properties, and newly recorded species in the country.

However, the five studies also state that there are still undiscovered and undocumented species of mushrooms in the country.

7.3 MUSHROOMS FROM CENTRAL LUZON

Seventy-six species of macrofungi were recorded in six Central Luzon Aeta communities from the provinces of Tarlac, Pampanga, and Zambales (De Leon et al. 2013). The identification was done using both morphological and molecular methods. The macrofungi recorded were classified into 41 genera and 23 families. However, due to synonymous taxa, some species have been listed under certain families. Out of the 76 macrofungi species recorded, 23 were classified up to genus level only while 53 were classified at the species level. The species listing was performed for two localities per province to compare the presence of macrofungi in the ancestral domain of the Aetas and their resettlement areas. Based on the findings, 25 species were recorded in Pampanga, 41 in Tarlac, and 48 in Zambales (Table 7.1). Out of these, the family Polyporaceae represented the majority of reported macrofungi with 10 species.

Moreover, Dulay et al. (2020) conducted a study on the distribution and species of macrofungi in Sitio Canding in the province of Tarlac. Opportunistic and purposive samplings were used in the documentation and collection of macrofungi. The study was conducted over three months, from August to October of 2019. The collection resulted in 116 macrofungi, which were taxonomically classified under 27 families, 47 genera, and 72 species. The identification was based on morphological features. According to the data gathered in the study, the families Agaricaceae and Polyporaceae had the highest number of species, with 12 and 10 species, respectively.

TABLE 7.1
Macrofungi Species Recorded from Central Luzon.

Species	Location
Agaricaceae	
Agaricus comtulus	Tarlac (Dulay et al. 2020)
Agaricus placomyces	Nueva Ecija (Culala and Dulay 2018)
Agaricus trisulphuratus	Zambales (De Leon et al. 2013)
Agaricus xanthodermus	Tarlac (Dulay et al. 2020)
Chlorophyllum molybdites	Nueva Ecija and Tarlac (Dulay et al. 2020; Dulay and Maglasang 2017; Culala and Dulay 2018)
Leucocoprinus cepaestipes	Nueva Ecija (Undan et al. 2016; Culala and Dulay 2018)
Macrolepiota procera	Pampanga (De Leon et al. 2013)
Macrolepiota rhacodes	Aurora and Tarlac (Tadiosa et al. 2011; De Leon et al. 2013)
Xanthagaricus flavosquamosus	Tarlac (Dulay et al. 2020)
Agaricus sp.	Tarlac (De Leon et al. 2013; Dulay et al. 2020)
Lepiota aspera	Aurora (Tadiosa et al. 2011)
Lepiota cristata	Nueva Ecija and Aurora (Tadiosa et al. 2011; Culala and Dulay 2018)
Leucocoprinus birnbaumii	Nueva Ecija (Culala and Dulay 2018)
Leucocoprinus sp.	Tarlac (Dulay et al. 2020)
Lycoperdon sp.	Tarlac (Dulay et al. 2020)
Hymenegaricus sp.	Tarlac (Dulay et al. 2020)
Macrolepiota sp.	Tarlac (De Leon et al. 2013)
Macrolepiota rhacodes	Aurora (Tadiosa et al. 2011)
Xanthagaricus sp.	Tarlac (Dulay et al. 2020)

TABLE 7.1 ***Continued***
Macrofungi Species Recorded from Central Luzon.

Species	Location
Amanitaceae	
Amanita cokeri	Tarlac (De Leon et al. 2013)
Amanita fulva	Aurora (Tadiosa et al. 2011)
Amanita onusta	Nueva Ecija (Dulay and Maglasang 2017; Culala and Dulay 2018)
Limacella illinita	Nueva Ecija (Culala and Dulay 2018)
Auriculariaceae	
Auricularia auricula	Nueva Ecija, Aurora, Pampanga, Tarlac, and Zambales (Tadiosa et al. 2011; De Leon et al. 2013; Culala and Dulay 2018)
Auricularia delicata	Aurora (Tadiosa et al. 2011)
Auricularia fuscosuccinea	Nueva Ecija (Sibounnavong et al. 2008)
Auricularia mesenterica	Tarlac (De Leon et al. 2013)
Auricularia polytricha	Nueva Ecija, Aurora, Pampanga, Tarlac, and Zambales (Tadiosa et al. 2011; De Leon et al. 2013; Dulay and Maglasang 2017; Dulay et al. 2020)
Auricularia tenuis	Tarlac (De Leon et al. 2013)
Auricularia sp.	Pampanga and Tarlac (De Leon et al. 2013; Dulay et al. 2020)
Auricularia auricular-judae	Nueva Ecija and Tarlac (Dulay et al. 2020; Dulay and Maglasang 2017)
Bankeraceae	
Phellodon niger	Pampanga, Tarlac, and Zambales (De Leon et al. 2013)
Bolbitaceae	
Agrocybe sp.	Tarlac (De Leon et al. 2013)
Conocybe sp.	Aurora and Tarlac (Tadiosa et al. 2011; Dulay et al. 2020)
Conocybe lactae	Nueva Ecija (Culala and Dulay 2018)
Panaeolus sp.	Nueva Ecija, Aurora, and Tarlac (Tadiosa et al. 2011; De Leon et al. 2013; Culala and Dulay 2018)
Paneolus antillarium	Nueva Ecija (Culala and Dulay 2018)
Paneolus foenisecci	Nueva Ecija (Lopez et al. 2016; Culala and Dulay 2018)
Panaeolus semiovatus	Nueva Ecija (Dulay and Maglasang 2017)
Panaeolus papilionaceous	Tarlac (De Leon et al. 2013)
Boletaceae	
Strobilomyces strobilaceus	Aurora (Tadiosa et al. 2011)
Cantharellaceae	
Cantharellus minor	Nueva Ecija (Sibounnavong et al. 2008)
Cantharellus sp.	Aurora and Bulacan (Tadiosa et al. 2011; Liwanag et al. 2017)
Cortinariaceae	
Gymnopilus lepidotus	Tarlac (Dulay et al. 2020)
Gymnopilus sp.	Tarlac (Dulay et al. 2020)
Crepidotaceae	
Crepidotus sp.	Bulacan (Liwanag et al. 2017)
Dacrymycetaceae	
Dacryopinax spathularia	Nueva Ecija, Aurora, Pampanga, Tarlac, and Zambales (Tadiosa et al. 2011; De Leon et al. 2013; Dulay and Maglasang 2017; Dulay et al. 2020)
Entolomataceae	
Clitopilus prunulus	Tarlac (Dulay et al. 2020)
Enteloma serrulatum	Aurora (Tadiosa et al. 2011)
Fomitopsidaceae	
Daedalea flavida	Tarlac (Dulay et al. 2020)
Daedalea ambigua	Bulacan (Liwanag et al. 2017)
Ganodermataceae	
Amauroderma rogusom	Aurora (Tadiosa et al. 2011)

TABLE 7.1 *Continued*
Macrofungi Species Recorded from Central Luzon.

Species	Location
Ganoderma australe	Tarlac (Dulay et al. 2020)
Ganoderma applanatum	Aurora, Pampanga, Tarlac, Bulacan, and Zambales (Tadiosa et al. 2011; De Leon et al. 2013; Liwanag et al. 2017)
Ganoderma lobatum	Aurora (Tadiosa et al. 2011)
Ganoderma gibbosum	Tarlac (Dulay et al. 2020)
Ganoderma lucidum	Nueva Ecija, Aurora, Pampanga, Tarlac, and Zambales (Tadiosa et al. 2011; De Leon et al. 2013; Undan et al. 2016; Dulay and Maglasang 2017; Culala and Dulay 2018; Dulay et al. 2020)
Ganoderma neo-japonicum	Nueva Ecija, Zambales, and Tarlac (De Leon et al. 2013; Dulay and Maglasang 2017; Dulay et al. 2020)
Ganoderma sinense	Nueva Ecija, Pampanga, and Zambales (De Leon et al. 2013; Dulay and Maglasang 2017; Culala and Dulay 2018)
Ganoderma sp.	Nueva Ecija and Bulacan (Liwanag et al. 2017; Dulay and Maglasang 2017)
Geastraceae	
Geastrum schmidelli	Tarlac (Dulay et al. 2020)
Hygrophoraceae	
Cantharocybe sp.	Tarlac (Dulay et al. 2020)
Hygrocybe coccinea	Aurora (Tadiosa et al. 2011)
Hygrocybe miniata	Aurora (Tadiosa et al. 2011)
Hymenochaetaceae	
Coltricia perennis	Nueva Ecija (Sibounnavong et al. 2008)
Hymenochaete rubiginosa	Aurora (Tadiosa et al. 2011)
Phellinus linteus	Nueva Ecija (Dulay and Maglasang 2017)
Phellinus pini	Nueva Ecija (Sibounnavong et al. 2008)
Phellinus sp.	Aurora and Bulacan (Tadiosa et al. 2011; Liwanag et al. 2017)
Lyophyllaceae	
Termitomyces bulborhizus	Tarlac (Dulay et al. 2020)
Termitomyces microcarpus	Tarlac (Dulay et al. 2020)
Termitomyces clypeatus	Nueva Ecija (Dulay and Maglasang 2017; Culala and Dulay 2018)
Lycoperdaceae	
Calvatia gigantea	Tarlac (De Leon et al. 2013)
Lycoperdon perlatum	Pampanga and Zambales (De Leon et al. 2013)
Lycoperdon echinatum	Aurora (Tadiosa et al. 2011)
Vascellum pratense	Pampanga (De Leon et al. 2013)
Marasmiaceae	
Marasmiellus ramealis	Tarlac (De Leon et al. 2013)
Marasmius androsaceus	Tarlac (De Leon et al. 2013)
Marasmius rotula	Pampanga and Tarlac (De Leon et al. 2013)
Marasmius siccus	Tarlac (De Leon et al. 2013)
Marasmius palmivorus	Tarlac (Dulay et al. 2020)
Marasmius candidus	Tarlac (Dulay et al. 2020)
Marasmius sp.	Pampanga, Tarlac, and Zambales (De Leon et al. 2013; Dulay et al. 2020)
Marasmius oreades	Nueva Ecija (Dulay and Maglasang 2017)
Chaetocalathus sp.	Tarlac (Dulay et al. 2020)
Hydropus marginellus	Tarlac (Dulay et al. 2020)
Gerronema keralense	Tarlac (Dulay et al. 2020)
Meripilaceae	
Meripilus giganteus	Nueva Ecija (Undan et al. 2016)
Meruliaceae	
Cymatoderma elegans	Tarlac (Dulay et al. 2020)

TABLE 7.1 *Continued*
Macrofungi Species Recorded from Central Luzon.

Species	Location
Cymatoderma africanum	Aurora (Tadiosa et al. 2011)
Gloeoporus dichrous	Nueva Ecija (Sibounnavong et al. 2008)
Podoscypha bolleana	Aurora (Tadiosa et al. 2011)
Nidulariaceae	
Cyathus sp.	Tarlac (Dulay et al. 2020)
Cyathus striatus	Tarlac (De Leon et al. 2013)
Sphaerobolus stellatus	Tarlac (De Leon et al. 2013)
Omphalotaceae	
Gymnopus sp.	Tarlac (Dulay et al. 2020)
Paxillaceae	
Paxillus sp.	Zambales (De Leon et al. 2013)
Phallaceae	
Aseroe rubra	Aurora (Tadiosa et al. 2011)
Dictyophora duplicata	Zambales (De Leon et al. 2013)
Dictyophora indusiata	Nueva Ecija (Dulay and Maglasang 2017)
Mutinus caninus	Tarlac (De Leon et al. 2013)
Phallus duplicatus	Tarlac (De Leon et al. 2013; Dulay et al. 2020)
Phallus indusiatus	Tarlac (Dulay et al. 2020)
Phallus multicolor	Zambales (De Leon et al. 2013)
Phallus sp.	Tarlac (Dulay et al. 2020)
Physalacriaceae	
Armillaria sp.	Tarlac (Dulay et al. 2020)
Oudemansiella canarii	Nueva, Ecija, Aurora, and Tarlac (Tadiosa et al. 2011; Culala and Dulay 2018; Dulay et al. 2020)
Pleurotaceae	
Pleurotus cystidiosus	Nueva Ecija (Dulay and Maglasang 2017)
Pleurotus porrigens	Tarlac (De Leon et al. 2013)
Pleurotus pulmunarius	Zambales (De Leon et al. 2013)
Pleurotus sajor-caju	Pampanga and Zambales (De Leon et al. 2013)
Pleurotus sp.	Nueva Ecija, Tarlac, and Zambales (De Leon et al. 2013; Dulay and Maglasang 2017)
Hohenbuehelia petaloides	Nueva Ecija (Sibounnavong et al. 2008)
Hohenbuehelia sp.	Tarlac (Dulay et al. 2020)
Pleurotus djamor	Nueva Ecija and Tarlac (Dulay and Maglasang 2017; Culala and Dulay 2018; Dulay et al. 2020)
Pluteaceae	
Volvariella dunensis	Tarlac (Dulay et al. 2020)
Pluteus multiformis	Tarlac (Dulay et al. 2020)
Volvariella volvacea	Nueva Ecija, Aurora, Pampanga, and Zambales (Tadiosa et al. 2011; De Leon et al. 2013; Dulay and Maglasang 2017; Culala and Dulay 2018)
Polyporaceae	
Coriolus versicolor	Nueva Ecija (Dulay and Maglasang 2017)
Daedaleopsis confragosa	Zambales (De Leon et al. 2013)
Lentinus cladopus	Tarlac (De Leon et al. 2013)
Lentinus sajor-caju	Nueva Ecija and Zambales (De Leon et al. 2013; Culala and Dulay 2018)
Lentinus swartzii	Nueva Ecija (Lopez et al. 2016)
Lentinus tigrinus	Nueva Ecija, Pampanga, Tarlac, and Zambales (De Leon et al. 2013; Dulay and Maglasang 2017; Culala and Dulay 2018)
Lentinus squarrosulus	Nueva Ecija, Pampanga, and Zambales (De Leon et al. 2013; Lopez et al. 2016; Culala and Dulay 2018)
Lenzites sp.	Bulacan (Liwanag et al. 2017)

TABLE 7.1 ***Continued***
Macrofungi Species Recorded from Central Luzon.

Species	Location
Lentinus sp.	Zambales and Tarlac (De Leon et al. 2013; Dulay et al. 2020)
Polystictus xanthopus	Zambales (De Leon et al. 2013)
Pycnoporus sanguineus	Aurora, Bulacan, Pampanga, and Zambales (Tadiosa et al. 2011; De Leon et al. 2013; Liwanag et al. 2017)
Polyporous grammocephalus	Nueva Ecija and Zambales (De Leon et al. 2013; Dulay and Maglasang 2017)
Polyporus sp.	Tarlac and Bulacan (Dulay et al. 2020; Liwanag et al. 2017)
Earliella scabrosa	Tarlac (Dulay et al. 2020)
Favolus emerici	Tarlac (Dulay et al. 2020)
Favolus sp.	Tarlac (Dulay et al. 2020)
Hexagonia tenuis	Aurora, Zambales, Tarlac, and Bulacan (Tadiosa et al. 2011; De Leon et al. 2013; Dulay et al. 2020; Liwanag et al. 2017)
Trametes cervina	Zambales (De Leon et al. 2013)
Trametes versicolor	Nueva Ecija, Bulacan, Pampanga, and Zambales (Sibounnavong et al. 2008; De Leon et al. 2013; Liwanag et al. 2017; Culala and Dulay 2018)
Trametes hirsuta	Tarlac (Dulay et al. 2020)
Trametes pubescens	Zambales (De Leon et al. 2013)
Trametes corrugata	Bulacan (Liwanag et al. 2017)
Trametes sp.	Tarlac (Dulay et al. 2020)
Microporus sp.	Nueva Ecija and Bulacan (Undan et al. 2016; Liwanag et al. 2017)
Microporus vernicipes	Aurora and Bulacan (Tadiosa et al. 2011; Liwanag et al. 2017)
Microporus xanthopus	Aurora and Bulacan (Tadiosa et al. 2011; Liwanag et al. 2017)
Polyporus picipes	Aurora (Tadiosa et al. 2011)
Polyporus hirsitus	Aurora and Bulacan (Tadiosa et al. 2011; Liwanag et al. 2017)
Coprinellus aureogranulatus	Nueva Ecija (Lopez et al. 2016)
Coprinellus disseminatus	Pampanga, Tarlac, and Zambales (De Leon et al. 2013; Dulay et al. 2020)
Coprinellus pakistanicus	Tarlac (Dulay et al. 2020)
Coprinellus sp.	Tarlac (Dulay et al. 2020)
Coprinellus clastophylla	Tarlac (Dulay et al. 2020)
Coprinellus micaceus	Nueva Ecija (Dulay and Maglasang 2017)
Coprinellus musae	Tarlac (Dulay et al. 2020)
Coprinus atramentarius	Aurora (Tadiosa et al. 2011)
Coprinus comatus	Nueva Ecija (Dulay and Maglasang 2017)
Coprinus dessiminatus	Aurora (Tadiosa et al. 2011)
Coprinus plicatilis	Nueva Ecija and Aurora (Tadiosa et al. 2011; Dulay and Maglasang 2017)
Coprinus niveus	Nueva Ecija and Aurora (Tadiosa et al. 2011; Culala and Dulay 2018)
Psathyrella sp	Tarlac (Dulay et al. 2020)
Pterulaceae	
Pterula sp.	Tarlac (Dulay et al. 2020)
Sarcoscyphaceae	
Cookeina tricholoma	Bulacan (Liwanag et al. 2017)
Schizophyllaceae	
Schizophyllum commune	Nueva Ecija, Aurora, Tarlac, and Bulacan (Tadiosa et al. 2011; Lopez et al. 2016; Liwanag et al. 2017; Dulay and Maglasang 2017; Culala and Dulay 2018; Dulay et al. 2020)
Serpulaceae	
Serpula similis	Tarlac (Dulay et al. 2020)
Stereaceae	
Stereum hirsitum	Nueva Ecija (Dulay and Maglasang 2017)
Strophariaceae	
Deconia coprophila	Tarlac (Dulay et al. 2020)
Naematoloma fasciculare	Nueva Ecija (Culala and Dulay 2018)

TABLE 7.1 ***Continued***
Macrofungi Species Recorded from Central Luzon.

Species	Location
Pholiota sp.	Pampanga (De Leon et al. 2013)
Psilocybe coprophila	Nueva Ecija (Dulay and Maglasang 2017)
Stropharia semiglobata	Tarlac (De Leon et al. 2013)
Stropharia squamosal	Nueva Ecija (Culala and Dulay 2018)
Tremellaceae	
Pseudohydnum gelatinosum	Nueva Ecija (Culala and Dulay 2018)
Tremella sp.	Tarlac (Dulay et al. 2020)
Tremella foliacea	Nueva Ecija (Dulay and Maglasang 2017)
Tremella mesenterica	Nueva Ecija (Dulay and Maglasang 2017)
Tricholomataceae	
Collybia sp.	Pampanga (De Leon et al. 2013)
Phyllotopsis nidulans	Nueva Ecija (Culala and Dulay 2018)
Mycena sp.	Pampanga and Zambales (De Leon et al. 2013)
Termitomyces robustus	Zambales (De Leon et al. 2013)
Termitomyces sp.	Zambales (De Leon et al. 2013)
Xylariaceae	
Daldinia concentrica	Aurora, Bulacan, Pampanga, Tarlac, and Zambales (Tadiosa et al. 2011; De Leon et al. 2013; Liwanag et al. 2017)
Xylaria papulis	Nueva Ecija (Undan et al. 2016)
Xylaria polymorpha	Tarlac and Bulacan (Liwanag et al. 2017; Dulay et al. 2020)
Xylaria sp.	Tarlac (Dulay et al. 2020)

Additionally, 21 species of macrofungi were identified from the collection conducted in Angat Watershed Reservation, Bulacan, Philippines (Liwanag et al. 2017). The identified species were characterized morphologically and classified under 10 families and 16 genera. Out of these, 13 mushrooms were identified down to the species level while eight were identified down to the genus level only. The majority of identified macrofungi belonged to the family Polyporaceae. Similarly, Tadiosa et al. (2011) also reported a high number of macrofungi from the family Polyporaceae in the preliminary study conducted in Aurora Province.

Furthermore, Dulay and Maglasang (2017) conducted a species listing in an agroecosystem in Cuyapo, Nueva Ecija, in June 2016. Aside from collection and photo-documentation, the method of sampling was not specified in the study. They reported 30 species of mushrooms and classified them under 21 genera and 18 families based on their micro- and macromorphological features. Similarly, the collection, identification, and phylogeny analysis of different macrofungi were conducted in Mt. Bangkay, Cuyapo, and Nueva Ecija in August 2015 (Undan et al. 2016). The identification and phylogeny analysis were done through a molecular approach via amplification and sequencing of the DNA of the samples' internal transcribed (ITS) region. In summary, five samples of macrofungi were collected, identified, and subjected to phylogenetic analysis. On the same note, the ribosomal DNA-ITS of some collected macrofungi in areas of Jaen, Nueva Ecija, were also used for identification and analysis. The collection sites were from Barangay Dampula, Langla, and Putlod in Jaen, Nueva Ecija. The study duration was from April to May 2015. The researchers collected and identified seven samples, and three of them belonged to the species *Lentinus swartzii* (Lopez et al. 2016).

In addition, a macrofungal survey was conducted in Puncan, Carranglan, Nueva Ecija, in April 2008 (Sibounnavong et al. 2008). Analysis of morphological characteristics was performed by comparing the collected specimens with published literature. The analysis resulted in the identification of seven species and genera classified under five families, namely, Auriculariaceae, Polyporaceae, Tricholomataceae, Hymenochaetaceae, and Cantharellaceae. Also, a species listing of naturally occurring mushrooms was performed by Culala and Dulay (2018) at the Central Luzon State University campus in the Science City of Munoz, Nueva Ecija. The study was conducted in October 2017. Collected mushrooms were analyzed and identified based on observation of morphological characteristics and morphometric features. A total of 35 mushroom species were identified and classified under 26 genera and 15 families. Out of these, 26 were identified down to the species level while nine were identified down to the genus level only. Table 7.1 shows the list of species reported from some of the provinces in Central Luzon.

7.4 MUSHROOM BIOACTIVITIES

Many macrofungi have the potential to be utilized as functional foods or pharmaceuticals for the treatment and prevention of a variety of chronic diseases, including diabetes, cardiovascular disease, cancer, etc. Aside from investigating antimicrobial activity, in the Philippines, several studies have described the biological activities of different macrofungi found in the country. Macrofungi, particularly mushrooms, have gained growing attention in recent years due to their bioactivities. These biological activities are commonly associated with potential therapeutic capabilities (Figure 7.2). Since macrofungi are recognized to have a variety of biomolecules with nutritional properties, they have been identified as functional foods in addition to being sources of pharmaceuticals and nutraceuticals, whether they are wild or cultivated fruiting bodies.

In the forests of Benguet and Mt. Province, the pharmacological importance of wild macrofungi was examined in order to aid in the treatment of a variety of health issues, including oxidative stress, bacterial infections, and thrombosis. In this study by Culliao et al. (2020), six wild macrofungi were extracted with chloroform and ethanol, and their crude extracts were evaluated for their total biological activities. These includes the *Albatrellus ellissi*, *Daedaleopsis confragosa*, *Lenzites betulina*, *Postia fragilis*, *Scleroderma citrinum*, and *Termitomyces erurrhizus*. Aside from investigating antibacterial properties, this study assessed the total phenolic content, lethality,

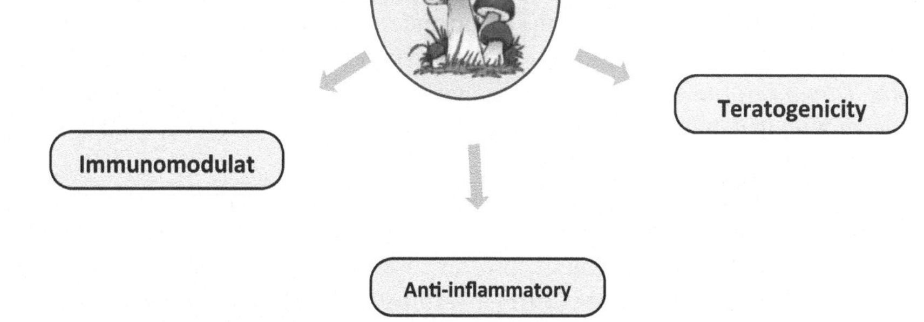

FIGURE 7.2 Common biological activities of substances derived from mushrooms.

antioxidant, and thrombolytic activity of the collected mushrooms. They reported that the highest total phenolic content was found in the ethanol extract of *D. confragosa*, which had 49.28 ± 0.30 μg gallic acid equivalents (GAE)/mg. The *S. citrinum* and *P. fragilis* ethanol extracts, on the other hand, possessed the most potent median effective concentrations (EC50) of 431.01 ± 17.82 and 469.63 ± 15.25 g/mL, respectively. The *D. confragosa* ethanol extract was the only extract that showed minimal toxicity, and none of the other test extracts were harmful or toxic. The highest percent clot reduction values were found in chloroform and ethanol extracts of *T. eurrhizus*, at 35.19 ± 0.13% and 32.41 ± 0.17%, respectively. The thrombolytic activity of macrofungal extracts from the Philippines had never been reported, and this study was the first to do so. These findings made by Culliao et al. (2020) suggested that the wild macrofungi examined may be used to make pharmaceutical formulations or nutraceuticals as a source of natural bioactive chemicals. Meanwhile, another study showed the biological activity of macrofungi from the Philippines, specifically the antioxidant property of mushrooms that are leaf-litter degrading and ligninolytic. Reyes and Nair (2016) evaluated five ligninolytic macrofungi, namely, *Ganoderma lucidum*, *Pleurotus florida*, *Pleurotus ostreatus*, *Schizophyllum commune*, and *Lentinus sajor-caju*. *Volvariella volvacea*, a leaf-litter degrading macrofungi, was included together with the ligninolytic macrofungi for *in vitro* assessment of the antioxidant activities of these cultured macrofungi. This study reported that the antioxidant activity of *G. lucidum* methanol and hot-water extracts, with absorbances of 0.75 and 0.54 nm, respectively, were higher than those of the other examined ligninolytic edible mushrooms and conventional compounds (vitamin C and tert-butylhydroquinone [TBHQ]). The absorbances of vitamin C and TBHQ were 0.49 and 0.39 nm, respectively. Methanol extracts of *L. sajor-caju*, on the other hand, displayed the highest bioactivity, with an absorbance of 0.80 nm. Similarly, antioxidant activities were found in the partially and fully opened stages of the basidiocarps of *V. volvacea*, regardless of strain. Furthermore, *V. volvacea* hot-water extracts exhibited stronger antioxidant activity than methanol extracts. This work adds to the body of knowledge, since the results of the experiment indicate that five of the Philippine's ligninolytic and leaf-litter degrading mushrooms have antioxidant properties. Finding novel natural sources, such as macrofungi, containing antioxidants that may trap free radicals and be employed to combat various illnesses related to oxidative stress is a significant step forward in revealing these macrofungi's further potential and bioactivity.

Moreover, there are other studies in the Philippines designed to disclose several bioactivities of certain macrofungi. For instance, a study by Mendoza et al. (2020) describes the bioactivities of *Pycnoporus sanguinues*, a Philippine basidiomycete, in a time-course submerged culture. This study revealed that the mycelial extracts' radical-scavenging efficacy increased over the incubation period, while the phenolic content was at its peak after 20 days of fermentation. Also, the researchers reported nine groups of mycochemicals from the 20-day-old mycelia of *P. sanguinues*. These mycochemicals are probably responsible for the aforementioned biological functions of *P. sanguinues*. Likewise, several bioactivities of a newly documented macrofungus, *Xylaria papulis*, from Paracelis, Mt. Province, were also assessed by De Leon et al. (2020b). The total phenolic content of *X. papulis* was found to be high, with 43.17 mg GAE/g, which indicates a good potential source of natural antioxidants. Moreover, the teratogenicity and cytotoxicity of *X. papulis* were also revealed. The teratogenic activity of *X. papulis* demonstrated a prominent growth retardation effect on developing *Danio rerio* (Zebrafish) embryos and led to head and tail deformities. Furthermore, the researchers found that embryos died at a much higher rate after being exposed to varying doses of *X. papulis* ethanol extract.

The lethal concentration 50 (LC_{50}) of the ethanolic extract of *X. papulis*, on the other hand, was found to be 218.507 ppm in a brine shrimp lethality test, indicating high toxicity. Another macrofungus with antioxidant and scavenging action was discovered in the Philippines. According to a study conducted by Dulay and Pamiloza (2018), *Lentinus strigosus*, a hairy sawgill mushroom gathered from the region, displayed functional activity aside from antibacterial activity. When compared to extracts of cultivated fruiting bodies, the extract of wild *L. strigosus* fruiting bodies demonstrated much stronger scavenging action (with an EC_{50} value of 534.67 µg/mL). The phenolic concentrations of the wild and cultivated samples were reported to be 79.70 and 78.80 mg ascorbic acid equivalent (mg AAE/g extract) samples, respectively.

Macrofungi indeed possess considerable functional biological activities, many of which have already been demonstrated, while more remain undiscovered. Among the biological activities and characteristics of macrofungi being studied in the Philippines, antibacterial activity is one of the most commonly assessed. Antibacterial resistance among harmful microorganisms has developed and is continuing to emerge over time. Over the last decade, it has become a critical concern all over the world. To address this significant issue, it is vital to explore new and local sources of active compounds that are not only effective in bacterial infection but also have a long-term effect by improving the body's immune system. One way to further analyze the potential of macrofungi is by screening their mycochemical components. Statistics show that there is a low percentage of investigated macrofungi compared to their total population on earth which is the same under the Philippine condition, that is why there is a need to further explore and investigate wild macrofungi to harness their possible pharmaceutical potential.

7.4.1 Antimicrobial Activity of Mushrooms against Gram-Positive Bacteria

To assess the antimicrobial activity of different mushroom extracts and compounds, various methodologies and a number of approaches have been used. Some of the developed techniques utilized over the years are broth macro- and microdilution methods; disc diffusion method; agar streak dilution method; concentration gradient test; other tests involving the introduction of the extract to the culture medium and colonial determination for the *in vitro* antimicrobial type of susceptibility testing. Some other studies have included *in vivo* tests for antimicrobial activity, depending on the type and purpose of the study. The variety of available tests is the reason why antimicrobial activity values are expressed in various units. In the Philippines, however, among the research published, the most common method used to assess antimicrobial activity is the disc diffusion method, which includes the planting of discs treated with extracts at different concentrations on top of the agar medium. The antimicrobial activity of the extract is determined by measuring the clear circular area around the disc, which is called the zone of inhibition (ZOI) and is usually presented in millimeters.

The Philippines is home to a varied range of macrofungi that grow naturally on a variety of surfaces. To capitalize on the potential of these wild genetic macrofungi as sources of food and pharmaceutical compounds, investigations on many different bioactivities, including antimicrobial activity of mycelia and fruiting bodies, are frequently included in evaluations. In the Philippines, several mushroom extracts, including those from wild and exotic, cultured varieties, were said to exhibit antibacterial action against Gram-positive bacteria (Table 7.2). For example, the ethanolic and acetone extracts of wild strain *Coprinus comatus* and *Pleurotus cystidiosus* from central Luzon exhibited antibacterial activity against *Staphylococcus aureus.* A study by Kalaw and Albinto (2014) revealed that the ethanolic extract of *C. comatus* generated a larger ZOI (mean of 14.09 mm) than its acetone extract, which generated a 13.16 mm mean ZOI. The acetone extract of *P. cystidiosus*, on the other hand, had a greater ZOI (15.25 mm) than its ethanol extract, which yielded a ZOI of 13.16 mm. The acetonitrile extracts of the Philippine wild and exotic mushrooms *Lentinus tigrinus* and *Pleurotus djamor*, respectively, also demonstrated to have antibacterial activity against *S. aureus*; however hexane extract of both *L. tigrinus* and *P. djamor* did not show any inhibition activity against the Gram-positive bacteria *S. aureus*, and Gram-negative *Escherichia coli* (Dulay et al. 2017). These findings regarding *L. tigrinus* are consistent with those of Dulay et al. (2014), who reported that an ethanol extract, lyophilized hot-water extract, and immobilized secondary mycelia of *L. tigrinus* exhibited antibacterial action against *S. aureus* but not against *E. coli.* Ethanol extracts of *Ganoderma lucidum*, *Ganoderma applanatum*, and *Polyporus* sp. also showed ZOI diameters ranging in size from 6.55–7.43 mm against *S. aureus* (Estebal and Gaylan 2017). Meanwhile, *Collybia reinakeana* collected from the northern Luzon produced its widest ZOI (14.07 mm) among different dilutions with 40 mL sterile distilled water added to 0.1 mL of 12-hour bacterial suspension in nutrient broth. This is the

TABLE 7.2
Macrofungal Extracts with Antimicrobial Activity against Gram-Positive Bacteria.

Microorganism	Macrofungus	Extract used	Result	Reference(s)
Staphylococcus aureus	*Coprinus comatus*	Ethanol extract	ZOI = 14.09 mm	Kalaw and Albinto (2014)
		Acetone extract	ZOI = 13.16 mm	Kalaw and Albinto (2014)
	Pleurotus cystidiosus	Ethanol extract	ZOI = 13.43 mm	Kalaw and Albinto (2014)
		Acetone extract	ZOI = 15.25 mm	Kalaw and Albinto (2014)
	Lentinus tigrinus	Acetonitrile extract	ZOI = 9.48 mm	Dulay et al. (2017)
		Ethanol extract	ZOI = 12.7 mm	Dulay et al. (2014)
		Lyophilized hot-water extract	ZOI = 8.00 mm	Dulay et al. (2014)
		Immobilized secondary mycelia	ZOI = 20.3 mm	Dulay et al. (2014)
	Pleurotus djamor	Acetonitrile extract	ZOI = 9.37 mm	Dulay et al. (2017)
	Ganoderma applanatum	Ethanol extract	ZOI = 7.12 mm	Estebal and Gaylan (2017)
	Ganoderma lucidum	Ethanol extract	ZOI = 7.43 mm	Estebal and Gaylan (2017)
	Polyporus sp.	Ethanol extract	ZOI = 6.55 mm	Estebal and Gaylan (2017)
	Collybia reinakeana	Mycelial disc	ZOI = 14.07 mm	Reyes et al. (2017)
	Tyromyces chioneus	Hot-water extract	ZOI = 15.67 mm	Capistrano et al. (2018)
	Auricularia	Extracted chitosan	ZOI = 66.17 mm	Chang et al. (2019)
	Polyporus pinistus	Ethanol extract	ZOI = 6.55 mm	Gaylan et al. (2017)
	Daedaleopsis confragosa	Ethanol extract	MIC = 12.50 µg/mL	Gaylan et al. (2017)
		Chloroform extract	ZOI = 11.33 mm	Culliao et al. (2020)
	Fomitopsis rosea	Ethanol extract	MIC = 2500 µg/mL	Gaylan et al. (2017)
	Microporus xanthopus	Ethanol extract	MIC = 2500 µg/mL	Gaylan et al. (2017)
	Hericium americanum	Mycelial disc	ZOI = 16.10 mm	Julian et al. (2018)
	Hericium erinaceus	Mycelial disc	ZOI = 14.33 mm	Julian et al. (2018)
	Xylaria papulis	Ethanol extract	ZOI = 18.87 mm	De Leon et al. (2020b)
	Pycnoporus sanguineus	Ethanol extract	ZOI = 8.00 mm	Mendoza et al. (2020)
	Schizophyllum commune	Dichloromethane extract	ZOI = 8.26 mm	Torres and Reyes (2020)
		Ethanol extract	ZOI = 6.37 mm	Torres and Reyes (2020)
		Methanol extract	ZOI = 6.38 mm	Torres and Reyes (2020)
	Lenzites betulina	Ethanol extract	ZOI = 13.33 mm	Culliao et al. (2020)
Bacillus cereus	*Fomes sp.*	Ethanol extract	ZOI = 8.00 mm	Estebal and Gaylan (2017)
	Ganoderma applanatum	Ethanol extract	ZOI = 7.43 mm	Estebal and Gaylan (2017)
	Ganoderma lucidum	Ethanol extract	ZOI = 7.00 mm	Estebal and Gaylan (2017)
	Polyporus sp.	Ethanol extract	ZOI = 7.06 mm	Estebal and Gaylan (2017)
Streptococcus iniae	*Volvariella volvacea*	Crude extract	ZOI = 9.51 mm	Serrano and Divina (2016)
	Pleurotus sajor-caju	Crude extract	ZOI = 11.07 mm	Serrano and Divina (2016)
Enterococcus faecalis	*Schizophyllum commune*	Dichloromethane extract	ZOI = 8.59 mm	Torres and Reyes (2020)
		Ethanol extract	ZOI = 6. 79 mm	Torres and Reyes (2020)

MIC = minimum inhibitory concentration
ZOI = zone of inhibition

first and currently the only report on the antibacterial activity of *C. reinakeana* in the Philippines (Reyes et al. 2017). *Tyromyces chioneus*, another macrofungus collected from Bambang, Nueva Vizcaya, in a location occupied by the Isinai tribe, exhibited antibacterial properties; its hot-water extract yielded a 15.67 mm diameter ZOI against *S. aureus*. This result, however, was significantly different from that of the positive control, streptomycin sulfate, which yielded a mean ZOI value of 40.21 mm against *S. aureus* (Capistrano et al. 2018). On the other hand, the chitosan extract of *Auricularia* yielded a very high activity of inhibition, with a 66.17 mm diameter ZOI at a 1000 μg/mL concentration; this shows that *Auricularia* has significant inhibitory activity against *S. aureus* even compared to that of the positive control streptomycin sulfate, which only generated a 26.5 mm ZOI in the study of Chang et al. (2019). Other assessed macrofungi collected from the Philippines that were observed to have antibacterial activity against the pathogenic, Gram-positive *S. aureus* include *Polyporus pinistus, Daedaleopsis confragosa, Fomitopsis rosea, Microporus xanthopus, Hericium americanum, Hericium erinaceus, Xylaria papulis, Pycnoporus sanguineus, Schizophyllum commune*, and *Lenzites betulina*, which are mostly from the mountainous regions and mushroom centers of Luzon. These macrofungi exhibited ZOI ranging from 6.37–18.87 mm among different extracts, but mostly ethanol extracts of the mushroom were assessed (Gaylan et al. 2017; Julian et al. 2018; De Leon et al. 2020b; Mendoza et al. 2020; Torres and Reyes 2020; Culliao et al. 2020).

Moreover, aside from *S. aureus*, which is commonly used as a test organism representing Gram-positive bacteria, macrofungi showed inhibitory activity against different Gram-positive bacteria. In the study of Estebal and Gaylan (2017) the ethanol extracts of *Fomes* sp., *G. applanatum, G. lucidum*, and *Polyporus* sp. showed antibacterial activity against *Bacillus cereus* ranging from a 7.00–8.00 mm mean diameter of ZOI. Moreover Serrano and Divina (2016) did an *in vitro* and *in vivo* antimicrobial assay of *Pleurotus sajor caju* and *Volvariella volvacea* against *Streptococcus iniae* in Nile Tilapia (*Oreochromis niloticus* L.). The result showed that *in vitro* analysis of extracts from *V. volvacea* and *P. sajor-caju* showed the widest inhibition activity of 9.51 mm and 11.07 mm, respectively, after 24 hours of incubation. On the other hand, *in vivo* assay wherein the Nile Tilapia were infected with the bacteria (*S. iniae*) showed that the percentage mortality of fish given with Pleurotus and Volvariella was lower than that of the negative control but higher than that of the positive control. These two macrofungi showed effective antibacterial properties, as they drastically reduced the death/mortality rate of infected fish compared to that in the negative control, where 80% of fish died after 96 hours. Furthermore, *S. commune* collected from Mt. Umubi in Nueva Vizcaya was also found to exhibit growth inhibition activity on *S. aureus* and *Enterococcus faecalis*. The dichloromethane and ethanol extracts of *E. faecalis* yielded ZOI values of 8.59 and 6.79 mm, respectively (Torres and Reyes 2020).

It can be pointed out from the data presented that among the extraction solvents used in macrofungal studies, the most commonly used is ethanol. Ethanol extraction is a very effective method, and the use of ethanol as an extraction solvent is thought to have numerous benefits over other organic solvents, including the fact that it is reasonably safe according to Ermi Hikmawanti et al. (2021). Ethanol has been also the choice of solvent for those looking to enhance extraction capacity rapidly and for the least amount of money. Furthermore, depending on the process and intended end products, ethanol extraction can be done at either room or low temperature, which makes it more convenient and efficient to use. When it comes to the macrofungi presented from the gathered studies from the Philippines, most showed inhibitory activity against *S. aureus*, as the majority of the studies used *S. aureus* as their test organism representative of Gram-positive bacteria since it causes a number of clinical conditions. It is the most common cause of skin and soft tissue infections such as boils, furuncles, cellulitis, and food poisoning. It is very closely associated with the environment, human health, and diseases, making it an ideal test organism in the search of possible antibacterial activity from new natural sources. However, it is still crucial to conduct new research with various mushroom species, as well as other pathogenic microorganisms that are harmful to the health of various organisms.

7.4.2 Antimicrobial Activity of Mushrooms against Gram-Negative Bacteria

The antibacterial efficacy of mushroom extracts or compounds against Gram-negative bacteria can also be evaluated using the same approaches used for Gram-positive bacteria. Different *in vitro* and *in vivo* assays may be conducted in accordance with the objective of a particular study. Antimicrobial

TABLE 7.3
Macrofungal Extracts with Antimicrobial Activity against Gram-Negative Bacteria.

Microorganism	Macrofungus	Extract used	Result	Reference(s)
Escherichia coli	*Tyromyces chioneus*	Hot-water extract	ZOI = 16.58 mm	Capistrano et al. (2018)
	Auricularia	Extracted chitosan	MIC = 250 μg/mL	Chang et al. (2019)
	Hericium americanum	Mycelial disc	ZOI = 15.85 mm	Julian et al. (2018)
	Xylaria papulis	Ethanol extract	ZOI = 16.87 mm	De Leon et al. (2020b)
	Wild *Lentinusstrigosus*	Ethanol extract	ZOI = 10.93 mm	Dulay and Pamiloza (2018)
	Pycnoporus sanguineus	Ethanol extract	ZOI = 7.11 mm	Mendoza et al. (2020)
	Daedaleopsis confragosa	Ethanol extract	ZOI = 12.33 mm	Culliao et al. (2020)
		Chloroform extract	ZOI = 10.00 mm	Culliao et al. (2020)
Klebsiella pneumoniae	*Daedaleopsis sp.*	Ethanol extract	MIC = 625 μg/mL	Estebal and Gaylan (2017)
	Daedalea sp.	Ethanol extract	MIC = 10000 μg/mL	Estebal and Gaylan (2017)
	Fomes sp.	Ethanol extract	MIC = 5000 μg/mL	Estebal and Gaylan (2017)
	Ganoderma applanatum	Ethanol extract	MIC = 20 μg/mL	Estebal and Gaylan (2017)
	Ganoderma lucidum	Ethanol extract	MIC = 5000 μg/mL	Estebal and Gaylan (2017)
	Lenzites elegans	Ethanol extract	MIC = 625 μg/mL	Estebal and Gaylan (2017)
	Microporus xanthopus	Ethanol extract	MIC = 5000 μg/mL	Estebal and Gaylan (2017)
	Polyporus sp.	Ethanol extract	MIC = 156 μg/mL	Estebal and Gaylan (2017)
Pseudomonas aeruginosa	*Ganoderma applanatum*	Ethanol extract	ZOI = 8.20 mm	Estebal and Gaylan (2017)
	Ganoderma lucidum	Ethanol extract	ZOI = 6.81 mm	Estebal and Gaylan (2017)
	Polyporus sp.	Ethanol extract	ZOI = 7.89 mm	Estebal and Gaylan (2017)

activity studies of macrofungi against Gram-negative bacteria in the Philippines have assessed various extraction methods and concentrations; however unlike the case of Gram-positive assessments, several resulted in no inhibition against the Gram-negative test organism. Therefore, the data collected for macrofungal extracts with antimicrobial activity against Gram-negative bacteria are not as extensive as for the Gram-positive bacteria. For instance, in a study by Kalaw and Albinto (2014) both *Coprinus comatus* and *Pleurotus cystidiosus* extracts were tested against Gram-positive and Gram-negative bacteria; however, there was no observed inhibitory activity against the test organism for Gram-negative bacteria (*Escherichia coli*). The same situation was reported for the acetonitrile and hexane extracts of *Lentinus tigrinus* and *Pleurotus djamor*, which showed inhibitory effects on the Gram-positive test organism but not on the Gram-negative test organism used (Dulay et al. 2017). Macrofungi in the Philippines showing antimicrobial activity against Gram-negative bacteria are shown in Table 7.3.

In addition, *Tyromyces chioneus* collected from an indigenous area in Bambang, Nueva Vizcaya, exhibited an inhibitory activity against the Gram-negative *E. coli* bacterium, as well as against *S. aureus*, which is a Gram-positive strain. Furthermore, the inhibition activity against *E. coli* exhibited a wider diameter of ZOI than the Gram-positive *S. aureus*. This study was made by Capistrano et al. (2018), who concluded that even though *T. chioneus*' antimicrobial property was nothing compared to that of the established antibiotic streptomycin sulfate, *T. chioneus* hot-water extract still can be utilized as a broad-range antibiotic against both the bacterial species. Another macrofungi that showed antibacterial activity against *E. coli* is *Auricularia* sp. collected from Manila by Chang et al. (2019). Reported in their study that extracted chitosan from *Auricularia* sp. produced better antibacterial activity than commercial chitosan from China. In their study, the extracted chitosan yielded an MIC of 250 μg/mL, whereas commercial chitosan yielded an MIC of 500 μg/mL. This suggests that the extracted chitosan has better antibacterial activity than the commercial chitosan because it can produce considerable inhibitory action at lower concentrations. The exotic *Hericium Americanum*, which was successfully cultivated in the Philippines, was also assessed for antibacterial activity against

E. coli. Julian et al. (2018) reported that *H. americanum* displayed its largest ZOI of 15.85 mm after 12 hours of incubation at a bacterial concentration of 0.1 mL bacterial suspension in 80 mL of sterile distilled water. After 24 hours of incubation, it yielded a 12.30 mm diameter ZOI, which shows a slight decrease compared to that seen after 12 hours incubation but still shows the antimicrobial potential of *H. americanum* against Gram-negative bacteria. Other promising macrofungi were also found to have intermediate inhibition effect on Gram-negative bacteria such as the ethanol extract of *Xylaria papulis*, a newly documented macrofungal species from Paracelis, Mt. Province, was reported by De Leon et al. (2020b) who yield a ZOI of 16.87 mm against *E. coli*, showing an intermediate potency. Other macrofungal ethanol extracts showing antibacterial activity against *E. coli* include those of wild *Lentinus strigosus* and *Pycnoporus sanguineus*, both collected from Central Luzon. The ethanol extract of wild *L. strigosus* exhibited a larger ZOI (10.93 mm) than that of cultured *L. strigosusis* (9.91 mm). Even so, statistical analysis showed that at the 5% level of significance, there was no significant difference identified between the effects (Dulay and Pamiloza 2018). *Pycnoporus sanguineus* also exhibited ZOI diameters ranging from 7.11 to 7.38 mm against *E. coli*. The extracts of 10–25-day-old mycelia were evaluated for inhibitory effects. In spite of the fact that extract from 10-day-old mycelia had the largest diameter ZOI, the differences between all mycelial extracts were not statistically significant according to Mendoza et al. (2020). *Daedaleopsis confragosa* was also assessed in both ethanol and a chloroform extract. It was reported by Culliao et al. (2020) that both showed growth inhibition against *E. coli*, with the ethanol extract yielding a wider ZOI (12.33 mm) than the chloroform extract (10.00 mm). The researchers interpreted this activity as partially active. Aside from *E. coli*, other used test organisms for the antimicrobial assessment of macrofungi against Gram-negative bacteria among the collected studies in the Philippines include *Klebsiella pneumoniae* and *Pseudomonas aeruginosa*. In the study made by Estebal and Gaylan (2017), there were eight macrofungal species collected from Bataan that showed antimicrobial activity against *K. pneumoniae*. These macrofungal species were identified as *Daedaleopsis* sp., *Daedalea* sp., *Fomes* sp., *Ganoderma applanatum*, *Ganoderma lucidum*, *Lenzites elegans*, *Microporus xanthopus*, and *Polyporus* sp. The broth microdilution assay used in this study showed that in each well, all macrofungal ethanol extracts demonstrated inhibitory activity against *K. pneumoniae*, with MIC_{50} values ranging 20–10000 µg/mL. The maximum activity (20 µg/mL) against *K. pneumoniae* was seen in *G. applanatum*. Meanwhile, three of the eight collected species exhibited antimicrobial activity against *P. aeruginosa*. The ethanol extracts of *G. applanatum*, *G. lucidum*, and *Polyporus* sp. displayed ZOI diameters of 8.20, 6.81, and 7.89 mm, respectively. Although these extracts produced inhibition zones, the recorded values were significantly lower than those of the positive control ciprofloxacin. The researchers provided a possible explanation as to why other macrofungal extracts were not able to produce ZOI in the disc diffusion experiment, explaining that some extracts may have failed to diffuse into the agar. As in the collected data for Gram-positive bacteria, the most used extract is an ethanol extract of macrofungal species, probably due to the convenience and efficiency of this extraction solvent, as was already mentioned earlier.

According to the Center of Disease Control and Prevention (CDC) in 2011, Gram-negative bacteria are usually resistant to a variety of medications and are becoming progressively resistant to the majority of antibiotics available. They are commonly associated with illnesses such as pneumonia, meningitis, wound infections, and bloodstream infections. To reduce the spread of Gram-negative bacteria, greater knowledge of how to regulate them in a healthcare setting is essential. The antimicrobial assay is one of the techniques used to gain more knowledge about a certain bacterial strain and to find new sources of antimicrobials that could help in combating the challenge of antimicrobial resistance.

7.5 TERATOGENIC MUSHROOMS

A teratogen is anything a person/animal is exposed to or ingests during pregnancy that's known to cause fetal abnormalities. Teratogenicity is the ability of a teratogen to cause fetal abnormalities or deformities. Embryos of Zebra fish (*D. rerio*) are now used as standard models in toxicity testing due to their similarity to higher forms of vertebrates. But despite its negative effect, teratogenicity testing can be desirable property because many anticancer drugs are teratogenic and teratogens can be used as anticancer drugs. How can this be, in theory, selective teratogens may suppress cancer cells that reactivate embryonic pathways while sparing most normal cells (Blagosklonny, 2005). However, despite its great potential as a anti-cancer drug, it is noteworthy to understand its teratogenic consequences. Several researchers are now examining the teratogenic consequences of numerous basidiomycetes that have attracted considerable attention.

Different toxicological endpoint parameters must be considered in assessing the teratogenic effects of a compound at different stages of development, such as head and tail malformation, growth retardation, scoliosis or flexure, hypopigmentation, and limited movement. Moreover, according to studies by local researchers, the Philippine mushroom research provides benchmark data on strong biological activity depending on the mushroom part, extract concentration, and time of exposure. One study reported by De Leon et al. (2020b) reported significant teratogenic effects of the ethanol extract of *Fomitopsis feei* on developing *D. rerio* embryos, including growth retardation, malformation of the tail, yolk deformities, pericardial edema, curved body, scoliosis, and lack of pigmentation. Embryos exposed to concentrations below 1000 ppm had a significant death rate. The highest hatchability was seen at concentrations of 750 ppm. In terms of heartbeat, as the concentration of the extract increased, the heartbeat decreased. Furthermore, the study conducted by Bustillos et al. (2016) stated that *Panaeolus antillarium* and *Panaeolus cyanesccens* extracts exhibited teratogenic effects. The results showed that exposing embryos to high concentrations (3%) of both extracts for extended periods increased the mortality of zebrafish embryos. Embryos exposed to 1% and 3% *P. antillarium* extract had the lowest hatchability of 8.33%, and those exposed to 1% and 3% *P. cyanescens* extract exhibited 16.67% and 0% hatchability, respectively. The most prominent teratogenic effect of both extracts was growth retardation. However, only *P. antillarium*-treated embryos exhibited malformations such as twisted tail and yolk deformation. Moreover, another significant study by Lazaro (2018) reported that *Pleurotus djamor*, *Pleurotus citrinopileatus*, and *Pleurotus ostreatus* hot-water extracts exhibited toxic and teratogenic effects against *D. rerio*. The results of this analysis revealed that the lethal effects of the three mushroom extracts were dose- and exposure-time-dependent. The 3% (highest concentration examined) *P. djamor* extract had the highest percentage mortality at 12 hours post treatment application (hpta), with a mean of 83.33%, while *P. citrinopileatus* had the lowest, with a mean of 41.67%. After 48 hours of exposure, embryos treated with 1% or higher concentrations of extracts from the three *Pleurotus* species showed a 100% mortality rate. Embryos treated with 0.1% and 0.5% extracts of the selected mushrooms had considerably lower hatchability than the control embryos. In 1% or higher concentrations of *P. djamor* and *P. ostreatus* extract, as well as in 0.5% or higher concentrations of *P. citrinopileatus* extract, no hatched embryos were found. The most noticeable morphological anomaly in embryos at 72 hpta was tail malformation, which was clearly caused by 0.1% and 0.5% extracts. Embryos treated with 0.5% *P. djamor* and *P. ostreatus* extracts and 0.1% *P. citrinopiletus* extract showed growth retardation.

Another study conducted by De Castro et al. (2016) on *Termitocyces clypeatus* showed that the mortality and growth rates of hatched embryos treated with 0.1% and greater concentrations of mushroom extract were considerably higher than those of control embryos. The hatchability of embryos treated with 0.5% and greater doses of *T. cleapeatus* extract was considerably lower than that of control embryos. The impact of the extract on embryos seemed to be dose-dependent. Several phenotypic endpoints of embryos discovered included wavy somites, unhatched embryos with twisted tail tips, embryos with delayed development (still in the segmentation phase), embryos with deformed tails, and coagulated embryos. Moreover, the study of De Castro and Dulay (2015a) stated that against *D. rerio*, the ethanolic extract of *Lentinus sajor-caju* and *Pleurotus ostreatus* yielded delayed growth and tail malformations such as bent tail and hook-like tail as the most marked teratogenic effects. The hatching rate of embryos treated with 2.5% of *L. sajor-caju* and 1% of *P. ostraetus* were significantly lower.

Another significant study on zebrafish embryo conducted by Reneses et al. (2016) reported that *Lentinus sajor-caju* hot-water extract yielded delayed growth after 12 hours of postfertilization at 0.5%, 1%, and 3% concentrations and resulted in coagulation after 48 hours of exposure. After 48 hours of exposure, hatchability of 0.1%, 0.05%, and control embryos were noted, and malformations such as the bent tail, hook-like tail, scoliosis, yolk deformities, and light coloration were found. The *Lentinus strigosus* mushroom mycelia and fruiting body hot-water extract assay conducted by Dulay and Pamiloza (2018), showed that at 100 μg/mL, the delayed development was the major teratogenic effect on zebrafish embryo. Hatchability and heart rate after 48 hours of exposure were recorded at 100 μg/mL and lower concentrations. *Ganoderma lucidum* hot-water extract exhibited teratogenic effects that were dose- and time-dependent. Malformations of the tail and delayed development were the most prevalent teratogenic effects of the extract (Dulay et al. 2012). At specific quantities, *Lentinus tigrinus* hot water extract drastically lowered hatchability and pulse rate and produced delayed development. As growth-delay-related endpoints, malformations of the tail, pericardial edema, and underdeveloped organs were identified (Dulay et al. 2014). Furthermore, *Termitomyces clypeatus* extract induced toxicity and teratogenicity in zebrafish embryos, yielding embryos with wavy somites, unhatched embryos with twisted tail tips, delayed embryonic development (still in the segmentation phase), embryos with malformed tails, and embryos with coagulated tails (De Castro et al. 2016).

The evaluation of *Lentinus squarrosulus* hot-water extract conducted by Reyes et al. (2016) revealed different sublethal dysmorphic effects like malformation of the tail, pericardial edema, yolk deformity, and scoliosis seen after 48 hours of exposure; however, 100% growth retardation was the most observed effect at higher concentrations of 0.5% and 1%. In addition, the tail was not detached, coagulation, lack of somites, and lack of visual heartbeat were recorded as lethal effects of the extract. Moreover, De Leon et al. (2020a), assessed the biological activity of *Xylaria papulis* embryo water. Growth retardation at concentrations ≥ 750 ppm for 12 to 48 hours and at 500 ppm for 36 and 48 hours, as well as abnormalities of the head and tail, were the most prominent effects. Due to exposure to varying concentrations of *X. papulis* ethanol extract, embryo mortality rates were dramatically elevated. However, the hatchability of zebrafish embryos was observed after 48 hpta. There was no observed hatchability in concentrations ≥500 ppm due to coagulation or halted growth before even reaching the hatching phase. Percent hatchability was only observed in 250 ppm with 8.33%, 100 ppm with 33.33%, and the control with 100%.

Another notable study performed by Dulay et al. (2014) on the hot-water extract of *Lentinus tigrinus* revealed functional activity at 0.5%–10%, where treated embryos exhibited developmental delays like tail malformations, underdeveloped organs, pericardial edema as the major growth-delayed endpoints. The absence of a heartbeat in the later stages was caused by developmental defects that restricted embryonic activity. In line with this, the study of Dulay et al. (2017) showed that *Volvariella volvacea* pileus and stipes exhibited teratogenic activity on all embryos at 0.10%, with 0.05% hot-water extract yielding different malformations like twisted tail and body, hook-like tail, and bent tail tip. Furthermore, no mortality was observed in embryos at both extracts at 0.10% and lower concentration, and the hatchability was recorded 100% at 0.01% concentration. However, at the concentration of 0.10%, both pileus and stipe extracts yielded 100% delayed development.

Furthermore, the study of teratogenic activity of *Trichaleurina celebica* conducted by Sogan et al. (2018) showed that the extract of fruiting bodies caused a 100% mortality rate after 48 hpta, with coagulation as the most lethal effect. The embryos exposed to 0.5% and 1% extracts exhibited tail malformation by the presence of bending tails and kinks as the most marked teratogenic endpoints. Aside from this, some incubated embryos of *T. celebica* showed the presence of head abnormalities and growth retardation that might be attributed to the delayed hatching process resulting in abnormalities. Similar symptoms of teratogenic effects on zebrafish embryos were discovered for various mushrooms, including *Panaeolus* sp., *Lentinus tigrinus*, *Ganoderma lucidum*, *Letinus sajor-caju*, and *Pleurotus ostreatus* (Dulay et al. 2012; Dulay 2014; De Castro and Dulay 2015a; Bustillos et al. 2016).

Another study by Dulay et al. (2012) on the extract of *Ganoderma lucidum* (Lingzhi or reishi mushroom), recorded that the fatal effect was dose- and exposure-period-dependent. At 1% or lower concentrations of extract and embryo water, hatching was accomplished 48 hpta. Embryos treated with 5% or higher concentrations hatched at a considerably lower rate ($p > 0.05$) than the control. The most prominent morphological defect in embryos at 72 hpta was tail malformation, which was clearly caused by 1% extract (55.56% tail malformation) and was found in all embryos subjected to 5% extract. Growth retardation was observed in embryos exposed to 5%, 10%, and 20% of *G. lucidum* extract treatment concentrations. At 72 hpta, embryos treated with 5% (44.44% mortality) or greater concentrations of the extract had significantly higher mortality rates ($p > 0.05$) than control embryos.

Teratogenicity and developmental toxicity of *Ganoderma lucidum, Fomitopsis feei, Termitocyces clypeatus, Lentinus sajor-caju, Lentinus strigosus Lentinus tigrinus, Lentinus squarrosulus, Panaeolus antillarium, Panaeolus cyanesccens, Pleurotus ostreatus, Pleurotus djamor, Pleurotus citrinopileatus, Trichaleurina celebica, Volvariella volvacea,* and *Xylaria papulis* extract in zebrafish embryo model were presented. Generally, the varying quantities of the mushroom extracts altered the hatchability of embryos: as the concentration of the extract increased, the hatchability percentage fell. One of the crucial sublethal effects of the extracts was a delay in embryonic development, which might explain the limited hatchability. It is also important to note that several developmental disorders may impede the hatching process. Malformations of the tail, for instance, inhibit the rupture of the chorion, which usually permits the larvae to hatch. When embryos were exposed to *G. lucidum* extract and other chemicals, such as coumarin and warfarin, the hatching process was disrupted. Another essential sublethal metric is heart rate. At the pharyngula stage of each embryo, when the tail was clearly pigmented, the pulse rate was measured to establish that the heartbeat recorded was a result of the varying extract concentrations and not the delayed development. In addition, the developing pulse rate was not changed by the nominal concentrations in the various investigations presented. Indeed, these highlighted mushrooms serve vital tasks in establishing the pharmacological potential of other mushroom species closely related to them.

7.6 CYTOTOXICITY OF MUSHROOMS

Mushrooms contain many bioactive compounds that are beneficial to human health but may become toxic when taken in high amounts. In line with this, some mushrooms are fully edible and others are recommended to be consumed in small amounts, with or without benefits to human health. The study conducted by Elumba et al. (2013) showed that the *Ganoderma applanatum* mushroom is consumable in small amounts because it contains varied cytotoxicity based on their brine shrimp lethality assay. According to them, *G. applanatum* contains some secondary metabolites that are toxic to humans when taken in high doses, even if it has good health benefits for humans. They added that some of the secondary metabolites of *G. applanatum* have DNA-binding activity in salmon sperm DNA. According to the authors, this is a good indicator that *G. applanatum* can be a candidate anticancer drug since many studies show that the mode of action for anticancer drugs is DNA binding.

On the other hand, the study of Abad (2014) showed that the mushroom *Pycnoporus coccineus* has medicinal benefits to humans as an aid for sore throat, oral thrush, and sore lips, and is used for teething rings. However, it is considered toxic and not suitable for eating when investigated for its cytotoxicity. Studies showed that *P. coccineus* has an LC_{50} value of 488.28 μg/mL, which is considered toxic. In addition to the cytotoxic effect and the medicinal benefits of *P. coccineus*, it contains 15 different secondary metabolites that are documented as having good benefits to humans, but it is not recommended for consumption due to the fact that some secondary metabolites may cause cytotoxic effects according to the literature cited in the study of Abad (2014).

7.6.1 Non-Toxic and Low-Toxic Mushrooms in the Philippines

Numerous mushrooms have been documented in the Philippines that are considered non-toxic and can be edible or not edible or low-toxic, which can be edible in small amounts or not edible. The study of Culliao et al. (2020) documented six different species of mushroom, wherein five of them, namely *Albatrellus ellisii*, *Lenzites betulina*, *Postia fragilis*, *Scleroderma citrinum*, and *Termitomyces eurrhizus*, were non-toxic according to Clarkson's criteria of toxicity, having an LC_{50} value of greater than 1000 μg/mL. The remaining species, *Daedaleopsis confragosa*, is considered a low-toxic mushroom with an LC_{50} value of 565.90 μg/mL and is not recommended for consumption. Among the five non-toxic species of mushroom, only *A. ellisii* and *T. eurrhizus* are edible, and *L. betulina*, *P. fragilis*, *S. citrinum* are not recommended for consumption due to bitter taste and the presence of a bioactive compound that may cause gastrointestinal and bodily irritation.

Many mushrooms are studied for their optimal culture conditions for mass production, and the cytotoxicity is also studied. In the study by De Leon et al. (2020a) on *Fomitopsis feei*, the authors found a suitable culture condition for the mushroom to grow and documented that *F. feei* has an LC_{50} value of 534.676 ppm in the brine shrimp lethality assay, which, according to their definition, is mildly toxic. As stated by the authors, *F. feei* has the potential to be an anticancer drug due to the fact that the bioactive compounds of this mushroom showed cytotoxicity in the brine shrimp assay. The study by Mendoza et al. (2020) suggested that the cytotoxicity of mushrooms occurs due to prolonged incubation of the culture. In their study, they observed that *Pycnoporus sanguineus* exhibited cytotoxicity in brine shrimp with an LC_{50} value of 154.83 μg/mL, which is considered mildly toxic, after 10 days of incubation. However, they noticed that after 15, 20, and 25 days of incubation, the LC_{50} value decreased to 90.45, 33.50, and 16.02 μg/mL, respectively. Those LC_{50} values were considered highly toxic according to the toxicity criteria. The authors suggest that the incubation period may affect the toxicity of mushrooms, and that mushrooms that are considered mildly toxic and highly toxic can be used to formulate drugs for cancer.

7.6.2 Highly Toxic Mushrooms in the Philippines

Mushrooms can also be highly toxic and may be considered edible or non-edible and have potential medicinal benefits. The highly toxic mushrooms documented in the Philippines are *Xylaria papulis* (De Leon et al. 2020b), *Trametes elegans* (Nanglihan et al. 2018), and *Polyporus grammocephalus* (Aquino et al. 2018). In the study of De Leon et al. (2020b), *X. papulis* exhibited a cytotoxic effect on brine shrimp, with an LC_{50} value of 218.507 ppm, which is considered highly toxic and non-edible according the Aldahi criteria of toxicity. The mortality rate of brine shrimp was 40% and 90% at low and high concentrations, respectively. According to authors, the cytotoxicity of *X. papulis* is a good indicator that it can be a potential drug for cancer, since most of the literature states that a high toxicity towards brine shrimp can indicated high toxicity in some cell lines. Furthermore, a similar study by Nanglihan et al. (2018) on *T. elegans* extract showed a cytotoxic effect in the brine shrimp, with 70% to 100% mortality and an LC_{50} value of 32.57 μg/mL which is considered highly toxic according to the toxicity criteria of Clarkson. The authors conclude that *T. elegans* is not an edible mushroom, but it may be a potential source of an anticancer drug due to its cytotoxic effect in the brine shrimp.

There is a highly toxic mushroom named *Polyporus grammocephalus* that is still edible. According to the study by Aquino et al. (2018), this mushroom is a highly toxic edible species consumed by the Gaddang community and the Aeta tribe as food. In their investigation of the cytotoxic effects on brine shrimp, the authors found that *P. grammocephalus* was highly toxic, having an LC_{50} value of 73.78 μg/mL. According to them, the cytotoxicity of *P. grammocephalus* is due to 12 mycochemicals detected that are not toxic to the human body. In line with this, they concluded that *P. grammocephalus* is an edible, highly toxic species of mushroom that can be used as a potential anticancer, anti-inflammatory, and antimicrobial drug.

7.6.3 Mushrooms in the Philippines that Exhibit Cytotoxicity on Cell Lines

Numerous reports in the literature showed that mushroom extracts represent potential candidates for anticancer agents since they have a cytotoxic effect on different cancer lines. One example is the study by Buniel et al. (2018) on the cytotoxicity of *Pleurotus ostreatus* var. *florida* in H-29 colon cancer cell lines. In their study, the authors found out the ethanolic extract of *P. ostreatus* var. *florida* caused damage to the DNA of H-29 cell lines, with a significant difference between the treated and untreated HT-29 in terms of the presence of DNA tail, tail moment, and tail length, which are indicators of genotoxicity. Additionally, this study also revealed that the extract has a cytotoxic effect on the cell lines, with an IC_{50} value of 164 mg/mL. Based on their results, they concluded that *P. ostreatus* var. *florida* can be a potential source of anticancer agents due to its genotoxic and cytotoxic effect on HT-29 colon cancer cell lines.

Another similar study was conducted by Arnante et al. (2017) on the cytotoxicity of a water-soluble polysaccharide extract of *Lenzites betulina* in K562 leukemia cells. This study showed that a polysaccharide of *L. betulina* can be a source of anticancer agents due to its genotoxic and cytotoxic effect on K562 leukemia cells. Its genotoxic effects are confirmed by significant differences in % DNA in the tail, tail moment, and tail length between treated and untreated K562 cells. For cytotoxicity, the water-soluble polysaccharide of *L. betulina* exhibited strong cytotoxicity on K562 cells, with an IC_{50} value of 0.07533 µg/µL, which is a relatively low value.

Furthermore, the compounds ergosterol peroxide (1) and ergosterol (2) isolated from dichloromethane extracts of *Geastrum triplex* and *Termitomyces clypeatus*, respectively, had cytotoxic effects on HT-29 (colon cancer), MCF-7 (breast cancer), and H69PR (small lung cell carcinoma) cells (Ragasa et al. 2020). These compounds exhibited strong cytotoxic activities on HT-29 colon cancer cells, with an IC_{50} of 1.79 µg/mL for compound 1 and of 2.98 µg/mL for compound 2, and in MCF-7 breast cancer cells, with an IC_{50} of 4.13 µg/mL for compound 1 and of 4.20 µg/mL for compound 2. These were comparable to the IC_{50} value of zeocin, which was the positive control with an IC_{50} value of 3.68 and 4.89 µg/mL for MCF-7 and HT-29, respectively. On the other hand, compounds 1 and 2 showed more moderate cytotoxic activity on H69PR (small lung cell carcinoma), with an IC_{50} of 7.78 µg/mL for compound 1 and of 6.83 µg/mL for compound 2, which was comparable to the IC_{50} value of 9.81 µg/mL for zeocin. According to this study, compound 1 (ergosterol peroxide) and compound 2 (ergosterol) isolated from *G. triplex* and *T. clypeatus*, respectively, can be used in further studies as potential sources of anticancer agents due to their cytotoxic activities on different cell lines.

Moreover, the study conducted by Dulay et al. (2021) reported the cytotoxicity of *Gymnopilus purpureosquamulosus* extract against acute myeloid leukemia (KBU, OCI-AML3, and MOLM13), lymphoma (J45.01, Toledo, and U937), and multiple myeloma (RPMI8226, MM.1R, and MC/CAR) cell lines. In their study, they found that the extract of *G. purpureosquamulosus* was highly cytotoxic to all leukemia, lymphoma, and multiple myeloma cells, with IC_{50} values ranging between 12–15 ppm in leukemia cell lines and 13–41 ppm in both lymphoma and multiple myeloma cells. In line with this, the authors observed that the extract of *G. purpureosquamulosus* inhibited cell proliferation and activated cell apoptosis due to the fact that they observed an increase in annexin V-positive cells and in the proportion of cells in the sub-G0/G1 phase. Moreover, and the extracts also activated the apoptosis pathway through a mechanism involving the extensive cleavage of PARP1 and caspase 3. The said activation of cell apoptosis is because of the damage to mitochondria. Additionally, the authors observed substantial depolarization of the mitochondrial membrane in cells exposed to *G. purpureosquamulosus* extract, which apparently induced the leaking of pro-apoptotic mitochondrial factors into the cytoplasm, triggering the caspase-dependent cascade of events that leads to apoptosis. Furthermore, this study also revealed that *G. purpureosquamulosus* extract possibly altered cellular redox status due to the activation of the ASK1-MEK-SAPK/JNK and ASK1-P38 MAPK pathway as suggested by the decreased protein levels of peroxiredoxin, thioredoxin, and thioredoxin reductase. Moreover, antioxidant N-acetylcysteine alleviated the cytotoxicity of

G. purpureosquamulosus. Pharmacological inhibition of SAPK/JNK and P 38, alleviated the *G. purpureosquamulosus* mediated cytotoxicity. Their result shows that the extract activated apoptosis in leukemia and lymphoma patient cell samples but not in mononuclear cell from health donors further supporting the therapeutic values of *G. purpureosquamulosus* for hematologic malignancies.

7.7 CONCLUSIONS AND FUTURE PERSPECTIVES

The ongoing documentation of various macrofungi thriving in the Philippines has revealed a diverse community of macrofungi. The macrofungi that were collected and studied throughout the country were mostly isolated from mountainous and forest regions. Although there are already documented and reported macrofungal discoveries and distributions, there is still a lot left to discover. To conclude, there have been a total of 189 species of macrofungi documented and classified under 32 families from different regions in Central Luzon. Polyporaceae has the highest number of species documented. The most tested macrofungal species against several pathogenic microorganisms are *Ganoderma applanatum*, *G. lucidum*, and *Polyporus* sp. which were tested and exhibited antimicrobial activity against several pathogens, including *Staphylococcus aureus*, *Bacillus cereus*, *Klebsiella pneumoniae*, and *Pseudomonas aeruginosa*, which were all assessed under one study, indicating a broad antimicrobial activities against both Gram-positive and Gram-negative bacteria. The most commonly used macrofungal extract is ethanol extract, which yield mid-range antibacterial activity shown by a 6–18 mm ZOI. The commonly reported biological activities of macrofungi aside from antimicrobial effects are antioxidant, antidiabetic, teratogenic, anticancer, cytotoxic, and scavenging activities. *Ganoderma lucidum*, *Fomitopsis feei*, *Termitocyces clypeatus*, *Lentinus sajor-caju*, *L. strigosus L. tigrinus*, *L. squarrosulus*, *Panaeolus antillarium*, *P. cyanesccens*, *Pleurotus ostreatus*, *P. djamor*, *P. citrinopileatus*, *Trichaleurina celebica*, *Volvariella volvacea*, and *Xylaria papulis* extracts exhibited strong teratogenicity. Additionally, the different mushroom species found in the Philippines that exhibit cytotoxic activities on brine shrimp are classified as non-toxic, low-toxic, mildly toxic, and highly toxic. Furthermore, the chapter also presents mushrooms with the potential to be anticancer agents due to their medium to strong cytotoxic activities on different cancer lines. These mushrooms could be used in the development of anticancer drugs because many anticancer drugs are teratogenic and cytotoxic, and teratogens can be used as anticancer drugs.

REFERENCES

Abad, M. 2014. Chemical constituents, antibacterial properties, and cytotoxic activities of *Pycnoporus coccineus*. *Quirino State University Research Journal* 3(1).

Aquino, Y., L. Vega, N. Medrano & R.M. Dulay. 2018. Mycochemicals, antioxidant and cytotoxic activities of *Polyporus grammocephalus* Berk (BIL7749). *International Journal of Biology, Pharmacy and Allied Sciences* 7(6), 966–975.

Arnante, M., M. Clerigo, A. Paano & M. Enriquez. 2017. Cytotoxic and genotoxic activity of an extract from the mushroom *Lenzites betulina* against K562 leukemia cells. In *Proceeding of the De La Salle University Research Congress 2017*, De La Salle University, Manila, Philippines.

Buniel, P., H. Scheewe, C. Jr. Sanico, Z. Alima & C. Demayo. 2018. Assessing the genotoxic and cytotoxic responses of the H-29 cancer cell lines on the ethanolic extracts of the oyster mushroom, *Pleurotus ostreatus* var. Florida. *International Journal of Pharmaceutical Science and Research* 9(10), 4201–4209.

Bustillos, R., Z. Paguio & R.M. Dulay. 2016. Philippine coprophilous mushrooms (*Panaeolus antillarium* and *Panaeolus cyanescens*) exhibit toxic and teratogenic effects in zebrafish (*Danio rerio*) embryo model. *Advances in Environmental Biology* 10(3), 75–80.

Capistrano, J.G., M.V. Perez, C.M. Soriano, M.K. Sunga, S.G. Villajuan, K.G.D. Waing & A.M. De Leon. 2018. Mycochemical, antioxidant, antibacterial and teratogenic activity of *Tyromyces chioneus* collected from Bambang, Nueva Vizcaya, Philippines. *International Journal of Biology, Pharmacy and Allied Sciences* 7(7), 1263–1274.

Centers for Disease Control and Prevention. 2011. Gram-negative bacteria infections in healthcare settings. *Center for Disease Control and Prevention*. Retrieved May 16, 2022, from www.cdc.gov/hai/organisms/gram-negative-bacteria.html

Chang, A.K., R.R. Frias, L.V. Alvarez, U.G. Bigol & J.P. Guzman. 2019. Comparative antibacterial activity of commercial chitosan and chitosan extracted from *Auricularia* sp. *Biocatalysis and Agricultural Biotechnology* 17, 189–195.

Culala, J.M. & R.M.R. Dulay. 2018. Species listing of naturally occurring mushrooms in Central Luzon State University, Science City of Munoz, Nueva Ecija, Philippines. *International Journal of Biology, Pharmacy, and Allied Sciences* 7(10), 1890–1899.

Culliao, A., R. Lumang-ay, G. Kingat, T. Colallad & M. Canaria. 2020. Preliminary bioactivity screening of crude extracts of six wild macrofungi from pine forests in Benguet and Mt. Province, Philippines. *Manila Journal of Science* 13, 74–88.

De Castro, M.E.G. & R.M.R. Dulay. 2015a. Toxic and teratogenic effects of *Lentinus sajor-caju* and *Pleurotus ostreatus* ethanolic extracts in *Danio rerio* embryo model. *International Journal of Biology, Pharmacy, and Allied Sciences* 4(4), 2261–2269.

De Castro, M.E.G. & R.M.R. Dulay. 2015b. Macrofungi in multistorey agroforestry systems in Mt. Makiling Forest Reserve, Los Baños, Laguna, Philippines. *Journal of Chemical, Biological and Physical Sciences* 5(2), 1646–1655.

De Castro, M.E.G., R.M.R. Dulay & M. Enriquez. 2016. Toxic and teratogenic effects of medicinal and culinary mushroom, *Termitomyces clypeatus*, collected from the Termite Mound in Mt. Makiling Forest Reserve, Los Baños, Laguna, Philippines on developing embryos of zebrafish (*Danio rerio*). *Scholars Research Library* 8(5), 237–242.

De Leon, A.M., E.O. Diego, L.K.F. Domingo & S.P. Kalaw. 2020b. Mycochemical screening, antioxidant evaluation and assessment of bioactivities of *Xylaria papulis*: A newly reported macrofungi from Paracelis, Mountain Province, Philippines. *Current Research in Environmental & Applied Mycology (Journal of Fungal Biology)* 10(1), 300–318.

De Leon, A.M., A.R. Dulay, A.L. Villanueva & S.P. Kalaw. 2020a. Optimal culture conditions and toxicity assessment of *Fomitopsis feei* (Fr.): A newly documented macro fungus from Philippines. *Studies in Fungi* 5(1), 491–507.

De Leon, A.M., J.J.D. Luangsa-ard, S.C. Karunarathna, K.D. Hyde, R.G. Reyes & T.E.E. Dela Cruz. 2013. Species listing, distribution, and molecular identification of macrofungi in six Aeta tribal communities in Central Luzon, Philippines. *Mycosphere* 4(3), 478–494.

De Leon, A.M., M. Pagaduan, B. Panto & S.P. Kalaw. 2021. Species listing of macrofungi found in Paracelis Mountain Province, Philippines. *The CLSU International Journal of Science & Technology* 5(2), 22–40.

Dulay, R.M.R., M.C. Arenas, S.P. Kalaw, R.G. Reyes & E.C. Cabrera. 2014. Proximate composition and functionality of the culinary-medicinal tiger sawgill mushroom, *Lentinus tigrinus* (Higher basidiomycetes), from the Philippines. *International Journal of Medicinal Mushrooms* 16, 85–94.

Dulay, R.M.R., E.C. Cabrera, S.P. Kalaw & R.G. Reyes. 2020. Nucleotide sequencing and identification of wild mushrooms from the Philippines. *Biocatalysis and Agricultural Biotechnology* 27(1), 4–18.

Dulay, R.M.R., J. Carandang VI, S.P. Kalaw & R.G. Reyes. 2020. Distribution and species listing of wild macrofungi in Sitio Canding, Barangay Maasin, San Clemente, Tarlac Province, Philippines. *Journal of Applied Biology & Biotechnology* 8(5), 7–15.

Dulay, R.M.R., S.P. Kalaw, R.G. Reyes & N. Alfonso. 2012. Teratogenic and toxic effects of lingzhi or reishi medicinal mushroom, *Ganoderma lucidum* (w.curt.:fr.) p. karst. (higher basidiomycetes), on zebrafish embryo as model. *International Journal of Medicinal Mushrooms* 14(5), 507–512.

Dulay, R.M.R., S.P. Kalaw, R.G. Reyes & E. Cabrera. 2014. Embryo-toxic and teratogenic effects of Philippine strain of *Lentinus tigrinus* (tiger sawgill basidiomycetes) extract on zebrafish (*Danio rerio*) embryos. *Annals of Biological Research* 5(6), 9–14.

Dulay, R.M.R. & C. Maglasang. 2017. *Species Listing of Naturally Occurring Mushroom in Agro-Ecosystem of Barangay Bambanaba*, Cuyapo, Nueva Ecija, Philippines.

Dulay, R.M.R., L.A. Miranda, J.S. Malasaga, S.P. Kalaw, R.G. Reyes & C.T. Hou. 2017. Antioxidant and antibacterial activities of acetonitrile and hexane extracts of *Lentinus tigrinus* and *Pleurotus djamour*. *Biocatalysis and Agricultural Biotechnology* 9, 141–144.

Dulay, R.M.R. & D.G. Pamiloza. 2018. Proximate composition and bioactivities of hairy sawgill mushroom, *Lentinus strigosus* (Bil 1324) from the Philippines. *International Journal of Biology, Pharmacy, and Allied Sciences* 7(3), 361–369.

Dulay, R.M.R., A. Rivera & M. De Castro. 2017. Wild paddy straw mushroom (*Volvariella volvacea*) exhibit toxic and teratogenic effects in zebrafish (*Danio rerio*) embryo. *International Journal of Biology, Pharmacy adn Allied Sciences* 6(5), 910–917.

Dulay, R.M.R., B. Valdez, Y. Li, S. Chakrabarti, B. Dhillon, S.P. Kalaw & E. Cabrera. 2021. Cytotoxicity of *Gymnopilus purpureosquamulosus* extracts on hematologic malignant cells through activation of the SAPK/JNK signaling pathway. *PLoS ONE* 16(5), 0252541.

Elumba, Z., F. Teves & R. Malaluan. 2013. DNA-binding activity and *in vivo* cytotoxicity of *Ganoderma applanatum* (Pers.) Pat. supercritical-CO2 extracts. *African Journal of Microbiology Research* 7(3), 202–210.

Ermi Hikmawanti, N.P., S. Fatmawati & A.W. Asri. 2021. The effect of ethanol concentrations as the extraction solvent on antioxidant activity of Katuk (*Sauropus Androgynus* (L.) merr.) leaves extracts. *IOP Conference Series: Earth and Environmental Science* 755(1), 012060.

Estebal, J.C. & C.M. Gaylan. 2017. *Antimicrobial and Antioxidant Activities of Ten Wild Mushroom Species Collected from Bataan*, Philippines. DSpace Repository. Retrieved February 17, 2022, from http://dspace.cas.upm.edu.ph:8080/xmlui/handle/123456789/903

Gaylan, C.M., J.C. Estebal, O.A. Tantengco & E.M. Ragragio. 2017. Anti-staphylococcal and antioxidant properties of crude ethanolic extracts of macrofungi collected from the Philippines. *Pharmacognosy Journal* 10(1), 106–109. https://doi.org/10.5530/pj.2018.1.19

Jacob, J., E. Romorosa & S.P. Kalaw. 2017. Species listing of macroscopic fungi in Isabela State University, Isabela as baseline information. *International Journal of Agricultural Technology* 13(7.1), 1199–1203.

Julian, A.V., C.A. Wright & R.G. Reyes. 2018. Prelude to successful cultivation of *Hericium* in the Philippines: Understanding its mycelial growth response on different culture media and its antibacterial activity. *International Journal of Pharmaceutical Research & Allied Sciences* 7(2), 1–7.

Kalaw, S.P. & R.F. Albinto. 2014. Functional activities of Philippine Wild Strain of *Coprinus comatus* (O. F. Müll.: Fr.) pers and *Pleurotus cystidiosus* O. K. Miller grown on rice straw based substrate formulation. *Mycosphere* 5(5), 646–655.

Kinge, T.R., N.A. Apalah, M.N. Theobald, N. Acha, & A.M. Mih. 2017. Species richness and traditional knowledge of macrofungi (Mushrooms) in the awing forest reserve and communities, Northwest Region, Cameroon. *Journal of Mycology* 24, 1–9. http://doi.org/10.1155/2017/2809239

Lazaro, J. 2018. Teratogenic activity of exotic mushrooms on the developing embryos of zebrafish (*Danio rerio*). *Ascendens Asia Journal of Multidisciplinary Research Abstracts* 2(5).

Lindequist, U., T.H. Niedermeyer & W.D. Jülich. 2005. The pharmacological potential of mushrooms. *Evidence-Based Complementary and Alternative Medicine* 2(3), 285–299.

Liwanag, J., E. Santos, F. Flores, R. Clemente & R.M.R. Dulay. 2017. Species listing of macrofungi in Angat watershed reservation, Bulacan province, Luzon island, Philippines. *International Journal of Biology Pharmacy and Allied Sciences* 6(5), 1060–1068.

Lopez, A.B.V., J.D.C. Aquino, J.Q. Undan, K.G.D. Waing & J.R. Undan. 2016. Molecular identification and phylogeny of some wild microscopic fungi from selected areas of Jaen, Nueva Ecija, Philippines. *Advances in Environmental Biology* 10(12), 153–158.

Lu, H., H. Lou, J. Hu, Z. Liu & Q. Chen. 2020. Macrofungi: A review of cultivation strategies, bioactivity, and application of mushrooms. *Comprehensive Reviews in Food Science and Food Safety* 19(5), 2333–2356.

Mendoza, W., R.M.R. Dulay, M.J.G. Valentino & R.G. Reyes. 2020. Mycelial biomass and biological activities of Philippine mushroom *Pycnoporus sanguineus* in time-course submerged culture. *Journal of Applied Biology and Biotechnology* 8(5), 8–3.

Nacua, A., H. Pacis, J. Manalo, C. Soriano, N. Tosoc, R. Padirogao & C. Deocaris. 2018. Macrofungal diversity in Mt. Makiling forest reserve, Laguna, Philippines: With floristic update on roadside samples in Makiling Botanic Gardens (MBG). *Biodiversitas Journal of Biological Diversity* 19(4), 1579–1585.

Nanglihan, K., R.M.R. Dulay & S.P. Kalaw. 2018. Myko-actives and functional activities of Philippine wild mushroom *Trametes elegans*. *International Journal of Biosciences* 13(5), 402–408.

Paguirigan, J., B. David, R. Elsisura, A. Gamboa, R. Gardaya, J. Ilagan, J. Mendiola, P. Pineda, R. Samelin & M.A. Pangilinan. 2020. Species listing and distribution of macrofungi in Consocep Mountain Resort, Tigaon and Mount Isarog National Park, Goa, Camarines Sur. *Philippine Journal of Systematic Biology* 14(1), 1–9.

Ragasa, C., G. Oyong M. Tan, M. de Los Reyes & M. de Castro. 2020. Cytotoxic sterols from Philippine mushrooms. *Asian Journal of Chemistry* 32(5), 1197–1202.

Raneses, M., R.M.R. Dulay & A.M. De Leon. 2016. Proximate nutritive composition and teratogenic effect of *Lentinus sajor-caju* collected from Banaue, Ifugao Province, Philippines. *International Journal of Biology, Pharmacy and Allied Sciences* 5(7), 1771–1786.

Reyes, N., S.P. Kalaw & A.M. De Leon. 2016. Antioxidant screening, teratogenecity and antifungal property of *Lentinus squarrosulus* (Mont.) singer. *Asian Journal of Biochemical and Pharmaceutical Research*, 2231–2560.

Reyes, R.G., M.A.G. del Rosario, J.P.G. Padua, M.C. Malonzo, A.J. Barza, R. Sumi, M. Mori, M. Miyazawa & F. Eguchi. 2017. The first report on the mycelial growth performance and antibacterial activity of *Collybia reinakeana* RGR-FE—NSC strain, a Philippine endemic edible mushroom. *International Journal of Pharmaceutical Research & Allied Sciences* 6(1), 113–119.

Reyes, R.G. & M.G. Nair. 2016. Ligninolytic and leaf litter degrading mushrooms from the Philippines with antioxidant activities. *International Journal of Pharmaceutical Research and Allied Sciences* 5(4), 67–74.

Serrano, V.K.D. & C.C. Divina. 2016. In—Vitro and In-Vivo antimicrobial assay of Pleurotus sajor caju, Volvariella volvacea and Auricularia auricula against Streptococcus iniae in nile tilapia (Oreochromis niloticus L.). *International Journal of Agricultural Technology* 12(7.2), 2155–2164.

Sibounnavong, P., C.C. Divina, S.P. Kalaw, R.G. Reyes & K. Soytong. 2008. Some species of macrofungi at Puncan, Carranglan, Nueva Ecija in the Philippines. *Journal of Agricultural Technology* 4(2), 105–115.

Sogan, M.M., J.A.L. Maslang & R.M.R. Dulay. 2018. Myco-chemicals and teratogenic activity of wild mushroom *Trichaleurina celebica* from Mt. Palali, Quezon, Nueva Vizcaya, Luzon Island, Philippines. *The CLSU International Journal of Science & Technology* 3(2), 17–23.

Sridhar, K. & S. Deshmukh. 2019. *Advances in Macrofungi: Diversity, Ecology and Biotechnology*. Taylor & Francis, Routledge, CRC Press, London, ISBN 9780367780333.

Sumic, Z., A. Tepic, S. Vidovic, A. Vakula, J. Vladic & B. Pavlic. 2017. Process optimization of Chanterelle (*Cantharellus cibarius*) mushrooms vacuum drying. *Journal of Food Processing and Preservation* 41(2), 12.

Tadiosa, E.R., E.A. Agbayani & N.T. Agustin. 2011. Preliminary study on the macrofungi of Bazal-Baubo watershed, Aurora Province, Central Luzon, Philippines. *Asian Journal of Biodiversity* 2(1), 149–171.

Torres, M.L. & R.G. Reyes. 2020. Antibacterial potential of different crude extracts of *Schizophyllum commune* mycelia grown on coconut water. *International Journal of Agricultural and Environmental Research* 6(2), 134–141.

Torres, M.L., E. Tadiosa & R.G. Reyes. 2020. Species listing of macrofungi on the Bugkalot Tribal community in Alfonso Castañeda, Nueva Vizcaya, Philippines. *Current Research in Environmental & Applied Mycology (Journal of Fungal Biology)* 10(1), 475–493.

Undan, J.Q., D.O. Alfonso, R.M.R. Dulay, A.M. De Leon, S.P. Kalaw, J.R. Undan & R.G. Reyes. 2016. Molecular identification and phylogeny of different macrofungi in Mt. Bangkay, Cuyapo, Nueva Ecija, Philippines based on ITS nrDNA region. *Advances in Environmental Biology* 10(12), 35–42.

8 Uses of Mushrooms in Traditional Medicine in China

Pu Liu, Yu Li

8.1 INTRODUCTION

Medicinal mushrooms are an important part of traditional Chinese medicine, with a history going back thousands of years. Medicinal mushrooms are mostly large macrofungi used to treat diseases and have a high medicinal value. Research on traditional Chinese medicine shows that most medicinal fungi belong to the Ascomycotina and Basidiomycotina (Xu and Lan 1997). However, the definition of "medicinal mushroom" has always been controversial. Broadly speaking, medicinal mushrooms are all fungi used as a source of pharmaceuticals; colloquially speaking, medicinal mushrooms are large medicinal fungi (Li 2019).

Medicinal mushrooms improve human health and may have preventive, inhibitory, or therapeutic effects on diseases. A variety of medicinal compounds (Lin and Zhang 2004), such as amino acids, proteins, polysaccharides, flavonoids, sterols, glycosides, terpenes, alkaloids, minerals, organic acids, vitamins, antibiotics, and other active compounds, can be extracted from the fruiting bodies, mycelia, sclerotia, or spores of medicinal mushrooms (Liaw et al. 2017). The primary biological activities of medicinal mushrooms are related to their anti-tumor and anti-oxidation properties, along with their capability to improve immunity, provide vitamins, lower blood sugar levels, reduce blood lipids, enhance memory, regulate gastrointestinal function, and protect the liver.

Based on their use, medicinal mushrooms can be divided into two types. The first is the edible and medicinal type, which represents raw material for food supplements in the traditional Chinese diet. Examples include *Auricularia heimuer* (Figure 8.1), *Lentinula edodes* (Figure 8.2), *Flammulina filiformis*, *Hericium erinaceus*, *Agaricus brasiliensis*, and *Ophiocordyceps sinensis*. The second type includes those used specifically for medical purposes. Among these are *Ganoderma lucidum* (Figure 8.3), *Tramates versicolor*, *Sanghuangporus vaninii* (Figure 8.4), *Wolfiporia cocos*, *Polyporus umbellatus*, *Shiraia bambuslcola*, and *Claviceps purpurea*. Currently, the list of the most promising medicinal mushrooms encompasses more than 20 species, such as *Wolfiporia cocos*, *Polyporus umbellatus*, *Laccocephalum mylittae*, *Ganoderma lucidum* (Figure 8.3), *Tremella fuciformis* (Figure 8.5), *Tramates versicolor*, *Schizophyllum commune* (Figure 8.7), *Grifola frondosa*, *Ophiocordycepe sinensis*, *Hericium erinaceus*, *Sanghuangporus vaninii* (Figure 8.4), and *Inonotus obliquus* (Huang 2005).

8.1.1 History of Medicinal Mushrooms in China

As already indicated, medicinal mushrooms have a long history in China. About 2,500 years ago, mushrooms were first reported as being used for medical purposes. In the first medicinal book to be produced in China in the late Eastern Han Dynasty, which was entitled "The Classic of Herbal Medicine" and covered the subject of agricultural and medicinal plants, 10 different kinds of medicinal mushrooms were mentioned. These were such species as *Ganoderma lucidum* (Figure 8.3), *Wolfiporia cocos*, *Polyporus umbellatus*, *Isaria cicadae*, *Lasiosphaera fenzlii*, *Auricularia heimuer* (Figure 8.1), and *Laccocephalum mylittae*. Some of these are still in use today. In the "The Compendium of Materia Medica" compiled by the famous medical scientist Li during the Ming

DOI: 10.1201/b23190-8

FIGURE 8.1 Wild fruiting bodies of *Auricularia heimuer.*

FIGURE 8.2 Wild fruiting bodies of *Lentinula edodes.*

FIGURE 8.3 Wild fruiting bodies of *Ganoderma lucidum.*

FIGURE 8.4 Wild fruiting bodies of *Sanghuangporus vaninii.*

FIGURE 8.5 Wild fruiting bodies of *Tremella fuciformis.*

FIGURE 8.6 Wild fruiting bodies of *Stropharia rugosoannulata.*

FIGURE 8.7 Wild fruiting bodies of *Schizophyllum commune*.

Dynasty (1368–1644), more than 20 kinds of medicinal mushrooms were mentioned, including *Auricularia heimuer* (Figure 8.1), *Lentinula edodes* (Figure 8.2), *Lasiosphaera fenzlii*, *Ganoderma lucidum* (Figure 8.3), *Isaria cicadae*, *Wolfiporia cocos*, and *Polyporus umbellatus*. During the early Qing Dynasty, Wang first discussed *Ophiocordyceps sinensis* in his "Essentials of Materia Medica," and Zhao described the shape and pharmacology of *Ophiocordyceps sinensis* collected from Yunnan, Guizhou, Sichuan, and other places in "Supplements to Compendium of Materia Medica."

In 1915, Hu produced "Fungal Identification," which began the process of modern fungal research in China. In modern China, domestic scholars began to investigate edible and medicinal fungal resources using the concepts and approaches of modern mycology. For example, the first edition of "Medicinal Fungi in China" written by Liu in (1974) included 78 species of medicinal mushrooms, a number that increased to 117 in the second edition (Liu 1978) and to 121 in the third edition (Liu 1984). In 1987, 272 medicinal mushrooms were recorded in "Illustrations of Medicinal Fungi in China" by Ying et al. (1987), most of which were macrofungi. Xu and Lan (2002) listed 311 medicinal mushrooms with high medicinal value in "Chinese Medicinal Mycology." More and more people have recently paid more attention to studies or investigations of medicinal fungi in China. However, many nomenclatural inconsistencies have been found in the Chinese reports. Dai and Yang (2008) critically reviewed the publications on Chinese medicinal fungi and enumerated 473 medicinal fungi from China in the checklist. One year later, Dai et al. (2009) reviewed the species diversity and utilization of medicinal mushrooms and fungi in China, identifying 540 species. In 2013, 314 kinds of medicinal fungi were described in "Illustrations of Medicinal Fungi in China," edited by Dai et al. (2013). Two hundred and sixty species of medicinal macrofungi were recorded in "A Field Guide to Antitumor Medicinal Fungi in China," edited by Chen (2013). More than 650 species of medicinal fungi in China were systematically described in "Medicinal Fungi of China" edited by Wu et al. (2013). In 2019, Wu et al. (2019) reported the resource diversity of Chinese macrofungi, which encompassed edible, medicinal, and poisonous species—they identified a total of 692 kinds of medicinal fungi and 277 kinds of edible and medicinal fungi in China. There were about a dozen medicinal fungi actively being used for clinical purposes at that time.

8.2 CHEMICAL COMPOUNDS FOUND IN MEDICINAL MUSHROOMS

8.2.1 Polysaccharides

In the 1950s, fungal polysaccharides were found to exhibit anticancer activity. After separation and purification, the chemical structure, physicochemical properties, and pharmacological effects of fungal polysaccharides were determined. Fungal polysaccharides are polymers formed by the condensation of more than 10 monosaccharides through glycosidic bonds. They are a kind of natural polymer compound that can be separated from the fungal fruiting body, mycelium, or fermentation broth. Fungal polysaccharides exhibit multiple functions, such as lowering blood pressure, reducing blood lipids, strengthening the stomach, promoting the biosynthesis of nucleic acids and proteins, repairing damaged tissue cells, delaying aging, and protecting the liver; they also possess immune regulatory, anti-oxidative, antiviral, anticoagulant, and anti-radiation, properties, among others. The anti-tumor activity of fungal polysaccharides is the most important biological activity known thus far, and it is also the most active focus of the research. Immune cells such as phagocytes, natural killer cells, cytotoxic T lymphocyte cells, and lymphokine-activated killer cells activate the network endothelial system to produce phagocytose, remove aging cells and foreign bodies, and regulate the formation of antibodies and complements, thereby improving the body's anti-tumor immunity. In the development of medicine and health care products, a variety of medicinal mushroom polysaccharides have been widely used, such as those from *Morchella esculenta*, *Hericium erinaceus*, *Laccocephalum mylittae*, *Schizophyllum commune*, *Polyporus umbellatus*, *Tramates versicolor*, *Grifola frondosa*, *Ganoderma lucidum*, *Cordyceps militaris*, *Auricularia heimuer*, *Wolfiporia cocos*, and *Tremella fuciformis* (Zhang and Li 2014).

The natural oyster mushroom selenium polysaccharide (Se-POP-3) displays a degree of activity against gastric and colon cancers *in vitro*, along with anti-tumor activity (Zhang et al. 2022). *Ganoderma sichuanense* (= *Ganoderma lingzhi*) polysaccharides have various pharmacological activities that can improve immunity; blood oxygen supply capacity; DNA, RNA, and protein synthesis in the liver, bone marrow, and blood; and can prolong life (Guan and Cong 1981). *Ganoderma lucidum* polysaccharides can also help resist fatigue, improve endurance, and promote recovery from fatigue, and offers potential for the development of anti-fatigue drugs (Chang et al. 2015). Several studies have shown that *G. lucidum* polysaccharides exhibit good antioxidant activity and can scavenge free radicals, improve immune function, and delay aging to a certain extent (Cai et al. 2021). The macromolecular polysaccharides extracted from the mycelium of *Hirsutella sinensis* also have the effect of regulating the intestinal microflora, reducing weight gain in mice, and improving insulin resistance (Wu et al. 2019). *Sanghuangporus sanghuang* polysaccharides not only exhibit anticancer activity, but also can be used to prepare composite materials. For example, *Sanghuangporus sanghuang* polysaccharides can be used as a natural reductant to prepare silver nanoparticles. Porous sponge dressing can be obtained after the lyophilization of these silver nanoparticles, as can chitosan solution, which has the potential as an ideal material for wound dressing (Ran et al. 2018).

Morchella angusticeps polysaccharides show strong cholesterol-lowering activity under a high-cholesterol diet; these polysaccharides can be used for the prevention of hyperlipidemia (Li et al. 2017; Tang et al. 2019). Polysaccharides are one of the primary active components of *Hericium erinaceus*. They exhibit clear protective effects on gastric mucosa, along with anti-tumor properties, and blood lipid-lowering effect. These polysaccharides are widely used in the treatment of gastritis, gastric ulcers, and other gastric diseases in China (Ma et al. 2021). *Isaria cicadae* (= *Cordyceps cicadae*) has a high nutritional value, and it has been shown that the polysaccharides of *Isaria cicadae* (= *Cordyceps cicadae*) have antioxidant and anti-aging activities, which can be developed into new drugs to reduce aging (Zhu et al. 2020). Some fungal polysaccharides, such as those of *Ganoderma lucidum*, *Tremella fuciformis*, and *Auricularia heimuer*, also exhibit good protective effects against radiation and can be used for the development of radiation protection drugs to reduce the damage to normal tissues after radiation treatment. These polysaccharides represent a kind of natural radiation protective agent with a wide range of sources (Wang et al. 2019).

8.2.2 Terpenoids

Terpenoids refer to turpentine and many volatile oils that contain some unsaturated hydrocarbons, have the general formula $C_{10}H_{16}$, and can be divided into monoterpenes, diterpenes, disesquiterpenes, triterpenes, and polyterpenes according to their chemical composition. At present, most of the compounds separated from medicinal mushrooms are sesquiterpenes, diterpenes, and triterpenes, which mainly have anticancer activity.

Many studies have shown that terpenoids from medicinal mushrooms have tumor cell-inhibiting, immunoregulatory, blood lipid-lowering, hepatoprotective, anti-inflammatory, anti-oxidative, and antiviral properties (Yang et al. 2019). Terpenoid was first separated from the fruiting bodies of *Ganoderma lucidum* (Figure 8.3) in 1982. At present, more than 150 triterpenoids have been separated from *Ganoderma*, including ganoderic acid, lucidenic acid, and ganolactone (Liu et al. 2019). A new triterpenoid, ethyl lucidenates A, was obtained from the fruiting body of *Ganoderma lucidum* (Figure 8.3), and its anti-tumor activity has been tested (Li et al. 2013). Two new triterpenes were isolated from *Ganoderma cochlear*, namely fornicatin G and fornicatin H (Peng et al. 2012). A new terpenoid, lucidone D, was isolated from the fruiting bodies of *Ganoderma lucidum* (Figure 8.3) (Liu et al. 2013). Lanostane triterpenoids were isolated from *Ganoderma applanatum* and effectively inhibit the production of fat with potential therapeutic implications for obesity (Su et al. 2021). Lanostane triterpenoids are also found in *Ganoderma lucidum* (Figure 8.3) and have been found to exert anti-inflammatory activity (Su et al. 2020). Various compounds such as sesquiterpenes and morelsins isolated from *Morchella importuna* have inhibitory effects on pancrelipase (Yang et al. 2020).

8.2.3 Steroids

Steroids are natural products that exist in organisms. Ergosterol is an important part of the cell membranes of microorganisms and exhibits activities associated with anti-tumor, antioxidant, antibacterial, and other effects in medicinal fungi. Sterol compounds are found in *Claviceps purpurea*, *Ganoderma sichuanense* (alternatively, *Ganoderma lingzhi*), *Polyporus umbellatus*, *Ophiocordyceps sinensis*, *Hericium erinaceus*, and other medicinal fungi.

The steroid compounds in the mycelia of *Hericium erinaceus* have been studied, and the content of a steroid identified as ergosterol by spectral analysis was higher in an alcohol extract than in a water extract (Li et al. 2001). The total sterols of *Ganoderma* were extracted using high-speed shearing flash extraction technology and included ergosterol and ergosta-7, 22-dien-3β-ol (He et al. 2013). The fingerprints of liposoluble constituents in the spore powder of *Ganoderma sichuanense* have used to determine the content of ergosterol and to identify whether the cell wall of spore cells was broken; the time of cell wall breaking had significant effects on the dissolution of ergosterol and liposoluble constituents in the spore powder of this mushroom (Wang et al. 2018). The sterol contents in the different parts of *Ophiocordyceps sinensis* (stromae, larval hosts, and sterile stipes) were determined by high-performance liquid chromatography. The results showed that the sterol contents in *Ophiocordyceps sinensis* were the highest in the larval hosts (6.08 mg/g), followed by the sterile stipes (5.03 mg/g). The value for the stroma was the lowest (3.51 mg/g) (Qian et al. 2018). The contents of three kinds of sterols (ergosterol, cholesterol, and sitosterol) in five different development stages of *Ophiocordyceps sinensis*, namely mycelium, ossified larva, 1 cm stroma stage, 4–6 cm stroma stage, and sporulation stage, were detected and analyzed, and contents were the highest when the stroma length was 4–6 cm (Qian et al. 2019).

8.2.4 Proteins

Medicinal mushrooms are rich in amino acids, various types of which have anticancer activities and play a role in regulating immunity at high levels. *Cordyceps militaris* is rich in protein components, among which glutelin is the most abundant (43.11%), followed by albumin (36.47%) and globulin (17.94%); these can meet the amino acid needs of the human body. This protein can be effectively used as a food ingredient to improve nutrition and product variety (Yu et al. 2021). The content of protein in *Cordyceps militaris* can exceed about 40%, which is more than double that of *Ophiocordyceps sinensis*, and it has as many as 18 kinds of amino acids. Studies have shown that *Cordyceps militaris* protein has anti-tumor and antibacterial activities (Fan and Lin 2013). The cytotoxic activity of new nano-protein extracts from *Ganoderma lucidum* (Figure 8.3) and *Pleurotus ostreatus* on macrophages and osteocytes has been studied *in vitro*. The results showed that the nano-protein extract could inhibit the proliferation of macrophages and had no effect on the growth of osteocytes, which offers potential for the treatment of orthopedic diseases (Zhang et al. 2017). The ultrasonic-assisted extraction process of *Cordyceps militaris* protein was optimized by response surface analysis. The results showed that the protein extraction rate under ultrasonic-assisted conditions was 85.6%, which was 1.68 times than that of the control group without ultrasonication (Hou et al. 2018).

8.2.5 Other Compounds

There are also other compounds in medicinal mushrooms, such as nucleic acids, polyhydric alcohols, vitamins, organic acids, fatty acids, enzymes, furane derivatives, flavonoids, trace elements, and pigments.

Flavonoids in medicinal mushrooms exhibit anticancer, antiviral, anti-inflammatory, and other activities, along with strong antioxidant properties, which can reduce harmful free radicals into stable, harmless products and can effectively remove oxygen free radicals from the body (Li et al. 2015; Yang et al. 2014). The results of an inhibition test on breast cancer cells using four flavonoid extracts from edible fungi showed that the flavonoid extracts in *Auricularia heimuer* have anti-tumor activity (Miao et al. 2014); pigments from *Auricularia heimuer* (Figure 8.1) can significantly increase cell viability and improve the cell viability of LO2 cells pretreated with ethanol. There are indications that this pigment has a therapeutic effect on alcoholic liver injury (Hou et al. 2021).

Research on the flavonoids of *Stropharia rugosoannulata* (Figure 8.6) indicated that they have a certain inhibitory effect against *Penicillium* sp., *Escherichia coli*, and *Saccharomyces cerevisiae* (Chen et al. 2010). A comparative study of secondary metabolites and activities of fruiting bodies and mycelia of *Sanghuangporus sanghuang* showed that the total flavonoid content of mycelia in an alcohol extract was higher than of fruiting bodies, so the fermented mycelia are superior to fruiting bodies for anti-oxidation (Zhang et al. 2020). *Schizophyllum commune* (Figure 8.7) also contains phenolic acids, which have an analgesic effect and can be used as in mechanistic studies of neurasthenia (Yao et al. 2016). *Cordyceps militaris* contains a variety of pigments, such as zeaxanthin and carotenoids, among which the cordyxanthin content is the highest, accounting for 86.7% of the total cordycepin. Lou et al. (2018) provided optimal process parameters for the extraction of carotenoids from *Cordyceps militaris* (Dong et al. 2013; Lou et al. 2018). In addition, *Cordyceps militaris* is also rich in liposoluble vitamins such as vitamins A, D, and E and water-soluble vitamins such as vitamins B and C (Tu et al. 2020).

8.3 PHARMACOLOGICAL ACTIVITIES INVOLVING MEDICINAL MUSHROOMS

8.3.1 Anti-Tumor Activity

The anti-tumor substances in medicinal mushrooms are primarily polysaccharides, which are thought to improve immunity or activate the phagocytic ability of reticuloendothelial cells and macrophages, directly kill tumor cells, and promote the production of interferons and interleukins to inhibit the growth of tumor cells.

A submicron powder of *Grifola frondosa* fruiting body has anti-tumor activity (Liu et al. 2018). *Fomitopsis pinicola* has a high content of anti-tumor substances, and its anti-tumor activity is mainly related to the apoptosis signaling pathway through expression of BAX, BCL-2, cytochrome C, and PARPs (Li 2018). *Polyporus umbellatus* polysaccharides can inhibit the spontaneous metastasis of Lewis lung cancer and S180 sarcoma (Zhang et al. 2013). *Hericium erinaceus* polysaccharides may play a certain role in the clinical treatment of gastric cancer, esophageal cancer, and other cancers (Younis 2017). *Wolfiporia cocos* polysaccharides can inhibit the growth of tumor cells, improve the phagocytic ability of macrophages, and promote the formation of immunoglobulins (Jiang et al. 2021). Compared with the control group, the polysaccharides in the mycelia of *Phellinus igniarius* significantly increased the thymus index and spleen index of tumor-bearing mice, reduced the division index of S-180 ascites tumor cells, and prolonged the survival of tumor-bearing mice (Yang et al. 2006). *Sanghuangporus vaninii* polysaccharides have inhibitory effects on breast and colon cancer cells and can be used as natural polysaccharides for the treatment of tumors (Wan et al. 2020; Wang et al. 2018). Eight compounds, including ergosterol, were extracted and identified from *Inonotus hispidus* and shown to inhibit the proliferation of human hepatoma cells (Yang et al. 2019).

8.3.2 Immune Regulation

According to several modern pharmacological studies, the extracts of many medicinal fungi exhibit the function of regulating human immunity and have been developed into a variety of drugs for the adjuvant treatment of tumors (Wu et al. 2006; Xie et al. 2006). The sulfated derivatives of lentinan have anti-HIV and immunity-enhancing activities (Wang et al. 1999). Similarly, carboxymethylated *Wolfiporia cocos* polysaccharides can enhance immune responses in mice and inhibit tumor growth (Chen 2002). Polysaccharides are the most abundant active substances in *Ganoderma leucocontextum* and primarily play an immunoregulatory role, acting on a variety of immune cells to regulate the immune function of the human body. It is a good auxiliary drug (Gao et al. 2020; Ren et al. 2021). The new polysaccharide MIPW50–1 was isolated from *Morchella importuna* and is composed of N-acetylglucosamine, galactose, glucose, and mannose. The experiment showed that this polysaccharide has an immunomodulatory effect and can be used as a potential immune enhancer (Wen et al. 2019). The protein hydrolysate of *Grifola frondosa* has a good immune-enhancing activity and can promote the proliferation of splenocytes; it acts as a new iron supplement that can be used as an immune-enhancing agent (Yuan et al. 2019). Wang (2020) studied the conditioning effect of *Tremella fuciformis* polysaccharides on exercise-induced immunosuppression and showed that these polysaccharides can increase levels of serum immunoglobulin, regulate inflammation and anti-inflammatory cytokines, and play a certain conditioning role in post-exercise immunosuppression.

8.3.3 Antioxidant Activity

Anti-oxidation is the removal of antioxidant free radicals. In the physiological process of cellular metabolism, a large number of unpaired atoms or groups of electrons are generated and can attack cell membranes and react with serum antiprotease, thereby causing harm to the human body (Seifried et al. 2007). Several studies have shown that synthetic antioxidants in clinical use can cause additional damage to the human body when scavenging free radicals (Yuan et al. 2008; Sun and Kennedy 2010).

Studies have shown that fungal polysaccharides, such as those produced by *Grifola frondosa*, *Tremella fuciformis* (Figure 8.5), and *Ganoderma lucidum* (Figure 8.3), can scavenge hydroxyl radicals and peroxides (Liu 1997). Numerous studies have found that the active protein components of *Ganoderma lucidum* (Figure 8.3) have strong antioxidant properties (Du et al. 2006). The antioxidant capacities of polysaccharides in different medicinal mushrooms vary, with differences seen between *Lentinula edodes* (Figure 8.2), *Flammulina velutipes*,

FIGURE 8.8 Wild fruiting bodies of *Naematelia aurantialba*.

Tremella fuciformis (Figure 8.5), *Hericium erinaceus*, *Tramates versicolor*, *Ganoderma lucidum* (Figure 8.3), *Auricularia cornea*, and *Boletus edulis* (Zhang et al. 2011; He and Zhang 2013; Sun et al. 2010). Terpenoids and polyphenols of various medicinal fungi also have antioxidant capacities, as seen in *Ganoderma atrum*, *Pleurotus ostreatus*, *Boletus edulis*, *Lasiosphaera fenzlii*, *Naematelia aurantialba* (Figure 8.8), *Tremella fuciformis* (Figure 8.5), and *Auricularia heimuer* (Figure 8.1) (Wang et al. 2015; Liu et al. 2011; Chen et al. 2011).

In addition, the antioxidant substances in medicinal mushrooms can also achieve antioxidant effects by improving the activity of related cells and enhancing the immunity of the human body (Gao et al. 2013). The extraction process of triterpenoids from the mycelia of *Sanghuangporus sanghuang* has been further optimized, and the resulting products exhibit high scavenging ability for hydroxyl radicals, as well as high antioxidant activity (Cai et al. 2019). *In vitro* antioxidant tests have shown that *Tropicoporus linteus* (= *Phellinus linteus*) polysaccharides also have strong DPPH free radical-scavenging ability and dose-dependent antioxidant activity (Wang et al. 2014). *Morchella esculenta* (Figure 8.9) can effectively scavenge hydroxyl, DPPH, and superoxide free radicals, and thus can be used as a natural antioxidant (Cai et al. 2018). *Schizophyllan commune* polysaccharides have good antioxidant activity, and the antioxidant activity of sulfate-esterified *Schizophyllan commune* polysaccharides is 2.6 times that of unprocessed polysaccharides, which indicates that sulfate-esterified modification can improve the antioxidant activity of *Schizophyllan commune* polysaccharides (Zhang and Cai 2017).

8.3.4 Anti-Inflammatory Activity

Ganoderma lucidum polysaccharides and triterpenoids also exhibit high anti-inflammatory activity. *Ganoderma lucidum* sterols also have anti-inflammatory activities and can be made into natural anti-inflammatory agents to prevent and treat inflammatory diseases (Xu et al. 2021). Cell pharmacological studies have shown that the polysaccharides in *Tropicoporus linteus* have good anti-inflammatory activity and can reduce the stimulation of lipopolysaccharide on macrophage RAW264 and inhibit the production of inflammatory cytokines such as TNF-α, IL-1, and IL-6 by regulating the signaling pathway of PPARγ and MAPK, which suggests that *Tropicoporus linteus* has a potential value in the treatment of inflammatory diseases (Hu et al. 2018). The anti-inflammatory and antioxidant drugs derived from the mycelia of *Sanghuangporus sanghuang* have a therapeutic effect on acute lung

FIGURE 8.9 Wild fruiting bodies of *Morchella esculenta*.

injury and can potentially be used for the treatment and prevention of inflammatory diseases (Lin et al. 2017). The exopolysaccharides of *Schizophyllum commune* (Figure 8.7) have a triple helical conformation and exhibited intestinal anti-inflammatory activity in a dextran sulfate sodium-induced colitis mouse model (Du et al. 2017). An exopolysaccharide was obtained from the fermentation broth of *Ophiocordyceps sinensis* that exhibited significant anti-inflammatory activity *in vitro* and may be used as a new anti-inflammatory drug (Li et al. 2021). In addition, three polyphenolic compounds have been isolated from *Inonotus hispidus* and exhibit dose-dependent anti-inflammatory activities (Li et al. 2019).

8.3.5 Antibacterial and Antiviral Activity

In nature, fungi need to fight off the threat of other microorganisms, and medicinal fungi in particular have antibacterial components that help humans. Studies have shown that preparations of *Hericium erinaceus* (preparation liquid of fruiting bodies, deep fermented culture, and polysaccharides) have different inhibitory effects on bacteria, yeasts, and molds (Zhang et al. 2003). *Ganoderma lucidum* polysaccharides can resist the herpes simplex virus (HSV). Research has shown that *Ganoderma lucidum* polysaccharides can inhibit the replication of filtering pathogens and achieve an antiviral function by interfering with the adsorption of the virus into the target cell at an early stage (Liu et al. 2004). *Ganoderma lucidum* also contains benzolactone constituents ganodumones A–F (1–6), which are new terpenoid compounds that were shown to have antibacterial effects in a biological activity evaluation (Lu et al. 2020). The antibacterial activity of compounds in the methanol extract of *Inonotus hispidus* has been determined, and it has been found that the content of total flavonoids was the highest, and the antibacterial activity of total flavonoids was the highest (Chen et al. 2018). The measured antibacterial activity of *Inonotus hispidus* showed that the inhibition rate against *Escherichia coli* was 40.35% under a high dose of 70% ethanol extract (Zan et al. 2015). The fermentation broth of the mycelia from both *Ganoderma lucidum* and *Hericium erinaceus* have an inhibitory effect on bacteria, such as *Staphylococcus aureus*, *Escherichia coli*, and *Bacillus subtilis*. The antibacterial effect of *Hericium erinaceus* is stronger than that of *Ganoderma lucidum* (Luo et al. 2017). Similarly, the fermented liquid of

Wolfiporia cocos also has a good inhibitory effect on *Staphylococcus aureus* and *Candida albicans* (Wu et al. 2016).

The water extract of *Amauroder marudis* has an inhibitory effect on HSV-1 and enterovirus 71 (EV-71), with the advantage of low toxicity. This extract has the potential to be further developed into antiviral products (Cai et al. 2020; Yu et al. 2015). The antibacterial ability of *Auricularia* polysaccharides was studied and the results showed that *Auricularia cornea* var. *Li* (Figure 8.10) has a strong antibacterial ability against *Escherichia coli*, *Staphylococcus aureus*, and *Bacillus subtilis*, compared with the antibacterial abilities of *Auricularia polytricha* and *Auricularia auricula* (= *Auricularia heimuer*, Figure 8.1) (Luo et al. 2018).

8.3.6 Hypoglycemic and Hypolipidemic Activity

It has been found that the polysaccharides of medicinal fungi with hypoglycemic and hypolipidemic activities include those from *Ganoderma lucidum*, *Cordyceps militaris*, *Tramates versicolor*, *Hericium erinaceus*, *Auricularia heimuer*, *Grifola frondosa*, *Tremella fuciformis*, and *Morchella esculenta*.

FIGURE 8.10 *Auricularia cornea* var. *Li* (Photographed by Fang et al. 2018).

The polysaccharides and triterpenes contained in *Ganoderma lucidum* have the effect of lowering blood lipid levels. Studies have found that *Ganoderma lucidum* polysaccharides have good cholesterol- and triglyceride-lowering effects, and at the same time can improve the body's antioxidant capacity and prevent the occurrence of atherosclerosis induced by hyperlipidemia (Chen et al. 2005). *Cordyceps militaris* is a kind of tonic medicinal material in China. It has a high nutritional value, can enhance immunity, and has an obvious inhibitory effect on lung cancer and liver cancer. Relevant studies have shown that the mycelia of *Cordyceps militaris* have a lowering effect on blood lipid levels (Wu et al. 2008). The intracellular polysaccharides of *Trametes versicolor* (= *Coriolus versicolor*, Figure 8.11) can significantly reduce serum cholesterol, triglyceride, and low-density lipoprotein cholesterol levels in patients with hyperlipidemia and can increase the levels of high-density lipoprotein cholesterol (Rao and Tang 2007). The water-soluble polysaccharides in *Morchella esculenta* (Figure 8.9) and a homogenate from dry fruiting bodies of *Boletus aereous* both showed good hypolipidemic activity (Wang et al. 2004; Ming et al. 2009). The neutral polysaccharide SSIPS1 isolated and purified from mycelia of *Sanghuangporus sanghuang* has potential inhibitory activity on α-amylase and α-glucosidase, as well as a hypoglycemic effect (Cheng et al. 2020). The polysaccharide PPB-2 extracted from the fruiting bodies of *Sanghuangporus baumii* (= *Phellinus baumii*, Figure 8.12) also has the functions of lowering blood sugar and protecting the liver (Zhang et al. 2019).

8.3.7 Other Activities

In addition, medicinal mushrooms also regulate the endocrine system and metabolism, exhibit a synergistic effect on the nervous system, relieve cough and asthma, and possess anti-aging and hypoglycemic properties. Different doses of *Hericium erinaceus* extract were used to treat a model of D-galactose-induced subacute aging in mice, and the results showed that this extract could effectively delay the aging effect of mice (Liu and Li 2009). In another study, it was also found that macromolecular polysaccharides from the mycelia of *Ganoderma lucidum* (Figure 8.3) could reduce the body weight of mice, via the regulation of gut microbiota (Chen et al. 2009). According to the

FIGURE 8.11 Wild fruiting bodies of *Trametes versicolor*.

FIGURE 8.12 Wild fruiting bodies of *Sanghuangporus baumii.*

reported statistics, more than 100 million people in China suffer from diabetes, which is one of the countries with the largest number of diabetic patients (Yue et al. 2020).

Clinical studies have shown that traditional medicines have adverse effects on the liver, kidneys, and gastrointestinal tract of diabetic patients, while various compounds in medicinal mushrooms have good hypoglycemic effects (Du et al. 2018; Ma et al. 2019). The main hypoglycemic mechanism of the medicinal mushroom *Grifola frondosa* is regulation of enzymes that to improve insulin resistance and repair the damaged islet B cells (Xiao et al. 2015). *Schizophyllum commune* (Figure 8.7) can relieve exercise fatigue. For example, Zhang et al. (2020) studied the effects of the dry powder and fermented broth of fruiting bodies of the *Schizophyllum commune* on the contents of liver glycogen, muscle glycogen, and enzyme (succinate dehydrogenase and lactic dehydrogenase) of mice. The results showed that the anti-fatigue effect of a fermented broth of this species was weaker than that of the dry powder of fruiting bodies.

8.4 SPECIES OF IMPORTANT MEDICINAL MUSHROOMS AND THEIR PRIMARY USES IN CHINESE MEDICINE

Medicinal mushroom resources are rich in China and have great potential for development. With the development of technology, a series of medicines and healthcare products have been developed in China. Traditional pills and decoctions, injections, granules, tablets, capsules, syrups, healthcare oral solutions, drinks, and instant teas have also been developed (Li et al. 2007).

8.4.1 Ascomycetes

Family Shiraiaceae

(1) *Shiraia bambusicola* (medicinal)
Products: *Shiraia bambusicola* granules, capsules (Wu et al. 2020).

Family Morchellaceae

(2) *Morchella esculenta* (edible and medicinal)
Products: *Morchella esculenta* condiment (Fu et al. 2012).

Family Clavicipitaceae

(3) *Claviceps purpurea* (edible and medicinal)
Products: nicergoline (www.nmpa.gov.cn/).

Family Cordycipitaceae

(4) *Beauveria bassiana* (medicinal)
Products: raw powders of *Beauveria bassiana*, pesticide products (the main role is to kill pests, used in biological control) (Liu et al. 2003).

(5) *Cordyceps militaris* (edible and medicinal)
Products: *Cordyceps militaris* powders, compound oral solution, lung clearing capsules, powder capsules, tablets (Wu et al. 2020; Xiao et al. 2003).

(6) *Isaria cicadae* (medicinal)
Products: *Isaria cicadae* powders, tablets (Cheng 2016; Li et al. 2018).

Family Ophiocordycipitaceae

(7) *Ophiocordyceps sinensis* (medicinal)
Products: *Ophiocordyceps sinensis* capsules, eye drops, capsules for heart disease, *Cordyceps cephalosporium* mycelia (Wu et al. 2020; Xiao et al. 2003).

Family Xylariaceae

(8) *Daldinia concentrica* (medicinal)
Application: used for AIDS treatment.

(9) *Xylaria nigripes* (medicinal)

8.4.2 Basidiomycetes

Family Agaricaceae

(1) *Agaricus bisporus* (edible and medicinal)
Products: liver blood rehabilitation tablets, and liver health supplement tablets (www.nmpa.gov.cn/).

(2) *Agaricus brasiliensis* (edible and medicinal)
(3) *Calvatia gigantea* (medicinal)
(4) *Calvatia lilacina* (medicinal)

Family Fistulinaceae

(5) *Fistulina hepatica* (medicinal)

Family Lycoperdaceae

(6) *Lasiosphaera fenzlii* (medicinal)
Products: compound puffball granules, and compound puffball salicylic acid powders (Wu et al. 2020).

Family Omphalotaceae

(7) *Lentinula edodes* (edible and medicinal, Figure 8.2)
Products: *Lentinula edodes* polysaccharide injections, polysaccharide capsules, polysaccharide tablets, liver health supplement tablets, lentinan injections (Wu et al. 2020; Xiao et al. 2003).

(8) *Marasmiellus androsaceus* (medicinal)
Products: *Marasmiellus androsaceus* tablets, capsules, medicinal wine, extract, compound hydrochloride tablets (Wu et al. 2020).

Family Physalacriaceae

(9) *Armillaria mellea* (edible and medicinal)
Products: gastrodia tuder halimasch tablets, gastrodia tuder halimasch powder, *Tremella fuciformis*, and *Armillaria mellea* tablets (Wu et al. 2020).

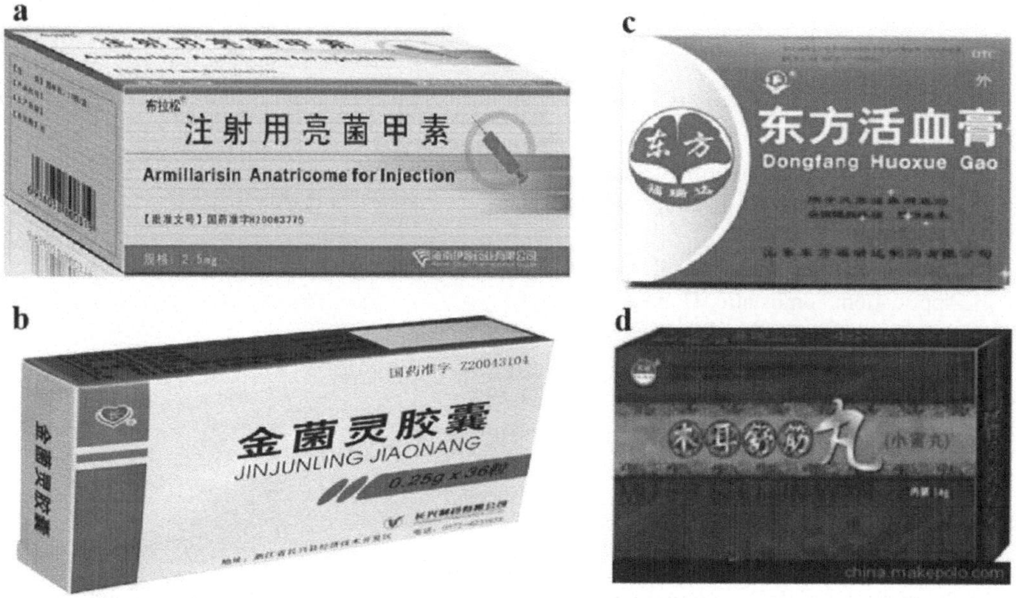

FIGURE 8.13 Primary types of medicinal mushroom medicines in China. (a) Armillarisin A injection, (b) *Flammulina filiformis* capsules, (c) Oriental blood-activating extract, (d) *Auricularia heimuer* muscle relaxant pills (Figures from (Li, 2019).

(10) *Armillariella tabescens* (edible and medicinal)
Products: *Armillariella* syrup, injections (Figure 8.13-a), combined sodium chloride injection, tablets, oral solution (Wu et al. 2020).

(11) *Flammulina filiformis* (edible and medicinal)
Products: *Flammulina filiformis* capsules (Figure 8.13-b), tablets, Oriental blood-activating extract (Figure 8.13-c) (Xiao et al. 2003; Li et al. 2009).

Family Pleurotaceae

(12) *Pleurotus ostreatus* (edible and medicinal)

Family Sarcomyxaceae

(13) *Sarcomyxa edulis* (edible and medicinal)

Family Schizophyllaceae

(14) *Schizophyllum commune* (Figure 8.7) (edible and medicinal)
Products: sizofiran (Xiao et al. 2003).

Family Strophariaceae

(15) *Stropharia rugosoannulata* (Figure 8.6) (edible and medicinal)

Family Tricholomataceae

(16) *Tricholoma matsutake* (edible and medicinal)
Products: compound *Tricholoma matsutake* capsules (Wu et al. 2020).

Family Geastraceae

(17) *Geastrum fimbriatum* (medicinal)
Application: hemostatic (Han et al. 2019).

(18) *Geastrum triplex* (medicinal)
Application: hemostatic (Han et al. 2019).

Family Phallaceae

(19) *Dictyophora indusiate* (=*Phallus indusiatus*, edible and medicinal)

Family Auriculariaceae

(20) *Auricularia heimuer* (edible and medicinal) (Figure 8.1)
Products: *Auricularia heimuer* muscle relaxant pills (Figure 8.13d) (Xiao et al. 2003).

(21) *Auricularia cornea* (Figure 8.14) (edible and medicinal)
Products: blood lipid-lowering tablets (Wu et al. 2020).

FIGURE 8.14 Wild fruiting bodies of *Auricularia cornea*.

Family Hymenochaetaceae

(22) *Inonotus hispidus* (medicinal)
Products: Inonotus hispidus tea.

(23) *Inonotus obliquus* (medicinal)
(24) *Sanghuangporus vaninii* (medicinal) (Figure 8.4)
Products: slices, tea, and superfine powders of *Sanghuangporus vaninii* (Tu et al. 2018).

(25) *Sanghuangporus baumii* (Figure 8.12) (medicinal)
Products: slices, tea, superfine powders of *Sanghuangporus baumii* (Tu et al. 2018).

(26) *Sanghuangporus sanghuang* (medicinal)
Products: slices, tea, superfine powders of *Sanghuangporus sanghuang* (Tu et al. 2018).

Family Fomitopsidaceae

(27) *Piptoporus betulinus* (medicinal)

Family Ganodermataceae

(28) *Ganoderma applanatum* (medicinal)
Products: anti-hepatitis soft capsules (Wu et al. 2020).

(29) *Ganoderma lucidum* (Figure 8.3) (medicinal)
Products: *Ganoderma lucidum* tea, spore powder capsules, oral solution, granules, spore powders, dispersible tablets, syrup, polysaccharide (Wu et al. 2020; Xiao et al. 2003).

(30) *Ganoderma sinense* (medicinal)

Family Laetiporaceae

(31) *Laetiporus sulphureus* (medicinal)

Family Meruliaceae

(32) *Irpex lacteus* (medicinal)
Products: kidney health supplement capsules (www.nmpa.gov.cn/).

Family Grifolaceae

(33) *Grifola frondosa* (edible and medicinal)
Products: *Grifola frondosa* polysaccharide lozenges, polysaccharide capsules (Wu et al. 2020; Song 2009).

Family Polyporaceae

(34) *Fomes fomentarius* (medicinal)
(35) *Laccocephalum mylittae* (medicinal)
Products: *Laccocephalum mylittae* capsules, tablets (Wu et al. 2020).

(36) *Polyporus umbellatus* (medicinal)
Products: *Polyporus umbellatus* polysaccharide injection, polysaccharide capsules (Wu et al. 2020; Xiao et al. 2003).

(37) *Trametes robiniophila* (medicinal)
Products: *Trametes robiniophila* granules (Wu et al. 2020; Xiao et al. 2003).

(38) *Trametes versicolor* (Figure 8.11) (medicinal)
Products: *Trametes versicolor* granules for liver health, capsules, corilan capsules, *Polystictus* glycopeptide tablets, polysaccharopeptide capsules (Wu et al. 2020; Xiao et al. 2003).

Family Fomitopsidaceae

(39) *Wolfiporia cocos* (medicinal)
Products: haw *Wolfiporia cocos* granules, polysaccharide oral solution, carboxymethyl pachyman (Wu et al. 2020; Xiao et al. 2003).

Family Incertae sedis

(40) *Taiwanofungus camphoratus* (medicinal)
Products: *Taiwanofungus camphoratus* oral solution, capsules (Huang et al. 2018; Li et al. 2013).

Family Hericiaceae

(41) *Hericium erinaceus* (edible and medicinal)
Products: *Hericium erinaceus* biscuits, tablets, powders, capsules for the stomach, compound granules (Xiao et al. 2003; Yang et al. 2003; Wang et al. 2008).

Family Tremellaceae

(42) *Tremella fuciformis* (Figure 8.5) (edible and medicinal)
Products: *Tremella* polysaccharide, polysaccharide capsules (Wu et al. 2020; Xiao et al. 2003).

Family Naemateliaceae

(43) *Naematelia aurantialba* (= *Tremella aurantialba*) (Figure 8.8) (edible and medicinal)

Products: *Tremella aurantialba* polysaccharide peptide capsule (Yang et al. 2021).

8.5 PERSPECTIVES

Mycomedicines are derived from specific parts of medicinal mushrooms, and the main medicinal parts include fruiting bodies, sclerotia, stroma, spore powders, fermented broth, and their processed products. With the development of modern technology, the number and types of mycomedicines are increasing day by day. The nomenclature of medicinal mushrooms should follow the principle of "one fungus, one Latin name," so that the development of mycomedicines can make great contributions to human healthcare (Bao et al. 2021). At present, there are still many medicinal mushrooms that remain in the stage of collecting wild resources and domestication, and the development and utilization of new drugs need to be further expanded.

With the rapid development of the economy in China, the concept of "big health" has become increasingly popular among the population, and more and more people pay attention to health preservation. Therefore, the development of medicinal mushrooms is of great significance to the healthcare industry (Wan et al. 2019). Medicinal mushrooms, as an important source of potential human food and pharmaceutical products, have considerable development potential and broad application prospects.

REFERENCES

Bao, H.Y., T. Bau & Y. Li. 2021. Development of the science of mycomedicine in China and its prospect. *Journal of Fungal Research* 19(1):12–18 (In Chinese).

Cai, C., S.W. Lei, S.D. Chen, Y.Z. Xie, H.H. Pan & G. Wang. 2020. Antiviral screening of extracta from 17 kinds of edible and medicinal fungi. *Acta Edulis Fungi* 27(1):75–84 (In Chinese).

Cai, C.S., J.X. Ma, C.R. Han, Y. Jin, G.Z. Zhao & X.W. He. 2019. Extraction and antioxidant activity of total triterpenoids in the mycelium of a medicinal fungus, *Sanghuangporus sanghuang*. *Scientific Reports* 9(1):7418–7428.

Cai, M., H.Y. Xing, B.M. Tian, J. Xu, Z.H. Li, H. Zhu, K. Yang & P.L. Sun. 2021. Characteristics and anti-fatigue activity of graded polysaccharides from *Ganoderma lucidum* separated by cascade membrane technology. *Carbohydrate Polymers* 269(1):118329.

Cai, Z.N., W. Li, S. Mehmood, W.J. Pan, Y. Wang, F.J. Meng, X.F. Wang, Y.M. Lu & Y. Chen. 2018. Structural characterization, in vitro and in vivo antioxidant activities of a heteropolysaccharide from the fruiting bodies of *Morchella esculenta*. *Carbohydrate Polymers* 195(1):29–38.

Chang, C.J., C.S. Lin, C.C. Lu, J. Martel, Y.F. Ko, M.O. David, S.F. Tseng, R.W. Tsung, Y.Y.M. Chen, J.D. Young & H.C. Lai. 2015. *Ganoderma lucidum* reduces obesity in mice by modulating the composition of the gut microbiota. *Nature Communications* 6:7489–7508.

Chen, C.X. 2002. The effect of carboxy methmethy pachymaran(CMP) on immunological function in mice. *Edible Fungi* 4:39–41 (In Chinese).

Chen, J.C., Y.B. Li, L. Wu, X.H. Zhou, Y.G. Duan & P.F. Lai. 2010. Extraction of flavonoids from stropharia rugoso-annulata farlow and the antibacterial activity. *Journal of Beijing Technology and Business University (Natural Science Edition)* 28(6):9–13 (In Chinese).

Chen, K.L. 2013. *A Field Guide to Anti-tumor Medicinal Fungi in China*. Beijing: Science Press (In Chinese).

Chen, L., W.F. Li, B. Ling & J. Ming. 2011. Extraction and antioxidant activity of polyphenols from *Tremella aurantialba*, *Tremella fucitormis*, and *Auricularia aurricula*. *Food Science* 32(20):52–56 (In Chinese).

Chen, W.Q., S.H. Luo, H.Z. Li & H. Yang. 2005. Effects of *Ganoderma lucidum* polysaccharides on serum lipids and lipoperoxidation in experimental hyperlipidemic rats. *China Journal of Chinese Materia Medica* 30(17):1358–1360 (In Chinese).

Chen, X.P., Y. Chen, S.B. Li & Y.G. Chen. 2009. Free radical scavenging of *Ganoderma lucidum* polysaccharides and its effect on antioxidant enzymes and immunity activities in cervical carcinoma rats. *Carbohydrate Polymers* 77(2):389–393.

Chen, Z.N., J.M. Xu, T. Ye, J.W. Zhou, Y.C. Sun, X.Y. Xie, J. Yu & Y. Liu. 2018. Identification of *Inonotus hispidus* and antioxidant and antimicrobial activities of different solvent extracts from its fruiting bodies. *Science and Technology of Food Industry* 39(23):99–104 (In Chinese).

Cheng, J.W., J.L. Song, H.L. Wei, Y.B. Wang, X.B. Huang, Y. Liu, N. Lu, L. He, G.Y. Lv, H.M. Ding, S.Z. Yang & Z.F. Zhang. 2020. Structural characterization and hypoglycemic activity of an intracellular polysaccharide from *Sanghuangporus sanghuang* mycelia. *International Journal of Biological Macromolecules* 164:3305–3314.

Cheng, X.H. 2016. Clinical observation of Wanyingchanhua powder combined with antibiotics in the treatment of vernal conjunctivitis. *Mod Diagn Treat* 27(18):3361–3362 (In Chinese).

Dai, Y.C., T. Bau, B.K. Cui & G.F. Qin. 2013. *Illustrations of Medicinal Fungi in China*. Haerbin: Northeast Forestry University Press (In Chinese).

Dai, Y.C. & Z.L. Yang. 2008. A revised checklist of medicinal fungi in China. *Mycosystema* 27(6):801–824 (In Chinese).

Dai, Y.C., Z.L. Yang, B.K. Cui, C.J. Yu & L.W. Zhou. 2009. Species diversity and utilization of medicinal mushrooms and fungi in China (Review). *International Journal of Medicinal Mushrooms* 11(3):287–302 (In Chinese).

Dong, G.Z., S.H. Wang, X.R. Ai, L. Yao, Z.W. Sun, C. Lei, Y. Wang & Q. Wang. 2013. Composition and characterization of cordyxanthins from *Cordyceps militaris* fruit bodies. *Journal of Functional Foods* 5(3):1450–1455.

Du, B., Y.D. Yang, Z.X. Bian & B.J. Xu. 2017. Molecular weight and helix conformation determine intestinal anti-inflammatory effects of exopolysaccharide from *Schizophyllum commune*. *Carbohydrate Polymers* 172(15):68–77.

Du, M., X.S. Hu & G.H. Zhao. 2006. Antioxidative activity and synergistic effect of different crude proteins in Se-enriched *Ganoderma lucidum*. *Science and Technology of Food Industry* 27(9):111–113 (In Chinese).

Du, Y., Y.F. Han, Z.M. Lu, X.G. Hua, X.Z. Hong & G. Yan. 2018. Glucose- and lipid-lowering activities of mycelial extracts from medicinal mushrooms. *Mycosystema* 37(5):656–666 (In Chinese).

Fan, H.T. & H.S. Lin. 2013. Advances on *Cordyceps militaris* constituents and pharmacological effect. *China Journal of Chinese Materia Medica* 38(15):2549–2552 (In Chinese).

Fang, R., P. Kirk, J.C. Wei, Y. Li, L. Cai, L. Fan, T.Z. Wei, R.L. Zhao, K. Wang, Z.L. Yang, T.H. Li, Y. Li, Phurbup-Dorji & Y.J. Yao. 2018. Country focus: China. In: K.J. Willis (ed.), *State of the World's Fungi. Report*. Kew: Royal Botanic Gardens, pp. 48–55.

Fu, T.Y., J.F. Ding, G.P. Xie, L. Chen & Y.H. Sun. 2012. Study on preparation technology of morchella seasoning. *China Condiment* 37(7):55–58 (In Chinese).

Gao, C.J., Y.H. Wang, C.Y. Wang & Z.Y. Wang. 2013. Antioxidant and immunological activity in vitro of polysaccharides from *Gomphidius rutilus* mycelium. *Carbohydrate Polymers* 92(2):2187–2192.

Gao, X., J.Y. Qi, C.T. Ho, B. Li, J.J. Mu, Y.T. Zhang, H.P. Hu, W.P. Mo, Z.Z Chen & Y.Z. Xie. 2020. Structural characterization and immunomodulatory activity of a water-soluble polysaccharide from *Ganoderma leucocontextum* fruiting bodies. *Carbohydrate Polymers* 249(1):1–43.

Guan, H.C. & Z. Cong. 1981. Effects of lingzhi Polysaccharide D6 on the Biosynthesis of nucleic acids and protein. *Beijing Medical College* 13(4):261–264 (In Chinese).

Han, D.X., J.L. You, M.H. Jin & X.L. Wu. 2019. Study on hemostatic effect and acute toxicity of DiXing. *Ginseng Research* 4:39–40 (In Chinese).

He, R.J., F. Zhou, Y.J. Zhao & P.L. Sun. 2013. Study on sterols extraction with high-speed shearing and antioxidation in *Ganoderma*. *Food and Fermentation Industries* 39(9):200–204 (In Chinese).

He, W.Q. & Z.F. Zhang. 2013. Antioxidant activities of polysaccharides of *Flammulina velutipes* obtained by ultrafiltration. *Agro Food Industry Hi-Tech* 24(4):53–56.

Hou, R.L., X. Liu, X.P. Wu, M.F. Zheng & J.S. Fu. 2021. The therapeutic effect of natural melanin from edible fungus *Auricularia auricula* on alcohol-induced liver damage in vitro and in vivo. *Food Science and Human Wellness* 10:514–522.

Hou, R.L., X. Liu, K.K. Xiang, L. Chen, M.F. Zheng & J.S. Fu. 2018. Optimization of ultrasonic-assisted extraction of *Cordyceps militaris* proteins by response surface and its nutritional assessment. *Biotechnology Bulletin* 34(11):198–204 (In Chinese).

Hu, T., Q.L. Lin, T. Guo, T. Yang, W.H. Zhou, X.F. Deng, J.K. Yan, Y. Luo, M.M. Ju & F.J. Luo. 2018. Polysaccharide isolated from *Phellinus linteus* mycelia exerts anti-inflammatory effects via MAPK and PPAR signaling pathways. *Carbohydrate Polymers* 200(15):487–497.

Huang, G.D., J.W. Peng, S.Y. Liang, J.H. Feng, X.M. Qu & X.F. Zhong. 2018. Study on the antialcoholism effects of antrodia camphorate Oral Liquid. *Modern Food Science and Technology* 34(7):56–60 (In Chinese).

Huang, N.L. 2005. The most promising main medicinal fungi in China. *Edible Fungi* 3–4 (In Chinese).

Jiang, Y.F., M.Y. Jin, & X.W. Zhou. 2021. Research progress on *Wolfiporia cocos* polysaccharides and their immunoregulatory functions. *Acta Edulis Fungi* 28(2):130–139 (In Chinese).

Li, B.K., L. Yu, C.G. Li & B. Wang. 2009. Clinical observation report of jinjunling capsule in treating 320 cases of cancer and chronic hepatitis. *Qilu Pharmaceutical Affairs* 28(4):245–246 (In Chinese).

Li, C., G.J. Peng, L. Jin, C.R. Li, M.Z. Fan & C.S. Sun. 2018. Evaluating the immunity function of Chanhuabao tablet. *Journal of Anhui Agricultural University* 45(1):161–165 (In Chinese).

Li, J.L., L. Lu, Y.H. Cang, C.C. Dai, J.Y. Qiu & K. Chen. 2001. Study on the sterols of *Hericium erinaceus* extracts. *Chinese Journal of Biochemical Phamaceutics* 22(3):124–126 (In Chinese).

Li, J.P., Y. Jin, P. Wang, M.R. Qin, X. Wu, J.W. Xian & X.W. Wang. 2013. Protective effect of *Taiwanofungus camphoratus* capsules on rat liver injury induced by tetrachloromethane. *China Pharmacist* 16(7):964–966 (In Chinese).

Li, L.Q., A.X. Song, W.T. Wong & J.Y. Wu. 2021. Isolation and assessment of a highly-active anti-inflammatory exopolysaccharide from mycelial fermentation of a medicinal fungus Cs-HK1. *International Journal of Molecular Sciences* 22(5):2450–2463.

Li, L.Y., H. Jin, J.Y. Zhang & R.C. Li. 2007. A brief account of Chinese medical fungi. *Journal of Microbiology* 27(2):57–61 (In Chinese).

Li, P., Y.P. Deng, X.X. Wei & J.H. Xu. 2013. Triterpenoids from *Ganoderma lucidum* and their cytotoxic activities. *Natural Product Research* 27(1):17–22.

Li, Q.J., H.Y. Bao, T. Bau & Y. Li. 2019. Chemical composition and anti-inflammatory activity of *Inonotus hispidus*. *Journal of Jilin University (Science Edition)* 57(1):177–181 (In Chinese).

Li, S.J., C.W. Liang, C.B. Cui, Y.J. Jin, X.S. Bai & Q. Duan. 2015. Study on extraction technology and antioxidant activity of total flavonoid from wild *Phellinus Vanini*. *Food Research and Development* 36(20):32–35 (In Chinese).

Li, X. 2018. Safety evaluation and anti-tumor mechanism of solid fermentation products of *Fomitopsis pinicola*. PhD dissertation, Jilin Agricultural University, 1–193.

Li, Y. 2019. History and development of Chinese medicinal mushrooms—excerpts from the speech delivered by Academician Li Yu of the Chinese academy of engineering at the 10th international conference on medicinal mushrooms. *Edible and Medicinal Mushroom* 27(6):357–362 (In Chinese).

Li, Y., Y. Yuan, L. Lei, F.H. Li, Y.Q. Zhang, J.L. Chen, G.H. Zhao, S.R. Wu, R. Yin & J. Ming. 2017. Carboxymethylation of polysaccharide from *Morchella angusticepes* Peck enhances its cholesterol-lowering activity in rats. *Carbohydrate Polymers* 172:85–92.

Liaw, C.C., S.J. Wu, C.F. Chen, M.N. Lai & L.T. Ng. 2017. Anti-inflammatory activity and bioactive constituents of cultivated fruiting bodies of *Xylaria nigripes* (Ascomycetes), a Chinese medicinal fungus. *International Journal of Medicinal Mushrooms* 19(10):1347–1351.

Lin, W.C., J.S. Deng, S.S. Huang, S.H. Wu, C.C. Chen, W.R. Lin, H.Y. Lin & G.J. Huang. 2017. Anti-inflammatory activity of *Sanghuangporus sanghuang* Mycelium. *International Journal of Molecular Sciences* 18(2):347–369.

Lin, Z.B. & H.N. Zhang. 2004. Anti-tumor and immunoregulatory activities of *Ganoderma lucidum* and its possible mechanisms. *Chinese Pharmacological Society* 25(11):1387–1395.

Liu, B. 1974. *Medicinal Fungi in China* (First edition). Tai Yuan: Shanxi People Press (In Chinese).

Liu, B. 1978. *Medicinal Fungi in China* (Second edition). Tai Yuan: Shanxi People Press (In Chinese).

Liu, B. 1984. *Medicinal Fungi in China* (Third edition). Tai Yuan: Shanxi People Press (In Chinese).

Liu, C., B.M. Li, J. Kang, H.Q. Wang & R.Y. Chen. 2013. A new terpenoid from *Ganoderma lucidum*. *Acta Pharmaceutica Sinica* 48(9):1450–1452 (In Chinese).

Liu, F. 1997. Free radical scavenging activities of mushroom polysaccharide extracts. *Life Sciences* 60(10):763–771.

Liu, H. & H. Li. 2009. Study on anti-aging effect of *Hericium erinaceus* extract. *Shandong Medical Journal* 49(16):37–38 (In Chinese).

Liu, J., H.Y. Bao & T. Bau. 2018. Chemical constituents and pharmacological activities of *Grifola frondose*. *Journal of Fungal Research* 16(3):150–157 (In Chinese).

Liu, J., H.Z. Chen & Z.H. Li. 2003. Process development and future for entomopathogenic fungus *Beauveria bassiana* production. *Chinese Journal of Biological Control* 19(2):86–90 (In Chinese).

Liu, J., F. Yang, L.B. Ye, X.J. Yang, K.A. Timani, Y. Zheng & Y.H. Wang. 2004. Possible mode of action of antiherpetic activities of a proteoglycan isolated from the mycelia of *Ganoderma lucidum* in vitro. *Journal of Ethnopharmacology* 95(2–3):265–272.

Liu, W., X.Z. Hu, L. Zhu, J.H. Gan, Y. Lu, N.P. Tao, X.C. Wang & C.H. Xu. 2019. Recent progress in research and application of *Ganoderma lucidum* triterpenoids. *Food Science* 40(5):309–315 (In Chinese).

Liu, X.Z., S.P. Nie, W.J. Li, Y. Chen & M.Y. Xie. 2011. Triterpenoids content and antioxidant activity of neutral components from *Ganoderma atrum*. *Journal of Nanchang University (Engineering &Technology)* 33(4):332–337 (In Chinese).

Lou, H.W., Y.H. Yu, Z.W. Ye, J.F. Lin, L.Q. Guo, H.B. Tang, X.Y. Wang, Y. Li & F. Yun. 2018. Optimization of ultrasonic-assisted extraction of carotenoids from *Cordyceps militaris*. *China Food Additives* 12:51–59 (In Chinese).

Lu, S.Y., Q.Q. Shi, X.R. Peng, L. Zhou, X.N. Li & M.H. Qiu. 2020. Isolation of benzolactones, Ganodumones A—F from *Ganoderma lucidum* and their antibacterial activities. *Bioorganic Chemistry* 98:103723.

Luo, J.W., F.L. Si, Z.X. Gu, L.L. Wang & X.X. Meng. 2018. Antioxidant and antimicrobial activities of polysaccharides from three species of *Auricularia*. *Food Science* 39(19):64–69 (In Chinese).

Luo, Q., Y.Z. Yang & G.X. Wang. 2017. Determination of the antimicrobial activity of *Ganoderma lucidum* and *Hericium erinaceus*. *Modern Animal Husbandry* 1(3):34–36 (In Chinese).

Ma, D.D., W.H. Wang, B. Xue, L. Li, L. Zhu & Z.D. Liu. 2019. The antitumor activity of natural components of edible and medicinal fungi. *Journal of Nanjing Normal University (Natural Science Edition)* 42(4):111–117 (In Chinese).

Ma, Q., Y. Yang, Z. Zhang, C.G. Wang & D. Wu. 2021. Progress on research and application of *Hericium erinaceus* Polysaccharides. *Acta Edulis Fungi* 28(6):199–216 (In Chinese).

Miao, Q.J., Y. Liu, F. Xu, S.X. Wang & S. Zhao. 2014. Research on total flavonoids bioactivity of edible fungus. *Food Science and Technology* 39(7):206–209 (In Chinese).

Ming, J., K.F. Zeng, G.H. Zhao, S.R. Wu, M.Y. Gui & Z.D. Chen. 2009. Hypolipidemic activity of water soluble polysaccharide PMEP-1 from *Morchella esculenta* (L.) Pers. *Food Science* 30(17):285–288 (In Chinese).

National Medical Products Administration. *Drugs*. [EB/OL]. (2022–01–01) [2022–05–01]. www.nmpa.gov.cn/. (In Chinese)

Peng, X.R., J.Q. Liu, J.J. Xia, Y.H. Yang & M.H. Qiu. 2012. Two new triterpenoids from *Ganoderma cochlear*. *Chinese Traditional and Herbal Drugs* 43(6):1045–1049 (In Chinese).

Qian, Z.M., M.T. Sun, W.Q. Li, G.R. Li, W.J. Li & Q.X. Mei. 2018. Analysis of sterols in different parts of *Cordyceps Sinensis*. *Lishizhen Medicine and Materia Medica Research* 29(12):2856–2858 (In Chinese).

Qian, Z.M., M.T. Sun, W.Q. Li, G.R. Li, W.J. Li & Q.X. Mei. 2019. Content analyses of three kinds of sterols in different development stages of Chinese *Cordyceps*. *Mycosystema* 38(4):539–544 (In Chinese).

Ran, L.X., Y.N. Zou, J.W. Cheng & F. Lu. 2018. Silver nanoparticles in situ synthesized by polysaccharides from *Sanghuangporus sanghuang* and composites with chitosan to prepare scaffolds for the regeneration of infected full-thickness skin defects. *International Journal of Biological Macromolecules* 15(125):392–403.

Rao, G. & X.W. Tang. 2007. Effect of *Coriolus versicolor* intracellular polysaccharide (CVICP) on blood lipid in patients with hyperlipidemia. *Chongqing Medicine* 36(13):1036–1037 (In Chinese).

Ren, L., J. Zhang & T.H. Zhang. 2021. Immunomodulatory activities of polysaccharides from *Ganoderma* on immune effector cells. *Food Chemistry* 340(15):127933.

Seifried, H.E., D.E. Anderson, E.I. Fisher & J.A. Milner. 2007. A review of the interaction among dietary antioxidants and reactive oxygen species. *The Journal of Nutritional Biochemistry* 18(9):567–579.

Song, M.X. 2009. Preparation of buccal tablets for *Grifola frondosa* polysaccharide. *Tianjin Pharmacy* 21(3):7–9 (In Chinese).

Su, H.G., X.R. Peng, Q.Q. Shi, Y.J. Huang, L. Zhou & M.H. Qiu. 2020. Lanostane triterpenoids with anti-inflammatory activities from *Ganoderma lucidum*. *Phytochemistry* 173:112256.

Su, H.G., Q. Wang, L. Zhou, X.R. Peng, W.Y. Xiong & M.H. Qiu. 2021. Functional triterpenoids from medicinal fungi *Ganoderma applanatum*: A continuous search for antiadipogenic agents. *Bioorganic Chemistry* 112:1–9.

Sun, Y.X. & J.F. Kennedy. 2010. Antioxidant activities of different polysaccharide conjugates (CRPs) isolated from the fruiting bodies of *Chroogomphis rutilus* (Schaeff.: Fr.) O. K. Miller. *Carbohydrate Polymers* 82(2):510–514.

Sun, Y.X., J.C. Liu & J.F. Kennedy. 2010. Purification, composition analysis, and antioxidant activity of different polysaccharide conjugates (APPs) from the fruiting bodies of *Auricularia polytricha*. *Carbohydrate Polymers* 82(2):299–304.

Tang, Y.W., Y.Q. Zhang, Y. Li, L. Lei, F.H. Li, J.C. Zhao, S.R. Wu & J. Ming. 2019. Regulation of cholesterol metabolism by sulfated polysaccharides from *Morchella angusticeps* Peck. *Food Science* 40(21):136–142 (In Chinese).

Tu, C.R., X.Q. Wei, H.Y. Zhang & T. Fan. 2018. Preparation of sanghuangporus sanghuang ultrafine powder and its effect on the dissolution rate of polysaccharides. *North Sericulture* 39(3):14–16 (In Chinese).

Tu, J.M., B. Wang & X. Xia. 2020. Research progress in bioactive components and pharmacological effects of *Cordyceps militaris*. *Edible Fungi* 42(6):4–7 (In Chinese).

Wan, H.Z., X.L. Xiao & G.X. Chai. 2019. Analysis on the health care function of edible and medicinal fungi under the general health trend. *Edible Fungi of China* 38(6):34–37 (In Chinese).

Wan, X.L., X. Jin, M.L. Xie, J. Liu, A.A. Gontcharov, H. Wang, R.N. Lv, D.R. Liu, Q. Wang & Y. Li. 2020. Characterization of a polysaccharide from *Sanghuangporus vaninii* and its antitumor regulation via activation of the p53 signaling pathway in breast cancer MCF-7 cells. *International Journal of Biological Macromolecules* 163(15):865–877.

Wang, D.B., S.Y. Lu, M. Geng, Y.C. Sun & K.J. Lian. 2008. Treatment of the Hp(+) functional dyspepsia with Hontoujun tablet in effect observation. *Medical Journal of Chinese Peoples Health* 20(13):1435–1436 (In Chinese).

Wang, D.L. 2020. Regulatory effect of *Tremella fuciformis* polysaccharides on basketball players exercise-induced immunosuppression. *Edible Fungi China* 39(8):199–203 (In Chinese).

Wang, J.Y., L.Y. Chang, Y.Y. Jia, H. Dong, Z.W. Zhou, Y.M. Sun & W. Long. 2019. Research progress on radio-protective effect of fungal polysaccharides. *Food and Drug* 21(3):235–238 (In Chinese).

Wang, J.Y., N. Feng, Y.F. Liu, P.L. Yan & Q.J. Tang. 2018. Ergosterol and fingerprint of liposoluble constituent in the *Ganoderma lingzhi* basidiospore. *Mycosystema* 37(9):1215–1223 (In Chinese).

Wang, S.C. & J.N. Fang. 1999. Preparation and structural analysis of sulfated derivatives of lentinan. *Acta Biochimica et Biophysica Sinica* 31(5):594–597 (In Chinese).

Wang, T.T., J.K. You, M. Yan, S.R. Wu & M.Y. Gui. 2015. In vitro antioxidant activity of polyphenols extracted from four wild edible fungus. *Food and Fermentation Industries* 41(9):148–152 (In Chinese).

Wang, W.H., Y. Yang, L.N. Zhu, W. Jia, J.S. Zhang, Y.F. Liu, M.Q. Yan, X.L. Zhao, K. Zhang, H.N. Zhang & H. Fan. 2018. Inhibitory effects of ethanol extract of *Sanghuangporus sanghuang* fruiting bodies on SW620 colon cancer cells. *Mycosystema* 37(12):1751–1760 (In Chinese).

Wang, Y.X., G.Z. Yang, Y. Di & Y.G. Su. 2004. Effects of *Boletus aereous* on blood lipids and antioxidant capacity in hyperlipidemia rats. *Chinese Journal of Preventive Medicine* 38(5):355 (In Chinese).

Wang, Z.B., J.J. Pei, H.L. Ma, P.F. Cai & J.K. Yan. 2014. Effect of extraction media on preliminary characterizations and antioxidant activities of *Phellinus linteus* polysaccharides. *Carbohydrate Polymers* 109(30):49–55.

Wen, Y., D. Peng, C.L. Li, X.J. Hu, S.X. Bi, L.Y. Song, B. Peng, J.H. Zhu, Y.Y. Chen & M.R. Yu. 2019. A new polysaccharide isolated from *Morchella importuna* fruiting bodies and its immunoregulatory mechanism. *International Journal of Biological Macromolecules* 137(15):8–19.

Wu, F., L.W. Zhou & Z.L. Yang. 2019. Resource diversity of Chinese macrofungi: Edible, medicinal, and poisonous species. *Fungal Diversity* 98:1–76.

Wu, Q.P., Y.Z. Xie, S.Z. Li, D.P. La Pierre, Z.Q. Deng, Q. Chen, C. Li, Z. Zhang, J. Guo, C.K.A. Wong, D.Y. Lee, A. Yee & B.B. Yang. 2006. Tumor cell adhesion and integrin expression affected by *Ganoderma lucidum*. *Enzyme and Microbial Technology* 40(1):32–41.

Wu, S.L., C.X. Shao, S.J. Tang, M.H. Qin, Y. Yang & H.Z. Liu. 2016. Antibacterial and antitumor activities of *Wolfiporia cocos* spent culture fluid. *Acta Edulis Fungi* 23(1):63–66 (In Chinese).

Wu, S.R., M. Yan, X. Chen, Y.Y. Deng, Y.R. Liu & D.F. Sun. 2020. Current status of the development of edible and medicinal fungi medicine in China. *Edible Fungi of China* 39(7):1–9 (In Chinese).

Wu, T.R., C.S. Lin, C.J. Chang, T.L. Lin, J. Martel, Y.F. Ko, D.M. Ojcius, C.C. Lu, J.D. Young & H.C. Lai. 2019. Gut commensal Parabacteroides goldsteinii plays a predominant role in the anti-obesity effects of polysaccharides isolated from *Hirsutella sinensis*. *Gut Microbiota* 68:248–262.

Wu, X.L., T. Bau & X.L. Mao. 2013. *Medicinal Fungi of China*. Beijing: Science Press (In Chinese).

Wu, Y.M., W.F. Huang & X. Lin. 2008. The effect of the mycelia of *Cordyceps militaris* on the regulation of high blood lipid. *Journal of South China Normal University (Natural Science Edition)* 2:111–117 (In Chinese).

Xiao, C., Q.P. Wu, Y.Z. Xie, J.M. Zhang & J.B. Tan. 2015. Hypoglycemic effects of *Grifola frondosa* (Maitake) polysaccharides F2 and F3 through improvement insulin resistance in diabetic rats. *Food & Function* 6(11):3567–3575.

Xiao, J.H., N.H. Jiang, Z.Q. Liang & A.Y. Liu. 2003. Research and application prospect of polysaccharides of *Cordyceps sinensis*. *Journal of Mountain Agriculture and Biology* 22(1):70–76 (In Chinese).

Xie, Y.Z., S.Z. Li, A. Yee, D.P. La Pierre, Z.Q. Deng, D.Y. Lee, Q.P. Wu, Q. Chen, C. Li, Z. Zhang, J. Guo, Z.D Jiang & B.B. Yang. 2006. *Ganoderma lucidum* inhibits tumor cell proliferation and induces tumor cell death. *Enzyme and Microbial Technology* 40(1):177–185.

Xu, J.T. & J. Lan. 1997. *Chinese Medicinal Mycology*. Beijing: Peking Union Medical College Press (In Chinese).

Xu, J.T. & J. Lan. 2002. Chinese medicinal mycology. *Beijing: Joint Publishing House of Peking Medical College and Peking Union Medical College* 31(1):29–34 (In Chinese).

Xu, J.T., C.M. Xiao, H.S. Xu, S.X. Yang & X.Q. Wu. 2021. Anti-inflammatory effects of *Ganoderma lucidum* sterols via attenuation of the p38 MAPK and NF-κB pathways in LPS-induced RAW 264.7 macrophages. *Food and Chemical Toxicology* 150:112073.

Yang, C.L., Q.F. Meng, Y. Zhang, Y. Hu, S.J. Xiao, Y.Q. Zhang, J.H. Ju & S.B. Fu. 2020. Morelsins A—F, six sesquiterpenoids from the liquid culture of *Morchella importuna*. *Tetrahedron* 76:31–38.

Yang, L.L., R.C. Li, Y. Cao, M.J. Li, X.Y. Luo, Z.H. Shen & Q.Q. Lu. 2021. Research progress on the pharmacology and health care effects of *Naematelia aurantialba* and its polysaccharides. *Edible and Medicinal Mushrooms* 29(3):176–182 (In Chinese).

Yang, Q., X.G. Hu, Q. Wang & W.Q. Wang. 2006. Study on antitumor or effect of medicinal fungi *Phellinus igniarius* extracts. *China Journal of Chinese Materia Medica* 31(20):1713–1715 (In Chinese).

Yang, S.D., H.Y. Bao & H. Wang. 2019. Chemical components and anti-tumor compounds from *Inonotus hispidus*. *Mycosystema* 38(1):127–133 (In Chinese).

Yang, S.M., Z.J. Wang, H.W. Yang & J.X. Yan. 2014. Optimization of extraction technology of total flavonoids in *Cordyceps* flowers and the antioxidant effects. *Food Science and Technology* 39(12):248–254 (In Chinese).

Yang, Y.L., J.L. Ren & H. Zhang. 2019. Research progress of terpenoids and bioactivities in edible mushroom. *Science and Technology of Food Industry* 40(1):305–310 (In Chinese).

Yang, Y.L., C.Y. Zhou, Y.Q. Bai, W. Jia & J.S. Zhang. 2003. Curative effect of "hougu stomachic capsule" on chronic superficial gastritis. *Acta Edulis Fungi* 10(3):17–21 (In Chinese).

Yao, H.M., G. Wang, Y.P. Liu, M.Q. Rong, C.B. Shen, X.W. Yan, X.D. Luo & R. Lai. 2016. Phenolic acids isolated from the fungus *Schizophyllum commune* exert analgesic activity by inhibiting voltage-gated sodium channels. *Chinese Journal of Natural Medicines* 14(9):661–670.

Ying, J.Z., X.L. Mao, Q.M. Ma, S.C. Zong & H.A. Wen. 1987. *Illustration of Medicinal Fungi in China*. Beijing: Science Press (In Chinese).

Younis, A.M. 2017. Anticancer potential of *Hericium erinaceus* extracts against particular human cancer cell lines. *Microbial Biosystems Journal* 2(1):9–20.

Yu, X.T., J.G. Huang, Y.Y. Han, Y.Z. Xie & H.H. Pan. 2015. Effect of aqueous extracts of *Amauroderma rudis* fruit bodies on herpes simplex virus infection of vero cells in vitro. *Acta Edulis Fungi* 22(3):55–59 (In Chinese).

Yu, X.Y., Y. Zou, Q.W. Zheng, F.X. Lu, D.H. Li, L.Q. Guo & J.F. Lin. 2021. Physicochemical, functional, and structural properties of the major protein fractions extracted from *Cordyceps militaris* fruit body. *Food Research International* 142(3):110211.

Yuan, B., C. Zhao, C. Cheng, D.C. Huang, S.J. Cheng, C.J. Cao & G.T. Chen. 2019. A peptide-Fe(II) complex from *Grifola frondosa* protein hydrolysates and its immunomodulatory activity. *Food Bioscience* 32:100459.

Yuan, J.F., Z.Q. Zhang, Z.C. Fan & J.X. Yang. 2008. Antioxidant effects and cytotoxicity of three purified polysaccharides from *Ligusticum chuanxiong* Hort. *Carbohydrate Polymers* 74(4):822–827.

Yue, X.M., J.H. Wu, Z. Ruan, M.L. Wolden, L.T. Li & Y. Lin. 2020. The burden of hypoglycemia in patients with insulin-treated diabetes mellitus in China: Analysis of electronic medical records from 4 tertiary hospitals. *Value in Health Regional Issues* 21:17–21.

Zan, L.F., R.J. Liang & H.Y. Bao. 2015. Antioxidant and antimicrobial activities of *Inonotus hispidus* extracts. *Northern Horticulture* 5:151–155 (In Chinese).

Zhang, A.Q., N.N. Xiao, P.F. He & P.L. Sun. 2011. Chemical analysis and antioxidant activity in vitro of polysaccharides extracted from *Boletus edulis*. *International Journal of Biological Macromolecules* 49(5):1092–1095.

Zhang, J.C. & R.C. Li. 2014. Overview of Chinese medicinal fungi research. *Journal of Anhui Agriculture* 42(18):6118–6120 (In Chinese).

Zhang, J.F., Z. Zhang, W.H. Wang, H.N. Zhang, N. Feng, D. Wu, Y.F. Liu & Y. Yang. 2020. Comparison of active secondary metabolites between mycelia and fruiting bodies of *Sanghuangporous sanghuang*. *Mycosystema* 39(2):398–408 (In Chinese).

Zhang, J.Y., L.L. Sun, P.A. Zapata, M. Arias, L. Atehortuac & T.J. Webster. 2017. Anti-inflammatory bone protective effects of nano-protein extracts from mushroom species: *Ganoderma lucidum* and *Pleurotus ostreatus*. *Journal of Nanoscience and Nanotechnology* 17(8):5884–5889.

Zhang, M.X., Y.W. Li, D.S. Zhang, P.W. Zhuang & Y.Y. Zhang. 2013. Inhibition of pachymaran on spontaneous lung metastasis in Lewis lung cancer of mice and its mechanism. *Drugs & Clinic* 28(6):842–846 (In Chinese).

Zhang, S.S. & H.Z. Cai. 2017. Preparation and antioxidative activities of the sulfated *Schizophyllan* polysaccharide. *Food Research and Development* 38(8):17–21 (In Chinese).

Zhang, T.T., J.H. Pan & S.X. Zhu. 2003. A study on the antimicrobial activities of *Hericium erinaceus* preparations. *Journal of Anhui Normal University (Natural Science)* 26(2):159–160 (In Chinese).

Zhang, X.H., L. Liu & N.N. Shang. 2020. The recovery effect of *Schizophyllum commune* on aerobic exercise fatigue. *Edible Fungi of China* 39(2):70–72 (In Chinese).

Zhang, Y.S., Z.M. Zhang, H. Liu, D. Wamg, J.H. Wang, M. Liu, Y.J. Yang & S. Zhong. 2022. A natural selenium polysaccharide from *Pleurotus ostreatus*: Structural elucidation, anti-gastric cancer and anti-colon cancer activity in vitro. *International Journal of Biological Macromolecules* 201(15):630–640.

Zhang, Z.F., G.Y. Lu, J.W. Cheng, W.M. Cai, L.F. Fan & L.X. Miao. 2019. Characterization and biological activities of polysaccharides from artificially cultivated *Phellinus baumii*. *International Journal of Biological Macromolecules* 129(15):861–868.

Zhu, Y.L., X.F. Yu, Q. Ge, J. Li, D.J. Wang, Y. Wei & Z. Ouyang. 2020. Antioxidant and anti-aging activities of polysaccharides from *Cordyceps cicadae*. *Biological Macromolecules* 15(157):394–400.

9 Diversity and Ecology of Wild Mushrooms in Ethiopia

Tatek Dejene, Busha Teshome, Wubalem Tadesse, Pablo Martín-Pinto

9.1 INTRODUCTION

One of the most varied groups of organisms in terrestrial habitats are the fungi, which are essential to the growth, operation, and stability of terrestrial ecosystems. In forest systems, fungi occur either as symbionts or as saprotrophs (Ferris et al. 2000; Nick et al. 2006; Nordén et al. 2004). Since fungi cannot produce their own food, they depend on other food sources for sugars and carbohydrates. Mycorrhizal species feed by adhering to living plants and developing beneficial symbiotic interactions with the plant's roots and surrounding tissues (Ferris et al. 2000; Dickie 2007; Lindahl et al. 2007; Wardle 2006). In this relationship, the fungi extend as auxiliary root systems, improving plants' ability to take up mineral nutrients and water from the soil and increasing pathogen resistance (Dahlberg et al. 2001; Dalong et al. 2011). Similarly, mycorrhizal fungi can obtain carbohydrates from the trees with which they have symbiotic connections (Lindahl et al. 2007). These saprotrophic fungi play a crucial role in the cycling of nutrients in forest ecosystems by decomposing decaying organic waste (Ferris et al. 2000). These properties cause the fungi to influence the diversity, production, and ecological processes of plant species (Pietras et al. 2013; van der Heijden et al. 2008). Moreover, fungi can also serve as signs of ecological stability in forest systems (Butler et al. 2002).

The preservation and management of forests, as well as the associated ecological processes, depend heavily on fungi (Bonet et al. 2014). Fungi have been utilized by humans as food and medicine since the earliest times (Boa 2004). Additionally, valuable wild mushrooms are used by rural populations all over the world to supplement their livelihoods (Cai et al. 2011; Sarma et al. 2010). It is common for wild mushrooms to be collected as valuable non-timber forest products (NTFPs) (Boa 2004; Chang and Lee 2004). Mushrooms provide economic benefits for the forest sector (Bonet et al. 2014). They help rural populations to avoid poverty and reinforce their livelihoods by providing a reliable source of income, as there has been an increasing demand for valuable mushrooms, particularly the edible varieties, all over the world (Pettenella et al. 2007); this demand can occasionally be exploited by forest managers for economic benefit (Martín-Pinto et al. 2006). In addition, forest fungi interact with other organisms in a variety of ways, as they provide food and shelter for various invertebrates, including mollusks, coleopterans, and others (Jonsell and Nordlander 2000).

9.2 GENERAL BACKGROUND OF MYCOLOGY IN ETHIOPIA

In Ethiopia, fungi are poorly studied components of forest ecosystems (Dejene et al. 2017a; Megersa et al. 2017; Sitotaw et al. 2015). Due to this, their taxonomy and ecology are poorly understood. However, according to Pegler and Rayner (1969) and Pegler (1977), the East African

 DOI: 10.1201/b23190-9

region that includes Ethiopia is rich in fungal species. The high diversity of the plant species in the forest ecosystem contributes to the high diversity of fungal species in Ethiopia (Sitotaw et al. 2015). Although the fungal diversity in the country is yet not estimated and not included in the biodiversity databases of the country, many Ethiopian fungi are supposedly widespread across Africa (Pegler 1977). The poor knowledge of fungi in Ethiopia indicated the urgent need for documentation to ensure the conservation and protection of the resources, as fungi are highly sensitive to habitat degradation, which is rife across the country (Badege 2001; Goldmann et al. 2015; IBC 2005).

Despite the poor knowledge of fungal resources, valuable wild mushroom collection and utilization have been common practices for the different ethnic groups in rural areas of Ethiopia (Semwal et al. 2014; Tuno 2001). Wild fungi have been used for food and medical purposes, and are also involved in local mythology (Abate 2014; Tuno 2001). Wild mushrooms have also been used as a valuable food during food shortage periods (Alemu et al. 2012; Lulekal et al. 2011). They are also commonly available in some local markets, allowing the local people to earn some economic benefits and support their household economy (Abate 2014). Although there have been some attempts to document the ethnomycological and socioeconomic elements of Ethiopia's wild mushrooms, the information that has been gathered is determined to be minor in light of the resource's rich abundance and ethnic diversity. The presence of anthropogenic impacts on wild mushrooms necessitates quick action in documenting, producing, and utilizing the species in the country, given that the conservation and maintenance of wild mushrooms is frequently a primary priority for current forest management efforts.

9.3 DIVERSITY OF VALUABLE WILD MUSHROOMS

There are many diverse ecosystems in Ethiopia, and the country's wide variation in elevation, rainfall patterns, and soil quality all play a role in the occurrence of various floral and faunal life forms (Bongers and Tenngkeit 2010; Friis et al. 2010). High macro- and micro-climatic variability has also facilitated the emergence of a variety of vegetation types across the country. According to Friis et al. (2010), there are 12 different types of flora in Ethiopia, ranging from the Afro-alpine formations to savanna, scrubland, and deserts. This vegetation's abundance in valuable NTFPs is a key characteristic (Adilo 2007; Seyoum 2007; Sultan 2009). The most significant NTFPs include bamboo, gum arabic, resin, honey, beeswax, and bamboo (Asfaw and Tadesse 2001; Lulekal et al. 2011).

The collection of wild mushrooms is a traditional cultural activity that has historically been carried out by rural people in Ethiopia. However, there is not much information available on this subject. As a result, the traditional knowledge and cultural background are being lost as habitat degradation and biodiversity loss occur in the country (Dejene et al. 2017b). Furthermore, publications on the variety of fungi and lists of significant species are also uncommon in the country (Dejene et al. 2017b). This is brought on by the lack of research facilities and experts in fungal ecology and taxonomy in the country. The lack of understanding of Ethiopia's fungi is also concerning because fungi are extremely vulnerable to habitat disturbances, specifically the numerous anthropogenic problems that occur there (IBC 2014). Knowledge about the country's fungal flora that can be collected by evaluating the literature available provide a foundation for future mycological research and will assist in raising awareness for resource protection and wider utilization. The body of information that exists on the diversity and availability of valuable fungal species, whether edible or medicinal (Table 9.1), from different habitats is likely to be much larger than what has been reported thus far. This is something that further complementary studies could confirm.

TABLE 9.1
Summary of Valuable Wild Mushrooms with Their Uses and Associated Habitats in Ethiopia.

List of valuable species	Use	Habitats	Referred sources
Agaricus arvensis Schaeff.	F	GA	Abate 2008, 2014; Alemu 2013; Muleta et al. 2013; Sitotaw et al. 2015
Agaricus augustus Fr.	F	NF	Dejene et al. 2017b; Alem et al. 2021
Agaricus bitorquis (Quél.) Sacc.	F	NF	Dejene et al. 2017b; Alem et al. 2021
Agaricus campestris L.	F	NF	Dejene et al. 2017b; Alem et al. 2021
Agaricus campestroides Heinem. and Gooss. -Font.	F	NF, PT, GA	Alemu 2013; Abate 2014; Sitotaw et al. 2015; Dejene et al. 2017a
Agaricus moelleri Wasser	F	NF	Dejene et al. 2017b; Alem et al. 2021
Agaricus murinaceus Bull.	F	NF	Dejene et al. 2017b; Alem et al. 2021
Agaricus subedulis Heinem.	F	NF, PT	Dejene et al. 2017b, 2017c; Dejene et al. 2021
Agrocybe pediades (Fr.) Fayod	F	NF	Dejene et al. 2017b, 2017c; Dejene et al. 2021
Ampulloclitocybe clavipes (Pers.) Redhead, Lutzoni, Moncalvo, and Vilgalys	F	NF	Dejene et al. 2017b; Alem et al. 2021
Armillaria heimii Pegler	FM	NF	Abate 2008, 2014; Osarenkhoe et al. 2014; Dejene et al. 2017b
Auricularia auricula-judae (Bull.) Quél.	FM	NF	Abate 2008, 2014; Osarenkhoe et al. 2014; Dejene et al. 2017b
Calvatia cyathiformis (Bosc) Morgan	F	NF	Dejene et al. 2017b; Alem et al. 2021
Calvatia gigantea (Batsch) Lloyd	F	NF	Dejene et al. 2017b; Alem et al. 2021
Calvatia rubroflava (Cragin) Lloyd	FM	NF	Dejene et al. 2017b
Calvatia subtomentosa Dissing and M. Lange	F	PT	Dejene et al. 2017b; Alem et al. 2021
Clitocybe carolinensis H.E. Bigelow and Hesler	F	NF	Dejene et al. 2017b; Alem et al. 2021
Clitocybe cistophila Bon and Contu	F	NF	Dejene et al. 2017b; Alem et al. 2021
Clitocybe foetens Melot	F	NF	Dejene et al. 2017b; Alem et al. 2021
Clitocybe fragrans (With.) P. Kumm.	F	NF	Dejene et al. 2017b; Alem et al. 2021
Clitocybe geotropa (Bull. ex DC.) Quél.	F	NF	Dejene et al. 2017b; Alem et al. 2021
Coprinellus domesticus (Bolton) Vilgalys, Hopple, and Jacq. Johnson	F	NF, PT	Dejene et al. 2017b
Coprinopsis nivea (Pers.) Redhead, Vilgalys, and Moncalvo	F	NF	Dejene et al. 2017b
Coprinus comatus (O.F. Müll.) Pers.	F	NF	Dejene et al. 2017b; Alem et al. 2021
Coprinus lagopus (Fr.) Fr.	F	NF	Dejene et al. 2017b; Alem et al. 2021
Coprinus micaceus (Bull.) Fr.	F	NF	Dejene et al. 2017b; Alem et al. 2021

TABLE 9.1 ***Continued***
Summary of Valuable Wild Mushrooms with Their Uses and Associated Habitats in Ethiopia.

List of valuable species	Use	Habitats	Referred sources
Coprinus niveus (Pers.) Fr.	F	NF	Dejene et al. 2017b; Alem et al. 2021
Coprinus pseudoplicatilis Voglino	F	NF	Abate 2008, 2014; Dejene et al. 2017b
Craterellus ignicolor (R. H. Petersen) Dahlman, Danell, and Spatafora	F	NF	Dejene et al. 2017b; Alem et al. 2021
Crepidotus applanatus (Pers.) P. Kumm.	F	NF	Dejene et al. 2017b; Alem et al. 2021
Crepidotus mollis (Schaeff.) Staude	F	NF	Dejene et al. 2017b; Alem et al. 2021
Gymnopilus pampeanus (Speg.) Singer	F	PT	Dejene et al. 2017b; Alem et al. 2021
Hygrocybe chlorophana (Fr.) Wünsche	F	NF	Dejene et al. 2017b; Alem et al. 2021
Hygrophoropsis aurantiaca (Wulfen) Maire	F	NF, PT	Dejene et al. 2017b, 2017c; Dejene et al. 2021
Hymenagaricus fuscobrunneus D. A. Reid and Eicker	F	NF	Dejene et al. 2017b, 2017c; Dejene et al. 2021
Laetiporus sulphureus (Bull.) Murrill	FM	NF	Tuno 2001; Alemu 2013; Muleta et al. 2013; Abate 2014; Osarenkhoe et al. 2014; Dejene et al. 2017b
Lentinellus cochleatus (Pers.) P. Karst.	FM	NF	Tuno 2001; Osarenkhoe et al. 2014
Lepista sordida (Schumach.) Singer	F	PT	Dejene et al. 2017b; Alem et al. 2021
Lepista sordida var. lilacea (Quél.) Bon	F	PT	Dejene et al. 2017b; Alem et al. 2021
Leucoagaricus americanus (Peck) Vellinga	F	NF	Dejene et al. 2017b; Alem et al. 2021
Leucoagaricus holosericeus (J. J. Planer) M. M. Moser	F	NF, PT	Dejene et al. 2017b, 2017c; Dejene et al. 2021
Leucoagaricus leucothites (Vittad.) Wasser	F	NF, PT	Dejene et al. 2017b, 2017c; Dejene et al. 2021
Leucoagaricus purpureolilacinus Huijsman	F	NF	Dejene et al. 2017b; Alem et al. 2021
Leucoagaricus rubrotinctus (Peck) Singer	F	NF, PT	Dejene et al. 2017b, 2017c; Dejene et al. 2021
Leucocoprinus birnbaumii (Corda) Singer	F	NF	Dejene et al. 2017b, 2017c; Dejene et al. 2021
Leucocoprinus cepistipes (Sowerby) Pat.	F	NF, PT	Dejene et al. 2017b, 2017c; Dejene et al. 2021
Lycoperdon perlatum Pers.	F	PT	Dejene et al. 2017b, 2017c; Dejene et al. 2021
Lycoperdon umbrinum Pers.	F	PT	Dejene et al. 2017b; Alem et al. 2021
Macrolepiota africana (R. Heim) Heinem.	F	PT	Dejene et al. 2017b; Alem et al. 2021
Macrolepiota procera (Scop.) Singer	F	NF	Abate 2008, 2014; Alemu 2013

TABLE 9.1 ***Continued***

Summary of Valuable Wild Mushrooms with Their Uses and Associated Habitats in Ethiopia.

List of valuable species	Use	Habitats	Referred sources
Morchella anatolica Isiloglu, Spooner, Alli, and Solak	F	PT	Dejene et al. 2017b; Alem et al. 2021
Morchella cf americana Clowez and C. Matherly	F	PT	Dejene et al. 2017b; Alem et al. 2021
Omphalotus illudens (Schwein.) Bresinsky and Besl	F	PT	Dejene et al. 2017b; Alem et al. 2021
Pholiota aurivella (Batsch) P. Kumm.	F	NF	Abate 2008, 2014; Alemu 2013
Pleurotus luteoalbus Beeli	F	NF	Dejene et al. 2017b; Alem et al. 2021
Pleurotus populinus O. Hilber and O.K. Mill.	F	NF	Dejene et al. 2017b; Alem et al. 2021
Pleurotus pulmonarius (Fr.) Quél.	F	NF	Dejene et al. 2017b; Alem et al. 2021
Polyporus badius (Pers.) Schwein.	F	PT	Dejene et al. 2017b; Alem et al. 2021
Polyporus tenuiculus (P. Beauv.) Fr.	F	PT	Dejene et al. 2017b; Alem et al. 2021
Polyporus tuberaster (Jacq. ex Pers.) Fr.	F	PT	Dejene et al. 2017b; Alem et al. 2021
Schizophyllum commune Fr.	FM	PT	Tuno 2001; Alemu 2013; Osarenkhoe et al. 2014
Suillus luteus (L.) Roussel	F	PT	Dejene et al. 2021; Alem et al. 2021
Termitomyces aurantiacus (R. Heim) R. Heim	F	TM	Sitotaw et al. 2015
Termitomyces clypeatus R. Heim	F	NF, TM	Sitotaw et al. 2015
Termitomyces clypeatus R. Heim	F	TM	Alemu 2013; Muleta et al. 2013; Sitotaw et al. 2015; Alem et al. 2021
Termitomyces eurrhizus (Berk.) R. Heim	F	TM	Sitotaw et al. 2015
Termitomyces letestui (Pat.) R. Heim	F	TM	Sitotaw et al. 2015
Termitomyces microcarpus (Berk. and Broome) R. Heim	FM	NF, TM	Alemu 2013; Muleta et al. 2013; Sitotaw et al. 2015; Alem et al. 2021
Termitomyces robustus (Beeli) R. Heim	FM	NF, TM	Alemu 2013; Muleta et al. 2013; Sitotaw et al. 2015; Alem et al. 2021
Termitomyces schimperi (Pat.) R. Heim	F	NF, TM	Sitotaw et al. 2015
Tricholoma portentosum (Fr.) Quél.	F	NF, PT	Dejene et al. 2017b; Alem et al. 2021
Tricholoma saponaceum (Fr.) P. Kumm.	F	NF, PT	Dejene et al. 2017b; Alem et al. 2021
Tylopilus niger (Heinem. and Gooss. -Font.) Wolfe	F	PT	Dejene et al. 2021
Vascellum depressum (Bonord.) F. Šmarda	F	GA	Alemu 2013; Abate 2014

Note: GA: grazing area, NF: natural forest, TM: termite mounds, PT: plantation forest, F: food, and FM: food and medicine.

9.4 DISTRIBUTION OF VALUABLE WILD MUSHROOMS

Wild edible mushrooms are frequently associated with local vegetation. Therefore, knowing the ecology of the host species aids in identifying potentially related taxa in any ecosystem (Harkonen et al. 2003). According to Friis et al. (2010), Ethiopia is home to a variety of ecosystems with a high richness of species, including fungi (Sitotaw et al. 2015). However, the available research only concentrates on a small number of species rather than providing a comprehensive picture of the nation's mycofloral profile. Here, we provide a summary and overview of Ethiopia's wild mushrooms and their associated environments (Table 9.1). The topics covered in the discussion are divided into four categories: (1) mushrooms in native forests, (2) mushrooms in grazing areas, (3) mushrooms in termite mounds, and (4) mushrooms in exotic tree plantations.

Indigenous forests are a typical part of the Ethiopian landscape (Friis et al. 2010), covering a range of environments. The occurrence of valuable wild mushrooms in these forests is widespread during the rainy season (Abate 2014). Some genera, such as *Lentinus* sp., are unique as they fruit during the dry season (Tuno 2001). Hjortstam and Ryvarden (1996) reported 15 species in the family *Corticiaceae* from the indigenous forest ecosystems of Ethiopia, of which *Mycoacia brunneofusca* and *Vuilleminia obducens* were new to science. Similarly, (Decock et al. 2005) reported taxa from the highland forest region, which included *Fomitiporia tenuis* and *F. aethiopica*. Some other taxa have also been reported by Tuno (2001); Abate (2008); (2014); Alemu (2013); Muleta et al. (2013) from indigenous forests in different parts of the country. Most recently, Megersa et al. (2017) reported 49 fungal taxa from the natural forests of Degaga. Also, our research team collected 64 (Dejene et al. 2017b) and 188 (Alem et al. 2021) macrofungal species from the dry Afromontane forests ecosystems in a different region of the country, suggesting the presence of moderate fungal diversity in the dry Afromontane forests of the country. Interestingly, some taxa such as *Agaricus* sp., *Agrocybe* sp., and *Calvatia* sp. could be identified only to the genus level, while some others could not be identified at all, indicating the likely presence of species new to science. Based on a survey report of NTFPs from the Benishnagul Gumz Region, Alemu et al. (2012) also reflected a wide diversity of macrofungal species in western dryland forests.

All the taxa found in native forests described in various literature sources were saprophytic. However, Alem et al. (2021) also reported an abundance of ectomycorrhizal genera such as *Tricholoma*, *Rhizopogon*, and *Suillus*. According to (Roy et al. 2008; Smith and Read 2008), the presence of these species in the country's natural forest ecosystem may be attributable to the high level of plant diversity in the forests, which may give ectomycorrhizal fungi a very wide host range. As a result, there may be several trees that can serve as hosts for mycorrhizal fungi (Hailemariam et al. 2013; Wubet et al. 2003). It is interesting to note that some *Trichoderma* fungi have the potential to operate as biocontrol agents in forests (Vinale et al. 2008). Additionally, the overall landscape connectedness of surrounding fragmented natural forests and exotic tree plantations may also play a role in the occurrence of ectomycorrhizal fungi in the fungal community assemblage (Boeraeve et al. 2018; Peay and Bruns 2014; Vannette et al. 2016). Most ectomycorrhizal species common in African forests, such as those in the genus *Amanita* (Okhuoya et al. 2010), were also reported by (Alem et al. 2021).

Similarly, important fungi that are found on highland plateaus between 2000–3000 meters above sea level can also be found in upland grazing pastures. These regions have farming practices that emphasize raising livestock in addition to growing crops. Some taxa in these regions were mentioned by Abate (1999, 2014); Alemu (2013); Sitotaw et al. (2015). *Agaricus*, *Lycoperdon*, and *Macrolepiota* were the most prevalent genera identified in the country's highland regions thus far. These species are saprophytes. Additional complementary studies may prove demonstrate that the variety of fungal species in Ethiopian grazing pastures is far higher than what has been previously documented.

The other ecological niche where lucrative fungi might be discovered is in termite mounds. These sites offer a unique example of fungus and insects coexisting (Frøslev et al. 2003; Tibuhwa 2012b; Yamada et al. 2005). In this relationship, small nodules that the fungus develops are eaten by termites, together with the comb-shaped mounds of deteriorated substrate. When it rains, the mycelium that forms in the termite combs deteriorates and generates mushrooms, which break through the soil and termite nests to reach the surface and disperse their spores (Frøslev et al. 2003). Most *Termitomyces* species in Ethiopia have been found in the lowland regions of the country, where termite mounds are more prevalent (Abate 2014; Muleta et al. 2013; Sitotaw et al. 2015). Additionally, the *Termitomyces clypeatus*, *T. microcarpus*, *T. robustus*, and *T. schimperi* species were reported by Alem et al. (2021) from the dry Afromontane forests region of the country. Because of its flavor and odor, *Termitomyces* sp. is highly appreciated by locals, especially in southwest Ethiopia (Abate 2014). This might open up new possibilities for expanding the management and use of termite mounds and other lowland forest ecosystems for the production of NTFPs in Ethiopia.

Additionally, valuable edible mushrooms can be found in exotic tree plantation forests. Due to the continuous degradation of the natural forests in Ethiopia, the creation of plantations of fast-growing trees has become a major forestry practice. This strategy has led to a rapid expansion of exotic tree species, mainly involving members of the genera *Eucalyptus*, *Cupressus*, *Pinus*, and *Acacia* in the past decades (Bekele 2011; Tesfaye et al. 2020). According to (Lemenih and Bekele 2008), the majority of these plantations were planted as community forests and offer sustainable production of high-value wood and NTFPs (Dejene et al. 2017a). The widespread establishment and expansion of non-native trees in the country suggest the unintentional introduction of related foreign fungus species. Different ectomycorrhizal fungi from the Mediterranean and temperate climates fall into this category (Megersa et al. 2017). *Suillus luteus*, for instance, is widespread in most *Pinus* plantations (Abate 2008; Dejene et al. 2021). Some other species in the genera *Pisolithus*, *Ramaria*, *Rhizopogon*, *Scleroderma*, and *Tylopilus* were also reported by Dejene et al. (2021) from *Pinus radiata* plantations. Furthermore, Dejene et al. (2017a) also reported some ectomycorrhizal species from the genera *Geastrum*, *Scleroderma*, and *Tylopilus* from the *Pinus patula* plantations. Such wild edible fungal species have the potential to diversify the value of plantation forests through mycosilvicultural management approaches (Boa 2004; Peredo et al. 1983), and these fungi can be produced in high quantities. Pictures of some of the wild mushrooms of Ethiopia are provided in Figure 9.1.

FIGURE 9.1 Pictures of selected wild mushroom in the forest of Ethiopia: (a) *Agaricus campestroides*, (b) *Agaricus subedulis*, (c) *Armillaria heimii*, (d) *Morchella anatolica*, (e) *Phylloporus* sp., (f) *Pluteus* sp., (g) *Calvatia subtomentosa*, (h) *Coprinellus* sp., (i) *Gymnopus foetidus*, (j) *Suillus luteus*, (k) *Tylopilus niger*, (l) *Scleroderma* sp., (m) *Lepista sordida*, (n) *Lycoperdon perlatum*, (o) *Macrolepiota africana*, (p) *Macrolepiota procera*, (q) *Hygrophoropsis aurantiaca*, and (r) *Calvatia rubroflava*. (Photo by Dejene et al., unpublished data).

9.5 ETHNOMYCOLOGY ASPECTS OF WILD MUSHROOM

In many rural communities around the world, wild mushrooms provide essential sources of income (Cai et al. 2011; Sarma et al. 2010). Individuals frequently collect these mushrooms because they are valuable NTFPs (Boa 2004; Chang and Lee 2004) that assist people in avoiding poverty and maintaining their way of life by providing a steady source of income (Cai et al. 2011). Due to this, ethnomycology has become a field of study throughout the world (Boa 2004). In many societies, mushrooms are used as a source of food, medicine, enzymes, and other industrial substances (Boa 2004). Additionally, they are crucial ingredients in traditional cuisines and recipes (Durazzo et al. 2017; Marconi et al. 2018). Mushrooms can also be used recreationally and are important to the myths and beliefs of many societies (Wani et al. 2010). Mushrooms are a good substitute for meat and fish (Dejene et al. 2021; Mattila et al. 2001) since they are a significant source of proteins, vitamins, lipids, carbs, amino acids, and minerals (Bano et al. 1993; Mattila et al. 2001).

Similar to wild edibles, wild mushrooms are an important NTFP in Ethiopia, especially in the southern and southwestern parts of the country (Asfaw and Tadesse 2001; Dejene et al. 2017a; Sitotaw et al. 2020). People frequently refer to wild mushrooms in the country as "enguday" or "yejib tila" (shade of the hyena), suggesting that wild mushrooms grow in areas where hyenas cast their shadow and indicating that the growth of wild mushrooms is relatively mysterious (Abate 2014). Although wild mushrooms have been used in traditional medicine and as food (Asfaw and Tadesse 2001; Semwal et al. 2014), little emphasis has been given to indigenous knowledge among various tribes.

Furthermore, there is a dearth of documentation regarding mushroom collection, usage, and valuable species. Nevertheless, in areas where they are consumed as food, wild mushrooms together with other foraging resources are used as a source of sustenance during times of food scarcity (Dejene et al. 2017a; Lulekal et al. 2011; Semwal et al. 2014). The local people sell wild mushrooms in various local marketplaces to generate more revenue for their households (Abate 2014). Ethnic groups from the southern and southwestern regions of Ethiopia are described as mycophiles, and the majority of them have a detailed folk taxonomy for mushrooms, indicating that the local people in these regions are good at sharing ethnomycological knowledge among themselves (Tibuhwa 2012a). For instance, the Sidama ethnic groups from the southern region of the country can easily discern between different types of wild mushrooms using their unique, ancient methods (Zeleke et al. 2020). They identify edible mushrooms primarily using physical traits, odor, and environment. The oldest accounts of this behavior among tribes in undeveloped nations have been putatively reflected by various authors, including Dejene et al. (2017a) and Tuno (2001). In addition to these traits, the edibility of mushrooms is also determined by whether or not they have a strong, unpleasant smell (Tuno 2001). Therefore, it may be assumed that these traditional techniques are the result of old experiments and perhaps accidental discoveries made by the native people, despite the confounded genealogy of their cultural uses (Ekandjo and Chimwamurombe 2012). The Sidama ethnic groups have assigned names to characteristics of several mushroom species. For instance, a highly prized species known as "Meine" is delicious to eat because of its taste but is rare because of its heavy collection. Medicinally important mushrooms are known as "gadifuto", which translates to "Hyena's fart". The Amhara ethnic group in the northern part of the country appears to have limited ethnotaxonomic knowledge of the wild mushroom species found in their area. They classify mushrooms as "enguday", which in general corresponds to "fungi" in English. All mushrooms with caps are classified as "yejib tila". In some northern localities, however, the local people may also assign mushrooms with specific names based on the season in which they grow. For example, mushrooms that grow following the first rain of the season are generally called "gunfane", which means common cold. In general, this observation is echoed by Tuno (2001), who reported that the traditional methods of mushroom utilization by the different ethnic groups provide a unique opportunity for studying how people belonging to traditional tribes in Africa utilize mushrooms as foods (Dejene et al. 2017a, 2017b; Tuno 2001).

On the other hand, one study from Ethiopia's northern region found that residents there have little awareness of ethnomycology and do not consume wild mushrooms, although their parents or grandparents did in the past (Zeleke et al. 2020). Commercial mushroom species, however, are grown artificially in the area and are available for purchase at the supermarkets in major cities. There will eventually be a loss of traditional mycological knowledge among the many ethnic groups in the northern part of the country as a result of the non-consumption of mushrooms in this region (Kamalebo et al. 2018).

Wild mushroom fruiting and collection are only permitted during the rainy season in Ethiopia, which is primarily from June to September. However, collection takes place during the dry season for certain species, such as *Lentinus* sp., (Tuno 2001). *Agaricus arvensis*, *A. campestroides*, *Laetiporus sulphureus*, *Termitomyces microcarpus*, *T. clypeatus*, *Lentinus* sp., *Schizophyllum commune*, and *Dictyophora indusiata* are some of the most commonly used species (Abate 2014, 2008; Muleta et al. 2013; Tuno 2001) and are included in Table 9.1. *Laetiporus sulphureus* and *S. commune* are two of species that have been extensively used and documented for ethnomycological purposes in Ethiopia (Tuno 2001). A list of commonly utilized species and their habitat details are provided in Table 9.1. They grow mainly in natural forests, grazing areas, and termite mounds (Abate 2014). Women are usually involved in the collection of mushrooms and recognize more fungi species than men. Children are also involved in the collection of some specific taxa, such as *Termitomyces* (Abate 2014; Alemu 2013). Women have also basic knowledge of mushrooms in terms of habitats, niches, and associated substrates, and have the expertise to distinguish between edible and poisonous species. This might be because they are more often involved in preparing and cooking mushrooms than men. Although it differs from place to place, the most recognized traits used to classify mushrooms as edible and non-edible are the colors, shapes, and the presence or absence of a strong, bad smell (Tuno 2001). These all help in the successful collection of wild edible mushrooms in a given locality. The traditional knowledge of processing, including handling and cooking, is also well known primarily by women (Abate 2014; Muleta et al. 2013). The information from the literature is very limited and does not reflect the wider cultural usage in the country. Thus, because numerous benefits can be derived from traditional uses, further ethnomycological work is needed to maximize the knowledge and use of wild mushroom resources in Ethiopia.

9.6 MEDICINAL ASPECTS OF WILD MUSHROOMS

Various bioactive compounds are known to be abundant in wild mushrooms (Lindequist et al. 2005). Many of these have been employed in traditional medicine around the world (Hobbs 1995). Interestingly, Abate (2014) reports that the Kaffa ethnic group in southern Ethiopia uses mushrooms in traditional medicine, and *Laetiporus sulphureus* is said to be one species that is frequently used. Its powder is typically kept for long periods to use as medication during child delivery in the Kaffa area (Abate 2014). People of the Benihsnagul Gumz region of western Ethiopia also exhibits traditional knowledge of the use of mushrooms as medicine (Alemu et al. 2012); knowledge of medicinal practices is handed down generational lines via oral communication. In all cases, most of the knowledge on medicinal uses has been found to be confined to elderly people of the village or is handled secretly by traditional healers of the locality. Both are key informants for the identification of medicinal species since they use different mushrooms for their traditional medicinal practices (Abate 2008). Consequently, it is difficult to get a list of species that have been used traditionally for medicinal purposes from any ordinary person (Abate 2014). Thus, further documentation and ethnomedicinal studies will be important for enhancing the understanding of indigenous knowledge systems and developing a list of species used in folk medicine in Ethiopia. A list of species used for traditional medicinal purposes is provided in Table 9.1.

The use of various species by Ethiopia's traditional healers in their traditional medical practices makes these individuals crucial sources of information for the identification and description of medicinal species (Dejene et al. 2017a, 2017b). *Lycoperdon perlatum* Pers. and *Calvatia rubroflava* Fr. were said to be the most important and valuable medicinal wild mushroom species; they are also used as food. In rural areas of the country, these species are crucial in the treatment of wounds and skin conditions. Abate (2014) mentioned that some fungal species are commonly used as traditional medicine in some parts of Ethiopia, which is an interesting observation, as locals regularly preserve the powder of these species for extended periods. Interestingly, *Ganoderma* species are also used for medicinal purposes by the local people in Wondo Genet, the southern part of the country (Zeleke et al. 2020).

FIGURE 9.2 Attempts to produce different valuable mushrooms at the laboratory of EEFRI, Ethiopia.

9.7 CULTIVATION ASPECTS OF MUSHROOMS

Mushroom cultivation is a relatively new practice in Ethiopia and is largely limited to major cities (Abate 2014). The most widely cultivated mushrooms, *Agaricus bisporus*, *Lentinula edodes*, and *Pleurotus ostreatus*, have been produced using agricultural and agro-industrial waste on a limited scale and in some restricted areas (Dejene et al. 2017b). However, there is still a lack of knowledge and cultivation practices that are needed to realize the potential of mushrooms (Dejene et al. 2017b). However, there are some attempts in the research to cultivate valuable mushrooms in the laboratory (Figure 9.2). In general, cultivated mushrooms are now consumed more frequently than ever, especially in major cities, because of their flavor and nutritional value. Conversely, traditional eating practices are impeding the localization of cultivation technologies, especially in regions where the consumption of mushrooms is not widespread.

9.8 CONCLUSIONS

We have highlighted the existence of valuable wild mushrooms in Ethiopia and presented their ecological niches in this review. According to the information presented, it is also generally understood that Ethiopians often gather wild mushrooms for sustenance and utilize them in traditional medicine. In some regions, particularly in the south and southwest parts of the nation, wild mushrooms are also valued by the local population economically. However, we believe that the information currently available is inadequate and does not convey a ethnomycological picture of the country, as Ethiopia is characterized by many different vegetation types, and the people living in different areas might have long histories of wild mushroom usage with wider traditional knowledge in the culture. Thus, we encourage more research to document this information in greater depth throughout the country. Wild mushrooms are not as well known for their food and medicinal purposes in the northern part of the country as they are in southern Ethiopia and have suffered many years of neglect. Consequently, mycological resources have remained understudied and little used. Thus, further studies are needed for the effective popularization, potential utilization, and conservation of wild mushrooms in the country.

REFERENCES

Abate, D., 1999. *Agaricus campestris* in upland Ethiopia. *Mycologist* 13, 28. https://doi.org/10.1016/S0269-915X(99)80075-5

Abate, D., 2008. Wild mushrooms in Ethiopia and our eating habit, in *National Mushroom Conference, Faculty of Science, Addis Ababa University*, May 14–15. Addis Ababa, Ethiopia.

Abate, D., 2014. Wild mushrooms and mushroom cultivation efforts in Ethiopia. *WSMBMP Bull.* 11, 1–3.

Adilo, M., 2007. *The Contribution of Non-Timber Forest Products to Rural Livelihood in Southwest Ethiopia.* MSc Thesis. Wageningen University, the Netherlands.

Alem, D., Dejene, T., Oria-de-Rueda, J.A., Martín-Pinto, P., 2021. Survey of macrofungal diversity and analysis of edaphic factors influencing the fungal community of church forests in Dry Afromontane areas of Northern Ethiopia. *For. Ecol. Manage.* 496, 119391. https://doi.org/10.1016/j.foreco.2021.119391

Alemu, F., 2013. Assessment of wild mushrooms and wood decaying fungi in Dilla University, main campus, Ethiopia. *Int. J. Adv. Res.* 1, 458–467.

Alemu, H., Debela, N., Mamuye, A., Jano, M., 2012. *Wild Edible Plants by Gumuz Tribes as Nutritious and Sustainable Food Stuffs in the Metekel and Kamashi Zones of Benishangul-Gumuz Regional State.* An assessments report submitted to Tikuret Legumuz Hizb Limat Mahibr, Assosa Ethiopia.

Asfaw, Z., Tadesse, M., 2001. Prospects for sustainable use and development of wild food plants in Ethiopia. *Econ. Bot.* 55, 47–62.

Badege, B., 2001. Deforestation and land degradation in the ethiopian highlands: A strategy for physical recovery. *Northeast Afr. Stud.* 8, 7–26.

Bano, Z., Shasirekha, M., Rajarathnam, S., 1993. Improvement of the bioconversion and biotransformation efficiencies of the oyster mushroom (*Pleurotus sajor-caju*) by supplementation of its rice straw with oil seed cakes. *Enzyme Microb. Technol.* 15, 985–989.

Bekele, M., 2011. Forest plantations and woodlots in Ethiopia. *African For. Forum* 1, 52.

Boa, E., 2004. *Wild Edible Fungi: A Global Overview of Their Use and Importance to People*, Non-wood Forest Products N°17. FAO, Rome.

Boeraeve, M., Honnay, O., Mullens, N., Vandekerkhove, K., De Keersmaeker, L., Thomaes, A., Jacquemyn, H., 2018. The impact of spatial isolation and local habitat conditions on colonization of recent forest stands by ectomycorrhizal fungi. *For. Ecol. Manage.* 429, 84–92. https://doi.org/10.1016/j.foreco.2018.06.043

Bonet, J.A., González-olabarria, J.R., Aragón, J.M.D.E., 2014. Mushroom production as an alternative for rural development in a forested mountainous area. *J. Mt. Sci.* 11, 535–543. https://doi.org/0.1007/s11629-013-2877-0

Bongers, F., Tenngkeit, T., 2010. Degraded forests in Eastern Africa: Introduction, in: Bongers, F., Tenningkeit, T. (Eds.), *Degraded Forests in East Africa: Management and Restoration.* Earthscan Ltd., London, pp. 1–18.

Butler, J., Alexander, K., Green, T., 2002. Decaying wood: an overview of its status and ecology in the United Kingdom and continental Europe. *USDA For. Serv. Gen. Tech. Rep. PSW-GTR* 181, 11–19.

Cai, M., Pettenella, D., Vidale, E., 2011. Income generation from wild mushrooms in marginal rural areas. *For. Policy Econ.* 13, 221–226.

Chang, Y.S., Lee, S.S., 2004. Utilisation of macrofungi species in Malaysia. *Fungal Divers.* 15, 15–22.

Dahlberg, A., Schimmel, J., Taylor, A.F.S., Johannesson, H., 2001. Post-fire legacy of ectomycorrhizal fungal communities in the Swedish boreal forest in relation to fire severity and logging intensity. *Biol. Conserv.* 100, 151–161. https://doi.org/10.1016/S0006-3207(00)00230-5

Dalong, M., Luhe, W., Guoting, Y., Liqiang, M., Chun, L., 2011. Growth response of Pinus densiflora seedlings inoculated with three indigenous ectomycorrhizal fungi in combination. *Brazilian J. Microbiol.* 42, 1197–1203. https://doi.org/10.1590/S1517-838220110003000045

Decock, C., Bitew, A., Castillo, G., 2005. Fomitiporia tenuis and Fomitiporia aethiopica (Basidiomycetes, Hymenochaetales), two undescribed species from the Ethiopian highlands: taxonomy and phylogeny. *Mycologia* 97, 121–129. https://doi.org/10.3852/mycologia.97.1.121

Dejene, T., Oria-de-rueda, J.A., Martín-Pinto, P., 2017a. Fungal diversity and succession following stand development in Pinus patula Schiede ex Schltdl. & Cham. plantations in Ethiopia. *For. Ecol. Manage.* 395, 9–18. https://doi.org/10.1016/j.foreco.2017.03.032

Dejene, T., Oria-de-Rueda, J.A., Martín-Pinto, P., 2017b. Edible wild mushrooms of Ethiopia: neglected non-timber forest products. *Rev. Fitotec. Mex.* 40, 391–397.

Dejene, T., Oria-de-Rueda, J.A., Martín-Pinto, P., 2017c. Fungal diversity and succession following stand development in *Pinus patula* Schiede ex Schltdl. & Cham. plantations in Ethiopia. *For. Ecol. Manage.* 395, 9–18. https://doi.org/10.1016/j.foreco.2017.03.032

Dejene, T., Worku, E., Martín-Pinto, P., 2021. Retention of matured trees to conserve fungal diversity and edible sporocarps from short-rotation pinus radiata plantations in Ethiopia. *Plant. Ethiop. J. Fungi* 7, 702. https://doi.org/10.3390/jof7090702

Dickie, I., 2007. Host preference, niches and fungal diversity. *New Phytol.* 174, 230–233.

Durazzo, A., Lisciani, S., Camilli, E., Gabrielli, P., Marconi, S., Gambelli, L., Aguzzi, A., Lucarini, M., Maiani, G., Casale, G., Marletta, L., 2017. Nutritional composition and antioxidant properties of traditional Italian dishes. *Food Chem.* 218, 70–77. https://doi.org/10.1016/j.foodchem.2016.08.120

Ekandjo, L.K., Chimwamurombe, P.M., 2012. Traditional medicinal uses and natural hosts of the genus *Ganoderma* in North-Eastern parts of Namibia. *J. Pure Appl. Microbiol.* 6, 1139–1146.

Ferris, R., Peace, A.J., Newton, A.C., 2000. Macrofungal communities of lowland Scots pine (*Pinus sylvestris* L.) and Norway spruce (*Picea abies* (L.) Karsten.) plantations in England: Relationships with site factors and stand structure. *For. Ecol. Manage.* 131, 255–267. https://doi.org/10.1016/S0378-1127(99)00218-2

Friis, B., Demissew, S., Breugel, P., 2010. *Atlas of the Potential Vegetation of Ethiopia.* The Royal Danish Academy of Sciences and Letters, København, Denmark.

Frøslev, T., Aanen, D., Læssøe, T., Rosendahl, S., 2003. Phylogenetic relationships of *Termitomyces* and related taxa. *Mycol. Res.* 107, 1277–1286.

Goldmann, K., Schöning, I., Buscot, F., Wubet, T., 2015. Forest management type influences diversity and community composition of soil fungi across temperate forest ecosystems. *Front. Microbiol.* 6, 1300.

Hailemariam, M., Birhane, E., Asfaw, Z., Zewdie, S., 2013. Arbuscular mycorrhizal association of indigenous agroforestry tree species and their infective potential with maize in the rift valley, Ethiopia. *Agrofor. Syst.* 87, 1261–1272. https://doi.org/10.1007/s10457-013-9634-9

Harkonen, M., Tuomo, N., Leonard, M., 2003. *Tanzanian Mushrooms. Edible, Harmful and Other Fungi.* Botanical Museum, P.O. Box 7, FIN-00014, University of Helsinki, Finland, Helsinki, Finland. https://doi.org/10.1017/S0269-915X(05)23306-2

Hjortstam, K., Ryvarden, L., 1996. New and interesting wood-inhabiting fungi (Basidiomycotina—Aphyllophorales) from Ethiopia. *Mycotaxon* 60, 181–190.

Hobbs, C., 1995. *Medicinal Mushrooms-An Exploration of Tradition, Healing, and Culture*, 2nd ed. Botanica Press, Santa Cruz, CA.

IBC (Institute of Biodiversity Conservation), 2005. *National Biodiversity Strategy and Action Plan.* Addis Ababa, Ethiopia, December.

IBC (Institute of Biodiversity Conservation), 2014. *Ethiopia's Fifth National Report to the Convention on Biological Diversity.* Addis Ababa, Ethiopia.

Jonsell, M., Nordlander, G., 2000. Insects in polypore fungi as indicator species: a comparison between forest sites differing in amounts and continuity of dead wood. *For. Ecol. Manage.* 157, 101–118.

Kamalebo, H.M., Nshimba, H., Wa, S., Ndabaga, C.M., 2018. Uses and importance of wild fungi: traditional knowledge from the Tshopo province in the Democratic Republic of the Congo. *J. Ethnobiol. Ethnomed.* 14, 1–12. https://doi.org/10.1186/s13002-017-0203-6

Lemenih, M., Bekele, M., 2008. *Participatory Forest Management Best Practices, Lesson Learnt and Challenges Encountered: The Ethiopian and Tanzanian Experiences.* Farm Africa/SOS-Sahel, Ethiopia.

Lindahl, B.D., Ihrmark, K., Boberg, J., Trumbore, S.E., Högberg, P., Stenlid, J., Finlay, R.D., 2007. Spatial separation of litter decomposition and mycorrhizal nitrogen uptake in a boreal forest. *New Phytol.* 173, 611–620. https://doi.org/10.1111/j.1469-8137.2006.01936.x

Lindequist, U., Niedermeyer, T., Jülich, W., 2005. The pharmacological potential of mushrooms. Evid based complement. Evidence-based complement. *Altern. Med.* 2, 285–99.

Lulekal, E., Asfaw, Z., Kelbessa, E., Van Damme, P., 2011. Wild edible plants in Ethiopia: a review on their potential to combat food insecurity. *Afrika Focus* 24, 71–121.

Marconi, S., Durazzo, A., Camilli, E., Lisciani, S., Gabrielli, P., Aguzzi, A., Gambelli, L., Lucarini, M., Marletta, L., 2018. Food composition databases: considerations about complex food matrices. *Foods* 7, 2. https://doi.org/10.3390/foods7010002

Martín-Pinto, P., Vaquerizo, H., Peñalver, F., Olaizola, J., Oria-De-Rueda, J.A., 2006. Early effects of a wildfire on the diversity and production of fungal communities in Mediterranean vegetation types dominated by *Cistus ladanifer* and *Pinus pinaster* in Spain. *For. Ecol. Manage.* 225, 296–305. https://doi.org/10.1016/j.foreco.2006.01.006

Mattila, P., Konko, K., Eurola, M., Pihlava, J., Astola, J., Al, E., 2001. Contents of vitamins, mineral elements, and some phenolic compounds in cultivated mushrooms. *J. Agric. Food Chem.* 49, 2343–2348.

Megersa, S., Gure, A., Feleke, S., Alemu, M., 2017. Macrofungi species richness and diversity in Dagaga and Gambo plantation and natural forests of Arsi Forest Enterprise, Oromia, Ethiopia. *Imp. J. Interdiscip. Res.* 3, 1681–1686.

Muleta, D., Woyessa, D., Teferi, Y., 2013. Mushroom consumption habits of Wacha Kebele residents, south-western Ethiopia. *Glob. Res. J. Agric. Biol. Sci.* 4, 6–16.

Nick, B., Shonil, B., Sarah, W., 2006. Macrofungal diversity in fragmented and disturbed forests of the Western Ghats of India. *J. Appl. Ecol.* 43, 11–17.

Nordén, B., Ryberg, M., Götmark, F., Olausson, B., 2004. Relative importance of coarse and fine woody debris for the diversity of wood-inhabiting fungi in temperate broadleaf forests. *Biol. Conserv.* 117, 1–10. https://doi.org/10.1016/S0006-3207(03)00235-0

Okhuoya, J., Akpaja, E., Osemwegie, O., Oghnekaro, A., Ihayere, C., 2010. Nigerian mushrooms: Underutilized non-wood forest resources. *J. Appl. Sci. Environ. Manag.* 14, 43–54.

Osarenkhoe, O.O., John, O.A., Theophilus, D.A., 2014. Ethnomycological conspectus of West African mushrooms : an awareness document. *Adv. Microbiol.* 4, 39–54. https://doi.org/10.4236/aim.2014.41008

Peay, K.G., Bruns, T.D., 2014. Spore dispersal of basidiomycete fungi at the landscape scale is driven by stochastic and deterministic processes and generates variability in plant-fungal interactions. *New Phytol.* 204, 180–191. https://doi.org/10.1111/nph.12906

Pegler, D., 1977. A preliminary Agaric flora of East Africa. *Kew Bull. Add. Ser.* 6, 1–615.

Pegler, D., Rayner, R., 1969. A contribution to the Agaric flora of Kenya. *Kew Bull.* 23, 347–412.

Peredo, H., Oliva, M., Huber, A., 1983. Environmental factors determining the distribution of *Suillus luteus* fructifications in *Pinus radiata* grazing-forest plantations. *Plant Soil.* 71, 367–370.

Pettenella, D., Secco, L., Maso, D., 2007. NWFP&S marketing: lessons learned and new development paths from case studies in some European Countries. *Small-Scale For.* 6, 373–390. https://doi.org/10.1007/s11842-007-9032-0

Pietras, M., Rudawska, M., Leski, T., Karliński, L., 2013. Diversity of ectomycorrhizal fungus assemblages on nursery grown European beech seedlings. *Ann. For. Sci.* 70, 115–121. https://doi.org/10.1007/s13595-012-0243-y

Roy, M., Dubois, M.-P., Proffit, M., Vincenot, L., Desmarais, E., Selosse, M.-A., 2008. Evidence from population genetics that the ectomycorrhizal basidiomycete *Laccaria amethystina* is an actual multihost symbiont. *Mol. Ecol.* 17, 2825–2838. https://doi.org/10.1111/j.1365-294X.2008.03790.x

Sarma, T., Sarma, I., Patiri, B., 2010. Wild edible mushrooms are used by some ethnic tribes of Western Assam. *The Bioscan* 3, 613–625. https://doi.org/10.13140/RG.2.1.3531.1842

Semwal, K.C., Lemma, H., Dhyani, A., Equar, G., Amhare, S., 2014. Short communication mushroom: Nature's treasure in Ethiopia. *Momona Ethiop. J. Sci.* 6, 138–147.

Seyoum, A., 2007. Economic value of afromontane forests of Sheka: Forest of Sheka, in: Fetene, M. (Ed.), *Forest of Sheka*. Addis Ababa, Ethiopia, pp. 183–218.

Sitotaw, R., Lulekal, E., Abate, D., 2020. Ethnomycological study of edible and medicinal mushrooms in Menge District, Asossa Zone, Benshangul Gumuz Region, Ethiopia. *J. Ethnobiol. Ethnomed.* 16, 1–14. https://doi.org/10.1186/s13002-020-00361-9

Sitotaw, R., Mulat, A., Abate, D., 2015. Morphological and molecular studies on *Termitomyces* species of Menge District, Asossa Zone, Northwest Ethiopia. *Sci. Technol. Arts Res. J.* 4, 49–57.

Smith, S., Read, D., 2008. *Mycorrhizal Symbiosis*. Academic Press, Cambridge.

Sultan, M., 2009. *The Role of Non-Timber Forest Products to Rural Livelihoods and Forest Conservation: A Case Study at Harana Bulluk District Oromia National Regional State, Ethiopia*. MSc Thesis. Wondo Genet: Wondo Genet College of Forestry and Natural Resource, Hawassa University.

Tesfaye, M.A., Gardi, O., Anbessa, T.B., Blaser, J., 2020. Aboveground biomass, growth and yield for some selected introduced tree species, namely *Cupressus lusitanica*, *Eucalyptus saligna*, and *Pinus patula* in the Central Highlands of Ethiopia. *J. Ecol. Environ.* 44, 1–18. https://doi.org/https://doi.org/10.1186/s41610-019-0146-z

Tibuhwa, D.D, 2012a. Folk taxonomy and use of mushrooms in communities around Ngorongoro and Serengeti National Park, Tanzania. *J. Ethnobiol. Ethnomed.* 8, 1–12.

Tibuhwa, D.D., 2012b. Termitomyces species from Tanzania, their cultural properties and unequalled basidiospores. *J. Biol. Life Sci.* 3, 140–159. https://doi.org/10.5296/jbls.v3i1.1723

Tuno, N., 2001. Mushroom utilization by the Majangir, an Ethiopian tribe. *Mycologist* 15, 78–79. https://doi.org/10.1016/S0269-915X(01)80087-2

van der Heijden, M.G.A., Bardgett, R.D., van Straalen, N.M., 2008. The unseen majority: soil microbes as drivers of plant diversity and productivity in terrestrial ecosystems. *Ecol. Lett.* 11, 296–310. https://doi.org/10.1111/j.1461-0248.2007.01139.x

Vannette, R.L., Leopold, D.R., Fukami, T., 2016. Forest area and connectivity influence root-associated fungal communities in a fragmented landscape. *Ecology* 97, 2374–2383. https://doi.org/10.1002/ecy.1472

Vinale, F., Sivasithamparam, K., Ghisalberti, E.L., Marra, R., Woo, S.L., Lorito, M., 2008. *Trichoderma*—plant—pathogen interactions. *Soil Biol. Biochem.* 40, 1–10. https://doi.org/10.1016/j.soilbio.2007.07.002

Wani, B., Bodha, R., Wani, A., 2010. Nutritional and medicinal importance of mushrooms. *J. Med. Plant Res.* 4, 2598–2604.

Wardle, D., 2006. The influence of biotic interactions on soil biodiversity. *Ecol. Lett.* 9, 349–373.

Wubet, T., Kottke, I., Teketay, D., Oberwinkler, F., 2003. Mycorrhizal status of indigenous trees in dry Afromontane forests of Ethiopia. *For. Ecol. Manage.* 179, 387–399. https://doi.org/10.1016/S0378-1127(02)00546-7

Yamada, A., Inoue, T., Wiwatwitaya, D., Ohkuma, M., Kudo, T., Abe, T., Sugimoto, A., 2005. Carbon mineralization by termites in tropical forests, with emphasis on fungus combs. *Ecol. Res.* 20, 453–460. https://doi.org/10.1007/s11284-005-0062-9

Zeleke, G., Dejene, T., Tadesse, W., Agúndez, D., Martín-Pinto, P., 2020. Ethnomycological knowledge of three ethnic groups in Ethiopia. *Forests* 11, 875. https://doi.org/10.3390/f11080875

Zewdie, M., Olsson, M., Verwijst, T., 2009. Above-ground biomass production and allometric relations of *Eucalyptus globulus* Labill. coppice plantations along a chronosequence in the central highlands of Ethiopia. *Biomass Bioenergy* 33, 421–428. https://doi.org/10.1016/j.biombioe.2008.08.007

10 Wild Edible Mushroom Trade in Europe

Special Focus on Estonia

Surya Sudheer, Ain Raal, Aivar Jürgenson, Rajeev Bhat

10.1 INTRODUCTION

Mushrooms are popular foods with high medicinal value. Because of the health benefits associated with mushrooms, they are considered functional foods. Mushrooms have been a part of human culture and civilisation for thousands of years. Mushrooms and their importance are cited in ancient literature. The Greeks believed that mushrooms could provide strength for warriors in battle, whereas Romans considered them "food of the gods". For the ancient Chinese, mushrooms were an "elixir of life", and for the Egyptians, they were "plants of immortality". Wild mushroom foraging is popular in many European countries, but the identification of wild edible mushrooms is still a challenge and invites many risks for people. Edible mushrooms can be safely consumed by a population but require proper documentation about edibility or related information for a broader audience to know their uses. People's opinions may differ regarding taste, texture, and smell. Mushrooms are highly prized as health food in different parts of the world, but at the same time are approached with suspicion because of their potential toxicity. Mushrooms are considered a next-generation food for health and are one of the emerging sources of "sustainable food". In this chapter, we emphasise wild edible mushroom legislation in Europe, with a special focus on the distribution, trade, and ethnomedicinal details of Estonian wild edible mushrooms.

10.2 AN OVERVIEW OF WILD EDIBLE MUSHROOMS AND RELATED LEGISLATION IN EUROPE

Many countries in the European Union (EU) have a long history of collecting and eating mushrooms. The populations of these countries who are fond of mushroom foraging are labelled as mycophiles. Mycophilic populations are common in many parts of southern and eastern Europe, with extreme levels of mycophily seen in Poland. In contrast, mycophobic areas are found in central, northern, and north-western countries (Kotowski et al., 2019). Before the twenty-first century, the Swedish people were mycophobic, but recent studies show that mushroom gathering has now become a common and appreciated recreational activity among the Swedish people (Svanberg and Lindh, 2019). A relatively high number of mushrooms has been commercialised in the EU. For example, in 2013, total of 268 fungal taxa (including nine genera of the Ascomycota and 74 genera of the Basidiomycota) were listed as commercialised in 24 EU countries (Peintner et al., 2013). This number has changed since then. The largest genus of commercialised mushrooms belongs to the order Agaricales (36). It is expected that by 2026, the edible mushroom market will reach a value of USD $21.67 billion, as compared to USD $13.71 billion in 2018 (Market Data Forecast, 2022). An increasing consumer demand for sustainable food and meat substitutes is driving the market for mushroom cultivation. Therefore, the European mushroom cultivation market is surging upward, and many companies are now investing in the production of mushrooms. Among all fungal taxa in the region, there are about 60 mushroom species that are cultivated on commercial mushroom farms, and the rest of the species are collected from forests by mushroom pickers and

DOI: 10.1201/b23190-10

then sold in the market. Many European countries like Germany, the Netherlands, France, the United Kingdom, Italy, and Poland are importing dried mushrooms from Asian countries. Among them, France and Italy offer opportunities for wild-collected gourmet mushroom suppliers, as both countries prefer dried mushrooms in traditional home cooking (CBI, 2020). This emphasises the importance of mushroom legislation for safe commercialisation around the world. For the safe commercialisation of mushrooms, more than half of the EU countries have set guidelines and legislation. This list includes countries in Western Europe such as France, Italy, Portugal, Spain, Belgium, and Switzerland. In Eastern Europe, Belarus, Poland, Russia, Slovakia, and Ukraine are on the list (Peintner et al., 2013). Mushroom commerce is dynamic in countries with a mycophilic population. Therefore, it is equally important to control mushroom commercialisation via regulatory rules to avoid problems related to mushroom poisoning after their consumption. Mushroom guidelines are based on the present use of wild edible mushrooms in a given country. This means that the local mycophilic population is aware of the edible mushroom species in the forests and is also well-informed about the places and seasons for collection, as well as safety (Stryamets et al., 2022).

Indigenous people in Europe generally collect different species of wild mushrooms for consumption and recreation. They preserve these mushrooms in different ways and sell them in markets or give them away as gifts (Peintner et al., 2013). One risk factor associated with mushroom foraging and eating is mushroom intoxication. In the EU region itself, around 50 to 100 cases of mycetismus are reported yearly (Smith and Davis, 2016). Therefore, the commerce of fresh and preserved wild edible mushrooms should follow safety guidelines and enacted laws for consumption and commercialisation. Mushroom legislation not only provides a list of wild fungal taxa that could be commercialised in a given country but also gives detailed procedures for collection, including time, quantities, and allowed methods; these guidelines can incorporate environmental conservation measures as well. In the EU, only a few mushrooms are known to be poisonous among the 5,000 known species recorded (Wennig et al., 2020). Consumption of poisonous mushrooms may lead to gastrointestinal irritation, acute renal failure, or psychotropic effects and death (Erden et al., 2013). Mushroom poisoning in the EU and US is mostly (90%) associated with amatoxin as is detected by symptomatology (i.e., Phalloides syndrome) (Santi et al., 2012; Mengs et al., 2012).

Every country in the EU has its own list of mushroom species published with varying degrees of comprehensiveness. Italy has published a list of 150 species (Sitta and Floriani, 2008). An ongoing project in Italy has focused on the molecular identification of both cultivable and wild mushroom species in the Italian territory (Giusti et al., 2020). Peintner et al. (2013) performed an excellent study on the legislation and guidelines for wild mushroom commerce in European countries. Some of their observations are provided herein. As per their study, Serbia has listed 15 mushrooms, France has listed 122, and Switzerland has recognised 114 species. The European mushrooms list is very heterogeneous at the genus level. *Lactarius* (23 spp.), *Russula* (19 spp.), and *Agaricus* (14 spp.) are popular genera, with most species approved for commercialisation in the EU. Porcini (*Boletus edulis*) and chanterelle (*Cantharellus cibarius*) mushrooms are present in all the lists. *Lactarius deliciosus*, *Morchella esculenta*, *Boletus badius*, *Agaricus campestris*, and *Craterellus cornucopioides* are mushroom species authorised to be marketed in 70% of European countries. Western Europe is significantly higher in mushroom species richness than Eastern Europe, and populations there consume mushrooms belonging to the genera *Coprinus*, *Flammulina*, *Craterellus*, *Macrolepiota*, *Morchella*, *Tricholoma*, and *Stropharia*. Compared with countries in central and southwest Europe, fewer mushroom species with less diversity are authorised for commerce in the countries of southeast Europe (Garibay-Orijel et al., 2007). Populations from Eastern Europe like to consume *Suillus* spp., while people in the Nordic region prefer species of *Russula* (*R. claroflava*—yellow swamp brittlegill, *R. decolourans*—copper brittlegill, *R. vinosa*—darkening brittlegill), as well as *Suillus variegatus* (velvet bolete), *Leccinum vulpinum* (foxy bolete), and *Hygrophorus camarophyllus* (arched woodwax). Based on geographical separation, cultural differences, and collecting habits, there are significant differences in the choice of mushroom species preferred by a given population. In some regions, hypogenous fungi (soil fungi) are collected with the help of pigs and dogs (Peintner et al., 2013).

10.3 MUSHROOM TRADE IN EUROPE

Chanterelles are among of the most valued edible forest mushrooms, and their international commercial value exceeds a billion dollars annually. The world production of wild chanterelles was estimated to be 150,000 to 200,000 tonnes per year in the 1990s. In the case of black truffles, the worldwide supply was estimated to be more than 200 tonnes per year. The largest producers of mushrooms were Italy, Netherlands, France, Poland, and Spain at that time (Pilz et al., 2003; Peintner et al., 2013). The black truffle industry is growing rapidly in Europe, especially in France, Italy, and Spain. The truffle sector itself generates tens of millions of Euros annually and also promotes activities such as truffle tourism and interdisciplinary research activities (Oliach et al., 2021). New markets are opening up all the time, and new healthy habits such as "sustainable food" and "healthy eating", have led people to consume more mushrooms and mushroom-based products (Bell et al., 2022). Wild mushrooms have always been a major part of Italian culinary tradition and have flourished in Italy for centuries. Because of the changes in consumer behaviour in recent years, the global market has been reduced to small number of mushroom species (e.g., Porcini) (Peintner et al., 2013). The five top-selling edible fungi included in most European mushroom lists are *B. edulis*, *Lactarius deliciosus*, *C. cibarius*, *A. campestris*, and *Morchella esculenta*. Interestingly, two important wild mushroom species currently traded in Italy are *Calocybe gambosa* and *Armillaria mellea*, which are not on the list of mushroom species in several other European countries. As per the current EU treaty regarding food safety demands, a mushroom can be sold in a country where the population has specific knowledge about the preparation of this species, no matter whether such general knowledge is lacking in other countries on the same mushroom species. For example, false morel mushrooms are allowed to be sold in Finland but not in Denmark (Peintner et al., 2013; Lagrange and Vernoux, 2020). As per the updated list of the CBI Ministry of Foreign Affairs, Poland is the largest fresh button mushroom exporter in the world. In 2019, Poland exported over 800 tonnes of dried mushrooms, worth €12.3 million. In Bulgaria, around 330 tonnes of dried mushrooms worth €7.8 million were exported in 2019. In contrast, Romania exported 224 tonnes of mushrooms, worth €7.3 million, in the same year (CBI, 2020).

Most of the younger population in the EU is in support of sustainable food production and consumption that minimises global environmental impacts. This has given more opportunities for the mushroom industry to collect, cultivate, and market wild edible mushrooms. It also creates a positive image for emerging suppliers from developing countries. China is currently the leading competitor for shiitake and other Asian dried mushrooms. Eastern European countries are strong competitors for dried wild mushrooms such as porcini, chanterelle, black trumpets, or morels. If we consider EU suppliers for wild culinary mushrooms, the main competitors are Eastern European countries like Romania, Bulgaria, Serbia, Bosnia and Herzegovina, Turkey, and North Macedonia, followed by Germany and Italy. Poland and the Netherlands are the main competitors for dried *Agaricus* mushrooms in the EU. Germany and Italy import dried mushrooms from Poland. Although Germany is an insignificant producer, it is known to re-export dried mushrooms. On the other hand, Italy produced some quantities of wild-collected dried mushrooms, but the produced quantity is not sufficient, so Italian imports are much greater than its exports. Bulgaria and Romania are trying to increase the production and export of dried wild mushrooms with the help of collectors who forage for fresh mushrooms in forests and sell them to processing companies. Most mushroom species collected this way are chanterelles, morels, porcini, truffles, and black trumpets. Serbia and Bosnia and Herzegovina mostly export porcini, chanterelles, black trumpets, *Lactarius* (milky-caps), and truffles to a smaller extent (Radomir et al., 2018). The most ubiquitous mushrooms that have been the subject of large-scale trade in Serbia are *Boletus* spp. and *C. cibarius* (CBI, 2022).

10.4 WILD EDIBLE MUSHROOMS OF ESTONIA

Estonia is a country in the Baltic region of Northern Europe that borders the Baltic Sea and the Gulf of Finland. When it comes to mushrooms, Estonia has much to offer. Over 7,000 species have been

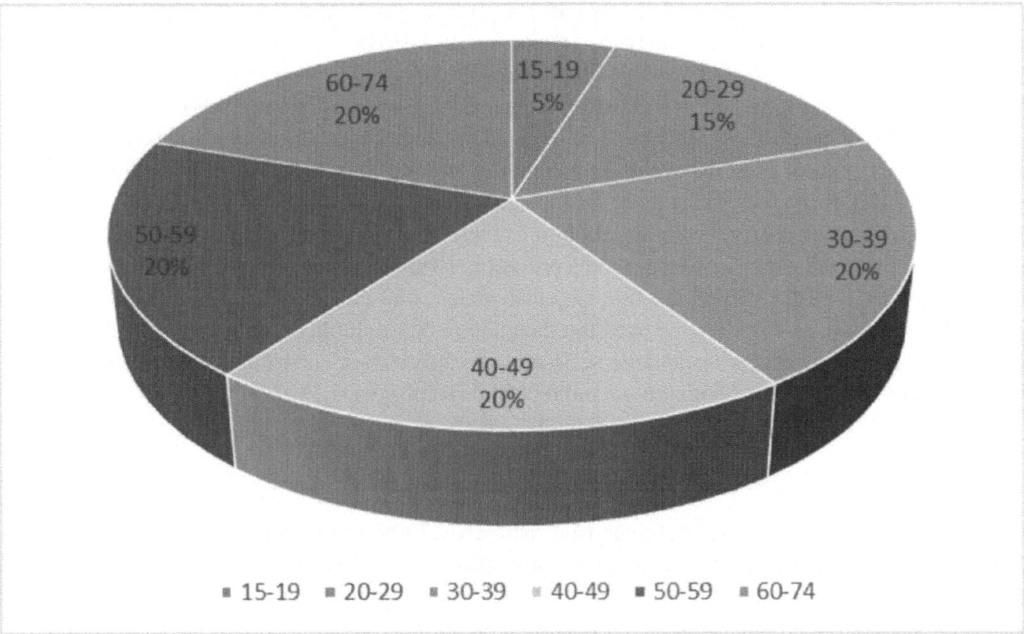

FIGURE 10.1 Mushroom and berry picking age groups in Estonia. Image retrieved and modified from Sepp and Lõhmus (2020) with permission by the Estonian Cooperation Assembly.

reported in Estonia. Of these, 400 are recognised as edible and about 200 as toxic (Kämärä, 2017). Estonia's knowledge and interest in mushrooms grew at the end of the nineteenth century. Popular activities for the Estonian population are often associated with nature. Mushroom and berry picking are among favoured activities, especially by rural residents, retired people, and people with a low income (Figure 10.1). A total harvest of ~8 million litres was estimated in 2011 (Kaldaru, 2011). A significant contribution has been made in the field of mycology by the well-known Estonian mycologist Erast Parmasto (1928–2012). His focus was mainly on the taxonomy of corticioid and polyporoid basidiomycetes, as well as the diversity of fungi in Estonia (Parmasto, 2012). In 2001, Parmasto documented the occurrence of nearly 360 species of basidiomycetes in a mycological "hotspot" at Soomaa National Park, Estonia (Parmasto, 2001; Põldmaa et al., 2019). The best season for mushroom picking in Estonia is from late August to mid-October. Rim (2016) reported that chanterelles and boletes are the most exported mushrooms from Estonia. The export volumes were high from 1995 to 2003, and then showed a decrease during the period of 2004 to 2009, and from 2010 the export of mushrooms started to rise again but still remained modest compared to before. About 292.5 tonnes were exported in the year 2000. The main countries to which mushrooms are exported are Finland and Sweden. Imports of mushrooms are lower than exports, and these indicators do not always reflect local mushroom consumption. Mushrooms are mainly bought in Estonia from Latvia and Lithuania and are resold from Estonia to Western Europe (Rim, 2016).

By the mid twentieth century, mushrooms had acquired a considerable role in the Estonian population's food consumption habits (Jürgenson, 2004). Aivar Jürgenson, an ethnologist from Estonia, wrote a book about the origin of mushrooms and their role in cultural history (Jürgenson, 2005). Estonian mycophiles are most interested in picking chanterelles, *Lactarius*, *Boletus*, and *Russula*. At least 50 different species of *Boletus* have been found and classified in Estonia (Nael, 2022). Chanterelles are inarguably the most sought-after mushrooms in Estonia and are marketed in fresh, blanched frozen, or blanched and salted forms. Other local favourite mushrooms include *Boletus* and *Russula*. Some *Lactarius* species such as *L. rufus*, *L. scrobiculatus,* and *L. deliciosus,* along with *Macrolepiota procera,* are also used culinarily or as medicine (Urb, 2021).

A detailed list of Estonian mushrooms has not yet been published. Saar et al. (2019) presented an updated red list of Estonian fungi to compare their assessment with the previous list, which was published in 1988 and 1998. The first red list comprised 12 non-lichenised fungal species (Kalamees, 1990), and the following Estonian Red Data Book (Lilleleht, 1998) included 91 fungal species (Järva et al., 1998; Järva et al., 1999). Threatened taxa in the last list belonged to the phyla Ascomycota and Basidiomycota in Estonia. Saar et al. (2019) listed all *Lactarius* species and selected *Thelephoroid* species; they also discussed the main habitats of threatened species and the major factors that endanger the distribution of these fungi in Estonia (Saar et al., 2019). These authors also listed the poisonous mushrooms present in Estonian forests, including *Amanita phalloides*, *A. virosa*, *Cortinarius rubellus*, *Gyromitra esculenta*, and *Inocybe erubescens* (Saar, 2019). In one of these studies, researchers reported the Estonian polypore biota to comprise over 260 species, of which 221 are verified species and the remaining are only molecularly documented. It is expected that more species are yet to be identified and reported (Runnel et al., 2021).

10.5 DISTRIBUTION OF WILD EDIBLE MUSHROOMS IN ESTONIAN FORESTS

Late mycologist Mall Vaasma published an article in 2009 about mushrooms in the forest in Estonia. Wild edible mushrooms and their availability are dependent on the weather and related conditions. South-eastern Estonia is popular for wild edible mushrooms, especially chanterelles. *L. rufus* (rufous milk cap) is found in Lahemaa, Estonia's largest national park (Figure 10.2). The most productive species in the forests are *L. rufus*, *Rozites caperata* (the gipsy mushroom), and *R. decolourans* (copper brittle gill). *Armillaria mellea* or honey fungus is a highly-produced species in Raplamaa. Estonian forests with the most diverse spectra of species might be found in Saaremaa and the Hiiumaa Islands, based on investigations published in the yearbook of the Estonian Naturalist Society (Kalamees, 2010). *L. rufus* grows near pine and birch trees and is found in Põlva County and Harju County. *L. scrobiculatus* is found in Jõgeva, Tartu, and Lääne-Virumaa counties. Similarly, *L. deliciosus*, known as the saffron milk cap or the red pine mushroom, is

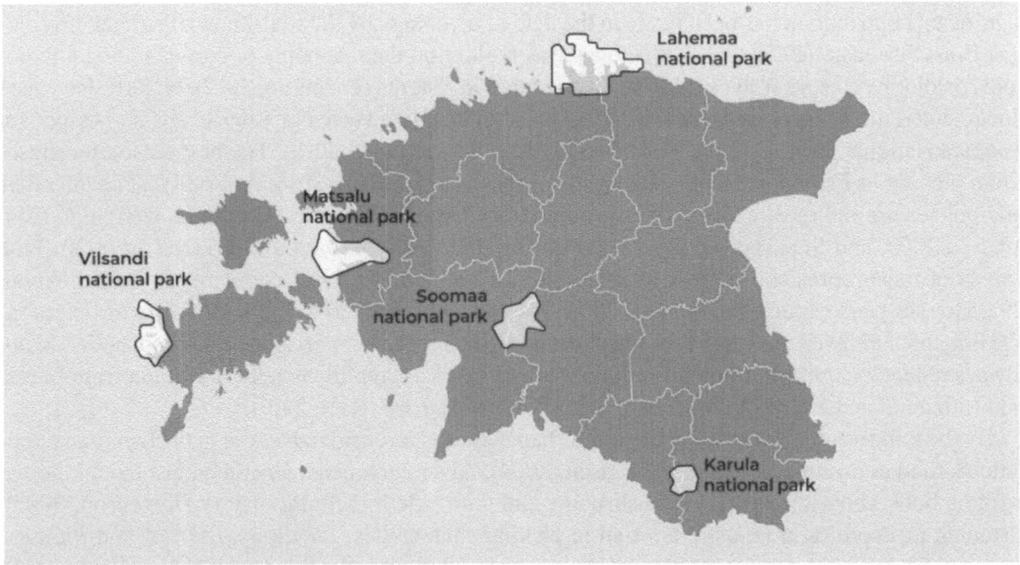

FIGURE 10.2 Natural sites in Estonia managed by the State Forest Management Centre where more number of visits reported. Image retrieved from Sepp and Lõhmus (2020) with the permission by the Estonian Cooperation Assembly.

FIGURE 10.3 Wild edible mushrooms for sales in Estonian market. Image courtesy Gyllenberg (2019).

found in Ida-Virumaa, Harju, and Võru counties. *Macrolepiota procera* is common in Harju, Pärnu, and Tartu counties, while *Russula* spp. can be found in Lääne, Jõgeva, and Võru counties. North Estonia alone lists around 60 species of mushrooms, mainly found along the Oandu nature trail (Tuberik, 2009).

10.6 A BRIEF OVERVIEW OF WILD EDIBLE/POISONOUS MUSHROOMS IN ESTONIA

10.6.1 Edible Mushrooms in Estonia

In Estonia, picking edible mushrooms has always been a popular activity, taking place mostly in autumn, but rarely in spring as well. About 400 species of edible mushrooms grow in Estonia, of which about 50 are considered good or very good. Usually, only 10–15 mushroom species are collected for food (Jaagomäe and Masso, 1966). The reason for this is the relatively limited knowledge of edible and poisonous mushrooms. To alleviate these problems, most families have at least one or more mushroom books in their home library. Several such manuals and catalogues were published during the period of Soviet Estonia before 1991 (Marland, 1956; Jaagomäe and Masso, 1966; Raitviir, 1972), as well as after Estonia's independence (Kuslap, 2004; Raitviir, 2010; Laux, 2011; Niiberg, 2016; Salo and Salo, 2017; Jürgenson, 2022).

In recent decades, people, especially younger people, have used the internet to identify and pick mushrooms. Several smartphone applications have been developed for this purpose, with which it is possible to browse photo galleries of mushrooms or take a photograph of a mushroom of interest and identify it through the image. Forest mushrooms are usually not collected for medicinal purposes in Estonia, and there is no mention of such mushrooms in any mushroom book.

10.6.2 Mushroom Poisoning in Estonia

Despite the abundance of readily available information about poisonous mushrooms, mushroom poisonings occur from time to time in Estonia. They first appear in the spring, when people eat the "brain mushroom" raw. Gyromitrin poisoning is mainly caused by three species: *Gyromitra esculenta* (brain mushroom), *G. gigas* (giant false morel), and *Discina ancilis* (pig's ears), which are edible mushrooms but are only safe when cooked and dried (Kalamees, 2011). The most dangerous mushroom poisoning in Estonia is associated with amanitin, which irreversibly damages the liver and kidneys. This substance is found in the most poisonous mushrooms in Europe, including the white destroying angel (*Amanita virosa*) and death cap (*A. phalloides*), as well as the deadly skullcap (*Galerina marginata*). More than 90% of all fatal mushroom poisonings are associated with the death cap (*A. phalloides*). Another poisonous compound found in mushrooms is muscarinic, which is reported in *Clitocybe* and *Inocybe* species (George and Hegde, 2013). The red fly agaric also contains a negligible amount of muscarine (0.0003%). In general, mushroom poisonings in Estonia, especially fatal poisonings, are not considered very frequent compared to those in Western Europe (Kalamees, 2011). One could emphasise the connections between mushroom poisoning and immigrants. For example, in the United States, green fly agaric poisonings have often affected immigrants from Asia, who confuse it with *Volvariella volvacea*, which is familiar to them from their homeland. Similarly, Estonians who have emigrated abroad, and also the immigrants in Estonia (such as the Caucasian Estonian diaspora and Ukrainian migrant workers in Estonia in recent years) can get easily confused with the species from their homeland (Millman, 2019).

10.7 WILD EDIBLE SPECIES IN ESTONIA

In the section that follows, brief descriptions are provided for some of the common, wild edible species of Estonia.

10.7.1 *Boletus* spp.

Members of the genus *Boletus* (phylum: Basidiomycota, class: Agaricomycetes, order: Boletales, family: Boletaceae) are some of the favourite mushrooms of Estonian mycophiles. Around 50 species of *Boletus* are recorded in Estonia (Nael, 2022). *B. edulis* Bull, called penny bun, ceps, or porcini, is the most popular one. This species are mainly encountered at the edges of clearings in broadleaf or coniferous forests. These fungi can form ectomycorrhizae with the roots of forest trees (Money et al., 2016). Healthy fruiting bodies of this fungus are collected from the forest to be cleaned, dried, packed, and finally marketed. The quality of such mushrooms is marked on the basis of their size, colour, flavour, and aroma. *Boletus* is an important ingredient in traditional dishes. *Boletus* mushrooms are good source of protein and dietary fibre. They also contain high amounts of iron, selenium zinc, essential fatty acids, vitamins, and manganese (Sande et al., 2019). They have been shown to possess high antioxidant and antimicrobial activities (Jaworska et al., 2015; Gałgowska and Pietrzak-Fiećko, 2020). The incorporation of *B. edulis* into cooked sausages has been shown to significantly improve firmness of frankfurters (Novakovic et al., 2020). Dried porcini mushrooms are marketed by many companies in Europe because of their robust scent and flavour (accessed from www.gurm.ee/en/a/funghi-porcini-secchi-special-dried-porcini-mushrooms, accessed in July 2022).

A targeted metabolite analysis was performed by Heleno et al. (2011) in six wild *Boletus* species and revealed proteins, carbohydrates, fatty acids (mainly linoleic acid) sugars (mainly mannitol and trehalose) and vitamins (tocopherols and ascorbic acid). Secondary metabolites, such as phenolic acids, were identified in *Boletus* and demonstrated to have antioxidant properties including free radical-scavenging activity, reducing power, and lipid peroxidation inhibition.

10.7.2 *Cantharellus cibarius* (Chanterelles)

Chanterelles (phylum: Basidiomycota, class: Agaricomycetes, order: Cantharellales, family: Cantharellaceae) are the most popular and widely marketed mushrooms throughout Europe. They often grow in coniferous or mixed forests, especially with tree species such as spruce, pine, and even oak. As chanterelles grow preferably in damp moss and grasses, finding them is often relatively easy.

Cantharellus cibarius has edible and medicinal value. This species is reported to have good radical-scavenging activity and antioxidant potential (Ebrahimzadeh et al., 2010). *C. cibarius* has been shown to exert significant antibacterial and anti-inflammatory effects (Nasiry et al., 2017). These mushrooms are yellow-golden in colour and are a great treat in a creamy mushroom sauce enjoyed along with pasta, potatoes, and grains. These mushrooms are mostly seen in South Estonia and on Saaremaa Island. Chanterelles are mostly available starting around the end of June every year. High-quality chanterelles are collected as per clients' requirements by companies in Estonia. Researchers from Poland conducted a multivariate analysis of 20 mineral elements in *Cantharellus* and found an abundance of essential micro- and macroelements based on samples from the fruiting bodies collected (Drewnowska and Falandysz, 2015). Chanterelles are thought to positively affect the immune system and help reduce swelling, and that also have a healing property that helps with helminth infections (Niiberg, 2016).

10.7.3 *Russula* spp.

Members of the genus *Russula* (phylum: Basidiomycota, class: Agaricomycetes, order: Russulales, family: Russulaceae) are commonly found in coniferous woodlands and are ectomycorrhizal. The young fruiting bodies can be found beneath pines and spruce. Most of the mushrooms have brightly coloured caps with white stipes. Nearly 50 species of *Russula* have been reported in Estonia (Nael, 2022). Some species are associated with one or more hosts in a range of habitats, while others are specific to a host or habitat, or both (Roberts, 2004). Some species are rich in proteins, phosphorus, calcium, and many other compounds. In 2007, an SU2 peptide from *R. paludosa* with positive effects on human health was isolated by Wang et al. (2007). Eating raw or improperly cooked *Russula* can cause vomiting and nausea. *R. xerampelina* is another edible species, which has a lightly fishy smell (O'Reilly, 2016).

10.7.4 *Lactarius* spp.

Approximately 62 species of the genus *Lactarius* (phylum: Basidiomycota, class: Agaricomycetes, order: Russulales, family: Russulaceae) were reported in Estonia by Kalamees (2008). Details on *Lactarius* species distribution in Estonia have been detailed in eight scientific literatures (Kalamees, 1962, 1978, 1979, 2008; Järva et al., 1998; Urbonas et al., 1986). *L. rufus* (milk cap) is found beneath pine, spruce, and rarely birch trees that grow in acidic soil. Milk caps growing in such acidic habitats are strong and acrid tasting mushrooms. *Lactarius* mushrooms produce a milky latex from the gills when they are cut or damaged. In Europe, about 110 *Lactarius* species are recognised (Leonardi et al., 2021). Being a wild ectomycorrhizal mushroom, *L. rufus* is a good source of beta glucans and has anti-inflammatory and anti-nociceptive potential (Ruthes et al., 2013). A new antifungal sesquiterpene from the fruiting bodies of *L. rufus* known as rufuslactone has been reported (Luo et al., 2005).

10.7.5 Truffles

Truffles (phylum: Ascomycota, class: Pezizomycetes, order: Pezizales, family: Tuberaceaeare) are the fruiting bodies of subterranean ascomycete fungi. They are ectomycorrhizal fungi, mainly found in association with roots of trees. They are dependent on mammals to spread their spores, and therefore have a specific smell to attract mammals (Lepp, 2012). Around 75 species of truffles have been reported from all around the world, among which five were reported by Estonian mycologists in 2011 as occurring in Estonian forests. The edibility of these mushrooms is not very clear (Tedersoo and Nadel, 2011). Later, in 2015, a group of mycologists found a new species of truffle, *Tuber aestivum* (locally named "summer truffle") with the help of a truffle-hunting dog in Valga County (Olle, 2015). Some truffle species are highly prized as food and known as "the diamonds of the kitchen" (Vidale, 2019). These truffles are used in French and Italian cuisine. At the present time, truffles are cultivated by companies and also harvested from natural environments.

10.7.6 Chaga Mushrooms (*Inonotus obliquus*)

Chaga mushrooms (phylum: Basidiomycota, class: Agaricomycetes, order: Polyporales, family: Hymenochaetaceae) have used by the people of Northern Europe and Russia for hundreds of years, who consume them in herbal teas. Estonia is a perfect land for growing chaga mushrooms. Chaga (*Inonotus obliquus*) is a popular drug in Estonia and has been used to enhance the immune system. It is also used as a cancer and rheumatism treatment in Estonian folk medicine (Drenkhan et al., 2022). Some companies in Estonia cultivate this mushroom on birch trees. For this, they choose birch trees that are at least 10 years old. In order to cultivate this mushroom, chaga dowels are planted into 4 cm drilled holes on the birch tree. An average of four to five planting holes are made per birch tree, and these are harvested three times in every 5–7 years. In the wild, chaga is found in Viljandi, Põlva, Valga, and Võru counties (Chaga Health, 2022). Studies have shown that chaga extract is able to protect liver tissue from the effects of tetra-butyl hydroperoxide, a chemical known to cause liver damage. It has high anti-oxidative properties and boosts immune defence (Won et al., 2011; Sak et al., 2014).

The medicinal potential of chaga mushrooms is well described in the scientific literature. This mushroom is highly praised in China, Korea, Japan, Russia, and the Baltics (Glamočlija et al., 2015; Fang et al., 2020). Chaga mushrooms have been demonstrated to have antioxidant, anti-inflammatory, anti-tumor, and hepatoprotective activities. Triterpenes and steroids are the main bioactive counterparts of this mushroom and are responsible for its medicinal properties (Nikitina et al., 2016; Drenkhan et al., 2022). About 40 triterpenes have been isolated from chaga mushrooms; of these, lanosterol and inotodiol were the first in chaga (Raal and Vilbaste, 2020). Steroids, polyphenols and polysaccharides also contribute to its medicinal properties (Nikitina et al., 2016; Zhang et al., 2015; Xu et al., 2015). Estonian-grown chaga shows its strongest cytotoxic effects against promyelocytic leukaemia and lung adenocarcinoma cells. Modest cytotoxic effects have also been found against colon adenocarcinoma, liver hepato-cellular carcinoma, oral epidermoid carcinoma, and prostate cancer cells (Nguyen et al., 2018).

10.7.7 *Tricholoma matsutake* (Matsutake)

Tricholoma matsutake (phylum: Basidiomycota, class: Agaricomycetes, order: Agaricales, family: Tricholomataceae), generally known as the matsutake mushroom or pine mushroom, is a highly prized and rare mushroom that grows in East Asia and Europe (Kalamees, 2010). It is an edible mycorrhizal mushroom and was found in Lahemaa National Park in Estonia (Mägi, 2019). This mushroom also has been reported in Finland, Canada, China, and Sweden. Matsutake mushrooms have a pungent, woody, spicy flavour and aroma. This mushroom is extremely rare as it needs very specific conditions to grow, and it is also difficult to spot in the forests since it is consumed by wild animals (Mägi, 2019). Matsutake mushrooms are rich in minerals, essential amino acids, protein, fibre, and carbohydrates. The presence of glutamic acid in this mushroom makes it useful for the supporting nervous system and brain function (Liu et al., 2010, 2020; Li et al., 2013).

10.8 CULTIVATED MUSHROOMS IN ESTONIA

The summer season in Estonia is favourable for growing mushrooms. Several edible and medicinal mushrooms can grow in vegetable gardens. Because of the growing number of mushroom enthusiasts, various training events and workshops are offered to allow enthusiasts to cultivate mushrooms in their homes. A number of companies are actively cultivating edible and medicinal mushrooms under controlled conditions. Edible varieties of mushrooms such as shiitake (*Lentinula edodes*, oyster (*P. ostreatus*), king trumpet (*P. eryngii*), lion's mane (*Hericium erinaceus*), and button mushrooms (*A. bisporus*) are cultivated in Estonia. The medicinal examples include lingzhi (*Ganoderma lucidum*) and chaga (*I. obliquus*). The oyster mushroom (*P. ostreatus*) has become well known in

Estonia and other Baltic countries and is available from commercial companies or can be cultivated as a stump mushroom. *P. ostreatus* has been developed most successfully in Estonia because it is considered the easiest to grow (Kivi, 2020). An active ingredient of *P. ostreatus*, lovastatin, is known to reduce bad cholesterol levels in blood (Niiberg, 2016). Another mushroom cultivated in Estonia is *Pholiota microspora* (Nameko) (Kivi, 2020). Health-promoting mushrooms known in Asia such as shiitake (*L. edodes*) or maitake (*Grifolia frondosa*) are little known in Estonia but are begin to gain interest. Finally, another type of cultivated fungus in Estonia is the tea mushroom or kombucha (*Medusomyces gisevi*), which is added to cool tea sweetened with sugar and is a symbiotic biological collection of yeast fungi (*Candida mycoderma*) and acetic acid bacteria (*Acobacter xylinum*). Various health properties are attributed to it (Niiberg, 2016).

10.9 USE OF MUSHROOMS IN ESTONIAN FOLK MEDICINE

As per the available report about 148,000 species of fungi have been described to date, in total, there may be 2.2–3.8 million of them. Up to 2,000 species of fungi are discovered every year (Jürgenson, 2022). Mushrooms are excellent resources of various bioactive compounds, especially polysaccharides and triterpenes, which contribute to their medicinal potential (Sudheer et al., 2019; Dasgupta and Acharya, 2019; Venturella et al., 2021). Studies show that traditional knowledge and modern research have verified the pharmacological potential of medicinal mushrooms used in European traditions (Gründemann et al., 2020). Mushrooms are well-known drugs in Estonia and have been used for medicinal purposes (Sak et al., 2014; Raal, 2010). Mushrooms have been used to treat different clinical conditions for years and are a part of folktales, fairy tales, folk remedies, and even in rituals (Youn et al., 2009; Dugan, 2008). About 100 different mushrooms are used in Chinese folk medicine, whereas only 30 species are known to have been used in European ethnomedicine. As per earlier reports, wild mushrooms have been used in the preparation of folk medicines in Estonia (Sak et al., 2014). Nearly 150 species of mushrooms with medicinal properties have been reported in Estonia, of which approximately 75% have antibiotic properties (Niiberg, 2016). It is known that in the 1930's, some wild mushroom species were sold in the markets for medicinal use, but these reports do not record exact species names. Among the different types of medicinal mushrooms in Estonia, the best-known mushroom is *I. obliquus* or chaga (Niiberg, 2016). As per the Estonian folk medicine database HERBA, chaga mushrooms are used to treat around 20 different illness including cancer. Since the sixteenth century, chaga mushrooms have been used as a remedy to treat cancer in Poland, Russia, and the Baltic countries. Sclerotia of this mushroom has been used in the preparation of folk medicine. A decoction prepared from chaga has been found to be effective in treating lung, uterine, colon, and liver cancer. It is also effective for leukaemia and melanoma (Won et al., 2011; Sak et al., 2014). Bioactive compounds such as triterpenoids, polyphenols, steroids, β-glucan, peptides, and polysaccharides are the main contributors to the medicinal potential of this mushroom (Hu et al., 2009).

Claviceps purpurea (ergot fungus), another folklore fungus, was used in obstetrics and gynaecology, and the alkaloid present in the fungus and known for the termination of pregnancy. However, overuse of this alkaloid results in severe poisoning and death (Raal, 2010). Poisonous *Bovista* (puffballs) and *Lycoperdon* mushrooms were also used in Estonian folk medicine. The beliefs about their therapeutic effect (as well as their danger) came from the peoples of the region (Wasson and Wasson, 1957). *Bovista* has been reported in homeopathic medicines and is used for treating cysts, ganglion, diabetes, eczema, tumours, and whitlow. Similarly, *Gomphidius* and *Phallus impudicus* were used in Estonian folk medicine. Another wild mushroom used for folk medicine was *A. muscaria* or fly agaric. It is a medicine for rheumatism and radiculitis, cancer, indigestion, liver problems, and toothache. It is believed that this medicine could also treat paralysis. Fly agaric has been marketed in Estonia as a rheumatism medicine since the 1930's (Jürgenson, 2022). Reishi (*G. lucidum*) with various medicinal properties is known to some Estonians as a dietary supplement (Niiberg, 2016).

10.10 CONCLUSIONS

Mushrooms were a significant part of the ancient culture of humans and have been utilised in folk medicine since ancient times. Species of mushrooms preferred in different cultures vary, and this relates to the mycophilic or mycophobic nature of a given culture. Because of the enormous number of reports on the edibility and medicinal potential of mushrooms, they are becoming favoured for culinary and medicinal purposes. Many countries in Europe have set certain legislation regarding the cultivation and marketing of mushrooms to control the illegal foraging and export of wild mushrooms. Outdated nomenclature, insufficient documentation of species, and the widespread use of different names to indicate taxa occurring on different continents has resulted in major confusion in mushroom classification and identification. The taxonomic conservation status of many species of mushrooms remains unreported. Estonian populations are among the mycophiles in Europe involved in mushroom picking activities in forests. The collection of wild mushrooms varies based on season and climate. Numerous species of mushrooms are in high demand in Estonia, including *Boletus*, *Lentinula*, chaga, and truffles. These mushrooms are pricy and are all-time favourites in Western cuisine. More and more members of younger generations are now changing to "sustainable food" habits that could lead to changes in their consumption behaviour, bringing more mushroom products onto the consumer's plate. A clear body of data relating to wild edible mushrooms and legislation of mushroom foraging and sale in Estonia is not yet available. It is expected that in the near future, more data on wild edible and medicinal mushrooms in Estonia and the Baltic regions will be available.

10.11 ACKNOWLEDGEMENTS

The authors (Surya Sudheer and Rajeev Bhat) acknowledge the ongoing project VALORTECH, which has received funding from the European Union's Horizon 2020 research and innovation program under grant agreement No. 810630. The authors would also like to acknowledge Kadri Pärtel, curator of the mycological collection at the Estonian University of Life Sciences, Estonia for her timely advice regarding literature on Estonian mushrooms.

REFERENCES

Bell, V., Silva, C.R.G., Guina, J. & Fernandes, T.H. 2022. Mushrooms as future generation healthy foods. *Frontiers in Nutrition 6*(9): 1050099. https://doi.org/10.3389/fnut.2022.1050099. PMID: 36562045; PMCID: PMC9763630.

CBI. 2020. *The European market potential for dried mushrooms. The market potential report*, 8 October 2020. https://www.cbi.eu/market-information/processed-fruit-vegetables-edible-nuts/dried-mushrooms/market-potential (visited in 25.8.2022).

Chaga Health. 2022. *Chaga cultivation*. https://chagahealth.eu/chaga-kasvatus/ (visited in 30.6.2022).

Dasgupta, A. & Acharya, K. 2019. Mushrooms: An emerging resource for therapeutic terpenoids. *3 Biotech 9*(10): 369. https://doi.org/10.1007/s13205-019-1906-2. PMID: 31588393; PMCID: PMC6760460.

Drenkhan, R., Kaldmäe, H., Silm, M., Adamson, K., Bleive, U., Aluvee, A., Erik, M. & Raal, A. 2022. Comparative analyses of bioactive compounds in *Inonotus Obliquus* Conks Growing on *Alnus* and *Betula*. *Biomolecules 12*(9): 1178. https://doi.org/10.3390/biom12091178

Drewnowska, M. & Falandysz, J. 2015. Investigation on mineral composition and accumulation by popular edible mushroom common chanterelle (*Cantharellus cibarius*). *Ecotoxicology and Environmental Safety 113*: 9–17. https://doi.org/10.1016/j.ecoenv.2014.11.028

Dugan, F. 2008. Fungi, folkways and fairy tales: Mushrooms & mildews in stories, remedies & rituals, from Oberon to the Internet. *North American Fungi*: 23–72. https://doi.org/10.2509/naf2008.003.0074

Ebrahimzadeh, M.A., Nabavi, S.M., Nabavi, S.F. & Eslami, S. 2010. Antioxidant and free radical scavenging activities of culinary-medicinal mushrooms, golden chanterelle *Cantharellus cibarius* and Angel's wings *Pleurotus porrigens*. *International Journal of Medicinal Mushrooms 12*(3): 265–272. https://doi.org/10.1615/IntJMedMushr.v12.i3.50

Erden, A., Esmeray, K., Karagöz, H., Karahan, S., Gümüşçü, H.H., Başak, M., Cetinkaya, A., Avcı, D. & Poyrazoğlu, O.K. 2013. Acute liver failure caused by mushroom poisoning: A case report and review of the literature. *International Medical Case Reports Journal 85*. https://doi.org/10.2147/IMCRJ.S53773

Fang, J., Gao, S., Islam, R., Teramoto, Y. & Maeda, H. 2020. Extracts of *Phellinus linteus*, bamboo (*Sasa senanensis*) leaf and chaga mushroom (*Inonotus obliquus*) exhibit antitumor activity through activating innate immunity. *Nutrients 12*: 2279.

Gałgowska, M. & Pietrzak-Fiećko, R. 2020. Mineral composition of three popular wild mushrooms from Poland. *Molecules 25*(16): 3588. https://doi.org/10.3390/molecules25163588

Garibay-Orijel, R., Caballero, J., Estrada-Torres, A. & Cifuentes, J. 2007. Understanding cultural significance, the edible mushrooms case. *Journal of Ethnobiology and Ethnomedicine 3*(1): 4. https://doi.org/10.1186/1746-4269-3-4

George, P. & Hegde, N. 2013. Muscarinic toxicity among family members after consumption of mushrooms. *Toxicology International 20*(1): 113–115. https://doi.org/10.4103/0971-6580.111559. PMID: 23833447; PMCID: PMC3702120.

Giusti, A., Ricci, E., Gasperetti, L., Galgani, M., Polidori, L., Verdigi, F., Narducci, R. & Armani, A. 2020. Molecular identification of mushroom species in Italy: An ongoing project aimed at reinforcing the control measures of an increasingly appreciated sustainable food. *Sustainability 13*(1): 238. https://doi.org/10.3390/su13010238

Glamočlija, J., Ćirić, A., Nikolić, M., Fernandes, Â., Barros, L., Calhelha, R.C., Ferreira I.C.F.R., Soković, M. & Griensven van L.J.L.D. 2015. Chemical characterization and biological activity of Chaga (*Inonotus obliquus*), a medicinal "mushroom". *Journal of Ethnopharmacology 162*: 323–332.

Gründemann, C., Reinhardt, J.K. & Lindequist, U. 2020. European medicinal mushrooms: Do they have potential for modern medicine? – An update. *Phytomedicine 66*: 153131. https://doi.org/10.1016/j.phymed.2019.153131

Gyllenberg, E. 2019. *Balti Jaama turg – the market you shouldn't miss in tallinn – Estonia.* https://www.rucksack.se/en/balti-jaama-turg-marknaden-du-inte-ska-missa-i-tallinn/ (visited in 2.5.2023).

Heleno, S.A., Barros, L., Sousa, M.J., Martins, A., Santos-Buelga, C. & Ferreira, I.C.F.R. 2011. Targeted metabolites analysis in wild Boletus species. *LWT – Food Science and Technology 44*(6): 1343–1348. https://doi.org/10.1016/j.lwt.2011.01.017

Hu, H., Zhang, Z., Lei, Z., Yang, Y. & Sugiura, N. 2009. Comparative study of antioxidant activity and antiproliferative effect of hot water and ethanol extracts from the mushroom *Inonotus obliquus. Journal of Bioscience and Bioengineering 107*(1): 42–48. https://doi.org/10.1016/j.jbiosc.2008.09.004

Jaagomäe, Õ. & Masso, S. 1966. *Seened Toidulaual.* Tallinn: Estonian SSR National Museum of Nature Publishing House, p. 89.

Järva, L., Kalamees, K., Kullman, B., Parmasto, E., Raitviir, A., Saar, I. & Vaasma, M. 1998. Red list of Estonian fungi. In: Perini, C. (Ed.). *Conservation of Fungi in Europe. Proceedings of the 4th Meeting of the European Council for the Conservation of Fungi, Vipiteno (Sterzing, Italy)*, 9–14 September 1997. Università degli Studi di Siena, Siena, Italy, pp. 136–138.

Järva, L., Kalamees, K., Kullman, B., Parmasto, E., Raitviir, A., Saar, I. & Vaasma, M. (Toim. E. Parmasto). 1999. Distribution maps of Estonian fungi. Eesti seente levikuatlas 2. In: *Protected Species and Species of the Estonian Red Data Book. Kaitsealused ja Eesti Punase raamatu liigid. Eesti Põllumajandusülikooli Zooloogia ja Botaanika Instituut*. Tartu.

Järva, L., Parmasto, E. & Vaasma, M. 1998. List of Estonian Fungi: With Host Index and Bibliography: Supplement 1 (1975–1990). Eesti seente koondnimestik. In Parmasto E. (Ed.). *Scripta Mycologica*, nr. 12. Tartu, Zooloogia ja Botaanika Instituut, Estonia, p. 183. ISBN 9985908112

Jaworska, G., Pogoń, K., Skrzypczak, A. & Bernaś, E. 2015. Composition and antioxidant properties of wild mushrooms Boletus edulis and Xerocomus badius prepared for consumption. *Journal of Food Science and Technology 52*(12): 7944–7953. https://doi.org/10.1007/s13197-015-1933-x

Jürgenson A. 2004. *Seentesse Suhtumine Meil Ja Mujal No. 48; Eesti Rahva Muuseumi Aastaraamat*, p. 107, https://ojs.erm.ee/index.php/ermer/article/view/119

Jürgenson, A. 2005. *Seened kultuuriloos*. Kultuur: Argo Publishing House, p. 358. ISBN 9949415306.

Jürgenson, A. 2022. *Seened kultuuriloos*. Edition: Elav teadus. Argo kirjastus. Tallinna Raamatutrükikojas, Argo publishing house, p. 15. ISBN 978-9949-688-88-3.

Kalamees, K. 1962. Species of the genus *Lactarius* occurring in the Estonian S.S.R. *Scripta Botanica* 2: 133–152.

Kalamees, K. 1978. A list of the Estonian Polyporales, Boletales, Russulales and Agaricales. *Scripta Mycologica* 8: 5–82.

Kalamees, K. 1979. The composition and seasonal dynamics of the fungal cover on mineral soils. *Scripta Mycologica* 9: 5–70.

Kalamees, K. 2008. Checklist of the species of the genus Lactarius (Phallomycetidae, Agaricomycetes) in Estonia. *Folia Cryptogamica Estonica 44*: 63–74.

Kalamees, K. 2010. Checklist of the species of the genus Tricholoma (Agaricales, Agaricomycetes) in Estonia. *Folia Cryptogamica Estonica 47*: 27–36.

Kalamees, K. 2011. *Mürkseened ja seenemürgitused. I Güromitriinimürgistus.* Eesti Loodus, 2011, 4; II Muskariinimürgistus Eesti Loodus, 6–7; III Amanitiinimürgistus. Eesti Loodus, 8, https://www.tartuloodusmaja.ee/wp-content/uploads/2020/03/Ettevaatust_seened.pdf

Kaldaru, H. 2011. *Marjade ja seente korjamine elanikkonnas.* Tallinn: Turu-uuringute AS, pp. 1–16, https://media.rmk.ee/files/Marjade%20ja%20seenete%20korjamine.pdf

Kämärä, M.-L. 2017. *The mushroom exhibition presents visitors with 150 mushroom species and mushroom worms*, 15 September 2017. https://opleht.ee/2017/09/seenenaitus-toob-kulastajate-ette-150-seeneliiki-ning-seeneussid/ (visited in 3.6.2022).

Kivi, K. 2020. Mükoloog: On üüratu hulk inimesi, kes tahvad kõike suhu pista. *Postimees*, 15 August 2020. https://leht.postimees.ee/7039787/mukoloog-on-uuratu-hulk-inimesi-kes-tahavad-koike-suhu-pista

Kotowski, M.A., Pietras, M. & Łuczaj, Ł. 2019. Extreme levels of mycophilia documented in Mazovia, a region of Poland. *Journal of Ethnobiology and Ethnomedicine 15*(1): 12. https://doi.org/10.1186/s13002-019-0291-6

Kuslap, E. 2004. *Korilase kokaraamat: seened*. Tallinn: Nõiaraamat Publisher, p. 143. ISBN 9985-9452-6-3.

Lagrange, E. & Vernoux, J.-P. 2020. Warning on false or true morels and button mushrooms with potential toxicity linked to hydrazinic toxins: An update. *Toxins 12*(8): 482. https://doi.org/10.3390/toxins12080482

Laux, H.E. 2011. *Seenelise teejuht. Kõik söögiseened ja nende mürgised teisikud.* Tallinn: Sinisukk Publisher, p. 720. ISBN 9789949148066.

Leonardi, M., Comandini, O., Sanjust, E. & Rinaldi, A.C. 2021. Conservation status of milkcaps (Basidiomycota, Russulales, Russulaceae), with notes on poorly known species. *Sustainability 13*(18): 10365. https://doi.org/10.3390/su131810365

Lepp, H. 2012. Spore release and dispersal. *Australian National Botanic Gardens*, 22 January 2012. https://web.archive.org/web/20161114222732/http://www.cpbr.gov.au/fungi/dispersal.html (visited in 5.8.2022).

Li, T., Zhang, J., Shen, T., Shi, Y.-D., Yang, S.-B., Zhang, T., Li, J.-Q., Wang, Y.-Z. & Liu, H.-G. 2013. Mineral element content in prized matsutake mushroom (*Tricholoma matsutake*) collected in China. *Chemical Papers 67*(6). https://doi.org/10.2478/s11696-013-0353-5

Liu, G., Wang, H., Zhou, B., Guo, X. & Hu, X. 2010. Compositional analysis and nutritional studies of *Tricholoma matsutake* collected from Southwest China. *Journal of Medicinal Plants Research 4*: 1222–1227.

Liu, J., Yang, F., Cheng, M. & Zhang, Y. 2020. Spinal cord injury can be relieved by the polysaccharides of *Tricholoma matsutake* by promoting axon regeneration and reducing neuroinflammation. *Neuroreport 31*(14): 1024–1029. https://doi.org/10.1097/WNR.0000000000001514

Luo, D.-Q., Wang, F., Bian, X.-Y. & Liu, J.-K. 2005. Rufuslactone, a new antifungal sesquiterpene from the fruiting bodies of the basidiomycete *Lactarius rufus*. *The Journal of Antibiotics 58*(7): 456–459. https://doi.org/10.1038/ja.2005.60

Mägi, K.A. 2019. An extremely rare mushroom grows in the forests of Estonia, the price of which exceeds one thousand euros per kilo. The news. *Rahageenius*, 30 July 2019. https://raha.geenius.ee/rubriik/uudis/eesti-metsades-kasvab-uliharuldane-seen-mille-kilohind-ulatub-ule-tuhande-euro/ (visited in 2.8.2022).

Market Data Forecast. 2022. *Europe edible mushroom market by type (Button, Shiitake, and Oyster), by application (Fresh mushrooms and processed mushrooms (dried, frozen, and canned)), and by region – industry analysis, size, share, growth, trends, and forecasts 2022 to 2027, January 2022*. https://www.marketdataforecast.com/market-reports/europe-edible-mushroom-market (visited 3.8.2022).

Marland A. 1956. *Eesti NSV tähtsamad söögi- ja mürkseened.* Tallinn: Eesti Riiklik Kirjastus, p. 51.

Mengs, U., Torsten Pohl, R. & Mitchell, T. 2012. Legalon® SIL: The antidote of choice in patients with acute hepatotoxicity from amatoxin poisoning. *Current Pharmaceutical Biotechnology 13*(10): 1964–1970. https://doi.org/10.2174/138920112802273353

Millman, L. 2019. *Fungipedia: A Brief Compendium of Mushroom Lore*. Princeton: Princeton University Press, p. 200.

Money, P. 2016. Fungal diversity. In: S.C. Watkinson, L. Boddy & N.P. Money (Eds.), *The Fungi*. Amsterdam: Elsevier, pp. 11–36, https://doi.org/10.1016/B978-0-12-382034-1.00001-3

Nael, I. 2022. Mushrooms. *Loodusveeb*, 13 January 2022. https://loodusveeb.ee/en/themes/gifts-nature/mushrooms (visited in 8.6.2022).

Nasiry, D., Khalatbary, A.R. & Ebrahimzadeh, M.A. 2017. Anti-inflammatory and wound-healing potential of golden chanterelle mushroom, cantharellus cibarius (Agaricomycetes). *International Journal of Medicinal Mushrooms 19*(10): 893–903. https://doi.org/10.1615/IntJMedMushrooms.2017024674

Nguyen, H.T., Ho, D.V., Nguyen, P.D.Q., Vo, H.Q., Do, T.T. & Raal, A. 2018. Cytotoxic evaluation of compunds isolated from the aerial parts of *Hedyotis pilulifera* and methanol extract of *Inonotus obliquus*. *Natural Product Communications 13*: 939–941.

Niiberg, T. 2016. *Ravivad seened. Kukeseenest India mereriisini*. Sinisukk, Estonia, p. 312. ISBN 9789949344437.

Nikitina, S.A., Habibrakhmanova, V.R. & Sysoeva, M.A. 2016. Chemical composition and biological activity of triterpenes and steroids of chaga mushroom. *Biochemistry 10*: 63–69.

Novakovic, S., Djekic, I., Klaus, A., Vunduk, J., Đorđević, V., Tomovic, V., Koćić-Tanackov, S., Lorenzo, J.M., Barba, F.J. & Tomasevic, I. 2020. Application of porcini mushroom (*Boletus edulis*) to improve the quality of frankfurters. *Journal of Food Processing and Preservation 44*(8). https://doi.org/10.1111/jfpp.14556

Oliach, D., Vidale, E., Brenko, A., Marois, O., Andrighetto, N., Stara, K., Martínez de Aragón, J., Colinas, C. & Bonet, J.A. 2021. Truffle market evolution: An application of the delphi method. *Forests 12*: 1174. htps://doi.org/10.3390/f12091174

Olle, M. 2015. *Truffle hunting dog makes unexpected discovery in Valga forest*, 22 September 2015. https://news.err.ee/116788/truffle-hunting-dog-makes-unexpected-discovery-in-valga-forest (visited in 21.7.2022).

O'Reilly, P. 2016. *Fascinated by Fungi*. First Nature Publishing. https://www.first-nature.com/fungi/russula-emetica.php (visited in 15.8.2022).

Parmasto, E. 2001. Fungi as indicators of primeval and old-growth forests deserving protection. In: D. Moore et al. (Eds.), *Fungal Conservation. Issues and Solutions*. Cambridge: Cambridge University Press, pp. 81–88. https://doi.org/10.1017/CBO9780511565168.006

Parmasto, E. 2012. Bibliography of biological (mainly mycological) publications by Erast Parmasto. *Folia Cryptogamica Estonica 49*: 1–18.

Peintner, U., Schwarz, S., Mešić, A., Moreau, P.-A., Moreno, G. & Saviuc, P. 2013. Mycophilic or mycophobic? Legislation and guidelines on wild mushroom commerce reveal different consumption behaviour in European countries. *PLoS One 8*(5): e63926. https://doi.org/10.1371/journal.pone.0063926

Pilz, D., Norvell, L., Danell, E. & Molina, R. 2003. Ecology and management of commercially harvested chanterelle mushrooms. *Gen. Tech. Rep. PNW-GTR-576*. Portland, OR: U.S. Department of Agriculture, Forest Service, Pacific Northwest Research Station, p. 83.

Põldmaa, K., Spirin, V., Miettinen, O., Runnel, K., Savchenko, A., Pärtel, K. & Kõljalg, U. 2019. Fungi from Soomaa National Park: Rarities and species new for Estonia. *Folia Cryptogamica Estonica 56*: 127–136. https://doi.org/10.12697/fce.2019.56.13

Raal, A. 2010. Farmakognoosia. In: *Alternative medicine*. Tartu: Tartu Ülikooli Kirjastus, p. 480. ISBN 9789949194124.

Raal, A. & Vilbaste, K. 2020. Estonian medicinal plants V. *Sirp*, 19 June 2020. https://www.sirp.ee/s1-artiklid/c21-teadus/eesti-ravimtaimedest-3/ (visited in 28.7.2022).

Radomir, M., Mesud, A. & Žaklina, M. 2018. Conservation and trade of wild edible mushrooms of Serbia—history, state of the art and perspectives. *Nature Conservation 25*: 31–53. https://doi.org/10.3897/natureconservation.25.21919

Raitviir, A. 2010. *Maalehe väike seeneraamat*. Tallinn: Maalehe Raamat, p. 144. ISBN 9789985644157.

Rim, M. 2016. *Metsa kõrvalsaaduste eksport ja import Eestis aastail 1995–2015. Bakalaureusetöö*. Tartu: EMÜ. https://dspace.emu.ee/xmlui/handle/10492/2573

Roberts, C., Ceska, O., Kroeger, P. & Kendrick, B.W. 2004. Macrofungi from six habitats over five years in Clayoquot Sound, Vancouver Island. *Canadian Journal of Botany 82*(10): 1518–1538. https://doi.org/10.1139/B04-114

Runnel, K., Miettinen, O. & Lõhmus, A. 2021. Polypore fungi as a flagship group to indicate changes in biodiversity – A test case from Estonia. *IMA Fungus 12*(1): 2. https://doi.org/10.1186/s43008-020-00050-y

Ruthes, A.C., Carbonero, E.R., Córdova, M.M., Baggio, C.H., Santos, A.R.S., Sassaki, G.L., Cipriani, T.R., Gorin, P.A.J. & Iacomini, M. 2013. *Lactarius rufus* (1→3),(1→6)-β-d-glucans: Structure, antinociceptive and anti-inflammatory effects. *Carbohydrate Polymers 94*(1): 129–136. https://doi.org/10.1016/j.carbpol.2013.01.026

Saar, I., Oja, J., Põldmaa, K., Pärtel, K., Zettur, I. & Kõljalg, U. 2019. Red list of Estonian fungi. *Folia Cryptogamica Estonica 56*: 117–126. https://doi.org/10.12697/fce.2019.56.12

Sak, K., Jürisoo, K. & Raal, A. 2014. Estonian folk traditional experiences on natural anticancer remedies: From past to the future. *Pharmaceutical Biology 52*(7): 855–866. https://doi.org/10.3109/13880209.2013.871641

Salo, P. & Salo, U. 2017. *Meie Metsade Seened*, Varrak, Estonia, p. 325. ISBN 9789985341575.

Sande, D., Oliveira, G.P. de, Moura, M.A.F. e, Martins, B. de A., Lima, M.T.N.S. & Takahashi, J.A. 2019. Edible mushrooms as a ubiquitous source of essential fatty acids. *Food Research International 125*: 108524. https://doi.org/10.1016/j.foodres.2019.108524

Santi, L., Maggioli, C., Mastroroberto, M., Tufoni, M., Napoli, L. & Caraceni, P. 2012. Acute liver failure caused by *Amanita phalloides* poisoning. *International Journal of Hepatology*: 1–6. https://doi.org/10.1155/2012/487480

Sepp, K. & Lõhmus, A. 2020. 3 Looduskeskkond kui avalik hüve, 3.1 Kuidas inimesed Eesti looduskeskkonda kasutavad? *Eesti Inimarengu Aruanne* 2019/2020, pp. 157–169. https://inimareng.ee/print/EIA-2019-3.1.pdf

Sitta, N. & Floriani, M. 2008. Nationalization and globalization trends in the wild mushroom commerce of Italy with emphasis on porcini (*Boletus edulis* and *Allied Species*). *Economic Botany 62*(3): 307–322. https://doi.org/10.1007/s12231-008-9037-4

Smith, M.R. & Davis, R.L. 2016. Mycetismus: A review. *Gastroenterology Report 4*(2): 107–112. https://doi.org/10.1093/gastro/gov062

Stryamets, N., Mattalia, G., Pieroni, A. & Sõukand, R. 2022. "Mushrooms (and a cow) are a means of survival for us": Dissimilar ethnomycological perspectives among hutsuls and romanians living across the Ukrainian-Romanian Border. *Environmental Management*. https://doi.org/10.1007/s00267-022-01619-6

Sudheer, S., Alzorqi, I., Manickam, S. & Ali, A. (2019). Bioactive compounds of the wonder medicinal mushroom "Ganoderma lucidum". In: Mérillon, J.M. & Ramawat, K. (Eds.), *Bioactive Molecules in Food. Reference Series in Phytochemistry*. Cham: Springer. https://doi.org/10.1007/978-3-319-78030-6_45

Svanberg, I. & Lindh, H. 2019. Mushroom hunting and consumption in twenty-first century post-industrial Sweden. *Journal of Ethnobiology and Ethnomedicine 15*(1): 42. https://doi.org/10.1186/s13002-019-0318-z

Tedersoo, L. & Nadel, T. 2011. Truffles in Estonia. *EestiLoodus*. http://eestiloodus.horisont.ee/artikkel3691_3636.html (visited in 2.7.2022).

Tuberik, T. 2009. Mushroom week. *Looduskalender*, 7 September 2009. https://www.looduskalender.ee/vana/en/node/4961.html (visited in 7.7.2022).

Urb, A. 2021. Best berry and mushroom picking spots in Estonia. *In Category: South Estonia, West Estonia, North Estonia, Food & Drink & Nightlife, Nature & Wildlife*, 5 September 2021. https://www.visitestonia.com/en/why-estonia/best-berry-and-mushroom-picking-spots-in-estonia (visited in 7.7.2022).

Urbonas, V., Kalamees, K. & Lukin, V. 1986. *Conspectus florum agaricalium fungorum (Agaricales s.l.)*, 1st ed. Vilnius, Mokslas: Lithuaniae, Latviae, Estoniae.

Venturella, G., Ferraro, V., Cirlincione, F. & Gargano, M.L. 2021. Medicinal mushrooms: Bioactive compounds, use, and clinical trials. *Internatinal Journal of Molecular Sciences 22*(2): 634. http://doi.org/10.3390/ijms22020634

Vidale, E. 2019. Truffles: The Precious Mushroom. *In*: Pullanikkatil, D., Shackleton, C. (Eds.). *Poverty Reduction Through Non-Timber Forest Products*. Sustainable Development Goals Series. Springer, Cham, Switzerland, pp. 79–81. https://doi.org/10.1007/978-3-319-75580-9_12

Wang, J., Wang, H.X. & Ng, T.B. 2007. A peptide with HIV-1 reverse transcriptase inhibitory activity from the medicinal mushroom *Russula paludosa*. *Peptides 28*(3): 560–565. https://doi.org/10.1016/j.peptides.2006.10.004

Wasson, V.P. & Wasson, R.G. 1957. *Mushrooms, Russia and History*. New York: Pantheon Books, p. 247. https://www.newalexandria.org/archive/MUSHROOMS+RUSSIA+AND+HISTORY+Volume+1.pdf

Wennig, R., Eyer, F., Schaper, A., Zilker, T. & Andresen-Streichert, H. 2020. Mushroom poisoning. *Deutsches Ärzteblatt International 117*: 701–708. http://doi.org/10.3238/arztebl.2020.0701. PMID: 33559585; PMCID: PMC7868946.

Won, D.P., Lee, J.S., Kwon, D.S., Lee, K.E., Shin, W.C. & Hong, E.K. 2011. Immunostimulating activity by polysaccharides isolated from fruiting body of *Inonotus obliquus*. *Molecules and Cells 31*(2): 165–173. https://doi.org/10.1007/s10059-011-0022-x

Xu, X., Quan, L. & Shen, M. 2015. Effect of chemicals on production, composition and antioxidant activity of polysaccharides of *Inonotus obliquus*. *International Journal of Biological Macromolecules 77*: 143–150.

Youn, M.-J., Kim, J.-K., Park, S., Kim, Y., Park, C., Kim, E.S., Park, K.-I., So, H.S. & Park, R. 2009. Potential anticancer properties of the water extract of *Inontus obliquus* by induction of apoptosis in melanoma B16-F10 cells. *Journal of Ethnopharmacology 121*(2): 221–228. https://doi.org/10.1016/j.jep.2008.10.016

Zhang, N., Chen, H., Zhang, Y., Xing, L., Li, S., Wang, X. & Sun, Z. 2015. Chemical composition and antioxidant properties of five edible Hymenomycetes mushrooms. *International Journal of Food Science and Technology 50*: 465–471.

11 Diversity, Distribution, Ecology, and Utilization of Wild Mushrooms in Bangladesh

F. M. Aminuzzaman, J. F. Tanni, K. Das

11.1 INTRODUCTION

Mushrooms are macrofungi with a stem (stipe), a cap (pileus), and gills (lamellae) or pores on the underside of the cap. They belong to the phyla Ascomycota and Basidiomycota (Cheung 2008; Miles and Chang 2004). Mushrooms have long been prized not only for their deliciousness, health benefits, and nutritional value as culinary ingredients but also for their medicinal properties as pharmaceutical sources (Wasser 2011). It is believed that there are 150,000 different species of mushrooms on the earth, although only around 10% of these have been named thus far (Mueller and Schmit 2007; Kirk et al. 2008). Protein, vitamins, minerals, and great quantities of glucan, selenium, thiamine, riboflavin, niacin, pantothenic acid, and folic acid all occur in mushrooms (Poucheret et al. 2006; Ahmad et al. 2013). Bioactivities such as immunomodulatory, antioxidant, anti-inflammatory, anti-diabetic, antibacterial, antifungal, antiviral, anticancer, hepatoprotective, glucose- and cholesterol-lowering, and anticancer properties have made mushrooms a hot topic recently (El Enshasy and Hatti-Kaul 2013; Kalač 2013). In nature, nutrients (especially nitrogen), moisture, forest type, and disturbances all influence the makeup and variety of macrofungi (O'Hanlon and Harrington 2012; Pradhan 2013). Mushrooms produce fruiting bodies seasonally and fill a wide variety of ecological niches in the forest ecosystem (Stamets 2000). Because of their crucial roles as ectomycorrhizal fungi in both unstructured and structured ecosystems, mushrooms are an important part of reforestation efforts. Moreover, macrofungi can be used as bioindicators of ecological health (Andrew et al. 2013). Given the wide-ranging biological and economic effects of wild mushrooms, studies on the taxonomy and variety of macrofungi are gaining prominence. There have been a lot of people working on the wild mushroom crop. Monitoring the reason and consequences of natural and artificial disturbances requires knowledge of diversity at the community and species levels. Ecosystem planners and managers could benefit greatly from information on the distribution of mushrooms species throughout a variety of plant communities (Engola et al. 2007).

The forests of Bangladesh can be grouped into four broad categories depending on their location, nature, and type of management. These categories are tropical moist deciduous, tropical evergreen or semi-evergreen, mangrove, and social forests. The current climatic conditions and varied vegetation types offer favorable environments that allow for the growth and survival of different types of naturally occurring wild mushrooms. Delicate fungal regeneration and development patterns, which depend on a specific microclimate, might be negatively affected by the current environmental problems of global warming and climate change. Identification and utilization of wild edible mushrooms are essential for improving a country's socioeconomic well-being. Bangladesh is covered in forests, with 2,600,000 acres of forested land (17% of the country's total land area). The Bangladesh Forest Department (BFD) owns and manages about 61.52% of these forest lands (1,600,000 hectares), while 26.80% of the forest area is unclassified (697,000 hectares), and 10.38% is community-owned and -managed (270,000 hectares) as private woodland and community forests (Foster-Turley et al. 2016; Dey 2018). Bangladesh has one of the lowest per capita forest lands in the world (Mukul and Quazi 2009), despite its great biodiversity. This is mostly attributable to the country's high human

DOI: 10.1201/b23190-11

population density in urban areas. Furthermore, the country has one of the highest rates of deforestation in South Asia, at 2600 hectares per year (FAO 2015; Poffenberger 2000). The wild mushroom flora is an important component of the forest ecosystem, and biodiversity, distribution, ecology, and utilization studies relating to their different properties have been largely neglected, limited, and improperly documented in the forest regions of Bangladesh. Rubina et al. (2017) found six edible, nine medicinal, four inedible, mushroom species, as well as one of unknown utilization, in the National Botanical Garden, Dhaka. In another study, Joty et al. (2020) identified five edible and 12 medicinal wild mushroom species from the Gajni forest. Therefore, to gather knowledge, improve fungal industry development, and to support mushroom collectors and farmers, a comprehensive and updated program of research on wild mushroom resource diversity and subsequent evaluation on edibility, medicinal uses, and poisonous or toxic properties is urgently needed.

11.2 CRITERIA OF RESOURCE EVALUATION

In this study, we summarize the data on wild mushrooms that were first collected in July, 2013, from a tropical moist deciduous forest region of Bangladesh (Rumainul et al. 2015; Rumainul and Aminuzzaman 2016) as a foundation of information for a more comprehensive study. We incorporate data on taxa, diversity, habitat, and utilization for the Bangladeshi wild mushrooms collected from different forest regions up to our last publication, which appeared in 2020 (Joty et al. 2020; Mafia et al. 2020 and Sonchita et al. 2020). To investigate the wild mushroom diversity of Bangladesh, numerous studies have already been conducted in various forest locations and published between 2015 to 2020. These are summarized in this chapter.

11.3 CURRENT STATUS AND DISTRIBUTION OF WILD MUSHROOM RESOURCES IN BANGLADESH

To date, there have been a total of 79 recorded sampling sites and 632 collected wild mushroom samples from different forest regions, parks, gardens, and the National Zoo in Dhaka, Bangladesh. The sampling regions include tropical moist deciduous forests (TMDF), tropical evergreen and semi-evergreen forests (TESF), mangrove forests (MF), Gajni forest regions (GF), social forests (SF), and parks, gardens, and the National Zoo of Dhaka (PGZD). From TMDF regions, the number of recorded samples was 160. From TESF areas, 128 mushroom samples were gathered, as well as 208 from MF regions; 32 from SF regions, 24 from GF areas, and 80 from PGZD zones, as shown in Figure 11.1. The major areas from which wild mushrooms were harvested in TMDF regions were Dhaka, Pabna, Rajshahi, Gazipur, Bogura, Dinajpur, Jaipurhat, and Modhupur; the major areas described as TESF regions were Rangamati, Bandarban, Sylhet, and Moulovibazar; the areas covering MF regions were Khulna, Bagherhat, Satkhira, Barisal, Barguna, Patuakhali, Pirojpur, Jhalokathi, and Kushtia; the location described as GF was Sherpur; the representative SF regions were Matihash, Panditpukur, Magura, Deogram, and Nandigram in the Bogura district; and the major parks and gardens under PGZD were Ramna Park, Boldha Garden, the National Botanical Garden, the National Zoo, and Dhanmondi Lake.

In Plate 11.1, the 25 most frequent wild mushroom species collected from Bangladesh are listed. These are *Ganoderma applanatum*, *Agaricus campestris*, *Lepiota cristata*, *Marasmius oreades*, *Volvariella volvacea*, *Trametes versicolor*, *Daedaleopsis confragosa*, *Amanita bisporigera*, *Russula nobilis*, *Coprinus pulmonarius*, *Mycena epipterygia*, *Crepidotus variabilis*, *Schizophyllum commune*, *Pleurotus ostreatus*, *Boletus edulis*, *Polyporus alveolaris*, *Tuber aestivum*, *Hypholoma fasciculare*, *Pycnoporus sanguineus*, *Hebeloma crustuliniforme*, *Xylaria polymorpha*, *Steccherinum ochraceum*, *Lycoperdon perlatum*, *Cortinarius saturninus*, and *Coprinellus micaceus*.

The less frequent genera encountered were *Ramariopsis*, *Gymnopilus*, *Daldania*, *Dictyophora*, *Phlebopus*, *Entoloma*, *Tylopilus*, *Cantharellus*, *Cerrena*, *Auricularia*, *Calvulina*, *Gomphus*, *Clitopilus*, *Laccaria*, *Psathyrella*, *Lycoperdon*, *Flammulina*, *Psilocybe*, *Aleuria*, *Callistosporium*,

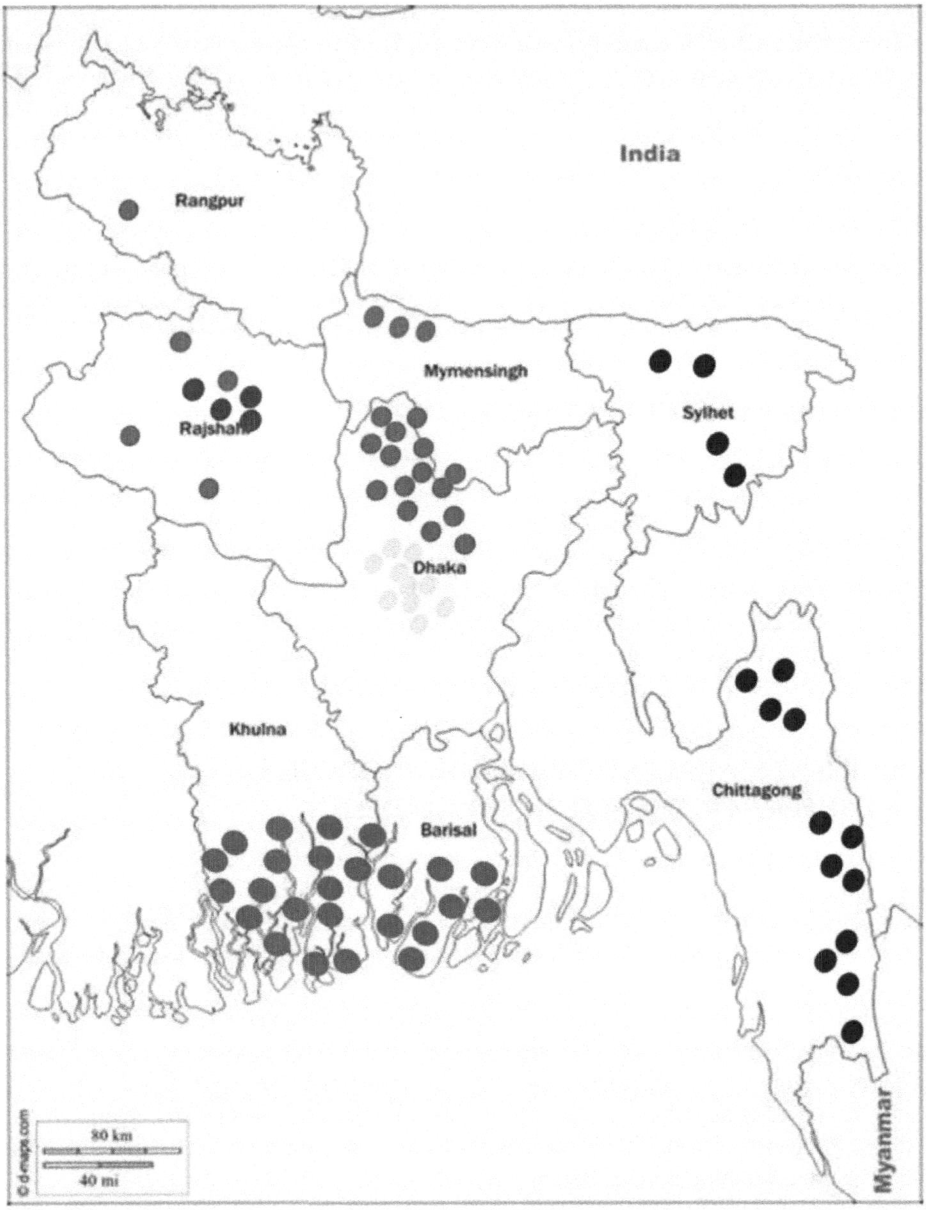

FIGURE 11.1 Sampling sites and number of collected mushroom samples from different forest regions of Bangladesh and parks and gardens of Dhaka city namely the tropical moist deciduous forest region (•), tropical evergreen or semi-evergreen forest region (•), Mangrove forest region (•), Gajni forest region (•), social forest region (•), and parks and gardens of Dhaka city (•). Each bullet represents eight number of samples collected from each sampling site.

Gymnopus, *Nolanea*, *Tyromyces*, *Innotus*, *Coltrica*, *Daedalea*, *Irpex*, *Gloephyllum*, *Megacollybia*, *Paneolus*, *Armillaria*, *Macrolepiota*, *Termitomyces*, *Chlorophyllum*, *Morganella*, *Stemonito*, *Cliopilus*, *Dacryopinax*, *Panellus*, *Hemiomyces*, *Craterellus*, *Ampulloclitocybe*, and *Collybia*. *Ganoderma* was the most frequently encountered genus in the collected samples, with a frequency of 27.20% and a density of 31.20% (Table 11.1). It was found in TMDF, TESF, MF, GF,

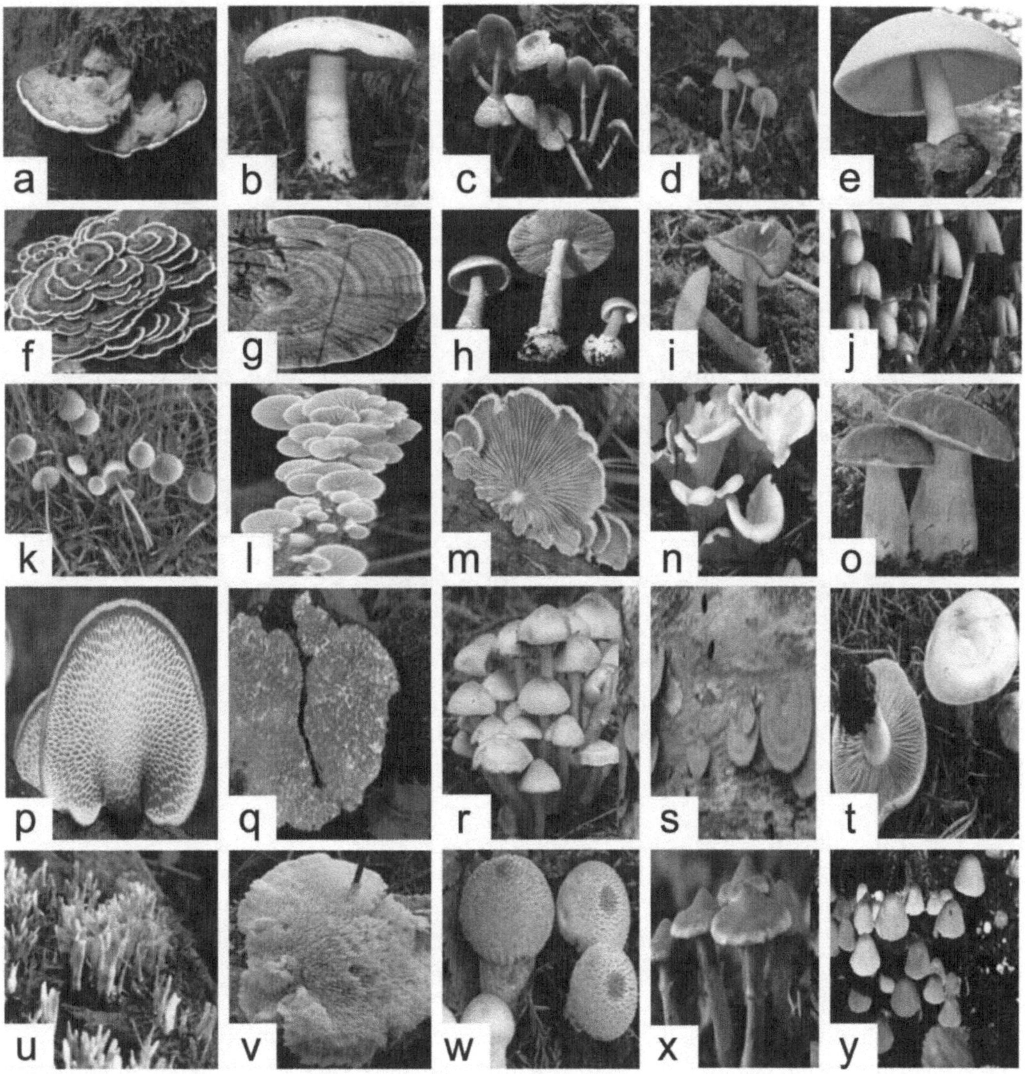

PLATE 11.1 The 25 most frequently occurring species of wild mushrooms collected from Bangladesh. (a) *Ganoderma applantaum*, (b) *Agaricus campestris*, (c) *Lepiota cristata*, (d) *Marasmius oreades*, (e) *Volvariella volvacea*, (f) *Trametes versicolor*, (g) *Daedaleopsis confragosa*, (h) *Amanita bisporigera*, (i) *Russula nobilis*, (j) *Coprinus pulmonarius*, (k) *Mycena epipterygia*, (l) *Crepidotus variabilis*, (m) *Schizophyllum commune*, (n) *Pleurotus ostreatus*, (o) *Boletus edulis*, (p) *Polyporus alveolaris*, (q) *Tuber aestivum*, (r) *Hypholoma fasciculare*, (s) *Pycnoporus sanguineus*, (t) *Hebeloma crustuliniforme*, (u) *Xylaria polymorpha*, (v) *Steccherinum ochraceum*, (w) *Lycoperdon perlatum*, (x) *Cortinarius saturninus*, (y) *Coprinellus micaceus*.

SF, and PGZD regions. The predominant species under this genus were *Ganoderma lucidum*, *Ganoderma tsugae*, *Ganoderma pfeifferi*, *Ganoderma lipsiense*, *Ganoderma applanatum*, and *Ganoderma zounatum*. The second most frequent genus recorded was *Agaricus*, whose frequency and density were 5.72% and 6.77%, respectively (Table 11.1). This genus was found in TESF, MF, and GF areas. The important species found under this genus were *Agaricus campestris*, *Agaricus xanthodermus*, *Agaricus silvicola*, *Agaricus aungustus*, *Agaricus arvensis*, and *Agaricus bitorquis*. The next genus was *Lepiota* whose frequency and density were calculated

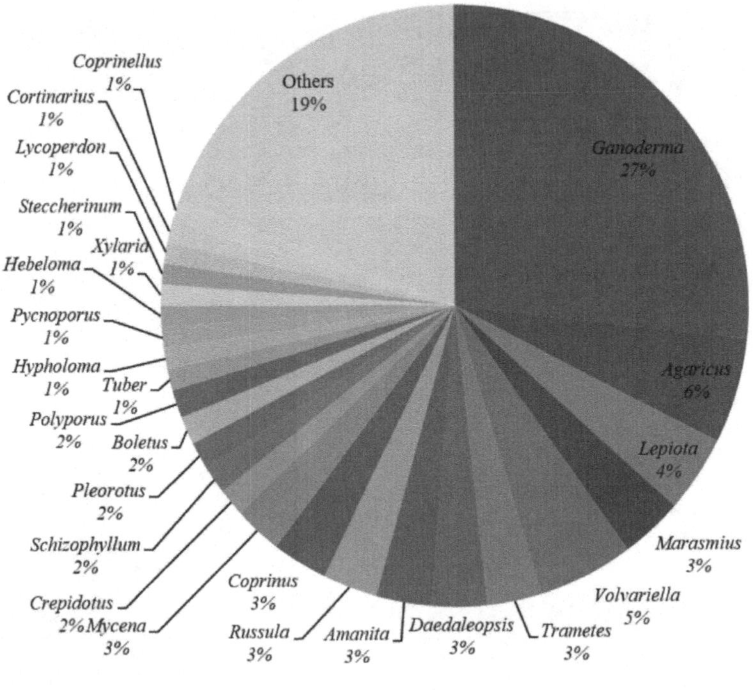

FIGURE 11.2 Wild mushroom genera collected from different forest regions of Bangladesh and parks, gardens, and the zoo of Dhaka; each sector represents one significant genus, and the remaining less frequent genera are combined and categorized as 'others'.

as 4.22% and 7.40%, respectively. This genus was found in TMDF, TESF, MF, and in PGZD. The next most frequent genera were *Marasmius, Volvariella, Trametes, Daedaleopsis, Amanita, Russula, Coprinus, Mycena, Crepidotus, Schizophyllum, Pleurotus, Boletus, Polyporus, Tuber, Hypholoma, Pycnoporus, Hebeloma, Xylaria, Steccherinum, Lycoperdon, Cortinarius,* and *Coprinellus*, as depicted in Table 11.1.

In the pie chart (Figure 11.2), the genera of collected wild mushroom samples are delineated. The genus *Ganoderma* covered the maximum area and represented 27% of the total number of samples collected. The second genus was *Agaricus*, which represented 6% of the total collection of samples. The third was *Volvariella*, representing 5% of the area of the graph. The next genera were *Lepiota* (4%), *Marasmius* (3%), *Trametes* (3%), *Daedaleopsis* (3%), *Amanita* (3%), *Russula* (3%), *Coprinus* (3%), *Mycena* (3%), *Crepidotus* (2%), *Schizophyllum* (2%), *Pleurotus* (2%), *Boletus* (2%), *Polyporus* (2%), *Hypholoma* (1%), *Tuber* (1%), *Hebeloma* (1%), *Pycnoporus* (1%), *Steccherinum* (1%), *Lycoperdon* (1%), *Cortinarius* (1%), and *Coprinellus* (1%). The remaining 43 genera were combined and indicated as others (19%). Each of these had a contribution of less than 1%.

The frequency of different families of macrofungi collected from different forest regions of Bangladesh is illustrated in a bar graph (Figure 11.3). The family with the highest frequency was the Ganodermataceae (26.79%), the second predominant family recorded was Agaricaceae (12.83%), and the third one was Polyporaceae (8.68%). The next most frequent family was the Pluteaceae (5.28%), followed by the Marasmiaceae (4.91%), Amanitaceae (3.02%), Russulaceae (3.02%), Mycenaceae (2.64%), Crepidotaceae (1.51%), Schizophyllaceae (1.51%), Boletaceae (1.51%), Tuberaceae (1.13%), Strophariaceae (1.13%), Hymenogastraceae (1.13%), Xylariaceae (1.13%), Steccherinaceae (1.13%), Lycoperdaceae (1.13%), Cortinariaceae (1.13%), and Psathyrellaceae (1.13%). The remaining less frequent families were sorted, combined, and labeled as others (19.25%).

TABLE 11.1
A Checklist of Diversified Wild Mushroom Genera of Bangladesh with Their Edibility Status, Habitat, Frequency, Density, and Distribution.

Genus	Edibility status	Number of species	Distribution	Hosts	Frequency (%)	Density (%)
Ganoderma	Edible, natural fungal medicine	14	TMDF, TESF, MF, GF, SF, PGZD	*Dalbergia sissoo*, *Swietenia macrophylla*, *Azadirachta indica*, *Leucaena leucocephala*, *Acacia nilotica*, *Shorea robusta*, *Terminalia arjuna*, *Terminalia bellireica*, *Albizia procera*, *Samanea saman*, *Bombax ceiba*, *Tectona grandis*	27.20	31.20
Agaricus	Edible, medicinal	6	TESF, MF, GF	Soil surface, *Cocos nucifera*, *Bambusa vulgaris*	5.72	6.77
Lepiota	Inedible, poisonous	3	TMDF, TESF, MF, PGZD	Soil debris, *Dalbergia sissoo*, *Terminalia arjuna*, *Cocos nucifera*	4.22	7.40
Marasmius	Edible, they are not generally used for medicinal purposes	6	TMDF, TESF, PGZD	Humus, *Santalum album*, *Artocarpus heterophyllus*, *Bambusa vulgaris*, *Dalbergia sissoo*, *Terminalia arjuna*	3.45	5.21
Volvariella	Edible, some are toxic, have antioxidant, anti-tumor properties	5	TMDF, TESF, MF, PGZD	Soil surface, *Macrophylla mahagoni*, *Azadirachta indica*, *Albizia saman*, *Bambusa vulgaris*, *Ziziphus jujube*, *Musa acuminate*, *Leucaena leucocephala*	5.36	7.34
Trametes	Medicinal	4	TMDF, TESF, MF, SF, PGZD	*Albizia lebbeck*, *Dalbergia sissoo*, *Cocos nucifera*, *Albizia saman*, *Terminalia arjuna*	3.65	4.32
Daedaleopsis	Inedible	2	TESF, MF, PGZD	Soil surface, *Borassus flabellifer*, *Diospyros ebony*, *Terminalia arjuna*, *Albizia saman*, *Leucaena leucocephala*	4.57	6.41
Amanita	Inedible, highly poisonous	3	TMDF, TESF, MF, PGZD	*Swietenia mahagoni*, *Dalbergia sissoo*, *Albizia saman*, *Leucaena leucocephala*, *Terminalia arjuna*, *Bambusa vulgaris*	5.64	7.21
Russula	Inedible	3	TMDF, TESF, GF, PGZD	Soil surface, *Terminalia arjuna*, *Andrographis paniculata*, *Albizia saman*	4.52	6.82
Coprinus	Edible but poisonous in combination with alcohol	2	TGEF, MF	Soil surface, *Swietenia mahagoni*, *Dalbergia sissoo*, *Leucaena leucocephala*, *Bambusa vulgaris*	5.32	7.26
Mycena	Edible	2	TMDF, TESF, MF	Soil surface, *Alstonia scholaris*, *Cocos nucifera*, *Dalbergia sissoo*, *Albizia saman*	3.54	5.17

TABLE 11.1 *(Continued)*
A Checklist of Diversified Wild Mushroom Genera of Bangladesh with Their Edibility Status, Habitat, Frequency, Density, and Distribution.

Genus	Edibility status	Number of species	Distribution	Hosts	Frequency (%)	Density (%)
Crepidotus	Inedible, poisonous	2	MF, PGZD	*Cocos nucifera, Bambusa vulgaris, Albizia richardiana, Dalbergia sissoo, Albizia saman*	3.20	5.31
Schizophyllum	Edible, medicinal, has anticancer, anti-tumor, immunomodulating properties	2	TESF, MF, PGZD	*Swietenia mahagoni, Albizia saman, Leucaena leucocephala*	2.67	4.24
Pleurotus	Edible, medicinal for blood sugar control, antioxidant capability	2	TESF, MF, PGZD	*Bambusa vulgaris, Neolamarckia cadamba, Swietenia mahagoni, Albizia saman*	4.12	6.71
Boletus	Edible, medicinal	2	TESF, MF, GF, SF	Dead wood, *Swietenia mahagoni, Dalbergia sissoo, Leucaena leucocephala, Terminalia arjuna,*	3.80	4.23
Polyporus	Edible, medicinal	2	TESF, MF, SF	Humus, *Ixora superba, Swietenia mahagoni, Leucaena leucocephala*	3.44	5.81
Tuber	Edible, superior anti-giogenic and anti-inflammatory activity	1	TESF, PGZD	Humus, *Dahlia, Albizia saman, Terminalia arjuna, Bambusa vulgaris*	2.32	3.24
Hypholoma	Inedible, poisonous	2	TESF, PGZD	*Ficus benghalensis, Cocos nucifera, Phoenix dactylifera, Terminalia arjuna, Swietenia mahagoni, Leucaena leucocephala*	2.51	3.72
Pycnoporus	Inedible, used for both industrial and medicinal purposes	1	TESF, MF, PGZD	*Heritiera fomes, Swietenia mahagoni, Leucaena leucocephala, Terminalia arjuna,*	3.23	4.23
Hebeloma	Edible	1	TMDF, TESF, GF	*Areca catechu, Dalbergia sissoo, Leucaena leucocephala*	4.10	5.11
Xylaria	Poisonous, medicinal, acts as diuretic, nerve-toxic agent, used in treating insomnia, trauma	3	TESF	*Leucaena leucocephala, Terminalia arjuna,*	4.10	4.56
Steccherinum	Inedible, medicinal value is not identified	1	TESF, MF	Humus, *Artocarpus chaplasha, Dalbergia sissoo, Leucaena leucocephala*	4.20	7.21
Lycoperdon	Edible at the tender stage but poisonous at maturity, medicinal	1	TMDF, TESF	Humus, *Azadirachta indica, Acacia nilotica, Swietenia mahagoni, Leucaena leucocephala*	3.80	4.21
Cortinarius	Edible, medicinal, anti-tumor qualities	1	TESF, MF	*Swetenia mahagoni, Terminalia arjuna, Leucaena leucocephala*	3.60	5.31

TABLE 11.1 ***(Continued)***
A Checklist of Diversified Wild Mushroom Genera of Bangladesh with Their Edibility Status, Habitat, Frequency, Density, and Distribution.

Genus	Edibility status	Number of species	Distribution	Hosts	Frequency (%)	Density (%)
Coprinellus	Edible, medicinal	2	TESF	Soil surface, *Terminalia arjuna, Leucaena leucocephala, Artocarpus heterophyllus, Dalbergia sissoo*	2.50	3.67
Ramariopsis	Edible	1	TMDF, PGZD	Soil surface, *Leucaena leucocephala*	2.70	11.28
Daldania	Poisonous	2	TMDF	*Acacia auriculiformis, Swietenia mahagoni*	2.70	2.25
Dictyophora	Edible	1	MF	*Bambusa vulgaris*	2.70	2.25
Phlebopus	Edible, ectomycorrizal mushroom	1	GF	On soil	7.14	2.14
Entoloma	Edible, some are poisonous	1	GF	On debris	70.00	15.25
Tylopilus	Inedible	1	TESF	*Albizia saman*	25.00	4.00
Cantharellus	Edible	2	TESF, PGZD	Soil surface, *Artocarpus heterophyllus*	15.00	2.00
Cerrena	Inedible	1	PGZD	*Albizia saman*	22.22	27.78
Auricularia	Edible	2	MF, PGZD	*Bambusa vulgaris, Albizia saman*	22.22	38.89
Calvulina	Edible, medicinal	1	TESF	Soil surface	30.56	16.67
Clitopilus	Edible	1	TESF	Soil surface	11.11	13.89
Laccaria	Edible	1	TESF	*Dalbergia sissoo*	11.11	2.78
Psathyrella	Edible, medicinal, have anticancer, anti-tumor, immunodulating properties	1	PGZD	Humus	2.51	5.00
Flammulina	Edible, potential medicinal uses	1	PGZD	Soil surface	3.21	5.01
Psilocybe	Inedible	1	PGZD	*Albizia saman*	4.21	5.00
Aleuria	Edible, medicinal	1	MF	Humus	11.11	2.78
Callistosporium	Inedible	1	TMDF, MF	*Bambusa vulgaris*	16.67	5.10
Gymnopus	Edible	1	MF	*Swietenia mahagoni*	11.11	10.00
Craterellus	Edible	1	TMDF	*Bambusa vulgaris*	2.11	3.02
Gomphus	Edible	1	TESF	*Albizia saman*	11.11	2.78
Nolanea	Edible, medicinal	1	MF	Humus	5.56	85.01
Tyromyces	Inedible, medicinal value is not identified	1	MF	*Cocos nucifera*	4.55	25

TABLE 11.1 *(Continued)*
A Checklist of Diversified Wild Mushroom Genera of Bangladesh with Their Edibility Status, Habitat, Frequency, Density, and Distribution.

Genus	Edibility status	Number of species	Distribution	Hosts	Frequency (%)	Density (%)
Innotus	Edible and medicinal with high nutritive value	1	MF	Goran, Gorjan	4.55	22.73
Coltrica	Inedible	1	MF	*Swetenia mahagoni*	25.00	21.43
Daedalea	Inedible, medicinal, acts as antioxidant	1	MF	*Dalbergia sissoo*	25.00	4.55
Irpex	Inedible	1	MF	*Mangifera indica*	50	13.64
Gloephyllum	Inedible	1	MF	Bohera	25	4.55
Marasmiellus	Edible and medicinal	1	MF	Coconut	75	106.75
Megacollybia	Edible	2	TMDF	Humus	2.31	3.21
Collybia	Edible	1	TMDF	Mahogony	1.12	4.32
Paneolus	Inedible	1	TMDF	Humus	3.21	5.21
Ampulloclitocybe	Inedible	1	TMDF	Soil surface, *Acacia auriculiformis*	4.21	3.11
Armillaria	Edible	2	TMDF	*Cocos nucifera, Swietenia mahagoni*	12.5	30.23
Macrolepiota	Edible	1	TMDF	*Swietenia mahagoni*	12.5	20.93
Termitomyces	Edible	2	TMDF	*Mangifera indica, Bambusa vulgaris*	12.5	2.70
Chlorophyllum	Poisonous	1	MF	*Cocos nucifera*	25.00	3.13
Morganella	Inedible	1	MF	*Oryza sativa*	25.00	3.13
Gymnopilus	Poisonous	2	TMDF, PGZD	Humus, Coconut	6.92	9.68
Stemonitis	Inedible	1	MF	*Swietenia mahagoni*	25.00	9.38
Cliopilus	Inedible	1	MF	*Albizia saman*	25.00	15.63
Dacryopinax	Edible	1	MF	*Areca catechu*	25.00	103.13
Panellus	Edible	1	MF	*Dalbergia sissoo*	25.00	18.75
Hemiomyces	Edible, medicinal	1	MF	*Cocos nucifera*	25.00	18.75

*TMDF = tropical moist deciduous forest region, TESF = tropical evergreen and semi-evergreen forest region, MF = mangrove forest region, GF = Ganji forest region, SF = social forest region, PGZD = parks, gardens, and the Zoo of Dhaka.

Note: Aminuzzaman and Das 2016; Das and Aminuzzaman 2017; Das et al. 2016; Mafia et al. 2020; Marzana et al. 2018; Rahaman et al. 2016; Joty et al. 2020; Rashid et al. 2016; Rashid et al. 2017; Rubina et al. 2017; Rumainul and Aminuzzaman 2016; Rumainul et al. 2015; Sonchita et al. 2020; Tanjim et al. 2019; Tanni et al. 2020.

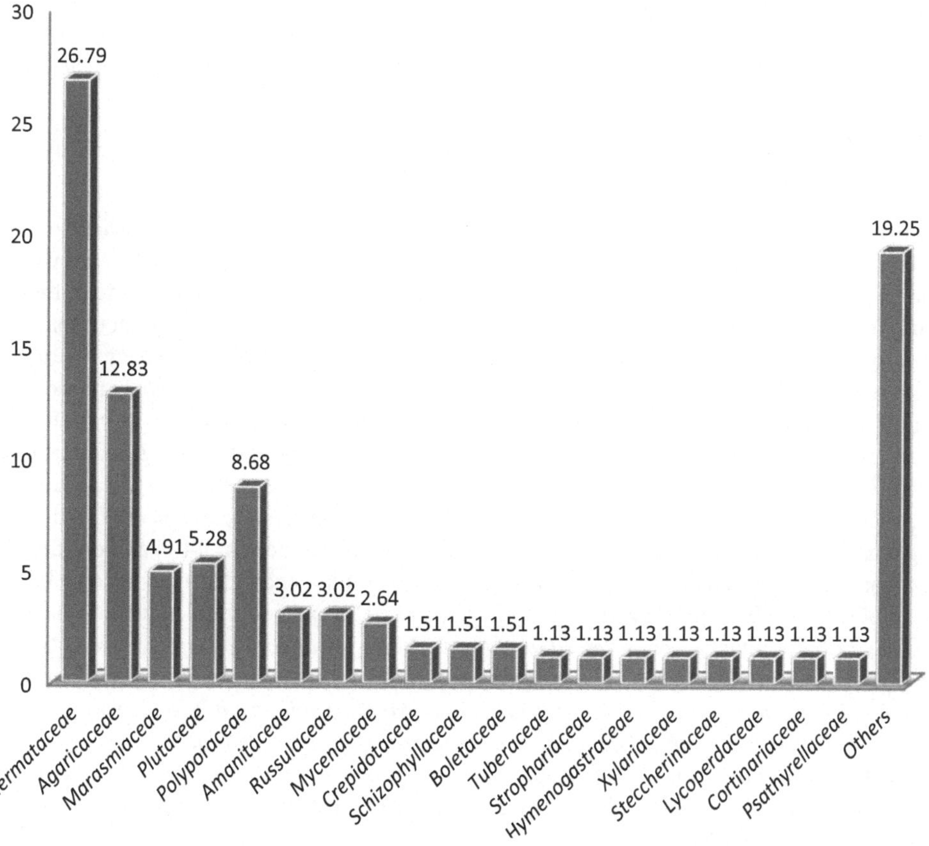

FIGURE 11.3 Different families of wild mushrooms collected from forest regions of Bangladesh and parks, gardens, and the zoo of Dhaka; each bar represents a particular family and less frequent families were combined and listed as 'others'.

11.4 WILD MUSHROOM DIVERSITY IN TROPICAL MOIST DECIDUOUS FORESTS

Mushroom flora is a crucial component of an ecosystem, but research on its biodiversity in the TMDF zones of Bangladesh has been mostly ignored. About 120,000 hectares, or about 0.81% of the total land area of Bangladesh, are covered by sal (*Shorea robusta*) tree forests, also known as moist deciduous forests. The country's greatest single mass of deciduous woodland is in the center of the area covered by these forests (Gittins and Akonda 1982). Investigation were carried out in eight separate locations, including Dhaka, Gazipur, Tangail, Bogura, Rajshahi, Pabna, Jaipurhat, and Dinajpur, within the tropical moist deciduous forest zones, and a total of 22 mushroom genera were identified from 160 collected field samples (Rumainul et al. 2015; Rumainul and Aminuzzaman 2016). In another survey, fourteen genera and 24 species of wild mushrooms were identified from this area (Rumainul et al. 2015). The most common genera were *Ganoderma*, *Lepiota*, *Marasmius*, and *Collybia*. In one survey, Rumainul and Aminuzzaman (2016) obtained eight genera and nine species of wild mushrooms from this forest area, where *Trametes*, *Daedaleopsis*, *Collybia*, and *Armillaria* were predominant. Pabna, Dhaka, and Rajshahi, located in the TMDF zones, are several hot sites where wood-decay fungus naturally flourish and from which a total of nine species belonging to the family Ganodermataceae were collected from a variety of hosts, including koroi (*Albizia procera*), aurjun (*Terminalia arjuna*), sissoo (*Dalbergia sissoo*), neem (*Azadirachta indica*), golden shower (*Acacia auriculiformis*), and the rain tree (*Albizia lebbeck*) (Mafia et al. 2020). The most frequent species were *Ganoderma applanatum* and *Ganoderma oregonense*, which had also been previously recorded (Rashid et al. 2017).

11.5 WILD MUSHROOM DIVERSITY IN TROPICAL EVERGREEN AND SEMI-EVERGREEN FORESTS

If temperature and moisture levels are favorable, mushrooms can occupy a variety of niches in natural forest ecosystems and are most prevalent during the rainy season. Hill forests, which make up most of Bangladesh's 680,000 hectares of mixed-evergreen forests, are primarily found in the country's northeast and southeast regions (Feeroz et al. 2014; Feeroz 2016). In addition, the tropical evergreen and semi-evergreen forest zones in the Sylhet and Moulovibazar districts, which are situated at 24.8917°N and 91.8833°E in Bangladesh's northeast, were also examined to assess the biodiversity and distribution of wild mushrooms, which naturally occur in various locations at various times of the year. Thus far, 30 genera have been identified from 128 collected samples in TESF regions. A survey was conducted in the hilly areas of Bangladesh, where 20 species of macrofungi belonging to 17 genera and 15 families were identified (Marzana et al. 2018) from the Rangamati hill forest, a TESF. The highest frequency of occurrence (44.44%) and density (55.56%) was recorded for *Xylaria polymorpha*. In another study, Tanjim et al. (2019) collected 53 macrofungal samples from TESF areas and a total of 15 genera and 25 species of wild mushrooms were identified. The predominant species were *Ganoderma tsugae*, *Ganoderma applanatum*, *Ganoderma bonninese*, *Agaricus bernardii*, *Coprinus disseminates*, *Daeleopsis confragosa*, *Schizophyllum commune*, *Pleurotus ostreatus*, *Pleurotus populinus*, *Mycena epiptergia*, *Steccherinum ochraceum*, and *Steccherinum ciliolatum*. *Ganoderma tsugae* and *Pleurotus ostreatus* when compared with other wild mushrooms in terms of occurance. The most common families found there were the Ganodermataceae, Agaricaceae, Polyporaceae, Schizophyllaceae, Pleurotaceae, Mycenaceae, and Steccherinaceae (Tanjim et al. 2019).

11.6 WILD MUSHROOM DIVERSITY IN MANGROVE FORESTS

A mangrove (Sundarban) representing the world's biggest single block of tidal halophytic forest is situated in Bangladesh's southwest region. The forest is located on the Ganges-Brahmaputra Delta at the point where it meets with the Bay of Bengal, a little south of the Tropic of Cancer between the latitudes of 21°30'N and 22°30'N and the longitudes of 89°00'E and 89°55'E. (Das et al. 2016). Along the Bay of Bengal coast, mangrove forests stretch over 801,700 hectares. About 601,700 hectares of the total coastal forest are natural mangroves, including the Sundarbans, the largest tract of productive mangrove forest in the world, and 200,000 hectares are coastal plantations (Payo et al. 2016; Khan 2011). Based on their ecological characteristics, mangrove forests provide a variety of hosts and habitats, including the sissoo (*Dalbergia sissoo*), coconut (*Cocos nucifera*), burflower (*Neolamarckia cadamba*), koroi (*Albizia procera*), rain (*Albizia saman*), mahogany (*Swietenia mahagoni*), and betel nut (*Areca cate*) trees. Altogether, 208 wild mushroom samples were harvested from this region, and 37 genera were identified. In one study, a total of 72 samples of fleshy wild macrofungi representing 21 genera and 32 species were identified from MF areas (Das et al. 2016). Predominant species included *Agaricus campestris*, *Agaricus xanthodermus*, *Agaricus silvicola*, *Agaricus aungustus*, *Agaricus arvensis*, *Agaricus bitorquis*, *Coprinus silvaticus*, *Coprinus plicatilis*, and *Marasamius siccus*. The most common families were the Agaricaceae, Marasmiaceae, Pluteaceae, Crepidotaceae, and Mycenaceae, which dominated the area. The largest collection of different xylotrophic macrofungi found in MF areas comprised 20 species belonging to 13 genera and seven families, including the Polyporaceae, Ganodermataceae, Hymenochaetaceae, Fomitopsidaceae, Xylariaceae, Steccherinaceae, and Gloeophyllaceae (Das and Aminuzzaman 2017). *Ganoderma*, *Trametes*, and *Inonotus* were the most common and predominant genera. In another study, a total of 16 mushroom species were identified, belonging to 10 genera and eight families (Rahaman et al. 2016). *Lepiota cristata* was found abundantly in the survey areas among other collected species and exhibited the maximum frequency of occurrence (25%). In addition, 24 different species belonging to 17 genera and 14 families were recorded, including members of the Amanitaceae (*Amanita bisporigera*), Pyronemataceae (*Aleuria aurantia*), Boletaceae (*Boletus subvelutipes*), Agaricaceae (*Agaricus* sp.), Tricholomataceae (*Callistosporium* sp.), Marasmiaceae (*Gymnopus* sp., *Ganoderma lucidum*, *Ganoderma applanatum*, *Ganoderma tsuage*) (Figure 11.2). *Amanita*, *Aleuria*, *Agaricus*, *Boletus*, *Mycena*, *Cortinarius*, *Nolanea*, and *Ganoderma* were the most common genera (Rashid et al. 2017) (Figure 11.3).

11.7 WILD MUSHROOM DIVERSITY IN THE GAJNI FOREST

The Gajni forest is located between 24°18' and 25°18'N latitudes and in between 89°53' and 90°91'E longitudes. It belongs to the Sherpur district, bounded by the Meghalaya state of India on the north and the Mymensingh and Jamalpur districts on the south, and is characterized by a wide range of vegetation and ecosystems. Among 24 harvested samples, seven genera were identified from the GF. Sonchita et al. (2020) recorded 12 species under seven different families from this region. The four major species that were identified belonged to the family Ganodermataceae and were *Ganoderma applanatum*, *Ganoderma lucidum*, *Ganoderma tropicum*, and *Ganoderma lobetum*. Another dominant genus was *Russla*. *Hebeloma*, *Boletus*, *Phlebopus*, and *Entoloma* were infrequently recorded. *Ganoderma applanatum* had the highest frequency (85.72%) among the taxa recorded, while *Phlebopus marginatus* had the lowest frequency (7.14%). Most of the wild species that were gathered were found on *Shorea robusta*, *Acacia auriculiformis*, soil, and detritus (Sonchita et al. 2020). In another study, 32 samples of mushrooms from the GF were gathered, and 28 species belonging to 11 genera and eight families were identified (Joty et al. 2020). *Ganoderma* sp. was widely distributed and had the highest frequency of occurrence (75%), while *Agaricus bitorquis* had the highest density (20.50%) and the sal tree (*Shorea robusta*) was the primary host. The predominant genera included *Amanita*, *Volvariella*, *Trametes*, *Agaricus*, and *Ganoderma*. The Ganodermataceae was the most common family among the wild mushrooms that were gathered, followed by the Polyporaceae, Agaricaceae, Amanitaceae, Rusullaceae, Pluteaceae, Marasmiaceae, and Strophariaceae (Figure 11.3). In addition, of the gathered species, five were edible, 11 were inedible or poisonous, and 12 were discovered to have a medicinal value (Joty et al. 2020).

11.8 WILD MUSHROOM DIVERSITY IN SOCIAL FORESTS

A survey of the biodiversity, distribution, and morphology of wild macrofungi was conducted in the SF of the Bogura (Former Bogra) district, and a total of six genera were identified from 32 harvested samples. A total of 16 species, belonging to two genera and seven families, were associated with the bark of the sissoo (*Dalbergia sissoo*) tree (Aminuzzaman and Das 2016). *Ganoderma* sp. (87.5% of the harvested samples) and *Polyporus* sp. (12.5%) were the two polypore genera recorded from the SF. *Ganoderma lucidum*, *Ganoderma multipileum*, *Ganoderma boninense*, and *Ganoderma* sp. displayed the highest frequency of occurrence (75%) and the highest density (66.67%), respectively (Aminuzzaman and Das 2016) (Figure 11.2).

11.9 WILD MUSHROOM DIVERSITY IN PARKS, GARDENS, AND THE ZOO OF DHAKA

Wild mushrooms have been surveyed in a variety of sites close to forested areas to study the diversity, ecology, and distribution of wild species in Bangladesh. The National Botanical Garden, National Zoo, Romna Park, Dhanmondi Lake, and Boldha Garden were the five parks and gardens in Dhaka, Bangladesh where a total of 22 genera were identified from 80 harvested samples. Tanni et al. (2020) identified 32 species of wild mushrooms belonging to 18 genera and 18 families. The genera *Ganoderma, Daedeleopsis, Ramariopsis, Crepidotus,* and *Daldinia* were the most frequently occurring genera. *Ganoderma lucidum* (9.46%) was the most frequent species followed by *Ganoderma applanatum* (8.1%), *Volvariella volvacea* (5.41%), *Agaricus bisporus* (5.41%), *Daedaleopsis confragosa* (4.05%), *Trametes versicolor* (4.05%), and *Ganoderma boninense* (4.05%) The species *Ramariopsis kunzei* (11.3%), *Ganoderma lucidum* (9.9%), *Crepidotus variabilis* (5.3%), and *Daedaleopsis confragosa* (3.76%) had the highest density of occurrence in the PGZD. The National Botanical Garden is located at 24°00' N (latitude), 90°00' E (longitude). The predominant species found in the National Botanical Garden is *Ganoderma applanatum*; in Ramna Park it is *Ganoderma lucidum*; in Dhanmondi Lake, it is *Ramariopsis kunzei*; in the Boldha Garden it is *Ganoderma lucidum*; and in the National Zoo it is *Amanita bisporigera*. The majority of the wild species are found in association with *Swietenia macrophylla, Acacia auriculiformis, Mangifera indica, Dalbergia sissoo, Acacia nilotica, Bambosa vulgaris, Azadirachta indica*, and various other plants (Tanni et al. 2020). In another study, 20 species belonging to 10 genera and 10 families were recorded from PGZD areas (Rubina et al. 2017). The predominant genera were *Ganoderma, Lepiota, Daedeleopsis, Russula, Psythyrella, Lycoperdon, Crepidotus, Psilocybe, Flammulina,* and *Cantharellus*. The koroi (*Albizia procera*), golden shower (*Acacia auriculiformis*), aurjun (*Terminalia arjuna*), *Mangifera indica*, bohera (*Terminalia bellirica*), neem (*Azadirachta indica*), white rangun (*Ixora superba*), rain (*Albizia lebbeck*), raj koroi (*Albizia richardiana*), and white chandan (*Santalum albus*) trees were the major hosts in the PGZD. The most common genera were *Cantharellus, Lepiota, Daedeleopsis, Russula, Lycoperdon, Crepidotus, Psilocybe,* and *Ganoderma*. According to the survey, just one species has unknown uses, while nine of the thirteen non-edible species have medicinal potential, and six species were edible (Rubina et al. 2017).

11.10 DISTRIBUTION OF WILD MUSHROOMS IN BANGLADESH

The frequency of the collected wild mushroom genera ranged from 2.32% to 70.00%, where the highest and lowest frequency was recorded for *Entoloma* and *Tuber*, respectively. On the other hand, the highest (106.75%) and lowest (2.00%) density were recorded for *Marasmiellus* and *Nolanea*, respectively (Table 11.1). Other collected genera exhibited medium values for frequency and density of occurrence. *Ganoderma* is distributed throughout the country and found in all forest regions, including PGZD areas. The next most frequent genera are *Lepiota, Volvariella, Amanita,* and *Russula*, distributed in TMDF, TESF, MF, and PGZD regions; *Trametes*, distributed in TESF, MF, SF, and PGZD areas; *Agaricus*, distributed in TESF, MF, GF zones; *Marasmius*, distributed in TMDF, TESF, PGZD regions; *Daedaleopsis*, distributed in TESF, MF, PGZD areas; *Mycena*, distributed in TMDF, TESF, MF locations; *Schizophyllum* and *Pleurotus*, distributed in TESF, MF, PGZD regions; *Hebeloma*, distributed in TMDF, TESF, GF areas; and *Pycnoporus*, distributed in TESF, MF, and PGZD zones. *Xylaria* and *Coprinellus* are only distributed in TESF areas. The distribution of wild mushrooms is mainly influenced by host, environmental, and edaphic factors, as reported by many researchers. In this review, the highest host range (12) was found for *Ganoderma*, whereas the lowest host range (2) was recorded for *Xylaria*. *Terminalia arjuna* is a common host for *Ganoderma, Lepiota, Marasmius, Trametes, Daedaleopsis, Amanita, Russula, Boletus, Tuber, Hypholoma, Pycnoporus, Xylaria, Cortinarius,* and *Coprinellus*. *Dalbergia sissoo* is a common host for *Ganoderma, Lepiota, Marasmius, Trametes, Amanita, Coprinus, Mycena, Crepidotus, Boletus Hebeloma, Steccherinum,* and *Coprinellus*. *Swietenia mahagoni* is a general host for *Ganoderma, Volvariella, Amanita, Coprinus, Schizophyllum, Pleurots, Boletus, Polyporus, Pycnoporus, Lycoperdon,* and *Cortinarius*. *Azadirachta indica* is a common host for *Ganoderma, Volvariella,* and *Lycoperdon*. *Leucaena leucocephala* is a common host for *Ganoderma, Volvariella, Daedaleopsis, Amanita, Coprinus, Schizophyllum, Boletus, Polyporus, Pycnoporus, Hebeloma, Xylaria, Steccherinum, Lycoperdon,* and *Coprinellus*. *Cocos nucifera* trees support such genera as

Agaricus, *Lepiota*, *Trametes*, *Crepidotus*, and *Hypholoma*. In a previous study, the morphology of 53 specimens of *Ganoderma lucidum* was examined in Norway, and a large variation among them was found to exist (Ryvarden 1994). The genus *Ganoderma* was also previously reported in China (Wang et al. 2012) and India (Dwivedi et al. 2012). Species of *Marasmius* were also recorded in Madagascar and Mascarenes (Antonin and Buyck 2006). Species of *Lepiota* were first reported in India (Dwivedi et al. 2012). As in our study, *Trametes* was previously reported from Sweden (Fries 1821), later in the Western Ghats region of India (Thiribhuvanamala et al. 2011), and then in and around Bangalore (Karnataka) of India; these mushrooms were found to be of medicinal importance (Pushpa and Purushothama 2012). The species was also reported in the Lagos State of Nigeria (Bankole and Adekunle 2012) and Chungcheong Province of Korea (Jo et al. 2013). Furthermore, the genus *Daedaleopsis* is a widely distributed genus containing six species (Kirk et al. 2008). *Daedaleopsis confragosa* was also reported in Pakistan (Farooq et al. 2013). This genus was first reported in Europe in 1791, where it was known to have antioxidant properties (Bolton 1791), and was also reported in the Central Luzon region of the Philippines as a saprophyte (De Leon et al. 2013). *Crepidotus ehrendorferi* was recorded in Slovakia (Ripkova and Glejdura 2010). The new species *Crepidotus pseudoantillarum* and *Crepidotus herrerae* were collected from Costa Rica and Mexico, respectively (Bandala et al. 2008). In our study, the *Schizophyllum commune* and other unidentified species of *Schizophyllum* were recorded from Sreemangal, Madhabkundu, and Jaintiapur. These were previously reported to occur on kala koroi (*Albizia lebbeck*) and kadam (*Anthocephalus chinensis*), respectively (Murray 2013; Chandulal et al. 2013). As in our study, the genera *Amanita* and *Russula* were previously reported in India (Mohanan 2011), northern Thailand (Lee et al. 2007), and North America (Stanis 1979; Kraigher et al. 1995), although *Amanita cinereovelata* was first identified from the sal forests of Bangladesh (Hosen et al. 2016). The Amanita species were also reported from other areas, including Switzerland (Breitenbach 1995), California (Desjardin et al. 2015; Thiers 1982), and North America (Jenkins 1986). As found in our study, *Russula* was previously harvested from the southern Kashmir Himalayas (Pala et al. 2012) and is widely distributed in North America (Kernaghan et al. 1997). *Russula firmula* and *Russula postiana* were isolated and characterized from the Himalayan moist temperate forests of Kashmir (Itoo et al. 2013). As in our study, a wild edible *Lycoperdon pyriforme* was reported in India (Vyas et al. 2014) and biochemically characterized in Turkey (Akatin 2013). *Boletus edulis* was previously described (Bulliard 1782; Thiers and A.H. Sm. 2016). The genus *Pycnoporus* contains five distinct species, which are saprophytic on the dead hardwoods (Eggert et al. 1998). It has been found in some other places in Europe and North America, including northeastern Ohio (Bunyard 2003). *Cortinarius* sp. was previously reported in North America and Europe (Breitenbach et al.1995).

11.11 UTILIZATION OF WILD MUSHROOMS

Mushrooms are a rich, low-calorie source of fiber, protein, and antioxidants. When consumed, these compounds can mitigate the risk of developing serious health conditions such as heart disease, cancer, and diabetes. *Ganoderma* is edible and well known for boosting immunity; fighting cancer, fatigue, and depression; promoting heart health; controlling blood sugar; and boosting antioxidant levels. *Ganoderma lucidum* contains bioactive compounds like ganoderic acids, ganodermanontriol, ganoderiol, polysaccharides, and β-glucan, providing health benefits including anti-metastatic, anti-tumor, antiviral, anti-HIV, and hepatoprotective properties and preventing cholesterol synthesis (Xu et al. 2010; Walton and Treasure 2014). This mushroom has a high nutritive and medicinal value and contributes to a healthy diet because of it is rich source of vitamins, minerals, and proteins (Garcha et al. 1993). Many species of *Agaricus* are edible and known to fight against physical and emotional stress, stimulate the immune system, reduce blood cholesterol, prevent osteoporosis and peptic ulcer, and support the circulatory and digestive systems. *Agaricus rubescens* and *Agaricus virgineoides* have been reported as edible and poisonous, respectively (Zhang et al. 2021). *Agaricus bisporus* has some bioactive compounds and also provides health benefits (Moro et al.

2011; Ndungutse et al. 2015; Lakhanpal and Rana 2005). The genus *Lepiota* can be divided into three groups: edible (*Lepiota procera*), inedible or poisonous (*Lepiota margani*), and non-poisonous (Ukwuru et al. 2018). *Marasmius* is edible, but the tough, fibrous stipes should be discarded. This mushroom is not generally used for medicinal purposes, but a ethanol extract of *Marasmius oreades* showed a moderate anticancer effect on the HT-29, MCF-7, and MDA-MB-231 cell lines (Shomali et al. 2019). *Volvariella* is edible but some species are toxic. This mushroom contains ergothioneine, a potent antioxidant that keeps the body safe from free radicles. *Volvariella volvacea* (paddy straw mushroom) is edible (Fui et al. 2018) and has the Fip-vvo immunomodulatory protein (Hsu et al. 1997). *Volvariella volvacea, Schizophyllum commune, Pleurotus* spp., *Auricularia* spp., and *Marasmiellus* spp. are mostly consumed by the indigenous people of Sabah as part of their daily diet (Fui et al. 2018). *Shizophyllum* is edible and has medicinal values which include anticancer, anti-tumor, and immunomodulatory properties. *Trametes* sp. is a medicinal mushroom (MM) with an impressive range of benefits. It contains a variety of powerful antioxidants and other compounds that help in boosting immune system against certain types of cancer. *Trametes* sp. improves the balance of gut bacteria, which positively impacts human immunity and fighting infection. In another study, Aravind et al. (2021) identified *Trametes versicolor, Daedaleopsis confragosa, Xylaria polymorpha, Xylaria hypoxylon, Coltricia cinnamomea*, and *Steccherinum* sp., as non-edible wild mushrooms, whereas *Polyporus resinosus, Ganoderma applanatum, Auricularia auricula*, and *Macrolepiota procera* were reported as edible. *Xylaria* has medicinal value and is used as a diuretic and nerve-toxic agent, treating insomnia and trauma. *Amanita* is inedible and highly poisonous although *Amanita vaginata, Amanita caojezong, Amanita cupreobrunneus, Amanita imazekii*, and *Amanita pseudoporphyria* have been reported as edible wild mushrooms (Aravind et al. 2021; Zhang et al. 2021). On the other hand, toxins of *Amanita muscaria* mushroom are successful in trapping or killing insects and other agricultural pests (Wang et al. 2021). Medicinal species in the genus *Polyporus* contain many bioactive compounds that are responsible for anti-tumor, anticancer, antioxidant, free radicle-scavenging, immune system enhancing, and antimicrobial activities. *Russula nobilis* and *Russula pubescens* were reported as poisonous (Zhang et al. 2021) but act as antibacterial agents by preventing blood clots and blood thickening. Some edible species such as *Russula aeruginea, Russula albonigra, Russula aurora, Russula cyanoxantha, Russula paludosa, Russula rosea, Russula turci, Russula vinosa, Russula virescens*, and *Russula stenocystidiata* were also reported from China (Zhang et al. 2021). *Coprinus* is edible but becomes poisonous in the presence of alcohol. However, it has anti-tumor and antiviral effects and helps in lowering blood sugar. *Coprinus comatus* is widespread and approved as an edible mushroom; it is often cultivated for consumption in many countries. Various studies have shown that it has many physiological activities such as antioxidant, anticancer, antiandrogenic, hepatoprotective, acetylcholinesterase-inhibitory, anti-inflammatory, antidiabetic, anti-obesity, antibacterial, antifungal, anti-nematode, and antiviral activity (Nowakowski et al. 2020). Some species of *Mycena* are edible while others contain toxins, but the edibility of most species is not known (Elkhateeb and Daba 2022). These authors also reported *Laccaria amethystina* as an edible mushroom, although it can absorb arsenic from the soil. *Laccaria laccata* and *Laccaria vinaceoavellanea* were also reported as edible species (Zhang et al. 2021). The parasol mushroom, *Macrolepiota procera*, is a choice edible species found on roadway edges (Elkhateeb and Daba 2022). *Crepidotus* is inedible and poisonous. *Pleurotus* is edible and has several health benefits. It lowers high blood pressure, builds strong bones, defends against neurodegenerative diseases, and has anti-inflammatory and anticancer properties. The white oyster mushroom, *Pleurotus florida*, contains the bioactive compound β-glucan, which shows antioxidant and antimicrobial activity (Ganeshpurkar et al. 2015; Menaga et al. 2012). In another study, Zhang et al. (2021) reported *Pleurotus giganteus* as mushroom of uncertain edibility. Mushrooms of the genus *Boletus* are edible and excellent sources of minerals such as selenium, manganese, zinc, copper, iodine, and molybdenum. *Boletus aereus, Boletus edulis, Boletus reticulatus*, and *Boletus pseudosulphureus* have been previously reported as edible mushrooms, whereas *Boletus satanas* is poisonous and two other species, *Boletus rhodopurpureus* and *Boletus eastwoodiae* have uncertain edibility (Zhang et al. 2021). The *Tuber aestivum* is edible and has superior anti-angiogenic as well as anti-inflammatory activities. *Hypholoma* is inedible and poisonous. Species of *Pycnoporus* are inedible but used for both industrial and medicinal purposes. *Hebeloma* is an edible mushroom, whereas *Steccherinum* is inedible. *Lycoperdon* is edible at a tender stage but becomes poisonous after maturity and has medicinal properties. *Cortinarius* and *Coprinellus* are both edible mushrooms and have medicinal uses. Eight species of *Cantharellus*, seven species of *Termitomyces*, two species of *Craterellus*, and one species of each of *Cortinarius*, *Auricularia*, *Phlebopus*, *Gomphus*, and *Cerrena* have been reported as edible, whereas *Tylopilus* microsporus, *Tylopilus neofelleus*, and *Psilocybe semilanceata* are reported as poisonous mushrooms (Zhang et al. 2021).

11.12 LIMITATIONS

Mushrooms are both common and highly diverse around the world; however, research on identifying wild mushrooms, as well as their future cultivation and preservation, is still experiencing some difficulties. The number of species of mushrooms is thought to be between 150,000 and 160,000, but only around 10% of the identified species are known to science (Wasser 2010; Hawksworth 2012; Kirk et al. 2008). In certain circumstances, up to 73% of the species found in diverse tropical regions have yet to be described (Bass and Richards 2011). According to recent studies, the actual number of fungal species that exist on the earth may be 50 times greater than the current statistics indicate. There may be up to five million different kinds of fungi, according to modern sequencing techniques (Blackwell 2011). Therefore, based on the current discovery pace of roughly 1,200 new species each year in the last 10 years of data, we would require more than 4,000 years to describe this total fungal diversity (Hibbett and Taylor 2013). Available data also reveal that just 10% of the world's mushroom biodiversity and about 1% of the world's fungi are known to humans (Wasser 2010; Chang and Wasser 2012). Numerous species of MMs are gravely misunderstood, and future research studies will be invalid without knowing their exact scientific nomenclature. DNA barcoding may be beneficial in conjunction with traditional taxonomical approaches for the accurate identification of MM species, including the examination of type material and standardized MM products. For instance, there is growing evidence that most species previously listed as lingzhi or reishi (*Ganoderma lucidum*) in most of the pharmacological research were misidentified. *Ganoderma lucidum* represents a taxonomic group, whose future subdivision calls for caution (Wasser et al. 2006). In addition, poisonous mushrooms, often known as toadstools, are harmful to humans and should not be consumed. Because they contain toxins or alkaloids (alpha-amantin), which are to blame for the negative and occasionally fatal consequences, these mushrooms are referred to as poisonous (Dutta 2008; Macrae et al. 1993). The ultimate function of the basidiocarp is protected by these secondary metabolites because the mycelium has expended a lot of energy and protoplasm to create a structure that can effectively distribute its spores (Kalu et al. 2013).

More effort must also be put into ensuring that mushrooms are grown on nutritious substrates with a hygienic environment to protect against the effects obtained from abiotic and biotic stresses, respectively. A recent review looked at the prevalence of 24 fungal diseases, three bacterial diseases, two viral diseases, several actinomycete molds and contaminants, and 11 non-infectious physiologic disorders in mushroom production farms (Aminuzzaman et al. 2022). Based on this and other studies, several infectious diseases threaten the mushroom industry and there are still obstacles to be overcome in terms of identification, preservation, and mushroom farming. These obstacles need to be addressed before this industry can continue expanding and contributing to Bangladesh's food security.

11.13 FUTURE PROSPECTS

Mushrooms are one of the few foods that may be consumed in unlimited quantities without negative health consequences. They do not perform miracles, but their consumption makes a big difference in how long and how well one lives. Mushrooms are a vegetarian food that is highly regarded by flexitarians and vegetarians alike (Thakur 2020). The proteins and peptides found in mushrooms are important natural bioactive sources (Zhou et al. 2019). Protein fragments called bioactive peptides (BAPs) are thought to contribute to health in a variety of ways (Bhandari et al. 2020). Mushrooms, which contain a lot of high-quality proteins, could be used as a BAP source in the future. Numerous endogenous peptides with varied biological activities, including those with antihypertensive, antioxidant, antibacterial, anticancer, and other properties, have recently been found in several types of mushrooms. Of the hundreds of species of mushrooms, roughly 2,000 are edible, about 200 have historically been gathered as food or for medicinal purposes (Erjavec et al. 2012), and fewer than half have been investigated (Erjavec et al. 2012). To learn more about the specific mushroom components that are good for the skin, many studies have been conducted. Mushrooms were shown to have a wide variety of bioactive metabolites, including lectins, phenolics, polysaccharides, polyphenolics, ergosterols, terpenoids, and volatile chemical compounds (Kalač 2013). The pharmaceutical industry is now offering mushroom polysaccharides such as schizophyllan, lentinan, polysaccharide krestin (PSK), and polysaccharide peptide (PSP). As an emulsifier, gel-forming agent, protector, and antibacterial component, mushroom chitosan has found a widespread application in the cosmetics industry (Badalyan 2014). The use of *Tremella* extracts in the treatment of neurodermatitis and scleroderma adds a touch of exoticism to otherwise all-encompassing skincare products (De Baets and Vandamme 2001). Products from biotransformation involving mushrooms, such as ceramides and lactic acid, have the potential to be employed in the cosmetics industry (Gao et al. 2004).

The worldwide community of mycologists is unanimous in its belief that an improved understanding of mushrooms has the potential to alleviate many types and stages of cancer. Consequently, it is important to investigate the untouched wilderness for rare species with therapeutic potential. There can be no long-term growth without the conservation and cloning of MMs. New anticancer and immune-stimulant chemicals need to be isolated, purified, and studied structurally. Thus far, research has uncovered several chemicals and defined the underlying mechanisms. However, future study is needed to explain the varied functions of many active substances and the pathways involved. Many wild mushrooms remained underutilized, despite their wide nutritional and bioactive potential (Kumar et al. 2021). Green techniques are being explored for the extraction of novel bioactive components from wild edible mushrooms.

11.14 CONCLUSIONS

Mushrooms are a superfood that can provide a wide range of health benefits. There is almost no room for the horizontal expansion of land for food, feed, and fiber production due to the rapidly increasing population, so vertical expansion of farming by integrating the optimal farming components is the only way to meet the growing need for food. Due to the losses caused by the threat of habitat degradation, research into the diversity of macrofungi is becoming more and more important. Mushrooms are one ecosystem component that not only makes use of vertical space but also contributes to solving problems related to food quality, health, and ecological sustainability. To keep up with the ever-evolving demand for food, it is important to encourage not only mushroom cultivation and mushroom consumption but also harvesting of more wild edible mushrooms. As a result, the mushroom industry has the potential to dramatically impact the world's economic and social landscape. Mushroom farming might get a significant boost from technological development in the areas of production, processing, value addition, and cost reduction, with enormous domestic market and export potential. Due to rising consumer expectations, mushroom research has exploded in recent years. Unemployed young people may make a decent living selling wild mushrooms, which is also beneficial for the economy. Thus, conservation can be achieved via correct identification, cultivation, genome preservation, raising awareness, and establishing national parks and forest reserve areas specifically to support the wild mushroom habitat. In addition, further research is required in different seasons and forest locations to identify and preserve newly introduced, non-indigenous mushroom species, which will serve as a comprehensive review of the mycota that is now present in Bangladesh.

REFERENCES

Ahmad, M.F., F.A. Ahmad, Z. Azad, A. Ahmad, M.I. Alam, J.A. Ansari & B.P. Panda. 2013. Edible mushrooms as health-promoting agent. *Advanced Science Focus* 1: 189–196. https://doi.org/10.1166/asfo.2013.1034.

Akatin, M.Y. 2013. Characterization of a β Glucosidase from an edible mushroom, *Lycoperdon pyriforme*. *International Journal of Food Properties* 16: 1565–1577. https://doi.org/10.1080/10942912.2011.600494.

Aminuzzaman, F.M. & K. Das. 2016. Morphological characterization of polypore macro fungi associated with *Dalbergia Sissoo* collected from Bogra district under the social forest region of Bangladesh. *Journal of Biology and Nature* 6(4): 199–212.

Aminuzzaman, F.M., S. Shahi, S. Thapa & K. Das. 2022. Mushroom diseases and their management: A review. In: Kulshreshtha, S., Ukaogo, P.O., Siddhant (Eds.), *Recent Advances in Mushroom Cultivation Technology and Its Application*. Delhi, India: Bright Sky Publications, pp. 1–27.

Andrew, E.E., T.R. Kinge, E.M. Tabi, N. Thiobal & A.M. Mih. 2013. Diversity and distribution of macrofungi (mushrooms) in the Mount Cameroon Region. *Journal of Environmental Microbiology* 3: 318–334. http://doi.org/10.5897/jene2013.0379

Antonín, V. & B. Buyck. 2006. *Marasmius* (Basidiomycota, Marasmiaceae) in Madagascar and the Mascarenes. *Fungal Diversity* 23: 17–50.

Aravind, S., D. Sharmah & D. Dhanavel. 2021. Exploitation and diversity of edible macrofungi resources in mount harriet national park, Andaman and Nicobar Islands. *Journal of Current Opinion in Crop Science* 2(3): 369–378.

Badalyan, S.M. 2014. Potential of mushroom bioactive molecules to develop healthcare biotech products. In *Proceedings of the 8th International Conference on Mushroom Biology and Mushroom Products*. New Delhi, India.

Bandala, V.M., L. Montoya & M. Mata. 2008. New species and records of *Crepidotus* from Costa Rica and Mexico. *Fungal Diversity* 32: 9–29.

Bankole, P.O. & A.A. Adekunle. 2012. Studies on biodiversity of some mushrooms collected in Lagos State, Nigeria using biotechnological methods. *Journal of Yeast and Fungal Research* 3(4): 37–48. http://doi.org/10.5897/jyfr12.008

Bass, D. & T.A. Richards. 2011. Three reasons to re-evaluate fungal diversity 'on Earth and in the ocean'. *Fungal Biology Reviews* 25: 159–164. https://doi.org/10.3732/ajb.1000298

Bhandari, D., S. Rafiq, Y. Gat, P. Gat, R. Waghmare & V. Kumar. 2020. A Review on bioactive peptides: Physiological functions, bioavailability, and safety. *International Journal of Peptide Research and Therapeutics* 26(1): 139–150. http://doi.org/10.1007/s10989-019-09823-5

Blackwell, M. 2011. The fungi: 1, 2, 3 . . . 5.1 million species? *American Journal of Botany* 98: 426–438.

Bolton, J. 1791. *A History of Funguses. Growing about Halifax*. Halifax, Canada: Husddersfield, p. 160.

Breitenbach, J. & F. Kränzlin. 1995. *Fungi of Switzerland. Volume 4: Agarics (2nd Part). Entolomataceae, Pluteaceae, Amanitaceae, Agaricaceae, Coprinaceae, Strophariaceae*. Verlag-Mykologia: Luzern, Switzerland, p. 368.

Bulliard, J.B.F. 1782. *Herbier de la France*. Vol 2 (in French). Paris, France: P.F. Didot, pp. 49–96.

Bunyard, B.A. 2003. A survey of fungal diversity in Northeast Ohio. *The Ohio Journal of Science* 103(2): 29–32.

Chandulal, K., C. Gopal & J. Priya. 2013. Studies on biodiversity of fleshy fungi in Navsari (south Gujarat), India. *International Journal of Biodiversity and Conservation* 5(8): 508–514. http://doi.org/10.5897/ijbc2013.0561

Chang, S.T. & S.P. Wasser. 2012. The role of culinary-medicinal mushrooms on human welfare with a pyramid model for human health. *International Journal of Medicinal Mushrooms* 1: 95–134. http://doi.org/10.1615/intjmedmushr.v14.i2.10

Cheung, P.C. 2008. *Mushrooms as Functional Foods*. Hoboken, NJ: John Wiley & Sons, pp. 1–34.

Das, K. & F.M. Aminuzzaman. 2017. Morphological and ecological characterization of xylotrophic fungi in mangrove forest regions of Bangladesh. *Journal of Advances in Biology & Biotechnology* 11(4): 1–15. http://doi.org/10.9734/JABB/2017/30971

Das, K., F.M. Aminuzzaman & N. Akhtar. 2016. Diversity of fleshy macro fungi in mangrove forest regions of Bangladesh. *Journal of Biology and Nature* 6(4): 218–241. www.researchgate.net/project/Ministry-of-Science-and-Technology-P-R-Bangladesh-No-39000000090206916-17-10-BS-63

De Baets, S. & E.J. Vandamme. 2001. Extracellular Tremella polysaccharides: Structure, properties and applications. *Biotechnology Letters* 23: 1361–1366.

De Leon, A.M., J.J.D. Luangsa-ard, S.C. Karunarathna, K.D. Hyde, R.G. Reyes & T.E.E. dela Cruz. 2013. Species listing, distribution, and molecular identification of macrofungi in six Aeta tribal communities in Central Luzon, Philippines. *Mycosphere* 4(3): 478–494. http://doi.org/10.5943/mycosphere/4/3/4

Desjardin, D.E., M.G. Wood & F.A. Stevens. 2015. *California Mushrooms: The Comprehensive Identification Guide*. Portland: Timber Press, p. 560.

Dey, T.K. 2018. *Introduction to the Wildlife of Bangladesh and Management Techniques*. Lalmatia, Dhaka, Bangladesh: Nature Conservation Society, p. 1148.

Dutta, A.C. 2008. *Botany for Degree Students*. New Delhi: Oxford University Press India.

Dwivedi, S., M.K. Tiwari, U.K. Chauhan & A.K. Pandey. 2012. Biodiversity of mushrooms of Amarkantak biosphere reserve forest of Central India. *International Journal of Pharmacy and Life Science* 3(1): 1363–1367. ISSN: 0976–7126

Eggert, C., P.R. LaFayette, U. Temp, K.E. Eriksson & J.F. Dean. 1998. Molecular analysis of a laccase gene from the white rot fungus *Pycnoporus cinnabarinus*. *Applied and Environmental Microbiology* 64: 1766–1772. https://doi.org/10.1128/AEM.64.5.1766-1772.1998

El Enshasy, H.A. & R. Hatti-Kaul. 2013. Mushroom immunomodulators: Unique molecules with unlimited applications. *Trends in Biotechnology* 31: 668–677. https://doi.org/10.1016/j.tibtech.2013.09.003

Elkhateeb, W.A. & G.M. Daba. 2022. The wild non edible mushrooms, what should we know so far? *International Journal of Advanced Biochemistry Research* 6(1): 43–50.

Engola, A.P.O., G. Eilu, J.D., Kabasa, L. Kisovi, P.K.T. Mun & D. Olila. 2007. Ecology of edible indigenous mushrooms of the Lake Victoria basin (Uganda). *Research Journal of Biological Sciences* 2(1): 62–68.

Erjavec, J., J. Kos, M. Ravnikar, T. Dreo & J. Sabotič. 2012. Proteins of higher fungi -from forest to application. *Trends in Biotechnology* 30(5): 259–273.

FAO. 2015. *Global Forest Resources Assessment 2015*. Rome, Italy: Food and Agriculture Organization of the United Nations.

Farooq, M., A. Akram, R. Afzal & K.S. Nazir. 2013. Ethno-morphological studies of mushrooms collected from Soon Valley. *IOSR-Journal of Pharmacy and Biological Sciences* 8(5): 5–11. http://doi.org/10.9790/3008-0850511

Feeroz, M.M. 2014. *Biodiversity of Chunati Wildlife Sanctuary: Fauna*. Dhaka, Bangladesh: Arannayk Foundation.

Feeroz, M.M. 2016. *Biodiversity of Inani Protected Forest*. Dhaka, Bangladesh: Arannayk Foundation, 168p.

Foster-Turley, P., R. Das, M.K. Hasan & P.R. Hossain. 2016. *Bangladesh Tropical Forests and Biodiversity Assessment*. United States Foreign Assistance Act, Section 118/119 Report. Washington, USA: Integra Government Service, LLC.

Fries, E.M. 1821. Systema *Mycologicum sistens* fungorum ordines, genera et species, huc usque cognitas, quas ad normam methodi naturalis determinavit. *Lundae, Ex Officina Berlingiana* 1: 1821–1832. https://doi.org/10.5962/bhl.title.5378

Fui, F.S., F.H. Saikim, J. Kulip, J. Seelan & S. Seelan. 2018. Distribution and ethnomycological knowledge of wild edible mushrooms in Sabah (Northern Borneo), Malaysia. *Journal of Tropical Biology and Conservation* 15: 203–222.

Ganeshpurkar, A., P. Pardhi, S.S. Bhadoriya N. Jain, G. Rai & A.P. Jain 2015. Antioxidant potential of White Oyster culinary-medicinal mushroom, *Pleurotus florida* (higher basidiomycetes). *International Journal of Medicinal Mushrooms*17: 491–498. http://doi.org/10.1615/IntJMedMushrooms.v17.i5.90

Gao, J.M., A.L. Zhang, H. Chen & J.K. Liu. 2004. Molecular species of ceramides from the ascomycete truffle Tuber indicum. *Chemistry and Physics of Lipids* 131: 2. https://doi.org/10.1016/j.chemphyslip.2004.05.004

Garcha, H.S., P.K. Khann & G.L. Soni. 1993. Nutritional importance of mushroom. In: Chang, S.T., Buswell, J.A., Chiu, S. (Eds.), *Mushroom Biology and Mushroom Products*. China: The Chinese University Press, pp. 227–235.

Gittins, S.P. & A.W. Akonda. 1982. What survives in Bangladesh? *Oryx* 16: 275–281.

Hawksworth, D.L. 2012. Global species number of fungi: Are tropical studies and molecular approaches contributing to a more robust estimate? *Biodiversity Conservation* 21: 2425–2433. http://doi.org/10.1007/s10531-012-0335-x

Hibbett, D.S. & J.W. Taylor. 2013. Fungal systematics: Is a new age of enlightenment at hand? *Nature Reviews Microbiology* 11: 129–133. http://doi.org/10.1038/nrmicro2963

Hosen, M.I., T.H. Li & W.Q. Deng. 2016. *Amanita cinereovelata*, a new species of *Amanita* section Lepidella from Bangladesh. *Mycological Progress* 14: 35. http://doi.org/10.1007/s11557-015-1058-7

Hsu, H.-C., C.-I. Hsu, R.-H. Lin, C.-L. Kao & J.-Y. Lin. 1997. Fip-Vvo, a new fungal immunomodulatory protein isolated from *Volvariella volvacea. Biochemistry Journal*323: 557–565. http://doi.org/10.1042/bj3230557

Itoo, Z.A., Z.A. Reshi & K.I. Andrabi. 2013. Characterization and identification of *Russula firmula* and *Russula postiana* from Himalayan moist temperate forests of Kashmir. *African Journal of Biotechnology* 12(23): 3643–3647. www.ajol.info/index.php/ajb/article/view/132069

Jenkins, D.T. 1986. *Amanita of North America*. Eureka, CA: Mad River Press, p. 197.

Jo, J.W., Y.N. Kwag, C.S. Kim & S.K. Han. 2013. Distribution of higher fungi in Chungcheong Province, Republic of Korea. *Journal of Asia-Pacific Biodiversity* 6(3): 347–359. https://doi.org/10.7229/jkn.2013.6.3.347

Joty, A.A., F.M. Aminuzzaman, N. Sultana, A. Tanjina, D.R.B. Sonchita & M.N. Islam. 2020. Diversity, distribution and morphology of wild mushrooms collected from gajni forest of Bangladesh. *International Journal of Environment* 9(2): 234–255. http://doi.org/10.3126/ije.v9i2.32843

Kalač, P. 2013. A review of chemical composition and nutritional value of wild-growing and cultivated mushrooms. *Journal of the Science of Food and Agriculture* 93: 209–218. https://doi.org/10.1002/jsfa.5960

Kalu, I.G., C.U. Nwachukwu, B.C. Ijioma & C.I. Evans-Kemka. 2013. *Problems Facing Mushroom Availability and Consumption in Owerri Municipal Council of Imo State Nigeria*. www.iiste.org

Kernaghan, G., R.S. Currah & R.J. Bayer. 1997. Russulaceous ectomycorrhizas of *Abies lasicarpa* and *Picea engelmanii. Canadian Journal of Botany* 75: 1843–1850.

Khan, M.M.H. 2011. *Tigers in the Mangrove: Reserarch and Conservation of the Tigers in the Sundarbans of Bangladesh*. Dhaka, Bangladesh: Arannayk Foundation, 191p.

Kirk, P.M., P.F. Cannon, J.C. David & J.A. Stalpers. 2008. *Ainsworth and Brisby's Dictionary of the Fungi*. 10th ed. Wallingford: Centre for Agriculture and Biosciences International.

Kraigher, H., R. Agerer & B. Javornik. 1995. Ectomycorrhizae of *Lactarius lygnyotus* on Norway spruce, characterized by anatomical and molecular tools. *Mycorrhiza* 5(3): 175–180. https://link.springer.com/article/10.1007/BF00203334

Kumar, K., R. Mehra, R.P.F. Guiné, M.J. Lima, N. Kumar, R. Kaushik, N. Ahmed, A.N. Yadav, & H. Kumar, 2021. Edible mushrooms: A comprehensive review on bioactive compounds with health benefits and processing aspects. *Foods* 10: 2996. https://doi.org/10.3390/ foods10122996

Lakhanpal, T.N. & M. Rana. 2005. Medicinal and nutraceutical genetic resources of mushrooms. *Plant Genetic Resources*3: 288–303. http://doi.org/10.1079/PGR200581

Lee, H.T., J. Nuytinck, A. Verbeken, S. Lumyong & D. Desjardin. 2007. *Lactarius* in northernspores and refractive hyphae in Amanita. *Mycotaxon* 52: 305–396.

Macrae, R., R.K. Robinson & M.J. Sadler. 1993. *Protein: Encyclopedia of Food Science and Food Technology and Nutrition*. London: Academic Press, 6, pp. 3792–3480.

Mafia, M.I., F.M. Aminuzzaman, M.S.M. Chowdhury & J.F. Tanni. 2020. Occurrence, diversity and morphology of poroid wood decay by *Ganoderma* spp. from tropical moist deciduous forest region of Bangladesh. *Journal of Agriculture and Natural Resources* 3(2): 160–174. http://dx.doi.org/10.3126/janr.v3i2.32498

Marzana, A., F.M. Aminuzzaman, M.S.M. Chowdhury, S.M. Mohsin & K. Das. 2018. Diversity and ecology of macrofungi in rangamati of chittagong hill tracts under tropical evergreen and semi-evergreen forest of Bangladesh. *Advances in Research* 13(5): 1–17. http://doi.org/10.9734/AIR/2018/36800

Menaga, D., P.U. Mahalingam, S. Rajakumar & P.M. Ayyasamy. 2012. Evaluation of phytochemical characteristics and antimicrobial activity of *Pleurotus florida* mushroom. *Asian Journal of Pharmaceutical and Clinical Research*2012(5): 102–106.

Miles, P.G. & S.T. Chang. 2004. *Mushrooms: Cultivation, Nutritional Value, Medicinal Effect, and Environmental Impact*, 2nd ed. Boca Raton, FL: CRC Press, pp. 1–26.

Mohanan, C. 2011. *Macrofungi of Kerala*. Kerala, India: Kerala Forest Research Institute, p. 597.

Moro, C., I. Palacios, M. Lozano, M. D'Arrigo, E. Guillamón, A. Villares, J.A. Martínez & A. García-Lafuente. 2012. Anti-inflammatory activity of methanolic extracts from edible mushrooms in LPS activated RAW 264.7 macrophages. *Food Chemistry*130: 350–355. http://doi.org/10.1016/j.foodchem.2011.07.049.

Mueller, G.M. & J.P. Schmit. 2007. Fungal biodiversity: What do we know? What can we predict? *Biodiversity Conservation* 16: 1–5. http://doi.org/10.1007/s10531-006-9117-7

Mukul, S.A. & S.A. Quazi. 2009. Communities in conservation: Changing protected area management and enhanced conservation in Bangladesh. In: Leslie, R.N. (Ed.), *The Future of Forests in Asia and the Pacific: Outlook for 2020*. Bangkok, Thailand: Food and Agriculture Organization of the United Nations (FAO-RAP Publication no 2008/3), pp. 143–159.

Murray, T. 2013. *Mushrooms and Fungi Photo Gallery*, pp. 1–48. https://pbase.com/tmurray74/mushrooms

Ndungutse, V., R. Mereddy & Y. Sultanbawa. 2015. Bioactive properties of mushroom (*Aaricus bisporus*) stipe extracts. *Journal of Food Processing and Preservation*39: 2225–2233. http://doi.org/10.1111/jfpp.12467.

Nowakowski, P., K. Sylwia, N.R. Markiewicz-Zukowska, M.H. Borawska & K. Socha. 2020. The two faces of *Coprinus comatus*—Functional properties and potential hazards. *Phytotherapy Research* 34: 2932–2944.

O'Hanlon, R. & T.J. Harrington. 2012. Macrofungal diversity and ecology in four Irish forest types. *Fungal Ecology* 5: 499–508. http://doi.org/10.1016/j.funeco.2011.12.008

Pala, S.A., A.H. Wani & R.A. Mir. 2012. Diversity of macrofungal genus *Russula* and *Amanita* in Hirpora Wildlife Sanctuary, Southern Kashmir Himalayas. *Biodiversitas* 13(2): 65–71. https://doi.org/10.13057/biodiv/d130203

Payo, A., A. Mukhopadhyay, S. Hazra, T. Ghosh, S. Ghosh & S. Brown. 2016. Projected changes in area of the Sundarban mangrove forest in Bangladesh due to SLR by 2100. *Climatic Change* 139(2): 279–291. https://link.springer.com/article/10.1007/s10584-016-1769-z

Poffenberger, M. 2000. *Communities and Forest Management in South Asia*. Philippines: IUCN, DFID, and Asia Forest Network.

Poucheret, P., F. Fons & S. Rapior. 2006. Biological and pharmacological activity of higher fungi: 20-Year retrospective analysis. *Cryptogam Mycol*ogy 27: 311–333.

Pradhan, B. 2013. A comparative study on the predictive ability of the decision tree, support vector machine and neuro-fuzzy models in landslide susceptibility mapping using GIS. *Computers and Geosciences* 51: 350–365. https://doi.org/10.1016/j.cageo.2012.08.023

Pushpa, H. & K.B. Purushothama. 2012. Biodiversity of mushrooms in and around Bangalore (Karnataka), India. *American-Eurasian Journal of Agriculture and Environmental Science* 12(6): 750–759.

Rahaman, M., F.M. Aminuzzaman, M.B. Hossain, S.N. Rashid & M.I. Rumainul. 2016. Biodiversity, distribution and morphology of mushrooms of south western region of Bangladesh. *International Journal of Advanced Research* 4(3): 60–79. http://doi.org/10.9734/JABB/2016/27711

Rashid, M., K. Akhter, M.S.M. Chowdhury & F.M. Aminuzzaman. 2017. Biodiversity, habitat and morphology of mushroom of different forest regions of Bangladesh. *International Journal of Advanced Research* 5(9): 945–957. http://doi.org/10.21474/IJAR01/5408

Rashid, S.N., F.M. Aminuzzaman, M.R. Islam, M. Rahaman & M.I. Rumainul. 2016. Biodiversity and distribution of wild mushrooms in the southern region of Bangladesh. *Journal of Advances in Biology & Biotechnology* 9(1): 1–25. http://doi.org/10.9734/JABB/2016/27711

Ripkova, S. & S. Glejdura. 2010. Crepidotus ehrendorferi in Slovakia and taxo-nomic notes on related species. *Czech Mycology* 61(2): 175–185.

Rubina, H., F.M. Aminuzzaman, M.S.M. Chowdhury & K. Das. 2017. Morphological characterization of macro fungi associated with forest tree of national botanical garden, Dhaka. *Journal of Advances in Biology & Biotechnology* 11(4): 1–18. http://doi.org/10.9734/JABB/2017/30970

Rumainul, M.I. & F.M. Aminuzzaman. 2016. Macro fungi biodiversity at the central and northern biosphere reserved areas of tropical moist deciduous forest region of Bangladesh. *Journal of Agriculture and Ecology Research International* 5(4): 1–11. http://doi.org/10.9734/JAERI/2016/22135

Rumainul, M.I., F.M. Aminuzzaman & M.S.M. Chowdhury 2015. Biodiversity and morphological characterization of mushrooms at the tropical moist deciduous forest region of Bangladesh. *American Journal of Experimental Agriculture* 8(4): 235–252. http://doi.org/10.9734/AJEA/2015/17301

Ryvarden, L. 1994. Can we trust morphology in Ganoderma? In: Buchanan, P.K., Hseu, R.S., Moncalvo, J.M. (Eds.), *Proceedings of Contributed Symposia 59A, Ganoderma – Systematics, Phytopathology and Pharmacology.* Fifth International Mycological Congress, Vancouver, 14–21 August 1994, pp. 19–24.

Shomali, N., O. Onar, B. Karaca, N. Demirtas, A.C. Cihan, I. Akata & O. Yildirima. 2019. Antioxidant, anticancer, antimicrobial, and antibiofilm properties of the culinary-medicinal fairy ring mushroom, *Marasmius oreades* (Agaricomycetes). *International Journal of Medicinal Mushrooms* 21(6): 571–582.

Sonchita, D.R.B., F.M. Aminuzzaman, A.A. Joty, J.F. Tanni, M.N. Islam & M. Rahaman. 2020. Diversity, distribution and morphological characterization of wild macro fungi from gajni Forest. *Asian Journal of Biology*: 19–32. http://doi.org/10.9734/AJOB/2020/v9i230084

Stamets, P. 2000. The role of mushroom in nature, culturing mushroom mycelium on agar media. In: *Growing Gourmet and Medicinal Mushrooms*. Hong Kong: Ten Speed Press, pp. 10–11.

Stanis, V.F. 1979. *Russula Species Occurring in the Boreal Region of Ontario and Quebec, Canada*. M.Sc. Thesis, Department of Botany. University of Toronto, Toronto, Ontario.

Tanjim, A., F.M. Aminuzzaman, M. Rahaman & J.F. Tanni. 2019. Biodiversity, distribution and morphological characterization of macrofungi in Sylhet and Moulvibazar under tropical evergreen and semi—evergreen forest regions of Bangladesh. *International Journal of Advanced Research* 7(11): 567–589. http://doi.org/10.21474/IJAR01/10047

Tanni, J.F., F.M. Aminuzzaman, M. Ahmed & M. Rahaman. 2020. Diversity and distribution of macro fungi in some selected parks and gardens of Dhaka City, Bangladesh. *Asian Journal of Biology* 9(1): 23–43. http://doi.org/10.9734/ajob/2020/v9i130076

Thakur, M.P. 2020. Advances in mushroom production: Key to food, nutritional and employment security: A review. *Indian Phytopathology* 73: 377–395. http://doi.org/10.1007/s42360-020-00244-9

Thiers & A.H. Sm. 2016. *Species List: Boletus and Other Tubed Mushrooms of Washington State (236)*, Mushroom observer section 501(c)(3).

Thiers, H.D. 1982. *The Agaricales (Gilled Fungi) of California. 1. Amanitaceae*. Eureka, CA: Mad RiverPress, p. 53.

Thiribhuvanamala, G., V. Prakasam, G. Chandrasekar, K. Sakthivel, S. Veeralakshmi, R. Velazhahan & G. Kalaiselvi. 2011. Biodiversity, conservation and utilization of mushroom flora from the western ghats region of India. *Proceedings of the 7th International Conference on Mushroom Biology and Mushroom Products (ICMBMP7)*. Tamil Nadu, India, pp. 155–164.

Ukwuru, M., A. Muritala & L. Eze. 2018. Edible and non-edible wild mushrooms: Nutrition, toxicity and strategies for recognition. *Journal of Clinical Nutrition and Metabolism* 2: 9.

Vyas, D., A. Chaubey & P. Dehariya. 2014. Biodiversity of mushroom in Patharia Forest of Sagar M.P.-111. *International Journal of Biodiversity and Conservation* 6(8): 600–607. http://doi.org/10.5897/IJBC2014.0681

Walton, E. & B. Treasure. 2014. Unlocking the secrets of medicinal mushrooms. *Biomedical Journal*37: 339–342. http://doi.org/10.4103/2319-4170.146538.

Wang, P., J. Liu, L. Xu, P. Huang, X. Luo, Y. Hu & Z. Kang. 2021. Classification of *Amanita* species based on bilinear networks with attention mechanism. *Agriculture* 11(5): 393.

Wang, X.C., R.J. Xi, Y. Li, D.M. Wang & Y.J. Yao. 2012. The species identity of the widely cultivated *Ganoderma*, '*G. lucidum*' (Lingzhi), in China. *PLoS ONE* 7(7): e40857. https://doi.org/10.1371/journal.pone.0040857

Wasser, S.P. 2010. Medicinal mushroom science: History, current status, future trends, and unsolved problems. *International Journal of Medicinal Mushrooms* 12: 1–16. http://doi.org/10.1615/IntJMedMushr.v12.i1.10

Wasser, S.P. 2011. Current findings, future trends, and unsolved problems in studies of medicinal mushrooms. *Applied Microbiology and Biotechnology* 89: 1323–1332. https://doi.org/10.1007/s00253-010-3067-4

Wasser, S.P., I.V. Zmitrovich, M.Y. Didukh, W.A. Spirin & V.F. Malysheva. 2006. *Morphological Traits of Ganoderma Lucidum Complex Highlighting G. Tsugae Var. Jannieae: The Current Generalization*. Ruggell, Liechtenstein: A.R.A. Gantner Verlag K.-G.

Xu, J.-W., W. Zhao & J.-J. Zhong. 2010. Biotechnological production and application of ganoderic acids. *Applied Microbiological Biotechnology* 87: 457–466. http://doi.org/10.1007/s00253-010-2576-5.

Zhang, Y., M. Mo, L. Yang, F. Mi, Y. Cao, C. Liu, X. Tang, P. Wang & J. Xu. 2021 Exploring the species diversity of edible mushrooms in Yunnan, Southwestern China, by DNA Barcoding. *Journal of Fungi* 7: 310. https://doi.org/10.3390/jof7040310

Zhou, R., Z. Liu, Y. Zhang, J.H. Wong, T.B. Ng & F. Liu. 2019. Research progress of bioactive proteins from the edible and medicinal mushrooms. *Current Protein & Peptide Science* 20(3): 196–219. http://doi.org/10.2174/1389203719666180613090710

12 Some Medicinal Fungi from the Forests Used by Tribal People in Central India

N. S. K. Harsh

12.1 INTRODUCTION

Macrofungi in forests influence the recycling of nutrients and provide many other ecological benefits. A recent study was conducted among the Baiga tribe living in sal (*Shorea robusta* Roth) forests of Balaghat, Bilsapur, Jabalpur, and Mandla districts in two states of central India, namely Chattisgarh and Madhya Pradesh, which exhibit 41% and 25% forest cover, respectively (India State of Forest Report 2019). The forests are tropical moist deciduous and tropical dry deciduous forests according to the Champion and Seth (1968) classification. Two dominant tree species are *Shorea robusta* Gaertn., and *Tectona grandis* L. f.

The Chattisgarh and Madhya Pradesh states have a fair degree of tribal peoples, representing 31% and 21% of the state populations, respectively (Tribal Population of India as per 2011 Census). The Baiga are a tribe with around 200,000 members that live in central India and are also known as the Bhuiya, Bhumaia, Bhumiaraja, Bhumij, Bhumija, and Bhumijan; they are a Munda or Kolarian tribe and have traditionally been slash-and-burn agriculturalists and hunters. Baiga means "sorcerer, medicine man." Other tribes in their area consider them priests with special knowledge about the soil (Hockings 1992). A majority of the Baiga tribe lives in and around forests and depends on the forests for their day-to-day needs, such as non-wood forest produce, food, medicine, and livelihood. They live away from the cities and depend largely on locally available materials. The modern system of healthcare is available to these people though government and non-government agencies, but due to their belief and traditions, they do not trust these services and resort to them only when the cases become critical. Initially, they take the help from local tribal medicine men, who use their knowledge and experience of forest plants and other items to treat diseases and disorders.

This traditional knowledge is considered the heritage of local peoples, and attempts are being made to document it the world over. However, extracting details from the custodians and practitioners is not an easy task because they keep it as a closely guarded secret and are not easily open to outsiders. In central India, a tribal medicine man only parts this knowledge with his sons and not daughters, as he considers them part of other families after marriage. If he does not have a son, the knowledge goes with him after his death. It is not incorrect to say that if a tribal medicine man dies without sharing his knowledge, it is as if a whole library is burned.

During the study, it was found that some macrofungi were used by the Baiga tribe as traditional medicine, and certain individuals knew about the habitat and period of occurrence of these fungi (Harsh et al. 1993, 1996, 1999). In the present chapter, information on medicinal fungi is documented.

12.2 MATERIALS AND METHODS

The study was conducted by visiting the places where Baiga medicine men were known to used forest produce in the treatment of human ailments. It was gathered through inquiries and

 DOI: 10.1201/b23190-12

personal approaches that the medicine men of the Baiga tribe use some higher fungi belonging to Basidiomycotina and Ascomycotina for their medicinal uses. These fungi were collected from the adjoining forest areas with the aid of knowledgeable tribal medicine men, and their uses in human diseases, methods of preparation of formulation, and doses were recorded. The specimens were brought to the laboratory for detailed taxonomical studies to confirm their identity by consulting standard monographs (Bakshi 1971; Bondartsev 1953; Coker and Couch 1969; Cunningham 1942; Ryvarden and Johansen 1980) and available expertise.

12.3 RESULTS

A total of nine fungi were identified as being used in tribal medicine. Details on the fungi, including their use for treating diseases, preparations, formulations, and doses are described herein.

12.3.1 *Astraeus hygrometricus* (Pers.) Morgan

Collected from sal (*Shorea robusta*) forests during early rains (July–August) where it forms ecto-mycorrhizae with the roots of the trees, unopened sporocarps are used as food and sold in the local markets. Opened sporocarps are used for the treatment of burns. The spore mass is mixed with mustard oil and applied over the burned skin (Harsh et al. 1999).

12.3.2 *Bovista coprophila* (Cooke and Massey) G. Cunn. (= *Bovista apedicellata* G. Cunn.)

Growing among grasses during early rains (July-August). The spore mass is mixed with mustard oil and applied for the healing of wounds.

12.3.3 *Bovista pusilla* (Batsch.) Pers. (= *Lycoperdon pusillum* Batsch.)

The sporocarps are collected from gardens, agricultural fields, grasslands, and freshly worked sandy soil throughout the rainy season (July-September). It is used to stop bleeding from deep cuts and wounds. The spore powder is mixed with mustard oil and applied over the wounds.

FIGURE 12.1 *Astraeus hygrometricus*—immature sporocarps being sold in the local tribal market as a food item and mature sporocarp.

12.3.4 *Cyathus striatus* (Huds.) Willd.

FIGURE 12.2 *Cyathus striatus.*

The sporocarps are found on wood and bamboo debris and collected during the mid-rainy season (August-September). The peridioles are crushed with clean water or rose petal extract and filtered through clean cotton; the filtrate is used as eye drops for eye redness, pain, and conjunctivitis (Harsh et al. 1999).

12.3.5 *Daldinia concentrica* (Bolten) Ces. and De Not

The sporocarps are collected from the dead wood and bark of the trees. For the treatment of chronic cough, sporocarps growing particularly on *Butea monosperma* trees or wood are used. The dried sporocarps are then powdered and a piece of old earthen vessel is also powdered. Both are mixed in equal proportions in honey and given to the patients at a dose of one teaspoon, twice daily.

FIGURE 12.3 *Daldinia concentrica.*

12.3.6 *Ganoderma lucidum* (Curtis) P. Karst.

FIGURE 12.4 *Ganoderma lucidum.*

The sporocarps are collected from the roots of living and dying trees and from stumps of various tree species during the rainy season (July-September). This fungus is used for the treatment of cataracts, asthma, and hydroceles by tribal medicine men.

For the treatment of cataracts, the sporocarps are dried over a flame and then finely powdered and mixed with coconut oil to prepare an ointment (5 g powder in 25 g oil). The ointment is applied around the eye socket twice daily for one and a half months (Harsh et al. 1999).

For the treatment of asthma, the sporocarps are finely powdered in the same manner as for the treatment of cataracts and then mixed in a locally-prepared liquor made from *Madhuca latifolia* flowers. The suspension is given at a dose of one teaspoon three times daily for two to three months to cure the disease (Harsh et al. 1999).

For the treatment of hydroceles, dried sporocarps are boiled with an equal amount of coconut oil, mustard oil, and linseed oil for about an hour. The cooled filtrate is applied around the scrotum three times a day for a month (Harsh et al. 1999).

12.3.7 *Microporus xanthopus* (Fr.) Kuntz.

FIGURE 12.5 *Microporus xanthopus.*

Sporocarps are collected from the dead twigs of fallen trees on the forest floor during the rainy season; sporocarps growing on the wood of *Terminalia tomentosa* are particularly preferred. This fungus is used for the treatment of ear pain, fever, vomiting.

For the treatment of ear pain, the dried sporocarps are ground with clean water and, after filtering through cotton, used as ear drops.

For the treatment of fever and vomiting, the sporocarps are ground with the resin of *Shorea robusta* trees and one teaspoon is given with water three times a day.

12.3.8 *Phallus impudicus* L.

FIGURE 12.6 *Phallus impudicus.*

The sporocarps are collected from leaf litter in old bamboo plantations during the middle of the rainy season (August). It is used for the treatment of typhoid and for labor pains in women. Sporocarps are crushed, and a suspension is made with jaggery and water; one teaspoon is given to patients three times a day for four days. The same dose is given to women experiencing labor pains to give relief.

12.3.9 *Xylaria polymorpha* (Pers.) Grev.

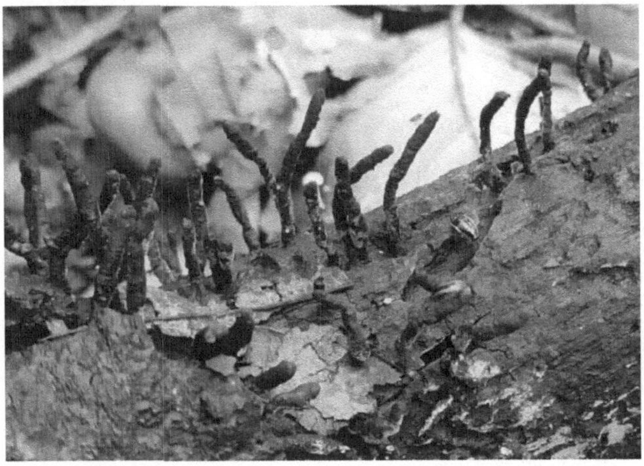

FIGURE 12.7 *Xylaria polymorpha.*

The sporocarps are collected from the roots and stumps of trees during the rainy season. This fungus is used for the activation of lactation after delivery. Dried sporocarps are powdered and mixed with old jaggery to make small, pea-size tablets. The prescription is one tablet twice a day with cow's milk before meals for five days (Harsh et al. 1999).

12.4 DISCUSSION

This documentation of the reported uses of macrofungi from forests in medicine by tribal medicine men has potential application for the benefit of mankind, particularly in the post-COVID global climate.

Polysaccharides isolated from the hot aqueous extract of *Astraeus hygrometricus* contain D-galactose, D-glucose, and D-galacturonic acid (Pramanik and Islam 2000). A water-soluble fraction was isolated from the aqueous extract of the fruit bodies of *Astraeus hygrometricus*; the structure of the repeating unit of glucan from fraction I (named ASQ-I) was determined to be an effective anti-tumor agent in other fungi (Chakraborty et al. 2004). The second polysaccharide fraction (AQS-II) was found to contain 63% polysaccharide and 35% protein. The polysaccharides include glucose, galactose, and fucose (Chakraborty et al. 2007). Three triterpenes have been isolated from the fruiting bodies of *A. astraeus*; these are 3-hydroxy-lanostane derivatives with a D-lactone in the side chain were named astrahygrol, 3-epi-astrahygrol, and astrahygrone (Takaishi et al. 1987). Basidiocarps of *A. hygrometricus* are rich in proteins, carbohydrates, minerals, crude fiber, and essential amino acids with a low concentration of fat. Experiments on animal models revealed the hepatoprotective, cardioprotective, anti-inflammatory, anticancer, and hypoglycemic efficacy of the mushroom (Dai et al. 2009; Biswas et al. 2017). However, the use of *A. hygrometricus* in the treatment of human diseases is not known.

Farrington (1887) claimed that *Bovista* spores restricted blood circulation through the capillaries and suggested uses to treat menstrual irregularity or trauma. He also mentioned that *Bovista* relieves some symptoms of suffocation and might be useful in remedying asphyxiation resulting from inhalation of charcoal fumes. Moreover, ailments suggested to be improved with the use of *Bovista* include "awkwardness in speech and action," "stuttering or stammering [in] children," "palpitation after a meal," diabetes mellitus, ovarian cysts, and "acne due to cosmetics" (Jain 1997).

Dai et al. (2009) reported *Cyathus striatus* to exhibit antibacterial and antifungal properties and suggested its efficacy in treating gastropathy. Fungal extracts prepared from *Cyathus striatus* showed significant inhibitory effects on the NF-kB activation pathway, suggesting activities worthy of investigation as cancer therapeutics (Petrova et al. 2007).

Dai et al. (2009) reported the use of *Daldinia concentrica* in treating infantile convulsion. Mo et al. (2018) reported it has antioxidant properties. *Daldinia concentrica* mushrooms were reported by tribal people to exhibit wound-healing activity, which was evaluated *in vivo* in a rat model (Thangaraj et al. 2017).

Medicinal uses of *Ganoderma lucidum* are well researched and documented; however, the ailments for which they were applied in this report have not been known before. Ayu et al. (2019) reported the use of *Ganoderma lucidum* by the Jakun people in Malaysia as an anticancer, stress-reducing, and cough-relieving medicine.

Ayu et al. (2019) reported the use of a water decoction of *Microporus xanthopus* for birth control and a treatment for aging, breast cancer, heart ailments, and gout by the Jakun people of Malaysia.

Phallus impudicus has a long history of use in Europe as an aphrodisiac and a treatment erectile dysfunction and other disorders (https://healing-mushrooms.net/phallus-impudicus). Though these traditional uses are not supported scientifically, the species showed promise in the treatment of certain cancers (Petrova 2008); it is known to have anti-tumor properties, and its extracts have also improved wound healing in experimentally-wounded rats and shown immunomodulating activity in a diabetic rat model (Vyacheslav et al. 2019).

Some interest has been generated regarding the use of *Xylaria polymorpha* in medicine, specifically as an antimicrobial agent. This fungus was shown to inhibit the growth of *Escherichia coli* in culture experiments (Hacioglue et al. 2011).

The traditional knowledge of simple application methods reveals the acceptance of local treatments among the tribal people who trust their medicine men, which, although not qualified doctors, receive an adequate following.

REFERENCES

Ayu, K.K., Alona, C.L., Mohd Ahmad, A., Patahayah, M., Mohd Salleh, S., Mohamed, M. & Lili, T. 2019. Documentation of macrofungi traditionally used by Jakun people in Johor, Malaysia in the treatment of various illnesses. *International Conference on Biodiversity 2018 IOP Conf. Series: Earth and Environmental Science* 269 (2019) 012013 IOP Publishing. http://doi.org/10.1088/1755-1315/269/1/012013.

Bakshi, B.K. 1971. *Indian Polyporaceae on Trees and Timber*. ICAR Pub, New Delhi, 246 p.

Biswas, G., Nandi, S., Kuila, D. & Acharya, K. 2017. A comprehensive review on food and medicinal prospects of *Astraeus hygrometricus*. *Pharmacognosy Journal* 9: 79–806. http://doi.org/10.5530/pj.2017.6.125

Bondartsev, A.S. 1953. *The Polyporaceae of the European USSR & Caucasia*. Keter Press, Jerusalem.

Census. 2011. Office of the Registrar General, India. Ministry of Tribal Affairs. https://tribal.nic.in/statistics

Chakraborty, I., Mondal, S., Pramanik, M., Rout, D. & Islam, S.S. 2004. Structural investigation of a water-soluble glucan from an edible mushroom, *Astraeus hygrometricus*. *Carbohydrate Research* 339(13): 2249–2254.

Chakraborty, I., Mondal, S., Rout, D., Chandra, K. & Islam, S.S. 2007. Structural investigation of a heteroglycan isolated from the fruit bodies of an ectomycorrhizal fungus *Astraeus hygrometricus*. *Carbohydrate Research* 342(7): 982–987.

Champion, H.G. & Seth, S.K. 1968. *A Revised Forest Types of India*. Manager of Publications, Government of India, Delhi.

Coker, W.C. & Couch, J.N. 1969 (Reprint). *The Gasteromycetes of Eastern United States and Canada*. 3301 Lehre, Verlag Von J. Cramer, New York, 201 p. 122.

Cunningham, G.H. 1942. *The Gasteromycetes of Australia and New Zealand*. John McIndoe, Donedin, New Zealand.

Dai, Y.C., Yang, Z.L., Cui, B.K., Yu, C.J. & Zhou, L.W. 2009. Species diversity and utilization of medicinal mushrooms and fungi in China (Review). *International Journal of Medicinal Mushrooms* 11: 287–302.

Farrington, E.A. 1887. *Lectures on Clinical Materia Medica*—Google Book Search. ISBN 978-81-8056-194-8. Retrieved 2008–11–25. https://books.google.co.ck/books?id=A2FNAAAAMAAJ&printsec=frontcover&source=gbs_ge_summary_r&cad=0#v=onepage&q&f=false

Hacioglu, N., Akata, I. & Dulger, B. 2011. Antimicrobial potential of *Xylaria polymorpha* (Pers.) Grev. *African Journal of Microbiology Research* 5(6): 728–730.

Harsh, N.S.K., Rai, B.K. & Soni, V.K. 1999. Some ethnomycological studies from Madhya Pradesh, India. In: Singh, J. & Aneja, K.R. (Eds.), *From Ethnomycological to Fungal Biotechnology: Exploiting Fungi from Natural Resources for Novel Products*. Plenum Publ. Corpn, New York, pp. 19–32.

Harsh, N.S.K., Rai, B.K. & Tiwari, D.P. 1993. Use of *Ganoderma lucidum* in folk medicine. *Indian Journal of Tropical Biodiversity* 1: 324.

Harsh, N.S.K., Tiwari, C.K. & Rai, B.K. 1996. Forest fungi in the aid of tribal women of Madhya Pradesh. *Sustainable Forestry* 1: 10–15.

Hockings, P. 1992. *Encyclopedia of World Cultures: South Asia*, edited by Paul Hockings, C.K. Hall & Company, USA, New York.

India State of Forest Report. 2019. A biennial publication published by Forest Survey of India, Dehradun and released by the Ministry of Environment, Forests and Climate Change, Govt. of India.

Jain, J.P. 1997. *Highlights of Homoeopathic Materia Medica*. Google Book Search. ISBN 978-81-7021-785-5. Retrieved 2008–11–30. https://books.google.co.in/books/about/Highlights_of_Homoeopathic_Materia_Medic.html?id=xzExF8ZLQd4C&redir_esc=y

Mo, G.Y., Wu, J.Q., Ren, Q.Y., He, X.S. & Huang, Y. 2018. Optimization of the extraction technology of anti-oxidant substances from *Daldinia concentrica* by orthogonal experiment. *Edible Fungi China* 37: 51–55.

Petrova, R. 2008. Substances from the Medicinal Mushroom *Marasmius oreades* Prevent iNOS Expression through NF-kB-Dependent and -Independent Mechanisms. *Thesis*.

Petrova, R., Mahajna, J., Reznik, A.Z. & Wasser, S.P. 2007. Fungal substances as modulators of NF-kB pathway. *Molecular Biology Reports* 34(3): 45–54. http://doi.org/10.1007/s11033-006-9027-5

Pramanik, A. & Islam, S.S. 2000. Structural studies of a polysaccharide isolated from an edible mushroom, *Astraeus hygrometricus*. *Indian Journal of Chemistry Section B-Organic Chemistry Including Medicinal Chemistry* 39(7): 525–529.

Ryvarden, L. & Johansen, I. 1980. *A Preliminary Polypore Flora of East Africa*, Fungiflora, Oslo, Norway.

Takaishi, Y., Murakami, Y., Ohashi, T., Nakano, K., Murakami, K. & Tomimatsu, T. 1987. Three Triterpenes from *Astraeus hygrometricus*. *Phytochemistry* 26(8): 2341–2344.

Thangaraj, R., Raj, S. & Regnganathan, K. 2017. The wound healing effect of King Alferd's mushroom (*Daldinia concentrica*) used by tribes of Sirumalai Hills, Tamilnadu, India. *International Journal of Pharmacy and Pharmaceutical Sciences* 9(7): 161.

Vyacheslav, B., Alexey, B., Aliaksandr, A. & Morox, V. 2019. Polysaccharides of mushroom *Phallus impudicus* mycelium: Immunomodulating and wound healing properties. *Modern Food Science and Technology* 35(9): 30–37.

13 The Current Status of Sri Lankan Mushroom Research

Mahesh C. A. Galappaththi, Samantha C. Karunarathna

13.1 INTRODUCTION

Mushrooms have a long association with humans and have had a profound biological and economic impact as nutritious food, medicinal mushrooms, and cosmetics. Since ancient times, mushrooms have been consumed by humans as a delicacy, probably for their pleasant taste and high levels of nutrients (Anusiya et al. 2021). Many mushrooms have been used in folk medicine for thousands of years (Hrudayanath and Sameer 2014). Mushrooms are a rich source of various bioactive substances and show a wide spectrum of health benefits because of their pharmacological properties (Anusiya et al. 2021).

Sri Lanka is a tropical island with an approximate land cover of 65,000 km^2 and was separated from the Indian mainland in the Miocene epoch. The central portion of the island has mountains with elevations up to 2,524 m. Sri Lanka, with its tropical humid climate (Silva 1988), is estimated to be home to more than 25,000 species of fungi, but only a little over 3,000 of these species have been documented (Adikaram and Yakandawala 2020). Although Sri Lanka is a potentially fungus-rich country, relatively few studies have been carried out on either fungi or lichens. Moreover, accurate estimates of the abundance of Sri Lankan fungi are further complicated due to the widely scattered available information and the synonyms used in different studies (Karunarathna et al. 2017).

Houttuyn (1783) described the first fungi (*Peziza ceylonische* and *P. Lembosa*) recorded from Sri Lanka. Tom Petch who worked at the Royal Botanical Gardens in Peradeniya, documented various plant pathogenic and other fungi (Petch 1908a, 1908b; 1909; 1910a, 1910b; 1913; 1915a, 1915b; 1916a, 1916b, 1916c; 1917a, 1917b; 1919; 1922; 1924a, 1924b, 1924c; 1925; 1926a, 1926b; 1927; 1928; 1945). In addition, Petch and Bisby (1950) also reported numerous taxa in the genera *Agaricus* (26 species), *Lepiota* (53 species) *Marasmius* (39 species), and *Mycena* (23 species). In addition, they also reported 30 genera of gasteromycetes and 17 genera of polypores from the wet zone of Sri Lanka (Wijesundera 1986). Coomaraswamy (1979), Coomaraswamy and de Fonseka (1981), and Coomaraswamy (1981) reported the results of three other significant studies in which approximately 100 genera of plant pathogenic fungi, 63 genera of soil fungi, and 53 genera of agarics were reported. Adikaram and Yakandawala (2020) published a checklist of 372 plant pathogenic fungi from Sri Lanka. The checklist provided the species names of the fungal pathogens and, in most cases, species names of the host plants.

A comprehensive overview of the taxonomy, diversity, ecology, ethnomycology, medicinal aspects, phytochemistry, and biotechnology of Sri Lankan mushrooms is lacking, and such an overview is essential for researchers and scientists working in the field of mycology. The present review provides information on the current status of all these aspects of mycology in Sri Lanka.

13.2 MUSHROOM TAXONOMY

The Dutch (1658–1796) and subsequent British (1796–1948) occupations can be considered important periods for macrofungal studies in Sri Lanka. Since the end of the colonial period, very few

 DOI: 10.1201/b23190-13

studies have been carried out by native Sri Lankans on the taxonomy, phylogeny, and cultivation of macrofungi in the country.

Houttuyn (1783) was probably the first westerner to write about the macrofungi of Sri Lanka. He was a Dutch naturalist and a systematic botanist (Wijnands et al. 2017). Berkeley and Broome (1871) provided descriptions for 23 species of mushroom belonging to the Entolomataceae (Pegler 1977). In this publication, Berkeley and Broome (1871) reported 403 species of agarics, 305 of which were new to science (Berkeley and Broome 1870, 1871, 1873; Karunarathna et al. 2017). A few years later, Cesati and Beccari (1879) reported a few species of agarics. Petch (1905–1925) published several valuable papers on Sri Lankan mushrooms (Karunarathna et al. 2012; Karunarathna et al. 2017).

Pegler (1972) published a revision of the genus *Lepiota*. Descriptive morphological features, habitats, and locations of 34 species of *Lepiota*, two species of *Macrolepiota*, and four species of *Leucocoprinus* were considered in this work. Pegler (1977) also carried out a revision of species in the Entolomataceae (Agaricales) from Sri Lanka and India, and he redescribed each species and provided line drawings. Coomaraswamy and Kumarasingham (1988) published a handbook of Sri Lankan macrofungi.

In 2011, *Lentinus giganteas* was transferred to *Pleurotus* on the basis of morphological features and ITS sequence analyses of collected specimens from Sri Lanka and Thailand (Karunarathna et al. 2011). Ediriweera et al. (2015) reported two mushrooms (*Panaeolus sphinctrinus* and *P. foenisecii*) from dry zone forests in Sri Lanka. Tipbromma et al. (2017) reported *Boletellus emodensis, Favolaschia manipularis, F. auriscalpium, Lentinus sajor-caju, Leucocoprinus cretaceus, Russula* cf. *virescens*, and *Tremella fuciformis* as new records for Sri Lanka. Recently, Ediriweera et al. (2021) reported nine new records of polypores from Sri Lanka based on both morphological and phylogenetic data.

People in Sri Lanka have consumed wild mushrooms since ancient times (Gamage and Ohga 2018). Tipbromma et al. (2017) documented edible wild mushrooms such as *Boletellus emodensis, Lentinus sajor-caju, Lentinus squarrosulus, Russula* cf. *virescens*, and *Tremella fuciformis* in Sri Lanka. Moreover, Gunasekara et al. (2021) reported the nutritional composition of three species of *Termitomyces* and four other wild edible mushrooms in Sri Lanka, while Adikaram et al. (2020a) documented the first report of the edible mushroom *Helvella crispa* from the island. In addition, Adikaram et al. (2020b) revisited *Leucocoprinus birnbaumii* after 100 years.

13.3 DIVERSITY AND ECOLOGY

Ediriweera et al. (2014) identified 49 species of macrofungi from the Sigiriya wilderness area, with 90% of them belonging to Basidiomycota. Thirty-nine percent of these were identified as polypores, 31% as agarics, and 8% as species of *Calocera*, *Tremella*, and *Auricularia*. Gnaneswaran and Wijayagunasekara (1999) reported nine insect pests associated with cultivated *Pleurotus ostreatus* in the Central Province. Also, they reported that only three species of insects (*Cyllodes bifacies, Drosophila funebris*, and *Bradysia papuera*) damaged *P. ostreatus*.

13.4 ETHNOMYCOLOGY

Local tribes and residents have considerable traditional knowledge of herbs, mushrooms, and other plants. The Sinharaja Forest Reserve is a UNESCO-recognized world heritage site in Sri Lanka with major ecological and biological importance (Hewage 2015). Natives of this region have a rich tradition of the usage of wild edible mushrooms for medicinal and culinary purposes, but a lack of scientific reports and proper research work on the ethnomycology of Sri Lanka has created a knowledge gap (Hewage 2015).

There are many vernacular names for mushrooms in Sri Lanka. "Hathu" is a general Sinhalese term for mushroom orchards, and mushroom picking is one of the villagers' preferred activities when visiting lowland forests (Ranawake 2021). When the mushroom season arrives, inhabitants gather edible mushrooms from the forests. A number of traditions are followed before entering the forests and picking mushrooms (Hewage 2015). Local people consume mushrooms such as hinvali hathu (*Termitomyces microcarpus/Entoloma microcarpus*), vali hathu (*Termitomyces alwisii*) (Hewage 2015), athuru-hathu/muukalan-hathu/pus-hathu (*Agaricus fulvoalbus*), indalolu (*Termitomyces eurrhizus*), kanda hathu (*Lentinus subnudes*), lena hathu (*Schizophyllum commune*), olu-weli-hathu (*Termitomyces* spp.), and ooru-hathu (*Lentinus giganteus*) (Ranawake 2021). Ancient villagers used to visit spots where there was a potential to find athuru-hathu (*Agaricus fulvoalbus*) during the fruiting season and sometimes waited a few days in the forest until they appeared (Ranawake 2021).

Kuḍava is a village located in the Sinharaja region, and villagers use "behet hathu" to treat bone injuries. They consider athuru-hathu (*Agaricus fulvoalbus*), urukammal hathu (*Lentinus giganteus*), and indalolu hathu (*Termitomyces eurrhizus*) are not given to menstruating women, breastfeeding mothers, or girls who recently underwent puberty since traditional Sri Lankan ayurvedic medicine considers these mushrooms make the body hot (Hewage 2015).

According to villagers who live in lowland tropical forests in Sri Lanka, thelka-hathu or wood-ear mushrooms (*Auricularia auricula* or *Auricularia polytricha*) are not edible. However, both species are considered edible in other parts of the world. Also, *Amanita muscaria* is not an edible species according to the villagers. In addition, puwak-bada-hathu and kaila-hathu are not edible, and their scientific names have not been confirmed (Ranawake 2021). There may be more varieties of medicinal and edible mushrooms in forests, but due to the lack of knowledge, they remain hidden. It is difficult to identify whether these mushrooms are edible or medicinal without traditional knowledge (Hewage 2015).

13.5 MEDICINAL ASPECTS AND PHYTOCHEMISTRY

Edible mushrooms are an exceptional source of protein and are exceedingly important for vegans. Although the medicinal properties of edible mushrooms are well documented in many countries, such information is inadequate in Sri Lanka. This has led many Sri Lankans to remain doubtful as to whether food and medicines obtained from fungal sources could hold any nutritional or therapeutic value. Therefore, there is an acute need to expand awareness about the medicinal and dietary importance of mushrooms.

Perera et al. (2001) revealed significant antibacterial activity of *Volvariella volvacea* against *Escherichia coli*. Ergosterol peroxide isolated from the ethyl acetate extract of *V. volvacea* showed significant activity against *E. coli*. The antibacterial activities of 21 basidiomycetes against *Klebiesiella*, *Pseudomonas*, *Staphylococcus*, and *Streptococcus* bacteria were investigated, and extracts of *Agaricus* spp. and *Schizophyllum* spp. exhibited significant antibacterial activities against *Klebsiella*. *Staphylococcus aureus* growth was relatively hindered by extracts of *Cantharellus* spp., *Clitocybe* spp., and two bracket fungi. Also, *Streptococcus* showed susceptibility to the extracts of *Amanita* spp. Menikpurage et al. (2009) studied the antifungal activity of *Pleurotus cystidiosus* against *Colletotrichum gloeosporioides*. An oxidized ergosterol, 3β, 5α, 6β-trihydroxyergosta-7,22-diene, (Figure 13.1) was identified and isolated from the fraction that exhibited the highest fungal activity. This compound is responsible for the activity of *P. cystidiosus* against *C. gloeosporioides* (Menikpurage et al. 2009).

An investigation of the antioxidant and cytotoxic activity of *P. cystidiosus* against the Hep-2 cancer cell line was carried out by Menikpurage et al. (2012) and revealed *P. cystidiosus* as a rich antioxidant that exhibited a cytotoxic effect against Hep-2 carcinoma cells. In addition, this research indicated that the usage of the entire mushroom as a therapeutic functional food may bring health benefits. *Anthracophyllum lateritium* is another terrestrial mushroom found in dry zone forest patches in Sri Lanka. Fernando et al. (2015b) examined the crude extract of *A. lateritium* and demonstrated significant antioxidant potential, antiproliferative activity, and apoptotic effect against RD

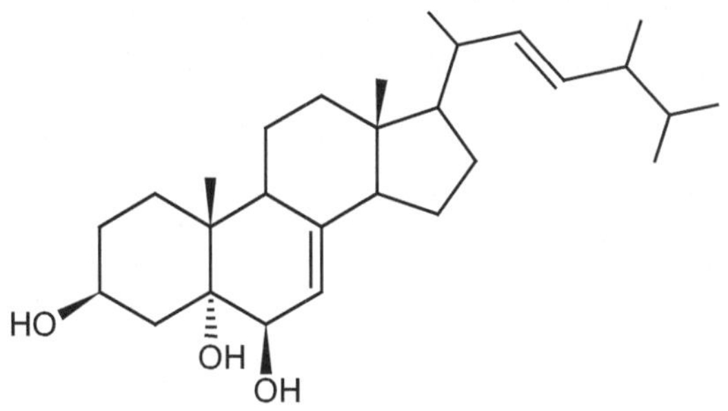

FIGURE 13.1 Structural formula of *3β, 5α, 6β*-trihydroxyergosta-7,22-diene.

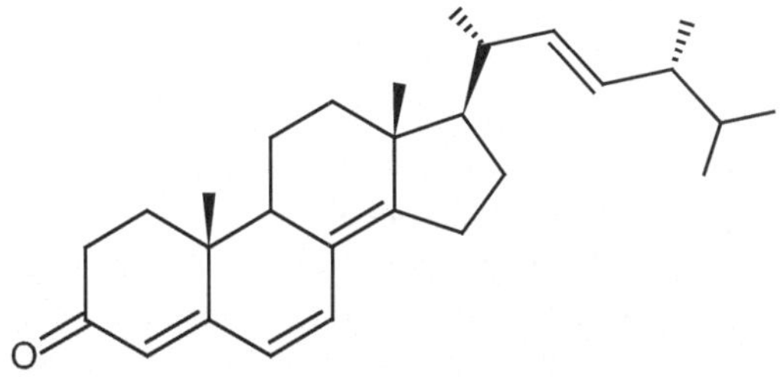

FIGURE 13.2 Structural formula of ergone.

cell lines and emphasized the ethnopharmacological use of this fungus to control cancer-causing oxidative damage. Fernando et al. (2015a) investigated the antioxidant activity of macrofungi collected in Sri Lankan dry zone forests. This investigation was based primarily on the contribution of flavonoid and phenol contents of macrofungi towards their antioxidant activity. The investigation revealed that all studied macrofungi found in the dry zone of Sri Lanka exhibited high antioxidant capacity. *Phellinus repandus* and *Inonotus porrectus* exhibited significant antioxidant activities, to which their considerable phenol and flavonoid contents contributed significantly. The terrestrial mushroom *Fulviformes fastuosus* is commonly found in the woodlands of Sri Lanka (Fernando et al. 2016a). Fernando et al. (2015c) examined the *in vitro* antiproliferative activity of *F. fastuosus* against Hep-2 and RD cell lines. This study further explained that the high cytotoxic and apoptotic effects of *F. fastuosus in vitro* can lead to its ethnopharmacological use to treat cancer. Fernando et al. (2016a) isolated a potential cytotoxic compound called ergone (Figure 13.2) from *F. fastuosus*. Ergone exhibits a powerful cytotoxic effect against RD cells with a high selectivity index compared to that for HepG2 cancer cells. The study highlighted that ergone can be used for the medical purpose of suppressing rhabdomyosarcoma (Fernando et al. 2016a).

White rots are wood-decay fungi that produce various polyphenolic secondary metabolites. *Flavodon flavus* is a common wood rot mushroom, and Fernando et al. (2016b) investigated the antioxidant capacity and total phenolic and flavonoid content of this terrestrial species, which is found in forest reserves in Dambulla and Mahiyanganaya, Sri Lanka. The study revealed that *F. flavus* possesses a very high antioxidant activity (EC_{50} 77.00 ± 0.18 μg/mL) and a high content of total phenols (55.7 ± 10.89 μg gallic acid/mg) and flavonoids (82.4 ± 4.0 μg epicatechine/mg); results suggested that these high contents of total phenols and flavonoids led to significant antioxidant ability. Rajapakse et al. (2018) carried out experiments to demonstrate the functional attributes of selected mushrooms (*Pleurotus ostresus*, *Ganoderma lucidum*, *Calocybe indica*, and *Calocybe* spp.) cultivated in Sri Lanka based on their antioxidant activities (free radical-scavenging activity).

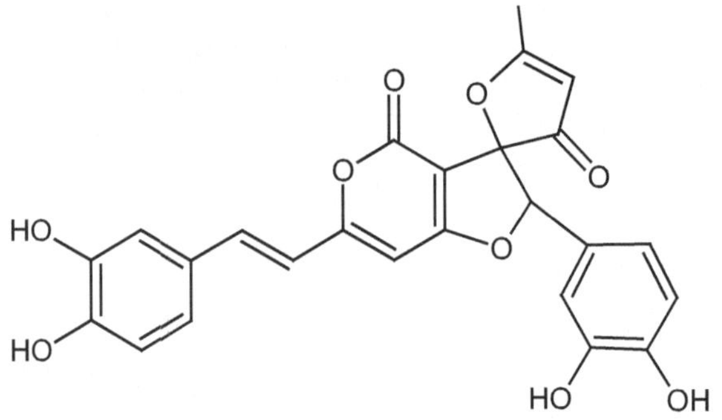

FIGURE 13.3 Structural formula of 2-(3,4-dihydroxyphenyl) -6-[(E)-2- (3,4-dihydroxyphenyl) ethenyl]-5'-methylspiro[2H-furo[3,2-c] pyran-3,2'-furan]-3',4-dione.

Antioxidant activity of all four mushrooms cultivated on three different substrates exhibited $IC_{50} \leq$ 100 µl/ml, indicating that all four mushrooms have moderate DPPH-scavenging activity compared to that of ascorbic acid and quercetin (Rajapakse et al. 2018).

The flavonoid compound 2-(3,4-dihydroxyphenyl)-6-[(E)-2-(3,4-dihydroxyphenyl) ethenyl]-5'-methylspiro [2H-furo[3,2-c] pyran-3,2'-furan]-3',4-dione (Figure 13.3) was isolated from *F. fastuosus* (Fernando et al. 2019). The compound showed significant anticancer activity on hepatoma cells with less cytotoxicity on normal CC-1 cells, indicating that this compound has high potential as a supportive therapeutic drug against hepatoma (Fernando et al. 2019).

Fernando et al. (2020) carried out an experiment to develop a novel method for isolating inoscavin A. *In vitro* cytotoxicity assays showed that inoscavin A intensely induced the apoptosis of human RD cells compared to CC-1 cells and suggested that the isolated compound can be used for the production of a drug to treat rhabdomyosarcoma (Fernando et al. 2020). Jayasuriya et al. (2020) investigated the anti-inflammatory potential of *Pleurotus ostreatus* and the underlying mechanisms of activity and suggested *P. ostreatus* as an applicable food that can control inflammation. They suggested that the possible mechanisms for the anti-inflammatory activity of *P. ostreus* were anti-histamine action, inhibition of cell migration to the site of inflammation, *in vitro* membrane stabilizing activity, and inhibition of NO production.

13.6 CULTIVATION AND BIOTECHNOLOGY

In 1985, the United Nations Development Program (UNDP) introduced commercial mushroom cultivation in Sri Lanka. (Rajapakse 2014; Karunarathna et al. 2017). Mushroom houses and spawn laboratories were built after the establishment of the Sri Lankan Export Development Board, and mushroom cultivation was improved by the Department of Agriculture through research and developmental activities (Karunarathna et al. 2017). Most indigenous mushrooms have not been domesticated in Sri Lanka, and artificial cultivation technologies can be used for the domestication of wild varieties. Udugama and Wickramarathna (1991) first introduced the cultivation of black ear mushrooms (*Auricularia* spp.) and milk mushrooms (*Callocybe indica*) in the country. Rupasinghe and Nandasena (2006) documented the potential of refuse tea as an alternative substrate for mushroom cultivation. Rajapakse et al. (2007) reported the efficiency of oyster mushroom (*Pleurotus* spp.) cultivation in different compost mixtures. Later, Pathmashini et al. (2008) discussed the efficiency of cultivating *Pleurotus ostreatus* on several types of sawdust media. Rajapakse (2011) introduced a new cultivation technology for *Volvariella volvacea.* Thanuejah et al. (2013) reported the potential of cultivating oyster mushrooms in Sri Lanka. Jeewanthi et al. (2017) discussed the growth and yield of *Ganoderma lucidum* (Curtis) P. Karst on different sawdust substrates. *Lentinus squarrosulus* and *Lentinus tuber-regium* are two edible wild mushroom species that can be found in Sri Lanka. Miriyagalla et al. (2019) investigated the domestication of these two wild species and suggested that both can be domesticated successfully and possibly introduced to mushroom farmers in Sri Lanka. The mycochemical composition of *Volvariella volvacea* cultivated on used palm oil bunch was discussed by Nasir and Siva (2019). Miriyagalla et al. (2022) successfully cultivated the wild edible mushrooms *Lentinus sajor-caju* and *Pleurotus tuber-regium* for the first time in Sri Lanka. In addition, Galappaththi et al. (2022) reported the nutritional composition of *Macrocybe gigantea* and successfully cultivated the species for the first time in Sri Lanka.

Mushroom biotechnology encompasses all facets of the application of mushrooms, such as the domestic cultivation of wild species, bioconversion of agricultural wastes, biodegradation resistance of substances, enhanced production of nutraceuticals, and bioremediation of xenobiotics (Singh et al. 2009). However, mushroom biotechnology in Sri Lanka is still in its infancy, except regarding domestication and cultivation, because of the absence of proper research. The lack of documented biotechnological studies related to mushrooms in Sri Lanka highlights the need for more advanced research as compared to that required in other countries.

13.7 CONCLUSIONS

Sri Lanka, with its diverse habitats with different ecological conditions, supports a wide range of mushrooms and other fungi with potentially rich nutritional and medicinal value. With the increasing human population and associated development across the landscape, the forest cover and microhabitats of fungi and lichens are shrinking in Sri Lanka, as is the case all over the world. Therefore, some unknown fungi and lichens might become extinct in their original habitats without humans ever knowing their value. Many mushrooms remain unreported, and their taxonomy, diversity, ecology, ethnomycology, medicinal aspects, phytochemistry, and biotechnological uses are unknown to us. The identification of mushrooms is an important part of mycological research in any country, and Sri Lanka is no exception. There is an urgent need for the establishment of a national collection of mushrooms and elucidation of their properties under several aspects, and a network of local and international working groups is required.

13.8 ACKNOWLEDGMENTS

Samantha C. Karunarathna thanks Qujing Normal University, Yunnan, China. Mahesh Galappaththi thanks the Postgraduate Institute of Science (PGIS), University of Peradeniya, Peradeniya 20400 Sri Lanka.

REFERENCES

Adikaram, N.K.B. & D.M.D. Yakandawala. 2020. A checklist of plant pathogenic fungi and Oomycota in Sri Lanka. *Ceylon Journal of Science* 49(1): 93–123.

Adikaram, N.K.B., D. Yakandawala & L. Jayasinghe. 2020a. The first report of *Helvella crispa* (Ascomycota, Pezizales), a rare fungal species in Sri Lanka. *Ceylon Journal of Science* 49(4): 485–489.

Adikaram, N.K.B., D. Yakandawala & L. Jayasinghe. 2020b. *Leucocoprinus birnbaumii* (Agaricales: Basidiomycota), attractive yellow houseplant mushroom, revisited after 100 years. *Ceylon Journal of Science* 49(2): 209–211.

Anusiya, G., U. Gowthama Prabu, N.V. Yamini, N. Sivarajasekar, K. Rambabu, G. Bharath & F. Banat. 2021. A review of the therapeutic and biological effects of edible and wild mushrooms. *Bioengineered* 12(2): 11239–11268. https://doi.org/10.1080/21655979.2021.2001183.

Berkeley, M.J. & C.E. Broome. 1870. On some species of the genus *Agaricus* from Ceylon. *Transactions of the Linnean Society of London* 27(2): 149–152.

Berkeley, M.J. & C.E. Broome. 1871. The fungi of Ceylon (Hymenomycetes from *Agaricus* to *Cantharellus*). *Botanical Journal of the Linnean Society* 11(56): 494–567. https://doi.org/10.1111/j.1095-8339.1871.tb00163.x.

Berkeley, M.J. & C.E. Broome. 1873. Enumeration of the fungi of Ceylon: Part 2. *Botanical Journal of the Linnean Society, Botany* 14(73): 29–140.

Cesati, V. & O. Beccari. 1879. *Mycetum in Itinere Borneensi Lectorum a cl. Od. Beccari*. Typis Regiae Scientarum Academiae, Napels, pp. 1–28.

Coomaraswamy, U. 1979. *A Handbook to the Fungi Parasitic on the Plants of Sri Lanka*. MAB-UNESCO Publication No 4, National Science Council of Sri Lanka, Colombo, Sri Lanka, pp. 1–167.

Coomaraswamy, U. 1981. *A Handbook to the Agarics of Sri Lanka*. MAB-UNESCO Publication No 5, National Science Council of Sri Lanka, Colombo, Sri Lanka, pp. 1–68.

Coomaraswamy, U. & R.N. de Fonseka. 1981. *A Handbook to the Soil Fungi of Sri Lanka*. MAB-UNESCO Publication No 7, National Science Council of Sri Lanka, Colombo, pp. 1–122.

Coomaraswamy, U. & S. Kumarasingham. 1988. *A Handbook to the Macrofungi of Sri Lanka*. MAB-UNESCO Publication No 17, Natural Resources, Energy and Science Authority of Sri Lanka, Colombo, Sri Lanka, pp. 1–124.

Ediriweera, S.S., C.M. Nanayakkara, O.V.D.S.J. Weerasena, S.C. Karunarathna, R.L.C. Wijesundera & M.A.S.U. Piyatissa. 2021. Morphology and phylogeny reveal nine new records of polypores from dry zone of Sri Lanka. *Chiang Mai Journal of Science* 48(3): 893–908.

Ediriweera, S.S., R.L.C. Wijesunderai, C.M. Nanayakkara & O.V.D.J.S. Weerasena. 2014. Macrofungi from the Sigiriya wilderness in Sri Lanka. *Journal of Mycopathological Research* 52(1): 47–51.

Ediriweera, S.S., R.L.C. Wijesundera, C.M. Nanayakkara & O.V.D.J.S. Weerasena. 2015. First report of *Panaeolus sphinctrinus* and *Panaeolus foenisecii* (Psathyrellaceae, Agaricales) on elephant dung from Sri Lanka. *Frontiers in Environmental Microbiology* 1(2): 19–23.

Fernando, D.M., A. Adhikari, C. Nanayakkara, E.D. de Silva, R. Wijesundera & P. Soysa. 2016a. Cytotoxic effects of ergone, a compound isolated from *Fulviformes fastuosus*. *BMC Complementary and Alternative Medicine* 16: 484. https://doi.org/10.1186/s12906-016-1471-8.

Fernando, D.M., A. Adhikari, N.H.K.S. Senathilake, E.D. de Silva, C.M. Nanayakkara, R.L.C. Wijesundera, P. Soysa & B.G.D.N.K. de Silva. 2019. In silico pharmacological analysis of a potent anti-hepatoma compound of mushroom origin and emerging role as an adjuvant drug lead. *Food and Nutrition Sciences* 10: 1313–1333.

Fernando, D.M., R.L. Wijesundera, P. Soysa, D. de Silva & C.M. Nanayakkara. 2015a. Strong radical scavenging macrofungi from the Dry Zone forest reserves in Sri Lanka. *Frontiers in Environmental Microbiology* 1(2): 32–38.

Fernando, D.M., R.L. Wijesundera, P. Soysa, D. de Silva & C.M. Nanayakkara. 2015b. Antioxidant potential, in vitro cytotoxicity, and apoptotic effect induced by crude organic extract of *Anthracophyllum lateritium* against RD sarcoma cells. *BMC Complementary and Alternative Medicine* 15: 398. https://doi.org/10.1186/s12906-015-0924-9.

Fernando, D.M., R.L.C. Wijesundera, P. Soysa, E.D. de Silva & C.M. Nanayakkara. 2015c. *In vitro* antiproliferative activity of the Sri Lankan mushroom *Fulviformes fastuosus*. *National Conference on Indigenous Systems of Medicine* 2015: 8.

Fernando, D.M., R.L. Wijesundera, P. Soysa, D. de Silva & C.M. Nanayakkara. 2016b. Antioxidant potential and content of the polyphenolic secondary metabolites of white rot macrofungi; *Flavodon flavus* (Klotzsch.) and *Xylaria feejeensis* (Berk.). *SDRP Journal of Plant Science* 1(1): 10–15.

Fernando, D.M., R.L. Wijesundera, P. Soysa, D. de Silva, C.M. Nanayakkara & A. Adhikari. 2020. *A Novel Method to Isolate Inoscavin A From Fulviformes Fastuosus and Medicinal Preparation Thereof to Treat Rhabdomyosarcoma Cancer Conditions*. Sri Lankan Patent, LK/P/1/18323. National Intellectual property office, Sri Lanka.

Galappaththi, M.C.A., Y. Lu, S.C. Karunarathna, N. Wijewardena, A. Karunarathna, M. Gamage & A.N. Ediriweera. 2022. First successful cultivation and nutritional composition of *Macrocybe gigantea* in Sri Lanka. *MycoAsia* 2022: 7.

Gamage, S. & S. Ohga. 2018. A comparative study of technological impact on the mushroom industry in Sri Lanka: A review. *Advances in Microbiology* 8(8): 665–686.

Gnaneswaran, R. & H.N.P. Wijayagunasekara. 1999. Survey and identification of insect pests of Oyster mushrooms (*Pleurotus ostreatus*) cultures in the central province of Sri Lanka. *Tropical Agricultural Research and Extension* 2(1): 21–25.

Gunasekara, N.W., C.M. Nanayakkara, S.C. Karunarathna & R.L.C. Wijesundera. 2021. Nutritional aspects of three *Termitomyces* and four other wild edible mushroom species from Sri Lanka. *Chiang Mai Journal of Science* 48(5): 1236–1246.

Hewage, D. 2015. Traditional knowledge of edible wild mushrooms in a village adjacent to the Sinharāja forest. *Journal of the Royal Asiatic Society of Sri Lanka* 60(1): 77–95. www.jstor.org/stable/44809414.

Houttuyn, M. 1783. *Handleiding tot de plant—en kruidkunde, benevens eene uitvoerige beschrijving der boomen, planten, heesters, kruiden, varens, mossens, bollen grasplanten, volgens het zamenstel van C. Linnaeus*. Volume 14, Lodewyk van Es: Amsterdam, pp. 585–698.

Hrudayanath, T. & K.S. Sameer. 2014. Diversity, nutritional composition and medicinal potential of Indian mushrooms: A review. *African Journal of Biotechnology* 13(4): 523–545. https://doi.org/10.5897/ajb2013.13446.

Jayasuriya, W.J.A.B.N., S.M. Handunnetti, C.A. Wanigatunge, G.H. Fernando, D.T.U. Abeytunga & T.S. Suresh. 2020. Anti-inflammatory activity of *Pleurotus ostreatus*, a culinary medicinal mushroom, in wistar rats. *Evidence-Based Complementary and Alternative Medicine* 2020: 6845383. https://doi.org/10.1155/2020/6845383.

Jeewanthi, L.A.M.N., K. Ratnayake & P. Rajapakse. 2017. Growth and yield of reishi mushroom [*Ganoderma lucidum* (Curtis) P. Karst] in different sawdust substrates. *Journal of Food and Agriculture* 10(1–2): 8–16.

Karunarathna, S.C., P.E. Mortimer, J. Xu & K.D. Hyde. 2017. Overview of research of mushrooms in Sri Lanka. *Revista Fitotecnia Mexicana* 40(4): 399–403.

Karunarathna, S.C., D. Udayanga, S.N. Maharachchikumbura, M. Pilkington, D.S. Manamgoda, D.N.N. Wijayawardene, H.A. Ariyawansa, A.R. Bandara, E. Chukeatirote, E.H.C. McKenzie & K.D. Hyde. 2012. Current status of knowledge of Sri Lankan mycota. *Current Research in Environmental & Applied Mycology* 2(1): 18–29.

Karunarathna, S.C., Z.L. Yang, O, Raspé, T.W.K. Ko, E.C. Vellinga, R.L. Zhao, A.H. Bahkali, E. Chukeatirote, J. Degreef, P. Callac & K.D. Hyde. 2011. *Lentinus giganteus* revisited: New collections from Sri Lanka and Thailand. *Mycotaxon* 118: 57–71.

Menikpurage, I.P., D.T. Abeytunga, N.E. Jacobsen & R.L. Wijesundara. 2009. An oxidized ergosterol from *Pleurotus cystidiosus* is active against anthracnose causing *Colletotrichum gloeosporioides*. *Mycopathologia* 167: 155. https://doi.org/10.1007/s11046-008-9158-4.

Menikpurage, I.P., S.S.S.B.D.P. Soysa & D.T.U. Abeytunga. 2012. Antioxidant activity and cytotoxicity of the edible mushroom, *Pleurotus cystidiosus* against Hep-2 carcinoma cells. *Journal of the National Science Foundation of Sri Lanka* 40(2): 107–114.

Miriyagalla, S.D., D.S. Manamgoda & D. Udayanga. 2022. Molecular characterization and cultivation of edible wild mushrooms, *Lentinus sajor-caju, L. squarrosulus* and *Pleurotus tuber-regium* from Sri Lanka. *Current Research in Environmental & Applied Mycology* 12(1): 28–43.

Miriyagalla, S.D., E.I.P. Silva, D. Udayanga & D.S. Manamgoda. 2019. Domestication of two edible wild mushrooms, *Lentinus squarrosulus* and *Lentinus tuber-regium* from Sri Lanka. *International Research Conference of Uva Wellassa University, Sri Lanka*, http://www.erepo.lib.uwu.ac.lk/handle/123456789/728.

Nasir, I.A. & R. Siva. 2019. Mycochemical composition of straw paddy mushroom (*Volvariella volvacea*) grown on used palm oil bunch. *Journal of Sustainability Science and Management* 14(6): 12–21.

Pathmashini, L., V. Arulnandhy & R.S.W. Wijeratnam. 2008. Cultivation of oyster mushroom (*Pleurotus ostreatus*) on sawdust. *Ceylon Journal of Science* 37(2): 177–182.

Pegler, D.N. 1972. A revision of the genus *Lepiota* from Ceylon. *Kew Bulletin* 27(1): 155–202.

Pegler, D.N. 1977. A revision of entolomataceae (Agaricales) from India and Sri Lanka. *Kew Bulletin* 32(1): 189–220.

Perera, W.A.S.W., D.T.U. Abeytunga & R.L.C. Wijesundera. 2001. Anti-bacterial activities of *Volvariella volvacea*. *Journal of the National Science Foundation of Sri Lanka* 29(1–2): 61–68.

Petch, T. 1908a. The genus *Endocalyx* Berkeley and Broome. *Annals of Botany* 22: 389–400.

Petch, T. 1908b. The Phalloideae of Ceylon. *Annals of the Royal Botanic Gardens, Peradeniya* 4: 139–184.

Petch, T. 1909. New Ceylon fungi. *Annals of the Royal Botanic Gardens, Peradeniya* 4: 299–307.

Petch, T. 1910a. Revisions of Ceylon fungi (Part II). *Annals of the Royal Botanic Gardens, Peradeniya* 4: 373–444.

Petch, T. 1910b. A list of the mycetozoa of Ceylon. *Annals of the Royal Botanic Gardens, Peradeniya* 4: 309–371.

Petch, T. 1913. Termite fungi: a résumé. *Annals of the Royal Botanic Gardens, Peradeniya* 5: 303–341.

Petch, T. 1915a. Horse-hair blights. *Annals of the Royal Botanic Gardens, Peradeniya* 6: 43–68.

Petch, T. 1915b. The pseudo-sclerotia of *Lentinus similis* and *Lentinus infundibuliformis*. *Annals of the Royal Botanic Gardens, Peradeniya* 6: 1–18.

Petch, T. 1916a. A preliminary list of Ceylon Polypori. *Annals of the Royal Botanic Gardens, Peradeniya* 6: 87–144.

Petch, T. 1916b. Ceylon lentini. *Annals of the Royal Botanic Gardens, Peradeniya* 6: 145–152.

Petch, T. 1916c. Revisions of Ceylon fungi (Part IV). *Annals of the Royal Botanic Garden, Peradeniya* 6: 153–183.

Petch, T. 1917a. Additions to Ceylon fungi. *Annals of the Royal Botanic Gardens, Peradeniya* 6(19): 195–256.

Petch, T. 1917b. Revisions of Ceylon fungi (Part V). *Annals of the Royal Botanic Gardens, Peradeniya* 6: 307–355.

Petch, T. 1919. Revisions of Ceylon fungi (Part VI). *Annals of the Royal Botanic Gardens, Peradeniya* 7: 1–14.

Petch, T. 1922. Additions to Ceylon fungi II. *Annals of the Royal Botanic Gardens, Peradeniya* 7: 279–322.

Petch, T. 1924a. Revisions of Ceylon fungi (Part VII). *Annals of the Royal Botanic Gardens, Peradeniya* 9: 119–181.

Petch, T. 1924b. Agaricaceae pleuropods Zeylanicae. *Annals of the Royal Botanic Gardens, Peradeniya* 9: 217–227.

Petch, T. 1924c. Ceylon pink-spored agarics. *Annals of the Royal Botanic Gardens, Peradeniya* 9: 201–216.

Petch, T. 1925. Additions to Ceylon fungi III. *Annals of the Royal Botanic Gardens, Peradeniya* 9: 313–328.

Petch, T. 1926a. Rhacophyllus B. & Br. *Transactions of the British Mycological Society* 11: 238–251.

Petch, T. 1926b. Additions to Ceylon fungi IV. *Annals of the Royal Botanic Gardens, Peradeniya* 10: 131–138.

Petch, T. 1927. Revisions of Ceylon fungi VIII. *Annals of the Royal Botanic Gardens, Peradeniya* 10: 161–180.

Petch, T. 1928. Tropical root disease fungi. *Transactions of the British Mycological Society* 13: 238253.

Petch, T. 1945. Ceylon fungi, old and new. *Transactions of the British Mycological Society* 27: 137–147.

Petch, T. & G.R. Bisby. 1950. *The Fungi of Ceylon*. Peradeniya Manual 6, Government Pub. Bureau, Colombo, Sri Lanka, pp. 1–111.

Rajapakse, J.C. 2014. Status of edible and medicinal mushroom research in Sri Lanka. In: Singh, M. (Eds.), *Proceedings of the 8th International Conference on Mushroom Biology and Mushroom Prodcuts*. Volume I & II, ICAR-Directorate of Mushroom Research, Solan, India, pp. 417–421.

Rajapakse, J.C., A. Karunatilaka, C. Nanayakkara & N. Punyasiri. 2018. Anti-oxidant attributes of four cultivated mushrooms in Sri Lanka. *International Conference on Nutrition, Health and Aging, September 26–27*, Frankfurt, Germany.

Rajapakse, J.C., P. Rubasingha & N.N. Dissanayake. 2007. The potential of using cost-effective compost mixtures for oyster mushroom (*Pleurotus* spp) cultivation in Sri Lanka. *Tropical Agricultural Research & Extension* 10: 29–32.

Rajapakse, P. 2011. New cultivation technology for paddy straw mushroom (*Volvariella volvacea*). *Proceedings of the 7th International Conference on Mushroom Biology and Mushroom Products, October 4–7, Arcachon, France* 1: 446–451.

Ranawake, A.L. 2021. The underutilized resources in the lowland wet zone forests of Sri Lanka and untapped Indigenous knowledge of peripheral households. *Journal of the University of Ruhuna* 9(2): 47–71.

Rupasinghe, K. & K. Nandasena. 2006. Potential exploitation of refuse tea as an alternative medium in mushroom cultivation. *Tropical Agricultural Research* 18: 399–404.

Silva, S.S.D. 1988. *Reservoirs of Sri Lanka and their fisheries*. Issue 298, Food and Agriculture Organization of the United Nations, Rome, pp. 1–128.

Singh, M.P., A.K. Srivastava, S.K. Vishwakarma, V.K. Singh & V.K. Pandey. 2009. Mushroom biotechnology. In: Singh, M.P., Agrawal, A. & Sharma, B. (Eds.), *Recent Trends in Biotechnology*. Volume 1, Nova Science Publishers Inc., New York, pp. 78–85.

Thanuejah, S., S. Karunakaran & V. Arulnandhy. 2013. Potential for oyster mushroom cultivation in Sri Lanka: A review. *2nd International symposium on minor fruits and medicinal plants for better life December 20, Kamburupitiya, Matara, Sri Lanka*, Faculty of Agriculture, University of Ruhuna, pp. 198–205.

Tipbromma, S., K.D. Hyde, R. Jeewon, S.S.N. Maharachchikumbura, J.K. Liu, D.J. Bhat, E.B.G. Jones, E.H.C. McKenzie, E. Camporesi, T.S. Bulgakov, M. Doilom, A.L. de Azevedo Santiago, K. Das, P. Manimohan, T.B. Gibertoni, Y.W. Lim, A.H. Ekanayaka, B. Thongbai & H.B. Lee et al. 2017. Fungal diversity notes 2017, 491–602: Taxonomic and phylogenetic contributions to fungal taxa. *Fungal Diversity* 83(1): 1–261. https://doi.org/10.1007/s13225-017-0378-0.

Udugama, S. & K. Wickramaratna. 1991. *Artificial Production of Naturally Occurring Lentinus Giganteus (Uru Paha), a Sri Lankan Edible Mushroom.* Horticultural Crop Research & Development Institute (HORDI), Gannoruwa, Peradeniya.

Wijesundera, R.L.C. 1986. Floristic studies and distribution of fungi in Sri Lanka. *Proceedings/Indian Academy of Sciences* 96(5): 375–377.

Wijnands, D.O., J. Heniger, J.F. Veldkamp, N. Fumeaux & M.W. Callmander. 2017. The botanical legacy of Martinus Houttuyn (1720–1798) in Geneva. *Candollea* 72(1): 155–198. http://doi.org/10.15553/c2017v721a11.

14 Bioactive Profile of *Xylaria spp.*—A Comparative Account

Tahir Syed, Kodandoor Sharthchandra, Gabbalkaje Harish Jithesh, Gopalappa Abhisheka, Shivannegowda Mahadevakumar, Kandikere Ramaiah Sridhar

14.1 INTRODUCTION

The cosmopolitan genus *Xylaria spp.* (Sordariomycetes; Xylariomycetidae, Xylariales, Xylariaceae) has been considered one of the largest among the ascomycetes (Li et al. 2017). It has been predicted that up to 300 species are distributed worldwide (Kirk et al. 2008). Carbonaceous stromata (upright straight or slender or globoid or irregular) with variations in shape, size, ornamentation, and pigmentation with apical apparatus in asci are considered the distinguishing features of *Xylaria spp.* (Roger 1979; Trierveiler-Pereira et al. 2009; Whalley 1996; Maharachchikumbura et al. 2016; Fournier et al. 2018). *Xylaria spp.* is found on a wide range of substrates including soil, leaf litter, woody litter, bark, tree trunks, seeds, fruits, dung, and termite mounds (Bayman et al. 1998; Davis et al. 2003; U'ren et al. 2009; Hsieh et al. 2010; Linnakoski et al. 2012; Rajulu et al. 2013; Jin et al. 2017). According to Rogers et al. (2005) and Ju and Hsieh (2007), nearly 20 species of *Xylaria spp.* are associated with termite nests. In addition, several *Xylaria spp.* are also found as endophytes on aroids, bromeliads, liverworts, orchids, palms, and various species of trees.

Xylaria spp. produces several valuable natural products and metabolites of medicinal and industrial significance (e.g., alkaloids, antioxidants, anticancer agents, cytochalasins, cytotoxins, enzymes, pigments, xyloketales, and other metabolites) (Whalley 1996; Stadler 2011; Rajulu et al. 2013; Ahmed and Jahan 2017; Divate et al. 2017; Abdel-Azeem et al. 2021; Deshmukh et al. 2021, 2022; Jagadish et al. 2021). They are known for their broad-spectrum antimicrobial activities (Thirumalesh et al. 2014; Ramesh et al. 2015; Elias et al. 2018; Devi et al. 2020). Many species of *Xylaria spp.* have been studied recently (Ma et al. 2013; Rebbapragada and Kalyanaraman 2016; Fernando et al. 2016; Divate et al. 2017; Zhou et al. 2018; Jagadish et al. 2021).

According to Debnath et al. (2018), *Xylaria spp.* is known all over the Indian subcontinent. Sixty species are known from southern India, representing about 10% of the globally known species (Mohanan 2011; Latha et al. 2015; Karun and Sridhar 2015; Patel and Krishnappa 2017; Dattaraj et al. 2020). As a result of a survey of different ecosystems of southwestern India, at least nine species of *Xylaria spp.* were recorded (Karun and Sridhar 2015). Studies in southwestern India have revealed the occurrence of 25 species of *Xylaria spp.*, with the highest number of species in Karnataka (17 species), followed by Maharashtra (13 species), Kerala (8 species), and Tamil Nadu (1 species). The present chapter is focused on the biochemical composition and antioxidant potential of *Xylaria arbuscula*, *X. grammica*, *X. minuta*, and *X. nigripes* occurring in woody litter, *Eucalyptus globulus* bark, woody litter, and termite mounds, respectively, across the scrub jungle expanses of the southwest region of India.

DOI: 10.1201/b23190-14

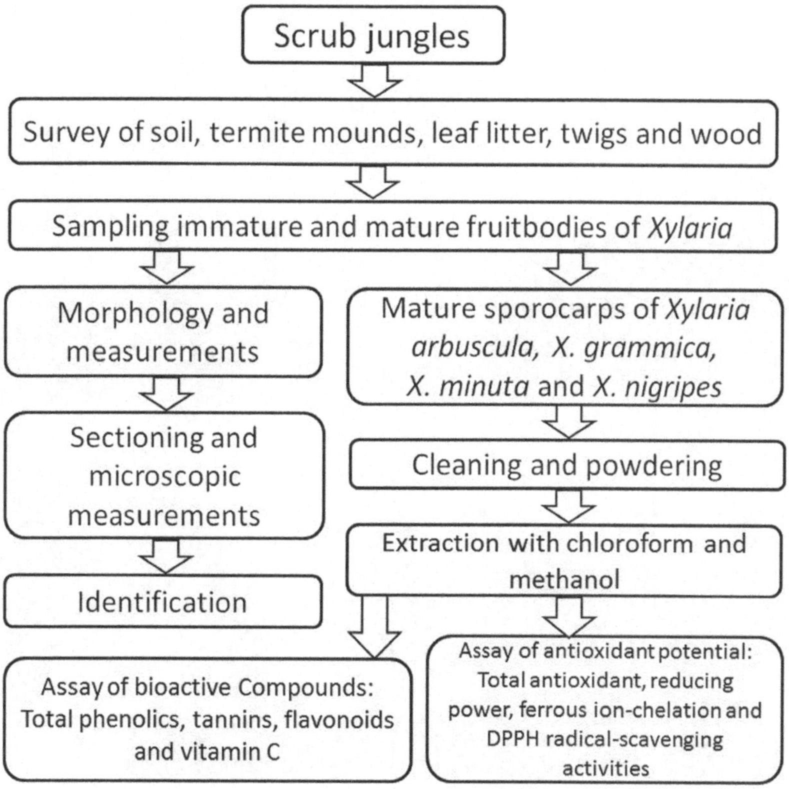

FIGURE 14.1 Schematic representation of the assessment of the biochemical and antioxidant potential of four *Xylaria spp.* species.

14.2 *XYLARIA SPP.* SPECIES AND PROCESSING

The scrub jungles of the southwest coast of India are the hub for species of *Xylaria spp.* that occur in different ecological niches (e.g., soil, humus, termite mounds, leaf litter, twigs, stubs, logs, and bark). During a survey of macrofungi on the Mangalore University Campus, located 5 km from the Arabian Sea beaches and adjacent to Mangalore City (Karnataka) (12°48'N, 74°55'E; 104–112 m above sea level), specimens of *Xylaria spp.* were collected. Four species of *Xylaria spp.* grown on different substrates were chosen for the study reported herein (Figure 14.1). Mature sporocarps of *Xylaria arbuscula were* found growing on the dead bark of *Pongamia pinnata*; those of *X. grammica* were growing on the bark of *Eucalyptus tereticornis,* those of *X. minuta* were growing on decaying unidentified fallen twigs, and those of *X. nigripes* were associated with termite mounds.

Mature sporocarps of four **species** *of Xylaria spp.* collected in triplicate from the scrub jungles of southwestern India were cleaned and dried at 55±2°C until they attained a constant weight. The samples were reduced to a powder using a mortar and pestle. Powdered replicate samples were extracted with chloroform and methanol on a shaker at room temperature overnight or up to 24 hr, air-dried, and then refrigerated for biochemical composition and antioxidant potential analyses. Assessment of the species of *Xylaria spp.* are schematically represented in Figure 14.2.

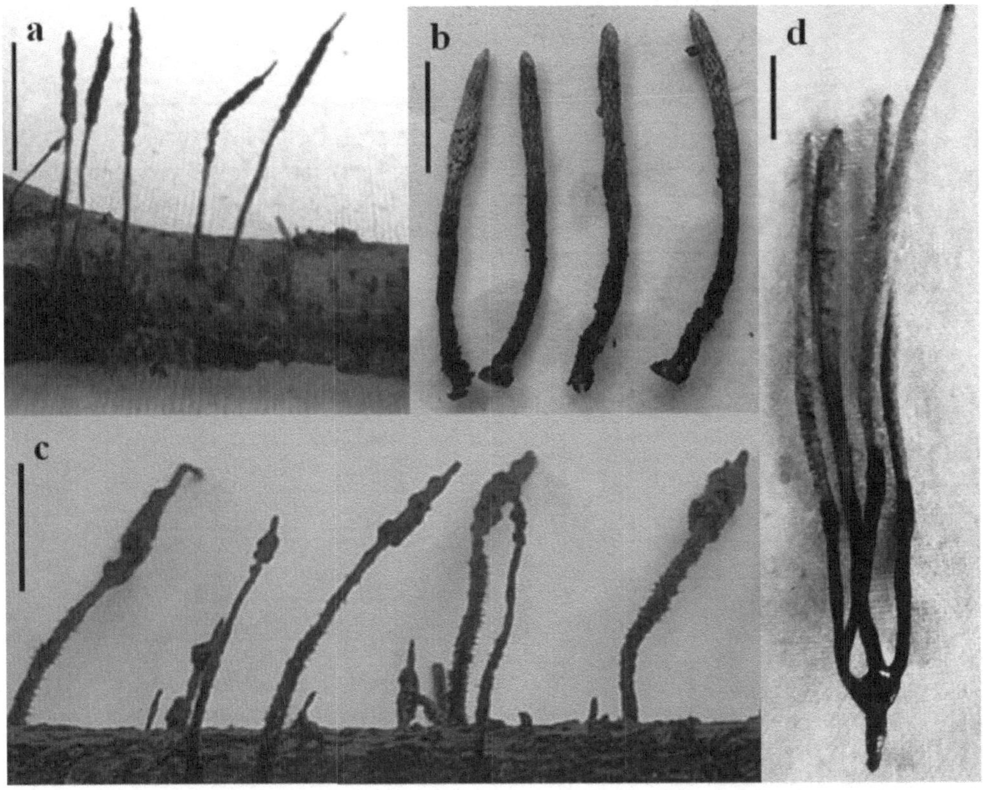

FIGURE 14.2 **Species of** *Xylaria spp.* studied for biochemical composition and antioxidant potential. (a) *Xylaria arbuscula* growing on *Pongamia pinnata* woody litter, (b) *Xylaria grammica* growing on *Eucalyptus tereticornis* bark, (c) *Xylaria minuta* growing on unknown woody litter, and (d) *Xylaria nigripes* associated with termite mounds (Scale bar: 10 mm).

14.3 BIOCHEMICAL ANALYSES

Powdered chloroform and methanol extracts of four species of *Xylaria spp.* were assessed for total content of phenolics, tannins, flavonoids, and vitamin C.

14.3.1 Total Phenolics

The total phenolics of sporocarp powder were evaluated according to Rosset et al. (1982). Sporocarp extracts were re-extracted with 50% aqueous methanol (5 ml) and distilled water (5 ml) in a water bath (95±1°C; up to 10 min) and centrifuged (101×*g*) to draw the supernatant. The rest of the material was extracted again. The extracts were pooled and distilled water was added to ensure 10 ml volumes. Aliquots (0.1 ml) (up to 1 ml with distilled water) were mixed with 2% Na_2CO_3 (in 0.1 N NaOH) (5 ml). After incubation (10 min) at laboratory temperature, Folin-Ciocalteau's reagent (1:2 v/v; 0.5 ml) was mixed in and the absorbance was measured (725 nm). Tannic acid served as a standard to define total phenolics as mg tannic acid equivalents (TAEs/g dry mass).

14.3.2 Tannins

To determine tannin levels in sporocarp extracts, the vanillin-HCl method was adapted (Burns 1971). Fifty milligrams of extracts were mixed with 5 ml of vanillin hydrochloride reagent (vanillin in methanol, 4%; concentrated HCl; ratio 1:1) and incubated at laboratory temperature (20 min). The absorbance was measured at 500 nm with a catechin standard (HPLC grade, 98%; Sigma Aldrich) to define results in mg catechin equivalents of tannin (mg CEs/g dry mass).

14.3.3 Flavonoids

The method by Chang et al. (2002) was adapted to find out the content of flavonoids in sporocarp powder. Aliquots of 0.5 ml of the extracts (1 mg/ml) were blended with 1.5 ml of methanol, 0.1 ml of 10% $AlCl_3$, 0.1 ml of 1 M KCH_3CO_2, and 2.8 ml of distilled water, and absorbance was measured at 415 nm after incubation at the laboratory (up to 30 min). Quercetin served as the standard to define flavonoid levels in mg quercetin equivalents (mg QEs/g dry mass).

14.3.4 Vitamin C

The quantity of vitamin C in sporocarp extracts was assessed according to Roe (1954) with a minor alteration. The flour (1 g) was extracted by 25 ml of trichloro acetic acid (TCA) (5%), an aliquot of extract (0.2 ml) was made up to 1 ml by 5% TCA, and 1 ml of chromogen reagent (0.6% copper sulfate, 5 ml; 5% thiourea, 5 ml; and 2.2% 2,4-dinitrophenylehydrazine, 90 ml) was added. After boiling for 10 minutes, samples were cooled to laboratory temperature and 65% H_2SO_4 (4 ml) was added, then samples were incubated at laboratory temperature for 30 minutes. The absorbance of samples with the standard was read at 540 nm. Ascorbic acid served as a standard. The quantity of vitamin C was defined in mg ascorbic acid equivalents (mg AAEs/g dry mass).

14.4 ANTIOXIDANT ASSAYS

Powdered chloroform and methanol extracts were assessed for antioxidant potential such as total antioxidant activity (TAA), reducing power (RP), ferrous ion-chelation capacity (FCC), and DPPH radical-scavenging activity (DPPH RSA).

14.4.1 Total Antioxidant Activity

Evaluation of the TAA of sporocarp extracts was carried out according to the procedure described by Prieto et al. (1999). The sporocarp flour extracts were mixed with sodium phosphate (28 mM) and ammonium molybdate (4 mM in 0.6 M sulfuric acid), incubated (95°C) up to 90 min, and cooled to attain the laboratory temperature; the absorbance was then assessed at 695 nm. Phosphomolybdenum complex served as a control, with methanol as a blank. The TAA was expressed as µM ascorbic equivalents (µM AAEs/g dry mass).

14.4.2 Reducing Power

The RP of sporocarp extracts was assessed using the technique described by Oyaizu (1986) with a minor alteration. Various concentrations (200–1,000 µg) were extracted using methanol and mixed with 2.5 ml of 0.2 M phosphate buffer (pH, 6.6). Then, 2.5 ml of potassium ferricyanide (1%) was added and samples were incubated after mixing (50°C) (up to 20 min). Later, 2.5 ml of trichloroacetic acid (10% w/v) was added, mixed and centrifuged (101 × g) (up to 10 min). The supernatant (2.5 ml) was blended with distilled water (2.5 ml). To this mixture, ferric chloride (0.1%; 5 ml) was added and the absorbance was assessed (700 nm). An increase in the absorbance of the mixture revealed an increase in RP.

14.4.3 Ferrous Ion-Chelation Capacity

The FCC of extracts of the sporocarp powder was evaluated using the procedure described by Hsu et al. (2003). To 1 ml of methanol extract, 2 mM $FeCl_2$ (0.1 ml) and 5 mM ferrozine (0.2 ml) were added, and the volume was increased to 5 ml with methanol. Samples were incubated at laboratory temperature for up to 10 min to measure the absorbance of a complex of Fe^{2+}-ferrozine at 562 nm. The reagents without sample served as a control for assessing the FCC.

$$\text{Ferrous ion-chelating capacity (\%)} = 1 - (A_{s562} / A_{c562}) \times 100$$

(where absorbance of the control, A_c; absorbance of sample, A_s).

14.4.4 DPPH Radical-Scavenging Activity

The DPPH RSA of the sporocarp extracts was evaluated using the procedure by Singh et al. (2002). A volume of 0.01 mM 2,2-diphenyl-1-picrylhydrazyl (DPPH) (4 ml) was added to 1 ml of the extracts, mixed, and allowed to react for up to 20 minutes at laboratory temperature. The reagents without the sample were used as a control, and the absorbance at 517 nm was measured to determine the DPPH RSA.

$$\text{DPPH free radical-scavenging activity (\%)} = [(A_{c517} - A_{s517})/(A_{c517})] \times 100$$

(where absorbance of the control, A_c; absorbance of sample, A_s).

14.5 DATA ANALYSIS

The differences in quantities of bioactive compounds and antioxidant activities between chloroform and methanol extracts of sporocarps of four species of *Xylaria spp.* were evaluated by Student's *t*-test using the software Statistica version #8.0 (StatSoft 2008).

14.6 DISCUSSION

14.6.1 Biochemical Composition

Among the four biochemical components in the four species of *Xylaria spp.*, the amount of total phenolics was highest, followed by flavonoids, tannins, and vitamin C (Figure 14.3). Total phenolics of all four species of *Xylaria spp.* species were higher in methanol than in chloroform extract, with the highest quantity in *X. nigripes* ($p<0.001$), followed by *X. arbuscula* ($p<0.001$), *X. minuta* ($p<0.001$), and *X. grammica* ($p<0.01$). The tannin content was higher in methanol than in the chloroform extract, *with the values in* descending order found for *X. arbuscula* ($p<0.001$), *X. minuta* ($p<0.01$), and *X. grammica* ($p<0.01$). Although the content of tannin was the highest in the methanol extract of *X. nigripes*, it was not significantly higher than in the chloroform extract ($p>0.05$). The flavonoid and vitamin C contents of all species of *Xylaria spp.* were higher in methanol than in chloroform extract ($p<0.001$). The relative abundance of both components in the methanol extract, in descending order, were *X. nigripes, X. minuta, X. grammica* and *X. arbuscula*.

The chloroform and methanol extracts of *X. nigripes* and *X. minuta* were found to possess lower quantities of total phenolics and flavonoids than the polar and non-polar solvent extracts of *X. nigripes* reported in Taiwan (Divate et al. 2017). Bioactive compounds from natural sources of health remedies will be preferable to synthetic compounds. *Traditional healers use Xylaria spp. to treat a variety of ailments (e.g., insomnia, trauma, and stomach ailments)* (Shortt 1867; Balakrishnan and Kumar 2001; Lin et al. 2013; Latha et al. 2015; Deshmukh et al. 2021). Similarly, these *Xylaria spp.* species are known for their antioxidant, immunomodulatory, anti-inflammatory, hepatoprotective, anti-tumor, and neuroprotective activities (Ma et al. 2013; Divate et al. 2017).

A qualitative assessment of 13 bioactive components present in chloroform and methanol extracts of *X. arbuscula* and *X. grammica* was carried out by Jithesh (2021). The total number of compounds

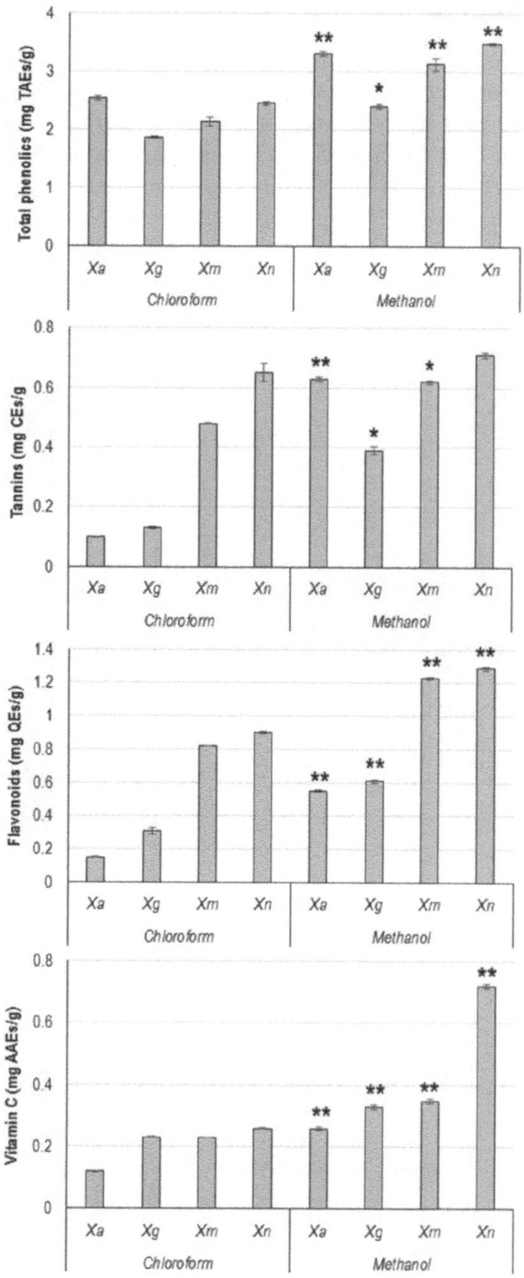

FIGURE 14.3 Total phenolics, tannins, flavonoids, and vitamin C in sporocarps of *Xylaria arbuscula, X. grammica, X. minuta,* and *X. nigripes.*

was higher in methanol than in chloroform extract. Alkaloids, flavonoids, phenols, and tannins were present in chloroform and methanol extracts of both species of *Xylaria spp.* Coumarins were confined to both extracts of *X. arbuscula*, while cardiac glycosides were confined to the methanol extracts of *X. arbuscula* and *X. grammica*. The seven remaining compound types were not detectable in *X. arbuscula* or *X. grammica* (carotenoids, glycosides, phycocyanins, quinone, quinone glycosides, saponins, steroids, and terpenoids).

14.6.2 Antioxidant Potential

TAA of all mushrooms was higher in methanol than in chloroform extract. In methanol extract, the TAA of *X. nigripes* was significantly high ($p<0.001$) followed by *X. grammica* ($p<0.01$) (Figure 14.4). It was also higher in *X. minuta* methanol extract than in chloroform extract ($p>0.05$). The TAA of the four species of *Xylaria spp.* in this study ranged from 20–40 μM AAEs/g in methanol extract, which is low compared to that reported for other species of *Xylaria spp.* (*X. escharoidea*, *X. hypoxylon*, *X. longipes*, and *X. polymorpha*) obtained from the scrub

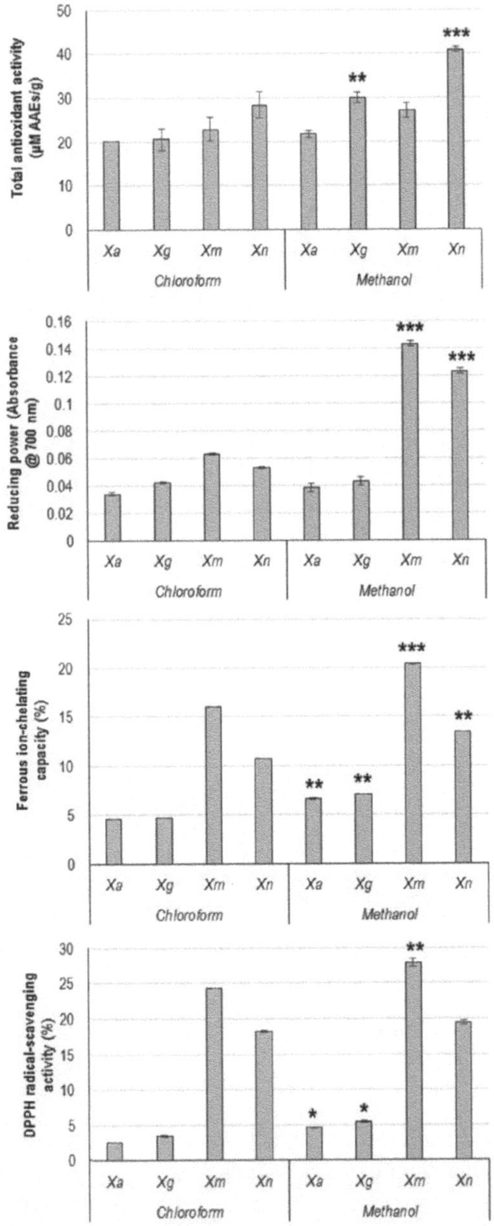

FIGURE 14.4 Total antioxidant activity, reducing power, ferrous ion-chelation capacity, and DPPH radical-scavenging activity of sporocarps of *Xylaria arbuscula*, *X. grammica*, *X. minuta*, and *X. nigripes*.

jungles of southwestern India (Jagadish et al. 2021). The TAA of polar and non-polar solvents of *X. nigripes* and *X. minuta* was lower than that of the aqueous extracts of other species of *Xylaria spp.* (*X. escharoidea*, *X. hypoxylon*, *X. longipes*, and *X. polymropha*) for which comparable data exist (Jagadish 2021).

The RP was higher in the methanol extract of *X. minuta* ($p<0.001$) than in that of *X. nigripes* ($p<0.001$). The FCC was higher in methanol than in chloroform extract, with the highest value seen in *X. minuta* ($p<0.001$), followed by *X. nigripes* ($p<0.01$), *X. grammica* ($p<0.01$), and *X. arbuscula* ($p<0.01$). As seen for TAA, the FCC was lower than the other three species of *Xylaria spp.* (*X. escharoidea, X. hypoxylon,* and *X. polymorpha*), while the FCC of *X. minuta* was comparable with that of *X. longipes* (Jagadish et al. 2021). The FCC of *Xylaria* spp. studied in chloroform and methanol was comparable to the aqueous extracts of other *Xylaria* spp. (*X. escharoidea*, *X. hypoxylon,* and *X. polymropha*), while it was higher than *X. longipes.*

DPPH RSA was the highest in the methanol extract of X. minuta ($p<0.01$), followed by that of *X. grammica* ($p<0.05$) and *X. arbusucla* ($p<0.05$). The DPPH RSA of *X. minuta* and *X. nigripes* methanol extracts were comparable to the reported values for other species of *Xylaria spp.*, such as *X. hypoxylon*, *X. longipes* and *X. polymorpha* (Jagadish et al. 2021). Although the reducing power of *X. minuta* and *X. nigripes* was less than that of *X. nigripes* from Taiwan, the DPPH RSA was higher (Divate et al. 2017). Similarly, the DPPH RSA of *X. nigripes* and *X. minuta* was higher than that of exopolysaccharides isolated from *X. nigripes* in China (Ma et al. 2013).

14.7 CONCLUSIONS

Four species of *Xylaria spp.* (*X. arbuscula, X. grammica, X. minuta,* and *X. nigripes*) growing in the scrub jungles of southwest India possess high amounts of total phenolics, flavonoids, tannins, and vitamin C. Most of the biochemical compounds and antioxidant potential (DPPH RSA, FCC, RP, and TAA) parameters were significantly higher in methanol than in chloroform extracts. Earlier studies offering qualitative assessment of other biochemical components found that levels were also higher in methanol than in chloroform extract. Owing to its bioactive components and antioxidant potential, *Xylaria spp.* is known in traditional medicine to treat many ailments (e.g., insomnia, trauma, and stomach ailments). Species of *Xylaria spp.* are well known for their immunomodulatory, anti-inflammatory, hepatoprotective, anti-tumor, and neuroprotective activities.

The scrub jungles of southwestern India, with their unique climate, geography, flora, and fauna, support many species of *Xylaria spp.,* which are responsible for recycling lignocellulosic plant remains. Human interference such as deforestation and the extraction of laterite, granite, and wood are the major causes of the depletion of *Xylaria spp.* Many termites mound-dependent *Xylaria spp.* are also threatened owing to the suppression of termite mounds or the use of pesticides. There is an urgent need to conserve and restore the scrub jungles of southwestern India in view of their economically valued macrofungi, including *Xylaria spp.,* and to harness the benefits of recycling lignocellulosic detritus, value for human nutrition, and application in human health management.

14.8 ACKNOWLEDGEMENTS

The authors are indebted to the Department of Biosciences, Mangalore University, for extending their laboratory facilities to carry out studies on *Xylaria spp.* growing in the scrub jungles.

REFERENCES

Abdel-Azeem, A.M., Yadav, A.N., Yadav, N. & Usmani, Z. 2021. *Industrially Important Fungi for Sustainable Development. Volume 1: Biodiversity and Ecological Perspective*. Springer, Cham, p. 390.

Ahmed, F.A. & Jahan, N. 2017. Optimization of culture conditions, phytoconstituents and cytotoxicity of *Xylaria hypoxylon* (L.) Grev. *Jahangirnagar University Journal of Biological Science* 6: 1–9.

Balakrishnan, V. & Kumar, N.A. 2001. 'Nilamanga' (*Sclerotium stipitatum*?)—A rare termite fungal sclerotia with medicinal properties known among the tribal and rural communities of Kerala. *Ethnobotany* 13: 9–14.

Bayman, P., Angulo-Sandoval, P., Báez-Ortiz, Z. & Lodge, D.J. 1998. Distribution and dispersal of *Xylaria* endophytes in two tree species in Puerto Rico. *Mycological Research* 102: 944–948.

Burns, R. 1971. Methods for estimation of tannins in grain sorghum. *Agronomy Journal* 63: 511–512.

Chang, C., Yang, M., Wen, H. & Chern, J. 2002. Estimation of total flavonoid content in propolis by two complementary colorimetric methods. *Journal of Food Drug Analysis* 10: 178–182

Dattaraj, H.R., Sridhar, K.R. & Jagadish, B.R. 2020. Diversity and bioprospect significance of macrofungi in the scrub jungles of southwest India. In: Öztürk, M., Egamberdieva, D. & Pešić, M. (Eds.), *Biodiversity and Biomedicine*. Elsevier. Academic Press, London, pp. 235–246.

Davis, E.C., Franklin, J.B., Shaw, A.J. & Vilgalys, R. 2003. Endophytic *Xylaria* (Xylariaceae) among liverworts and angiosperms: Phylogenetics, distribution and symbiosis. *American Journal of Botany* 90: 1661–1667.

Debnath, S., Majumdar, K., Das, P. & Saha, A.K. 2018. New distribution record of five species of *Xylaria* from Tripura, Northeast India. *Research & Reviews: Journal of Life Sciences* 8: 1–10

Deshmukh, S.K., Sridhar, K.R. & Gupta, M.K. 2021. Application of selected species of the genus *Xylaria* in traditional medicine. In: Sridhar, K.R. & Dehmukh, S.K. (Eds.), *Advances in Macrofungi: Pharmaceuticals and Cosmeceuticals*. CRC Press, Boca Raton, pp. 122–136.

Deshmukh, S.K., Sridhar, K.R., Saxena, S. & Gupta, M.K. 2022. Recent advances in the discovery of bioactive metabolites from *Xylaria* Hill ex Schrank. In: Arya, A. & Rusevska, K. (Eds.), *Biology, Cultivation and Applications of Mushrooms*. Springer Nature Singapore Pte Ltd, Singapore, pp. 47–116.

Devi, R., Kaur, T., Guleria, G., Rana, K.L., Kour, D. et al. 2020. Fungal secondary metabolites and their biotechnological application for human health. In: Rastegari, A.A., Yadav, A.N. & Yadav, N. (Eds.), *Trends of Microbial Biotechnology for Sustainable Agriculture and Biomedicine Systems: Perspectives for Human Health*. Elsevier, Amsterdam, pp. 147–161.

Divate, R.D., Wang, P.-M., Wang, C.-C., Chou, S.-T., Chang, C.-T. et al. 2017. Protective effect of medical fungus *Xylaria nigripes* mycelia extract against hydrogen peroxide-induced apoptosis in PC12 cells, Int. *Journal of Immunology and Pharmacology* 30: 105–112.

Elias, L.M., Fortkamp, D., Sartori, S.B., Ferreira, M.C., Gomes, L.H. et al. 2018. The potential of compounds isolated from *Xylaria* spp. as antifungal agents against anthracnose. *Brazilian Journal of Microbiology* 49: 840–847.

Fernando, M.D.M., Wijesundera, R.L.C., Soysa, S.S.B.D.P., de Silva, E.D. & Nanayakkara, C.M. 2016. Antioxidant potential and content of the polyphenolic secondary metabolites of white rot macrofungi; *Flavodon flavus* (Klotzsch.) and *Xylaria feejeensis* (Berk.). *SDRP Journal of Plant Science* 1: 1–6.

Fournier, J., Lechat, C. & Courtecuisse, R. 2018. The genus *Xylaria sensu lato* (*Xylariaceae*) in Guadeloupe and Martinique (French West Indies) III. Taxa with slender upright stromata. *Ascomycete.org* 12: 81–164.

Hsieh, H.M., Lin, C.R., Fang, M.J., Rogers, J.D., Fournier, J. et al. 2010. Phylogenetic status of *Xylaria* subgenus *Pseudoxylaria* among taxa of the subfamily Xylarioideae (Xylariaceae) and phylogeny of the taxa involved in the subfamily. *Molecular Phylogenetics and Evolution* 54: 957–969.

Hsu, C.L., Chen, W., Weng, Y.M. & Tseng, C.Y. 2003. Chemical composition, physical properties and antioxidant activities of yam flours as affected by different drying methods. *Food Chemistry* 83: 85–92.

Jagadish, B.R., Sridhar, K.R., Dattaraj, H.R., Chandramohana, N. & Mahadevakumar, S. 2021. Bioactive attributes of *Xylaria* species from the scrub jungles of southwest India. In: Abdel-Azeem, A.M., Yadav, A.N., Yadav, N. & Sharma, M. (Eds.), *Industrially Important Fungi for Sustainable Development, Volume 2: Bioprospecting for Biomolecule*. Springer Nature, Switzerland, AG, pp. 717–735.

Jin, Z., Li, D., Liu, T., Yu, F., Zhang, Z. et al. 2017. Cultural endophytic fungi associated with *Dendrobium officinale*: Identification, diversity estimation and their antimicrobial potential. *Current Science* 112: 1690–1697.

Jithesh, G.H. 2021. *Studies on the Biochemical and Antioxidant Attributes of Xylaria of the West Coast of Karnataka*. M.Sc. Dissertation, Department of Biosciences, Mangalore University, Mangalore, Karnataka, India, p. 23.

Ju, Y.-M. & Hsieh, H.-M. 2007. *Xylaria* species associated with nests of *Odontotermes formosanus* in Taiwan. *Mycologia* 99: 936–957. http://doi.org/10.3852/mycologia.99.6.936

Karun, N.C. & Sridhar, K.R. 2015. *Xylaria* complex in the South Western India. *Plant Pathology & Quarantine* 5: 83–96.

Kirk, P.F., Cannon, P.F., Minter, D.W. & Stalpers, J.A. 2008. *Dictionary of the Fungi*, 10th Edition. CABI, Egham, p. 2600.

Latha, K.P.D., Veluthoor, S. & Manimohan, P. 2015. On the taxonomic identity of a fungal morph used in traditional medicine in Kerala State, India. *Phytotaxa* 201: 287–295.

Li, Q.R., Liu, L.L., Zhang, X., Shen, X.C. & Kang, J.C. 2017. *Xylaria spinulosa* sp. nov. and *X. atrosphaerica* from southern China. *Mycosphere* 8: 1070–1079.

Lin, Y., Wang, X.Y., Ye, R., Hu, W.-H., Jiao, H.J. et al. 2013. Efficacy and safety of Wuling capsule, a single herbal formula, in Chinese subjects with insomnia: A multicenter, randomized, double-blind, placebo-controlled trial. *Journal of Ethnopharmacology* 145: 320–327.

Linnakoski, R., Puhakka-Tarvainen, H. & Pappinen, A. 2012. Endophytic fungi isolated from *Khaya anthotheca* in Ghana. *Fungal Ecology* 5: 298–308.

Ma, Y.P., Mao, D.B., Geng, L.J., Zhang, W.Y., Wang, Z. & Xub, C.P. 2013. Production optimization, molecular characterization and biological activities of exopolysaccharides from *Xylaria nigripes*. *Chemical and Biochemical Engineering* 27: 177–184.

Maharachchikumbura, S.S.N., Hyde, K.D., Jones, E.B.G., McKenzie, E.H.C., Bhat, D.J. et al. 2016. Families of sordariomycetes. *Fungal Diversity* 79: 1–317.

Mohanan, C. 2011. *Macrofungi of Kerala*. Handbook # 27. Kerala Forest Research Institute, Peechi, Kerala, India, p. 597.

Oyaizu, M. 1986. Studies on products of browning reactions: Antioxidative activities of products of browning reaction prepared from glucosamine. *Japanese Journal of Nutrition* 44: 307–315.

Patel, K.J.N. & Krishnappa, M. 2017. Diversity of Xylariaceae members in Sagara Taluk, Karnataka, India. *Journal of Mycology and Plant Pathology* 47: 447–452.

Prieto, P., Pineda, M. & Aguilar, M. 1999. Spectrophotometric quantitation of antioxidant capacity through the formation of a phosphomolybdenum complex: Specific application to the determination of vitamin E. *Analytical Biochemistry* 269: 337–341.

Rajulu, M.B.G., Thirunavukkarasu, N., Babu, A.G., Aggrawal, A., Suryanarayanan, T.S. & Reddy, M.S. 2013. Endophytic Xylariaceae from the forests of Western Ghats, Southern India: Distribution and biological activities. *Mycology* 4: 29–37.

Ramesh, V., Santosh, K., Pavunraj, M., Karunakaran, C. & Rajendran, A. 2015. *In vitro* antifungal and anticancer potential of *Xylaria curta* fruiting body fractions against human fungal pathogen and cancer cell lines. *Current Research in Environmental & Applied Mycology* 5: 20–26.

Rebbapragada, D. & Kalyanaraman, R. 2016. Evaluation and optimization of antioxidant potentiality of *Xylaria feejeensis* hmjau22039. *Asian Journal of Pharmaceutical and Clinical Research* 9: 269–273.

Roe, J.H. 1954. Chemical determination of ascorbic, dehydroascorbic and diketogluconic acids. In: Glick, D. (Ed.). *Methods of Biochemical Analysis*. Volume 1, InterScience Publishers, New York, pp. 115–139.

Roger, J.D. 1979. The Xylariaceae: Systematic, biological and evolutionary aspects. *Mycologia* 71: 1–41.

Rogers, J.D., Ju, Y.-M. & Lehmann, J. 2005. Some *Xylaria* species on termite nests. *Mycologia* 97: 914–923. http://doi.org/10.3852/ mycologia.97.4.914

Rosset, J., Bärlocher, F. & Oertli, J.J. 1982. Decomposition of conifer needles and deciduous leaves in two black Forest and two Swiss Jura streams. *International Review Gesamten Hydrobiologie* 67: 695–711.

Shortt, J. 1867. An account of *Sclerotium stipitatum* (Berk. et Curr.) of Southern India. *The Journal of the Linnean Society (Botany)* 9: 417–419. http://doi.org/10.1111/j.1095-8339.1867.tb01305.x

Singh, R.P., Murthy, C.K.N. & Jayaprakasha, G.K. 2002. Studies on antioxidant activity of pomegranate (*Punica granatum*) peel and seed extracts using in vitro methods. *Journal of Agricultural and Food Chemistry* 50: 81–86.

Stadler, M. 2011. Importance of secondary metabolites in the Xylariaceae as parameters for assessment of their taxonomy, phylogeny, and functional biodiversity. *Current Research in Environmental & Applied Mycology* 1: 75–133.

StatSoft. 2008. *Statistica Version # 8*. StatSoft, Tulsa.

Thirumalesh, B., Thippeswamy, B. & Krishnappa, M. 2014. Antibacterial activity of *Xylaria* species *in vitro* against *Xanthomonas campestris* PV. *mangiferaeindicae* isolated from bacterial black spot of mango fruit. *International Journal of Life Science & Pharma Research* 3: 125–130.

Trierveiler-Pereira, L., Romero, A.I., Baltazar, J.M. & Loguercio-leite, C. 2009. Addition to the knowledge of *Xylaria* (Xylariaceae, Ascomycota) in Santa Catarina, Southern Brazil. *Mycotaxon* 107: 139–156.

U'ren, J.M., Dalling, J.W., Gallery, R.E., Maddison, D.R., Davis, E.C. et al. 2009. Diversity and evolutionary origins of fungi associated with seeds of a neotropical pioneer tree: A case study for analysing fungal environmental samples. *Mycological Research* 113: 432–449.

Whalley, A.J.S. 1996. The xylariaceous way of life. *Mycological Research* 100: 897–922.

Zhou, J., Diao, X., Want, T., Chen, G., Lin, Q. et al. 2018. Phylogenetic diversity and antioxidant activities of culturable fungal endophytes associated with the mangrove species *Rhizophora stylosa* and *R. mucronata* in the South China Sea. *PLoS ONE* 13: e0197359. http://doi.org/10.1371/journal.pone.0197359

15 Some Uncommon Traditionally and Medicinally Important Wild Macrofungi of the Garhwal Himalayas, Uttarakhand, India

Vipin Parkash, Ranjna Kaundal, Supriti Paul, Anuj Chauhan, Ashok Aggarwal

15.1 INTRODUCTION

The Garhwal Himalayas of Uttarakhand, located in the western fraction of the central Himalayas, represent one of the most biodiversity-rich regions of India. This region is blessed with a large forest cover of about 34,650 km^2 (Botanical Survey of India report) that expands over an elevation range of 250 to 3500 m above sea level. This large forest area and range in elevation are responsible for the characteristic temperature, precipitation, relative humidity, and other favourable conditions of the region, permitted a rich diversity of higher flora and an associated high-level diversity of macrofungi (Joshi et al. 2012). The forests are dominated by tree genera such as *Cedrus*, *Quercus*, *Rhododendron*, *Pinus*, *Shorea*, *Cupressus*, *Abies*, *Picea*, *Ulmus*, *Juglans*, *Acer*, *Buxus*, and *Betula* (Vishwakarma et al. 2011). The macrofungi with fleshy-to-leathery, umbrella-to-shelf-like fruiting bodies fall under the categories of gilled mushrooms, bracket mushrooms, puffballs, coral mushrooms, jelly mushrooms, stinkhorns, and bird's nest mushrooms (Bates 2006). Mushrooms mainly belong to the Basidiomycetes, but some members of Ascomycetes also produce mushroom-like fruiting bodies (Atkinson 1961). Macrofungi are essential components of forest ecosystems and help to maintain biodiversity and ecosystem function by influencing carbon and nutrient cycling (Stametes 2000).

In the 21st century, macrofungi are known to have great ecological value, diversity, and economic importance. Macrofungi such as mushrooms are forest products that are considered vital components of food safety and security. They are low in caloric value but high in vitamin, mineral, and protein contents and thus are considered highly desirable foods (Singh et al. 2017; Bhatt et al. 2018). A great number of macrofungal species are not only edible and nutritious but also have therapeutic values, as they contain different secondary metabolites with antioxidant, anti-mutagenic, anticancer, and antimicrobial properties (Barros et al. 2007). Today, different compounds are extracted from these natural resources and suggested for use in inhibiting bacterial growth and further medical applications (Kothiyal and Singh 2022). The present chapter deals with some uncommon, medicinally important macrofungi of the Garhwal Himalayas in the Uttarakhand state of India.

15.2 TRUFFLES

15.2.1 Habitat

Truffles are hypogeal fungi that grow in almost all continents; however, most of the edible varieties have been identified in various parts of Europe (particularly in Spain, France, Italy, England, and

DOI: 10.1201/b23190-15

Croatia), Australia, New Zealand, North America (British Columbia and the Pacific Northwest) (Berch and Bonito 2016), Asia (Middle East and China), and Africa (Ferdman et al. 2005). Truffles are ectomycorrhizal and occur in nature at a depth of 4 to 60 cm, colonizing the roots of different deciduous trees such as oaks, pine, hazel, beech, birch, cherry, and other forest tree species.

In Uttarakhand, these macrofungi are found in the Jaunsar region of Chakrata, Dehradun, Uttarakhand. These mushrooms are found in humid forests, especially near sal forests/trees (*Shorea robusta* Gaertn. f.) in and around the time of the monsoon during the months of July and August, or the month of *Shravan* (as per the Hindu calendar). An interesting fact is that these mushrooms are not cultivated for commercial purposes. Truffles locally known as 'Jhanda' in the Jaunsar region of Uttarakhand and are used as a vegetable in traditional cuisine. Truffles are collected by digging them out from cracks or humps in forest soil.

15.2.2 Morphology

Truffles are found with various textures (smooth, wrinkled, bruised, and reticulate) and colour (black, brown, and white). Truffles are generally potato-like or lobed, which justifies their popular generic name, 'tuber.' Truffles are aromatic, slightly garlicky, pungent, and earthy in nature (see Figure 15.1).

15.2.3 Traditional and Medicinal Importance

Truffles are extremely expensive mushrooms and are highly prized because of their cooking value; they are considered 'the diamond of the kitchen' and 'black diamond of the earth.' They are rich in dietary fibre, protein, fat, and essential amino acids (methionine, threonine, isoleucine, phenylalanine, valine, and serine). Truffles are rich in aroma, as they contain different volatile organic compounds such as alcohol aldehydes, ascorbic acid, and ketones in various ratios (Federico et al. 2015). Studies have demonstrated that some truffles have the most widespread fungal sterol, ergosteroids, which can be transformed into vitamin D (Dogan and Aydın 2013).

Truffles have antimicrobial activity and have been shown to inhibit the growth of *Staphylococcus aureus* by 70–100% (Gouzi et al. 2011) and of *Pseudomonas aeruginosa* by 40% (Janakat et al. 2005). Truffles show antimicrobial activity against both Gram-positive and Gram-negative bacteria, and also against *Candida albicans* (Dogan and Aydın 2013). Truffles have been reported to have a high content

FIGURE 15.1 Truffles extracted from soil.

of phenolic compounds and antioxidants, as well as anti-inflammation activities (Chen et al. 2016; El Enshasy et al. 2013). Truffles have not been thoroughly studied for anticancer and immunomodulatory properties; however, they are rich in polysaccharides that have been reported to show immunomodulatory and anticancer properties (Patel and Goyal 2017). The polysaccharides derived from truffles through fermentation showed *in vitro* anticancer and anti-tumor activities on various cancer cell lines, such as HL-60 (Human promyelocytic leukemia), HepG2 (human hepatocellular carcinoma), SK-BR3 (human breast cancer), A549 (human lung adenocarcinoma), and HCT-116 (human colon carcinoma) cells (Zhao et al. 2014). Current unhealthy-food-dominated and stress-filled lifestyles are leading to some common problems such as depression and bipolar disorder (Rao et al. 2008). Truffles are considered a rich source of L-tyrosine, which acts as a precursor of neurotransmitters (catecholamines) as well as a critical constituent of thyroid hormones (Miot and Rousset 2000; Young et al. 2007). Dopamine and norepinephrine are catecholamines. These are important messengers of the neural circuits; hence, they improve functioning in the hippocampus and hypothalamus to counter psychological disorders and also improve serum levels of triiodothyronine (T3) and total thyrotropin (Wang et al. 2011; Young et al. 2007).

Traditionally, truffles have been used in local vegetable preparation and are considered aphrodisiac in nature due to the presence of rostenol, a steroidal pheromone an sexual enhancer. A study has shown that high levels of dopamine and serotonin neurotransmitters are attributed to high sexual arousal (El Enshasy et al. 2013).

15.2.4 Cultivation

As already discussed, truffles are not cultivated for commercial purposes. However, scientific cultivation of these macrofungi includes the process of planting and establishing host deciduous tree orchards and inoculating the host rhizosphere with truffle mycelia. Before inoculation, other contaminating soil fungi in the rhizosphere of the host plants are to be controlled to ensure the establishment of the truffle mycelia. Once the rhizosphere is cleared of contaminating fungi, the truffle mycelia culture is inoculated into the root zone of the host tree. Well-drained, argillaceous or calcareous soils with neutral or alkaline pH are perfect for truffle farming. Truffles do not grow in extremes of temperature or moisture. Truffles are very slow growing and as per the literature review, it requires 5–7 years to develop their mycorrhizal network, and only after that do the host plants permit the production of the fruiting bodies (Fischer et al. 2017).

15.3 MORELS

15.3.1 Habitat

Morchella species, which include the true morels, morel mushrooms, yellow/brown/black morels, and sponge morels, are among the most expensive macrofungi/mushrooms in the world due to their high medicinal and nutritional value. Due to their high market value, they are also known as the 'growing gold of mountains.' In India, they are also known as the 'Guchhi' mushroom and are most likely to be found in nature as a wild mushroom, growing during the rainy season in the northern region of India.

In Uttarakhand, the mushroom is found in the high mountain area of Sahiya village, Chakrata, Uttarakhand. The first author has also seen this fungus in the Benog Reserve Forest in Mussoorie, Dehradun. It is also has been reported in the Bharsar locality of the Garhwal Himalaya (Semwal et al. 2014). Morels grow naturally in humus-rich, loamy soil and have saprobic and mycorrhizal associations with different evergreen and deciduous conifers at elevations of 2200 to 3500 masl. Although it is generally known as the 'Guchhi,' in the Chakrata region, it is known as 'Juma' by the local people.

15.3.2 Morphology

Yellow morel mushrooms (collected macrofungal specimen, see Figure 15.2) consist of a cap and stipe, and the cap of the mushroom usually has a yellowish brown colour with a conical shape of

FIGURE 15.2 Yellow and black/brown morels.

about 3–8 cm in diameter and 5–12 cm in length. The stipe is a pale cream colour and is 3–12 cm long with a 1.5–6 cm diameter and is tapered towards the top (Hamayun et al. 2006). They vary in colour (blonde/grey), shape, and size (tight/loose), forming an imperfect growth pattern. The inner portion of the morel is hollow and whitish in colour, while the external portion converts into a bump-like structure. A large number of morels collected in the northwestern Himalayas have been assigned as yellow morels (*M. crassipes* and *M. spongiola*) and black morels (*M. elata*, *M. angusticeps*, and *M. gigas*), as reported by Kanwal et al. (2011).

15.3.3 Traditional and Medicinal Importance

Species of *Morchella* are well-known mushrooms/macrofungi throughout the globe and are also known for their high nutritional value, offering proteins (32.7%), carbohydrates (38%), vitamins (vits. B, C, and D), dietary fibre (17.6%), minerals, and trace elements (copper, manganese, cobalt, zinc, iron, calcium and magnesium) (Genccelep et al. 2009). The fruiting bodies of morel mushrooms are a good source of total tocopherol compounds that show strong antioxidant activity (Elmastas et al. 2006); phenolic compounds, beta-carotene, and linoleic acid present in mycelia also exhibit antioxidant activities (Mau et al. 2004; Heleno et al. 2013). Some crude polysaccharides extracted from *M. esculenta* exhibit antioxidant activity (Meng et al. 2010) and are potentially tumor-resistant (Li et al. 2013). Studies have reported that methanol, ethanol, and chloroform extracts of this mushroom show antibacterial activity against *Staphylococcus aureus*, *Salmonella typhimurium*, *Listeria monocytogenes*, *Escherichia coli*, and members of the family Enterobacteriaceae (Heleno et al. 2013). For more than 2000 years, species of *Morchella* have been used in traditional Chinese medicine as well as in Malaysia and Japan for the treatment of different diseases (Hobb1995). *Morchella* spp., with its broad variety of biomolecules, is commonly used to treat different types of indigestion and cardiac diseases (Ying et al. 1987). The powdered form of *M. esculenta* can also be used as an antiseptic and as a treatment for arthritis and general weakness (Mau et al. 2004; Mahmood et al. 2011). Other than that, *M. esculenta* is used as a remedy in colds, headaches, and hepatitis B. Previous studies show that it can reduce fatigue, regulate blood sugar levels, and tackle sleeping problems as well as blood cholesterol levels (Mahmood et al. 2011; Sher et al. 2011). Methanol extracts of *M. esculenta* contain various compounds that show strong anti-inflammatory activity and reduce pain (Christine et al. 2002).

In Uttarakhand, the Bhotia tribe residing in the high elevations of the Chamoli District of Uttarakhand use *M. esculenta* to make a decoction by boiling the fruiting bodies in the water. The local people in the Kullu district of Himachal Pradesh boil fruiting bodies in milk before they are consumed (Nautiyal et al. 2001). As already discussed, the local name for a morel in Uttarakhand is 'Juma/Guchhi', but in the Himachali dialect (Himachal Pradesh), it is known as 'Chunchuroo' (Semwal et al. 2014).

15.3.4 Cultivation

True morels are not cultivated like most grocery store mushrooms, as their farming is quite challenging. 'Gucchi' mushrooms cannot be cultivated commercially; instead, they grow wild only, as cultivation technology is not yet standardized. However, Ronald Owen of San Francisco patented a successful morel growing process in 1982, and growers in China were so successful that they could grow 6,000 morels on an acre of land. Morels are cool-season mushrooms that grow best when the weather transitions from winter to spring. In the United States of America, the most commercially cultivated species is *Morchella rufobrunnea*. The key techniques of cultivation include spawn production, land preparation, spawning, addition of exogenous nutrition, fruiting management, and harvest. The application of exogenous nutrition is one of the most important breakthroughs in cultivation, but the mechanism underlying its importance remains unclear. The environment for *M. esculenta* cultivation through spawn running requires temperatures of 21–24°C, relative humidity up to 85–90%, light for 8–10 hrs, and air exchange for 5–7 hrs over a duration of 10–14 days (Source: Kisan Suvidha app, www.kishansubidha.gov.in)

15.4 ENOKITAKE

15.4.1 Habitat

Enokitake is scientifically named *Flammulina velutipes*, syn. *Flammulina filliformis* (Wang et al. 2018). It is an edible mushroom belonging to the family Physalacriaceae of the division Basidiomycota. The wild mushroom is morphologically different from the cultivated varieties; thus, the cultivated mushrooms are recognized as a different fungal species. The wild enokitake mushroom is yellowish to dark brown in colour with short and thick stems/stipes, whereas the cultivated varieties are white in colour and have long and thin stipes, as they are grown without light in a carbon dioxide-rich environment. The mushroom grows in a wide range of habitats on woody trees in the winter season (e.g., September to March) and thus is also known as the 'winter mushroom.' It is found on tree trunks of *Cedrus deodara* Roxb. (Devdar) in Benog Wildlife Sanctuary, Mussoorie, Uttarakhand.

15.4.2 Morphology

Enokitake grows as a saprophyte on the stumps, logs, roots, and living wood of hardwood trees, growing in clusters and sometimes appearing terrestrial. The cap/pileus is 4–9 cm, convex in shape, becoming broadly convex to flat, moist and sticky when fresh, and bald. The colour is a fairly variable dark orange-brown to yellowish brown, often vanishing with maturity. Gills are broadly or narrowly attached to the stem/stipe, whitish to pale yellow, and crowded or close to each other. The stipe/stem is 40–130 × 4–12 mm, equal towards the base, tough, pale to yellowish brown or orange-brown when immature, and covered with a dark, rusty brown to the blackish velvety coating when it matures. The flesh is a soft tissue with a whitish to a yellowish tinge (see Figure 15.3).

15.4.3 Traditional and Medicinal Importance

The enokitake mushroom is edible but also possesses medicinal properties. The mushroom is very famous in Japanese and Chinese cultures and is a major part of their cuisine and gives a crunchy

FIGURE 15.3 *Flammulina velutipes* on a *Cedrus deodara* tree trunk.

texture while eating. The mushroom possesses antioxidant activity (Fu et al. 2002) and high nutritional content in terms of protein (Tang et al.2016). Jang et al. (2002) studied the inhibitory effect of enokitake mushroom extract against the browning of apples and observed positive results. The mushroom was also effective for inhibiting the melanosis of shrimp (Jang et al. 2003). The anti-melanosic property of the mushroom can be ascribed to the presence of ergothioneine. Tang et al. (2016) gave a complete description of different biological activities exhibited by enokitake mushrooms, along with the compounds responsible for these activities, which include anticancer, anti-tumor, hepatoprotective activity, anti-inflammatory, antimicrobial, cholesterol-lowering, and antihypertensive properties. Interestingly, carbon dots have also been synthesized from enokitake mushrooms (Pacquiao et al. 2018), as they are a good source of protein, vitamins, minerals, and nutrients, representing various heteroatoms that can yield highly fluorescent carbon dot nanoparticles (Pacquiao et al. 2018).

In the study area of Uttarakhand, the villagers cook this mushroom as a local vegetable with other green vegetables. After drying, the caps are ground into powder and taken with boiled milk to attenuate debility of the body.

15.4.4 Cultivation

The commercial cultivation of enokitake mushrooms is done on sawdust and rice bran (Nakamura 1981) through a solid-state fermentation process, usually in polypropylene bags or bottles. Sometimes, substrates such as sugarcane bagasse and cottonseed husk are also used for

cultivation (Chang and Miles 1989; Yang 1986). The spawn of the mushroom is usually prepared aseptically on cereal grain substrate or wood chips (Mata and Savoie 2005). It was observed that enokitake mushroom grew well in a medium supplemented with soybeans, yeast powder, and peptone as nitrogen sources, whereas nitrate or amine nitrogen sources did not give satisfactory results (Wang 2000).

15.5 CORDYCEPS

15.5.1 Habitat

Ophiocordyceps sinensis (= *Cordyceps sinensis* [L.] Link.) is an entomopathogenic and medicinal fungus that is widely acclaimed for its medicinal functions. *Cordyceps*, or caterpillar fungus, is thought to have been discovered about 1,500 years ago by sheep herdsmen in Tibet. *Cordyceps* species are found in restricted regions of high-altitude areas and thrive well in alpine and subalpine grasslands, fields, open dwarf scrublands, and mountain slopes of the Tibetan Plateau and high Himalayas (Baral et al. 2015). *Cordyceps sinensis* is found in alpine fields in India, Nepal, Bhutan, and China. In China, it is popularly known as 'Dong Chong Xia Cao,' and is 'Tockukasoin' in Japan, 'Yartsa Gunbu' in Tibet, 'Yarsagumba' in Nepal, and 'Keedajadi' in the Garhwal Himalayas of India (Belwal et al. 2019). Market accessibility in Nepal and the Tibetan Autonomous Region (TAR) of China has led to bulk trading of the fungus. In the northern state of India, Uttarakhand, moist alpine regions of the Dharchula and Munsyari blocks, the Bageshwar district, the Nanda Devi Biosphere Reserve, and the Chamoli district are hotspots for *Cordyceps* occurrence.

15.5.2 Morphology

Ophiocordyceps sinensis (Berk.), earlier known as *Cordyceps sinensis*, is a macrofungus that belongs to the Ophiocordycipitaceae family, division Ascomycetes, and is parasitic to underground-dwelling larvae of moths that belongs to the family Hepialidae (Zhong et al. 2016). The fungus uses the host's body as a substrate for mycelium formation, and the stroma (fruiting body) arises from the mouth of the dead host, which is finally converted into a sclerotium covered by the intact exoskeleton for survival in winter (Li et al. 2011). The stroma is a club-like structure that emerges about 4–7 cm above the ground and grows from the sclerotium, which is yellow to brown in colour; stipes bring the spores away from the host (see Figure 15.4).

15.5.3 Traditional and Medicinal Importance

Cordyceps species are used to treat disorders like diarrhoea, cough, headache, cardiovascular disorder, sexual dysfunction, rheumatism, asthma, pulmonary disease, and renal and liver diseases for centuries in Bhutanese indigenous and Chinese traditional medicine (Jiang and Yao 2002; Chakraborty et al. 2014; Belwal et al. 2019). Zhang et al. (2009) observed over 30 bioactivities of caterpillar fungus, including antioxidant, anti-tumor, anti-inflammatory, anticancer, and immunomodulatory properties. 'Yarsagumba' (*C. sinensis*) has been prized as an aphrodisiac (sexual enhancer) for the past 500 years by the Chinese. Since 1964, the species is classified as a drug in the Chinese Pharmacopoeia. It is also broadly used as a tonic and aphrodisiac (Zhou et al. 2009). Within a small period of time, the term 'Himalayan herbal Viagra' (aphrodisiac) has led to a rise in the price of caterpillar fungus. As per the report, the value ranges from USD $20,000–40,000 per kg in the international market (Winkler 2008, 2009; Negi et al. 2014). Chen et al. (2013) tested the effects of

FIGURE 15.4 *Ophiocordyceps sinensis* with stroma and sclerotium (black coloured) growing on a caterpillar.

water extracts of *C. sinensis* on an aging model in rats, and the results revealed enhanced resistance to the symptoms of induced aging in a dose-dependent manner.

In Uttarakhand, the 'Keedajadi' mushroom is sun-dried and taken in powder form with warm water during the daytime or with boiled milk to improve bodily strength and endurance. The local villagers also use 'Keedajadi' as tea with other herbal products.

15.5.4 Cultivation

The cultivation of *Ophiocordyceps sinensis* = *Cordyceps sinensis* in the laboratory is difficult, but another important species, *Cordyceps militaris*, can be grown and cultivated in laboratories. *Cordyceps* cultures are obtained from the stroma and stipe of fresh, healthy, sun-dried specimens. A small portion of *Cordyceps* tissues is transferred aseptically into petri plates containing media such as potato-dextrose agar, wheat grain extract, maize grain extract, and malt extract agar, incubated at 25°C for at least 8–10 days, and observed regularly for the appearance of mycelia. Mass inocula are prepared on different substrates (wheat grains, maize grains, and sorghum) for large-scale cultivation trials of *C. militaris*. Polypropylene bags containing sterilized maize grains, wheat grains, or sorghum grains are inoculated with a pure culture of *C. militaris*. These polypropylene bags are incubated at 25°C for 15 days in a dark room. The entire grain surface is then colonized by the fine thread-like mycelium of the fungus. When these polypropylene bags were opened and conditioned to different moisture, humidity, temperature, and light variations, it was noticed that bags subjected to moderate daylight (2–4 h), optimal temperature (25°C), and 50%–70% atmospheric humidity started showing vertical stromata which resulted in the formation of complete fruiting bodies in 50–60 days (Chen et al. 2013).

15.6 LION'S MANE

Lion's mane is a Basidiomycetes fungus belonging to the order Russulales and the family Hericiaceae. It is scientifically named *Hericium erinaceus* (Bull.) Pers. The common names are lion's mane, old man's beard, and monkey's mushroom, or, in local dialects, 'pom pom' and 'yamabushitake' (Thongbai et al. 2015).

15.6.1 Habitat

Lion's mane is a saprophytic fungus occurring mostly on dead wood. It has also been observed in cracks of living hardwoods and knotholes (Boddy et al. 2011). The fungus is found commonly distributed in North America (Canada), Central America, South America (Columbia), and Europe (Germany, Spain, and Netherlands), whereas its distribution is rare and scattered among Asian countries like China, India, Korea, and Japan (Kałucka and Olariaga Ibarguren 2019).

In Uttarakhand, it is found naturally growing on the stem/trunk of *Quercus leucotrichophora* A. Camus, and was collected from Chakrata region, Dehradun by the author and from Rudraprayag and Bageshwar of the Garhwal and Kumaun Himalayas by Bhatt et al. (2018).

15.6.2 Morphology

The fruiting body is tough, and white-coloured spines hang in a clump with a base attached to the tree with an appearance like a white beard, hence the common name, 'old man's beard.' The old fruiting bodies turn yellowish brown. Basidiomata are 60–100 × 40–70 mm, yellowish white, and stipitate (see Figure 15.5). Primary branches bear small spines (up to 4–5 mm long) all around, giving rise to secondary branches. Secondary branches are comparatively thinner and elongated (up to 6–10 mm) but have no spines and bear tertiary branches. Tertiary branches are fertile in nature and bear spines. Spines are dense, tapering, overhanging, and long; they are white in colour. Microscopically spores are smooth and globose.

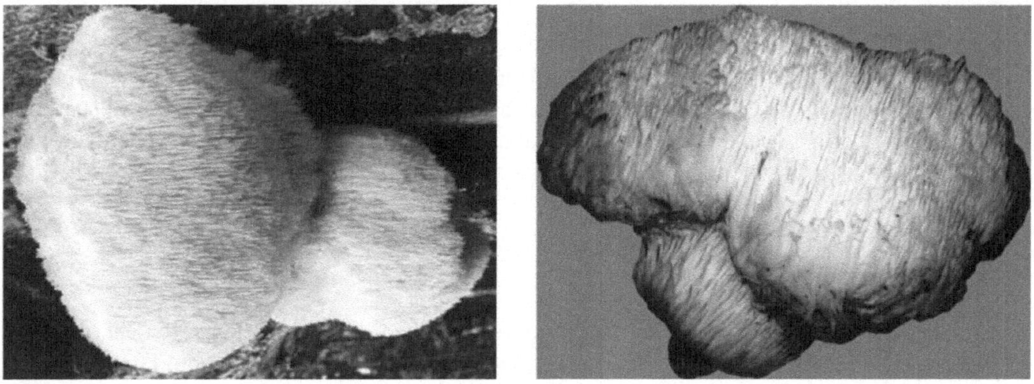

FIGURE 15.5 Lion's mane (*Hericium erinaceus*) fungus on a tree trunk and after removal from the tree trunk.

15.6.3 Traditional and Medicinal Importance

Hericium erinaceus fruiting bodies are known to contain anti-tumor compounds such as glucoxylans, xylans, heteroxyloglucans, and galactoxyloglucans (Mizuno et al. 1992). *Hericiumerinaceus* uniquely contains about 70 different secondary metabolites, among which five classes of organic compounds are described, namely aromatic compounds, erinacines, steroids, alkaloids, and lactones. Medicinal aromatic compounds called 'hericerins' are reported from this mushroom and are known to possess anticancer properties (Friedman 2015). Erinacines potentially have neuroprotective properties as a group of cyathane-type diterpenoids (Tang et al. 2015). Other medicinal properties include antibiotic abilities; attenuation of cognitive impairment; antioxidant, antidiabetic, antihypolipidemic, and anti-hypertensive effects; and beneficial effects againt ischemia, depression, and anxiety (Friedman 2015).

In the Chakrata region of Dehradun, Uttarakhand, this fungus is eaten by villagers as a local vegetable. The basidiomata or fruiting bodies are dried and boiled with milk/tea to strengthen the nervous system. *Hericium erinaceus* fungi are medicinally important due to the presence of therapeutic polysaccharide contents, which are composed of β-glucans. Some important antioxidants and immune-supporting β-glucan polysaccharides are also found in this fungal species. Lion's mane mainly contains erinacines and hericenones, the bioactive substances that give it its neuroprotective properties (Bhatt et al. 2018).

15.6.4 Cultivation

Sokol et al. (2015) reviewed the cultivation methodologies of *Hericium erinaceus* and categorized two types of cultivation systems being used, namely extensive and intensive cultivation. In extensive cultivation, no special/sophisticated equipment is required, and mushrooms are grown in uncontrolled conditions. They grow under high humidity conditions on wood logs/stumps inoculated with the mycelia of *H. erinaceus*. However, this method takes a long time to producing fruiting bodies that may extend up to one year. In intensive cultivation, on the other hand, sterilized substrates are used, and the culture is run in heat-resistant (for sterilization) polybags or bottles. A filter is used to facilitate aeration without contamination. In this method, better yield is observed in shorter cultivation times but with high input expenditure.

15.7 CONCLUSIONS AND FUTURE PROSPECTS

Macrofungi represent an imperative aspect of the biota in all forest ecosystems. Their ecological roles range from ectomycorrhizal associations with many of plant species to the key decomposition of dead plant material. Some macrofungal species are edible and suitable for human consumption in the form of various local cuisines and traditional remedies. During the past few decades, several workers have studied the diversity of mushrooms in India. Only a few reports are available in the literature with respect to wild edible and medicinal macrofungi from the Garhwal Himalayas. However, there are macromycoflora rich forests in the Garhwal Himalayas, which are yet to be explored thoroughly for the occurrence and diversity of wild edible and medicinal macrofungi/mushrooms. The present chapter is an effort to document some uncommon traditionally and medicinally important wild macrofungi of the Garhwal Himalayas of Uttarakhand, India. The conservation and multiplication of these species for sustainable usage by humankind is very much required as a prospect.

15.8 ACKNOWLEDGEMENTS

The senior author (VP) is thankful to the Director of the Forest Research Institute, Dehradun for facilitating the research activities in laboratories under the Forest Pathology Section. The authors are also thankful to Dr. Manoj Kumar, Forest Pathology Section, Forest Research Institute, Dehradun, for providing help with fungal photography.

REFERENCES

Atkinson, G.F. 1961. *Studies of American Fungi-Mushrooms, Edible, Poisonous, etc. (2)*. Hafner Publishing Company, New York.

Baral, B., B. Shrestha & J. Silva. 2015. A review of Chinese Cordyceps with special reference to Nepal, focusing on conservation. *Environmental and Experimental Biology* 13: 61–73.

Barros, L., M.J. Ferreira, B. Queiros, I.C. Ferreira & P. Baptista. 2007. Total phenols, ascorbic acid, β carotene and lycopene in Portuguese wild edible mushrooms and their antioxidant activities. *Food Chemistry* 103(2): 413–419.

Bates, S.C. 2006. A preliminary checklist of Arizona macrofungi. *Canotia* 2: 47–78.

Belwal, T., I.D. Bhatt, D. Kashyap, K. Sak, H.S. Tuli, R. Pathak, R.S. Rawal & S.M. Ghatnur. 2019. Ophiocordyceps sinensis. Nonvitamin and Nonmineral Nutritional Supplements. *Elsevier*: 527–537.

Berch, S.M. & G. Bonito. 2016. Truffle diversity (Tuber, Tuberaceae) in British Columbia. *Mycorrhiza* 26(6): 587–594. https://doi.org/10.1007/s00572-016-0695-2

Bhatt, R.P., U. Singh & P. Uniyal. 2018. Healing mushrooms of Uttarakhand Himalaya, India. *Current Research in Environmental and Applied Mycology* 8(1): 1–23.

Boddy, L., M.E. Crockatt & A.M. Ainsworth. 2011. Ecology of *Hericiumcirrhatum*, *H. coralloides* & *H. erinaceus* in the UK. *Fungal Ecology* 4(2): 163–173.

Chakraborty, S., S. Chowdhury & G. Nandni. 2014. Review on Yarsagumba (*Cordyceps sinensis*)—An exotic medicinal mushroom. *International Journal of Pharmacognosy & Phytochemical Research* 6: 339–346.

Chang, S.T. & P.G. Miles. 1989. *Edible Mushroom & Their Cultivation*. CRC Press Inc., Florida, USA, p. 345.

Chen, P.X., S. Wang, S. Nie & M. Marcone2013. Properties of *Cordyceps sinensis*: A review. *Journal of Functional Foods* 5(2): 550–569.

Chen, Y.F., W.W. Jiang, S.Q. Zhang, J.Q. Kan & Y. Liang. 2016. Antioxidant activity & characterization of one new polysaccharide obtained from Perigord Truffle (*Tuber huidongense*). *Evidence-Based Complementary & Alternative Medicine* 35(3): 71–93. https://doi.org/10.1155/2016/3537193

Christine, J.G.D., N. Pugh, D.S. Pasco & S.A. Ross. 2002. Isolation of a galactomannan that enhances macrophage activation from the edible fungus *Morchella esculenta. Journal of Agricultural and Food Chemistry* 50(20): 5683–5685.

Dogan, H.H. & S. Aydın. 2013. Determination of antimicrobial effect, antioxidant activity & phenolic contents of desert truffle in Turkey. *African Journal of Traditional, Complementary and Alternative Medicines* 10(4): 52–58. www.ncbi.nlm.nih.gov/pubmed/24146501.

El Enshasy, H., E.A. Elsayed, R. Aziz & M.A. Wadaan. 2013. Mushrooms & truffles: Historical biofactories for complementary medicine in Africa & in the middle East. *Evidence-Based Complementary & Alternative Medicine*: 620–451. https://doi.org/10.1155/2013/620451.

Elmastas, M., I. Turkekul, L. Ozturk, I. Glucin, O. Isildak & H.Y. Aboul-Enein. 2006. Antioxidant activity of two wild edible mushrooms (*Morchella vulgaris &Morchella esculenta*) from north Turkey. *Combinatorial Chemistry and High Throughput Screening* 9: 443–448.

Federico, V., T. Cosimo, P. Antonio, B. Nadia, L. Valentina, M. Stefano & A. Amedeo. 2015. Volatile organic compounds in truffle (*Tuber magnatum* Pico): Comparison of samples from different regions of Italy & from different seasons. *Scientific Reports* 5(1): 26–29. https://doi.org/10.1038/srep12629

Ferdman, Y., S. Aviram, N. Roth-Bejerano, J.M. Trappe & V. Kagan-Zur. 2005. Phylogenetic studies of *Terfezia pfeilii* & *Choiromyces echinulatus* (Pezizales) support new genera for southern African truffles: Kalaharituber & Eremiomyces. *Mycological Research* 109(2): 237–245. www. ncbi.nlm.nih.gov/pubmed/15839107.

Fischer, C.R., D. Oliach, J.A. Bonet & C. Colinas. 2017. *Best Practices for Cultivation of Truffles.* Forest Sciences Centre of Catalonia, Solsona, Spain. http://hdl.handle.net/10459.1/62828.

Friedman, M. 2015. Chemistry, nutrition, & health-promoting properties of *Hericium erinaceus* (Lion's Mane) mushroom fruiting bodies & mycelia & their bioactive compounds. *Journal of Agricultural & Food Chemistry* 63(32): 7108–7123.

Fu, H.Y., D.E. Shieh & C.T. Ho. 2002. Antioxidant & free radical scavenging activities of edible mushrooms. *Journal of Food Lipids* 9(1): 35–43.

Genccelep, H., Y. Uzun, Y. Tunccturk & K. Demirel. 2009. Determination of mineral contents of wild-grown edible mushrooms. *Food Chemistry* 113(4): 1033–1036.

Gouzi, H., L. Belyagoubi, K.N. Abdelali & A. Khelifi. 2011. *In vitro* antibacterial activities of aqueous extracts from Algerian desert truffles (Terfezia & Tirmania, Ascomycetes) against *Pseudomonas aeruginosa & Staphylococcus aureus. International Journal of Medicinal Mushrooms* 13(6): 553–558. www.ncbi.nlm.nih.gov/pubmed/22181843

Hamayun, M., S.A. Khan, H. Ahmad, D.H. Shin & I.J. Lee. 2006. Morel collection and marketing: A case study from the Hindu-Kush mountain region of Swat, Pakistan. *Lyonia A Journal of Ecology and Application* 11(2): 7–13.

Heleno, S.A., D. Stojkovic, L. Barros, J. Glamoclija, M. Sokovic, A. Martins, M.J.R.P. Queiroz & I.C.F.R. Ferreira. 2013. A comparative study of chemical composition, antioxidant & antimicrobial properties of *Morchella esculenta* (L.) Pers. From Portugal & Serbia. *Food Research International* 51(1): 236–243.

Hobb, C. 1995. *Medicinal Mushrooms.* Botanica Press, Santa Cruz, CA, p. 251.

Janakat, S.M., S.M. Al-Fakhiri & A.K. Sallal. 2005. Evaluation of antibacterial activity of aqueous & methanolic extracts of the truffle *Terfezia claveryi* against *Pseudomonas aeruginosa. Saudi Medical Journal* 26(6): 952–955. www.ncbi.nlm.nih.gov/pubmed/15983681.

Jang, M.S., A. Sanada, H. Ushio, M. Tanaka & T. Ohshima. 2002. Inhibitory effects of 'Enokitake' mushroom extracts on polyphenol oxidase & prevention of apple browning. *LWT-Food Science & Technology* 35(8): 697–702.

Jang, M.S., A. Sanada, H. Ushio, M. Tanaka & T. Ohshima. 2003. Inhibitory effect of enokitake extract on melanosis of shrimp. *Fisheries Science* 69(2): 379–384.

Jiang, Y., & Y.J. Yao. 2002. Names related to *Cordyceps sinensis* anamorph. *Mycotaxon* 84: 245–254.

Joshi, S., R.P. Bhatt & S.L. Stephenson. 2012. The current status of the family Russulaceae in the Uttarakhand Himalaya, India. *Mycosphere* 3(4): 486–501.

Kałucka, I.L. & I. Olariaga Ibarguren. 2019. Hericium erinaceus. *The IUCN Red List of Threatened Species* 2019: e.T70401627A70401637. https://doi.org/10.2305/IUCN.UK.2019-3.RLTS.T70401627A70401637.

Kanwal, H.K., K. Acharya, G. Ramesh & M.S. Reddy. 2011. Molecular characterization of *Morchella* species from the Western Himalayan region of India. *Current Microbiology* 62(4): 1245–1252.

Kothiyal, G. & K. Singh. 2022. Antimicrobial and phytochemical screening of selected wild mushrooms naturally found in Garhwal Himalayan region, Uttarakhand, India. *Journal of Advanced Biotechnology and Experimental Therapeutics* 5(2): 417–432.

Li, K., Q. Li, Y. Yuan, J.Y. Chen, L.Q. Huang & Z.G. Liu. 2011. Identification & detection of *Cordyceps sinensis* with LAMP. *Chinese Traditional & Herbal Drugs* 42(8): 1605–1608.

Li, S., Y. Sang, D. Zhu, Y. Yang, Z. Lei & Z. Zhang. 2013. Optimization of fermentation conditions for crude polysaccharides by *Morchella esculenta* using soybean curd residue. *Industrial Crops and Products* 50: 666–672.

Mahmood, A., R.N. Malik, Z.K. Shinwari & A. Mahmood. 2011. Ethnobotanical survey of plants from Neelum, Azad Jammu & Kashmir, Pakistan. *Pakistan Journal of Botany* 43: 105–110.

Mata, G. & J.M. Savoie. 2005. Wheat straw. In: Gush, R. (Ed) *Mushrooms Grower's Handbook 2*. Mushroom World, Seoul.

Mau, J.L., C.N. Chang, S.J. Huang & C.C. Chen. 2004. Antioxidant properties of methanolic extracts from *Grifola frondosa, Morchella esculenta & Termitomyces albuminosus* mycelia. *Food Chemistry* 87(1): 111–118.

Meng, F., B. Zhou, R. Lin, L. Jia, X. Liu, P. Deng, K. Fan, G. Wang, L. Wang & J. Zhang. 2010. Extraction optimization and in vivo antioxidant activities of exopolysaccharide by *Morchella esculenta* SO-01. *Bioresource Technology* 101(12): 4564–4569.

Miot, F., C. Dupuy, J. Dumont & B. Rousset. 2000. Chapter 2 thyroid hormone synthesis & secretion. *Endotext.* MDText.com www.ncbi.nlm.nih.gov/pubmed/25905405

Mizuno, T., T. Wasa, H. Ito, C. Suzuki & N. Ukai. 1992. Antitumor-active polysaccharides isolated from the fruiting body of *Hericium erinaceum*, an edible & medicinal mushroom called *Yamabushitake* or *Houtou. Bioscience, Biotechnology and Biochem*istry 56: 347–348.

Nakamura, K. 1981. *Mushroom Cultivation in Japan.* Asaki Publication House, Asaki, Japan.

Nautiyal, S., R.K. Maikhuri, K.S. Rao & K.G. Saxena. 2001. Medicinal plant resources in Nanda Devi Biosphere Reserve in the central Himalayas. *Journal of Herbs, Spices and Medicinal Plants* 8(4): 47–64.

Negi, C.S., M. Pant, P. Joshi & S. Bohra. 2014. Yartsa Gunbu [*Ophiocordyceps sinensis* (Berk.) G.H. Sung et al.]: The issue of its sustainability. *Current Science* 107: 882–887.

Pacquiao, M.R., M.D.G. de Luna, N. Thongsai, S. Kladsomboon & P. Paoprasert. 2018. Highly fluorescent carbon dots from enokitake mushroom as multi-faceted optical nanomaterials for Cr6 + & VOC detection & imaging applications. *Applied Surface Science* 453: 192–203.

Patel, S., A. Rauf, H. Khan, S. Khalid & M.S. Mubarak. 2017. Potential health benefits of natural products derived from truffles: A review. *Trends in Food Science & Technology* 70: 1–8. http://doi.org/10.1016/j.tifs.2017.09.009

Rao, T.S.S., M.R. Asha, B.N. Ramesh & K.S.J. Rao. 2008. Understanding nutrition, depression & mental illnesses. *Indian Journal of Psychiatry* 50(2): 77–82. https://doi.org/10.4103/0019-5545.42391

Semwal, K.C., S.L. Stephenson, V.K. Bhatt & R.P. Bhatt. 2014. Edible mushrooms of the Northwestern Himalaya, India: A study of indigenous knowledge, distribution and diversity. *Mycosphere* 5(3): 440–461.

Sher, H., M. Elyemeni, K. Hussain & H. Sher. 2011. Ethnobotanical & economic observations of some plant resources from the northern parts of Pakistan. *Ethnobotany Research and Application* 9: 27–41.

Singh, U., R.P. Bhatt, S.L. Stephenson, P. Uniyal & T. Mehmood. 2017. Wild edible mushrooms from high elevations in the Garhwal Himalaya—II. *Current Research in Environmental & Applied Mycology* 7(3): 208–226.

Sokol, S., I. Golak-Siwulska, K. Sobieralski, M. Siwulski & K. Górka. 2015. Biology, cultivation, & medicinal functions of the mushroom *Hericium Erinaceum. Acta Mycologica* 50(2). http://doi.org/10.5586/am.1069

Stametes, P. 2000. The role of mushroom in nature, culturing mushroom mycelium on agar media. In: *Growing Gourmet & Medicinal Mushrooms.* Ten Speed Press, Hong Kong, pp. 10–11.

Tang, C., P.C.X. Hoo, L.T.H. Tan, P. Pusparajah, T.M. Khan, L.H. Lee & K.G. Chan. 2016. Golden needle mushroom: A culinary medicine with evidenced-based biological activities and health promoting properties. *Frontiers in Pharmacology* 7: 474.

Tang, H.Y., X. Yin, C.C. Zhang, Q. Jia & J.M. Gao. 2015. Structure diversity, synthesis, & biological activity of cyathane diterpenoids in higher fungi. *Current Medicinal Chemistry* 22(19): 2375–2391.

Thongbai, B., S. Rapior, K.D. Hyde, K. Wittstein & M. Stadler. 2015. *Hericium erinaceus*, an amazing medicinal mushroom. *Mycological Progress* 14(10): 1–23.

Vishwakarma, M.P., R.P. Bhatt & S. Gairola. 2011. Some medicinal mushrooms of Garhwal Himalaya, Uttarakhand, India. *International Journal of Medicinal & Aromatic Plants* 1: 33–40.

Wang, D.C. 2000. Preliminary study on carbon & nitrogen sources for *Flammulina velutipes* under solid culture conditions. *Edible Fungi of China* 19(5): 38–39

Wang, L., G. Wang, J. Zhang, G. Zhang, L. Jia, X. Liu, P. Deng & K. Fan. 2011. Extraction optimization & antioxidant activity of intracellular selenium polysaccharide by *Cordyceps sinensis* SU-02. *Carbohydrate Polymers* 86: 1745–1750.

Wang, P.M., X.B. Liu, Y.C. Dai, E. Horak, K. Steffen & Z.L. Yang. 2018. Phylogeny & species delimitation of *Flammulina*: Taxonomic status of winter mushroom in East Asia & a new European species identified using an integrated approach. *Mycological Progress* 17(9): 1013–1030. https://doi.org/10.1007/s11557-018-1409-2

Winkler, D. 2008. Yartsa Gunbu (*Cordyceps sinensis*) and the fungal commoditization of Tibet's rural economy. *Economic Botany* 62: 291–305.

Winkler, D. 2009. Caterpillar fungus (*Ophiocordyceps sinensis*) production and sustainability on the tibetan plateau and in the Himalayas. *Asian Medicine* 5: 291–316.

Yang, X.M. 1986. *Cultivation of Edible Mushroom in China*. Agriculture Printing House, Beijing, pp. 489–510.

Ying, J.Z., X.L. Mao, Q.M. Ma, Y.C. Zong & H.A. Wen. 1987. *Icons of Medicinal Fungi from China*. Science Press, Beijing.

Young, S.N. 2007. L-tyrosine to alleviate the effects of stress? *Journal of Psychiatry & Neuroscience* 32(3): 224–225. www.ncbi.nlm.nih.gov/pubmed/17476368

Zhang, Y., L. Xu, S. Zhang, X. Liu, Z. An, M. Wang & Y. Guo. 2009. Genetic diversity of *Ophiocordyceps sinensis*, a medicinal fungus endemic to the Tibetan Plateau: Implications for its evolution & conservation. *BMC Evolutionary Biology* 9(1): 1–12.

Zhao, W., X.H. Wang, H.M. Li, S.H. Wang, T. Chen, Z.P. Yuan & Y.J. Tang. 2014. Isolation and characterization of polysaccharides with the antitumor activity from Tuber fruiting bodies & fermentation system. *Applied Microbiology & Biotechnology* 98(5): 1991–2002. https://doi.org/10.1007/s00253-013-5379-7

Zhong, X., S. Li, Q. Peng, J. Zhang, X. Kan, G. Zhang & X. Liu. 2016. A Polycephalomyces hyperparasite of *Ophiocordyceps sinensis* leads to shortened duration of production and reduced numbers of host ascospores. *Fungal Ecology* 21: 24–31.

Zhou, X., Z. Gong, Y. Su, J. Lin & K. Tang. 2009. *Cordyceps* fungi: Natural products, pharmacological functions & developmental products. *The Journal of Pharmacy & Pharmacology* 61: 279–291.

16 Hidden Treasure of Himalayan Forests
Mushrooms with Curative Potential

Priyanka Uniyal, Upendra Singh, Rajendra Prasad Bhatt

16.1 INTRODUCTION

Fungi represent one of the most diversified groups of living organisms on earth and form a wide variety of ecological associations with many other organisms. For quite a long time, fungi were thought to be close to plants, but recent studies suggest their close affinities with the animal kingdom. Either as saprophytes or in positive and negative ecological interactions, fungi play vital roles in different types of ecosystems and in the proper functioning of all ecosystems. Fungi are essential for maintaining the forest and plantation health, as plants are not able to sequestrate the phosphorus present in soil, and without fungi, this essential element for plant growth and survival comes under serious threat. This problem is overcome by soil fungi, which sequestrate the phosphorus for the plants involved in the mycorrhizal associations. Despite being a highly diverse and rich assemblage of organisms, fungi remain to be explored, as roughly 144,000 species have yet to be discovered. According to the report of Willis (2018), fungal exploration has increased in Asian countries, where the history of utilizing fungi as food and cures for various health conditions is very old. However, many other geographical regions like Africa, Australia, and South America need to be explored in terms of their mycoflora and to obtain a more accurate idea of fungal diversity and its ecological functions (Genevieve et al. 2019; Monkai et al. 2017; Stamets 2000; Willis 2018).

Fungi are very diverse in morphology, as they are known to occur in various shapes, sizes, and colours. Due to their unique appearance, wild fungi have been objects of curiosity and mystery. The somatic structure of fungi is the thread-like mycelium, which bears the spore-producing fruiting bodies. In higher fungi, spore-producing bodies are commonly referred to as carpophores or sporocarps (Figure 16.1). These aerial sporocarps of higher fungi are commonly known as 'mushrooms'. Mushrooms are multi-diverse in form and ecological association. Some common forms are morels, puffballs, stinkhorns, jelly, and bracket fungi (Figure 16.1). Chang and Miles (1992) refined the definition that is now universally accepted. They defined the mushroom as a 'macrofungus with a distinctive fruiting body which can be either epigeous or hypogeous and large enough to be seen with the naked eye and can be picked with hand'.

As human history suggests, mushrooms have been known to human beings for thousands of years. Mushrooms have a long history of use in Chinese and Japanese cultures, where they are believed to have first been cultivated, and even now form an integral part of the diet of these populations. The cultural history of the use of mushrooms is evident from the diversity of mushroom cuisines in these countries. However, mushrooms are used worldwide as food due to their rich flavour and texture. Apart from this, fruiting bodies are a rich source of vitamins, fibre, minerals, trace elements, protein, and a variety of polysaccharides. Another interesting feature of the mushrooms is the presence of toxins, which are responsible for frequent cases of mushroom poisoning worldwide due to the consumption of misidentified mushrooms. The presence of toxins is most common in poisonous and psychedelic/hallucinogenic mushrooms (Bhambri et al. 2022; Bhatt et al. 2018).

In recent times, mushrooms have gained much importance not only due to their edible or poisonous nature, but also for their therapeutic potential. Fruiting bodies and mycelia of mushrooms contain a wide spectrum of bioactive compounds, which result in bioactivities including antidiabetic,

 DOI: 10.1201/b23190-16

anti-oxidative, anticancer, anti-tumour, immunomodulatory, cytotoxic, neuroprotective, cardioprotective, antimicrobial, and antiparasitic activities. In addition, mushrooms also contain nutrients such as glutathione, ergothioneine, selenium, and vitamin D. Studies have suggested that various mushrooms can be used to reduce the adverse effects of radiotherapy and chemotherapy in cancer patients. Mushroom supplements can be used as neutraceuticals along with other drugs in such patients (Bhambri et al. 2022; Bhatt et al. 2018; Elkhateeb and Daba 2021).

The Himalayan state of Uttarakhand contains 1.63% of the total geographical area of India, accounting for about 53,483 km^2, and lies from 28°43' N to 31°28' N in latitude and 77°34' E to 81°03' E in longitude. According to Champion and Seth (1968), out of sixteen types of forests in India, eight are found in Uttarakhand (Figure 16.2). The vast elevation range has resulted in a variety of floral components in these forests. In temperate mixed forests, the main components are *Quercus*, *Myrica*, *Rhododendron*, *Alnus*, *Lyonia*, and *Aesculus* spp. Coniferous forests are formed of *Abies*, *Cedrus*, *Picea*, and *Pinus* spp. Tropical forests are composed mainly of *Dalbergia sissoo*, *Shorea robusta*,

FIGURE 16.1 Diversity of sporocarps in mushrooms. (a) Bracket fungi, (b) earth tongue, (c) puff balls, (d) cup fungi, (e) stinkhorns, (f) morels, (g) earth stars, (h) spinous fungi, (i) soral fungi, (j) jelly fungi, (k) dead man's fingers, (l) Earth balls, (m) moose antlers or rabbit ear fungi.

FIGURE 16.2 Different types of forests in the Uttarakhand Himalaya. (a) mixed deciduous forest, (b) *Picea* forest, (c) *Cedrus* forest, (d) broadleaf forest, (e) pine forest.

Acacia catechu, *Bauhinia* spp., *Terminalia* spp., and *Bombax ceiba* (Joshi et al. 2012; Sati and Bandooni 2018). Vast floral diversity in different types of forests in Uttarakhand supports the growth of various higher fungi. These fungi help to sustain these forests by providing different types of ecological services. Various investigators have explored the mycoflora of Uttarakhand, and discoveries of many new taxa from these forests in the past few decades indicate the rich macrofungal diversity of this region and reflect the need for more studies, including analysis of different aspects.

16.2 WILD TREASURE: MEDICINAL MUSHROOMS OF UTTARAKHAND

Uttarakhand is a mountainous state where still most of the rural population depends upon forest resources. Many reports (e.g., Bhatt et al. 2016; Semwal et al. 2014; Singh et al. 2017) on the edibility of wild mushrooms in Uttarakhand have reflected that knowledge about these wild resources is confined to particular groups of people, and many locals are still unaware of the benefits of the wild fungi. The present chapter deals with mushrooms from the Uttarakhand Himalaya with curative potential. Taxa are presented with detailed taxonomic descriptions (colour codes follow Kornerup and Wanscher 1978) and notes on medicinal properties.

16.2.1 *Aleuria aurantia* (Pers.) Fuckel, Jb. nassau. Ver. für Naturk. 23–24: 325 (1870) [1869–70] **Figure 16.3e**

Ascomata 7–70 mm in diam., cup-shaped, uplifted becoming applanate, sessile, yellowish orange (4B7–4B8) to dark orange, glabrous. Hymenophore yellowish white to orange-white (3A2–5A2), pruinose. Context very thin, solid, yellowish white to orange-white (3A2–5A2). Taste and odour are not distinct. Spore print not obtained.

Habit and habitat: Gregarious, on a sandy surface.
Localities: Chopta-Baniyakund, Jakhdhar (Rudraprayag district); Mukteshwar (Nainital district).
Medicinal properties: *Aleuria aurantia* is commonly known as the 'orange peel fungus'. A lectin (AAL) extracted from *A. aurantia* exhibits antifungal activity, particularly against Mucorracemosus. AAL has also been found effective in the treatment of pancreatic cancer (Amono et al. 2012; Kim et al. 2022).

16.2.2 *Armillaria mellea* (Vahl) P. Kumm., Führ. Pilzk. (Zerbst): 134 (1871) **Figure 16.3d**

Pileus 10–65 mm broad, subglobose, expanding convex to plano-convex, then flattened and centrally depressed or umbonate; honey yellow (5D6) to yellowish brown (5D8) to bronze-brown (5E5), with small, crowded, darker fibrillose squamules, surface viscid when moist, margin inrolled when young, becoming acute, entire, paler, striate. Lamellae adnate to sub-decurrent, whitish to yellow, moderately distant, lamellulae present. Stipe 30–110 × 5–20 mm, central, equal or slightly bulbous at base, tough, whitish to yellow at apex, brownish grey (4D2) to greyish brown (6D3) or dark brown (6F6) towards the base, longitudinally striate, covered with whitish fibrous scales. Annulus persistent, membranous, cottony, white to yellowish, fibres on the ventral side. Context white, woolly or fleshy. Taste mild. Odour sweet. Spore print white.

Habit and habitat: Clustered, on *Quercus semecarpifolia* in subalpine mixed forests.
Localities: Chopta-Baniyakund (Rudraprayag district); Pindari trek (Bageshwar district).
Medicinal properties: *Armillaria mellea* exhibits antidiabetic, anticancer, antioxidant, antimicrobial, immunostimulatory, and anti-platelet aggregation properties. It has the capacity to reducing whole blood viscosity and improve blood supply to brain cells. Several bioactive compounds are responsible for the anti-bacterial activities of this fungus; 4-O-methylmelleolide, judeol, and armillaric acid show antibacterial activity against Gram-positive bacteria. Moreover, sesquiterpenoids melleolides B, C, and D also exhibit antibacterial activity (Arnone et al. 1986; Donnelly et al. 1985; Gao et al. 2009; Lin et al. 1988; Strapáč et al. 2016; Wang et al. 2007). An et al. (2017) investigated the neuroprotective effects of *A. mellea* mycelium polysaccharides (AMPS), which can be used for the cure of Alzheimer's disease.

16.2.3 *Astraeus hygrometricus* (Pers.) Morgan, J. Cincinnati Nat. Hist. Soc. 11: 20 (1889) **Figure 16.3c**

Basidiomata 20–32 mm broad, subepigeous, globose or subglobose to slightly compressed, with whitish basal mycelium. Exoperidium whitish, fleshy, thick, hygroscopic, splitting into 7–14 acute rays and becoming star-shaped, rays recurved when moist, becoming inwards on drying, distinctly layered; outermost layer pale buff, lost with age; inner layer up to 12 mm wide, whitish or yellowish becoming dark brown, cracking deeply, thicker towards the base; pseudo-parenchymatous layer yellowish brown to dark brown, up to 10 mm thick, soon splitting into a characteristic

FIGURE 16.3 (a) *Cantharellus cibarius*, (b) *Coprinus comatus*, (c) *Astraeus hygrometricus*, (d) *Armillaria mellea*, (e) *Aleuria aurantia*.

reticulate pattern, hard when dry. Endoperidium sessile, subglobose, 10–25 mm wide, whitish at first, becoming pale buff, opening by an irregular, often longitudinal, apical tear. Peristome lacking. Columella lacking. Gleba pure white to light brown. Taste and odour are indistinct. Spore print not obtained.

Habit and habitat: Solitary to scattered on sandy ground.

Localities: Chirbatiya, Chopta-Baniyakund (Rudraprayag district).

Medicinal properties: *Astraeus hygrometricus* is commonly known as 'futfut' in Uttarakhand. Various bioactive compounds such as phenolics, tannins, flavonoids, vitamin C, phytic acid, lycopene, and β-carotene are present in this mushroom, providing anti-oxidative, antimicrobial, antiparasitic, antidiabetic, anti-inflammatory, hepatoprotective, cardioprotective, anticancer effects. Fe^{2+}-chelating capacity, reducing power, and DPPH radical-scavenging activity are associated with this mushroom. A heteroglucan present in *A. hygrometricus* causes tumour regression resulting in the anti-tumour capacity of the mushroom (Biswas et al. 2017; Mallick et al. 2010; 2016, Pavithra et al. 2016).

16.2.4 *Calocera viscosa* (Pers.) Fr., Syst. Mycol. (Lundae) 1: 486 (1821) Figure 16.5b

Basidiomata 35–90 × 20–40 mm, composed of cylindrical dichotomous branches, dendroid, bright yellow to yellowish orange, white near the base; branches cylindrical, 12–50 × 1–2 mm, often forked

FIGURE 16.4 (a) *Coprinellus micaceus*, (b) *Cordyceps militaris*, (c) *Schizophyllum commune*, (d) *Sparassis latifolia*, (e) *Tremella fuciformis*.

at apex; context concolourous to basidiomata, soft-cartilaginous. Taste and odour are indistinct. Spore print not obtained.

Habit and habitat: Solitary to scattered, growing on log or stumps of *Abies* spp.

Localities: Chopta-Baniyakund, Amkoti (Rudraprayag district); Mukteshar (Nainital district).

Medicinal properties:*Calocera viscosa* contains a variety of bioactive compounds such as alkaloids, tannins, flavonoids, sterols, glycosides, saponins, serotonin, terpenoids, and phenols. It exhibits antioxidant and antibacterial activities, showing inhibitory effects against *Klebsiella pneumonia*, *Escherichia coli*, and *Staphylococcus aureus* (Muszyήska et al. 2012a, b; Naik S et al. 2019).

16.2.5 *Cantharellus cibarius* Fr., Syst. Mycol. (Lundae) 1: 318 (1821)

Pileus 30–85 mm broad, convex at first, becoming funnel-shaped, pale yellow Figure 16.3a (2A3) to orange-yellow (4B8) or golden yellow (5B7), glabrous to tomentose or with few tiny, appressed fibrills; margin inrolled at first, wavy with age; context white, thin at the margin. Hymenophore in form of gill folds, decurrent, close to sub-distant, dichotomously forked, concolourous with pileus. Stipe 30–60 x 10–20 mm, more or less equal, sometimes narrowing slightly downwards, solid, smooth, yellowish white (2A2) or concolourous with the pileus, base whitish, bruising brownish; context white. Taste is not distinctive. Odour is fragrant and fruity. Spore print yellowish white (2A2) to pale yellow (2A3).

Habit and habitat: Gregarious, under *Quercus leucotrichophora* and *Cedrus deodara*.

Localities: Phedkhal, Nagdev-Jhandidhar, Adwani (Pauri Garhwal district); Gajar (Almora district), Hat-Kalyani (Chamoli district), Dhakuri (Bageshwar district).

Medicinal properties: Different extracts of *C. cibarius* have been reported to have various biological properties such as antimicrobial, insecticidal, scavenging, antioxidant, anti-aging, anti-hypoxic and anticancer activities. A chemo-preventive effect against colon cancer has been displayed by the polysaccharide extract of *C. cibarius*. A methanolic extract of the golden chanterelle is effective against a wide range of bacteria and shows antihypertensive activities. The polysaccharide fraction of *C. cibarius* is neuroprotective and hence can be promising in curing neurodegenerative diseases (Elkhateeb and Daba 2021; Khalili et al. 2014; Lemieszek et al. 2018).

16.2.6 *Coprinellus micaceus* (Bull.) Vilgalys, Hopple, and Jacq. Johnson, Taxon 50 (1): 234 (2001) Figure 16.4a

Pileus 10–30 mm broad, obtusely cylindrical when young, expanding to convex or campanulate with age, conspicuously striated up to the disc, surface covered by small, white to whitish, shiny, clustered mica-like granules at first then diffuse with age, becoming glabrous at maturity; yellowish brown (5E8) to light brown (6D5) at first, becoming dark brown (6F7) with maturity or when pileus deliquesces; margin irregularly wavy-lobed, becoming expanded with revolute or split, concolourous to disc at first, fading to light yellowish brown (5E6); context up to 1 mm thick, concolourous to surface. Lamellae adnexed or free, crowded, white at first, becoming dark brown (6F7), finally black with maturity; lamellulae numerous, of 2–3 lengths. Stipe 25–60 × 2–4 mm, equal or slightly tapering upwards, smooth to tomentose, white, greyish brown (6E3) on bruising at the base, hollow. Taste and odour are indistinct. Spore print black.

Habit and habitat: Gregarious, on the ground.
Localities: Chaubatta, Khirsu forest (Pauri Garhwal district).
Medicinal properties: Bioactive compounds present in *Coprinellus micaceus* exhibit antioxidant and antibacterial properties. According to Zahid et al. 2006, the methanolic extract of *C. micaceus* shows antibacterial activity against *Enterococcus faecalis*, *Staphylococcus aureus*, *Corynebacterium xerosis*, and *Escherichia coli*. In addition, *C. micaceus* also exhibits glutathione S-transferase (GST) inhibitory activity due to the presence of (Z, Z)-4-oxo-2,5-heptadienedioic acid, which can be promising in the treatment of cancers (Ata 2009, Selem et al. 2021; Zahid et al. 2006).

16.2.7 *Coprinus comatus* (O. F. Müll.) Pers., Ten. Disp. Meth. Fung. (Lipsiae): 62 (1797) Figure 16.3b

Pileus 12–55 mm in diam., cylindrical to oval when young, becoming bell-shaped or conic, surface milk white, light brown (5D8) to dark brown (8F6) near margin, then blackish with age, surface glabrous, light brown (5D8) to black, squamulose yellowish white to dark brown (8F6) scales; margin decurved, becoming inrolled with age, light brown (5D8) to dark brown (8F6) becoming 'ink' black with maturity; context white. Lamellae free or adnexed, crowded, white, turning reddish white, then black and inky with age, lamellulae abundant, with 3–4 series. Stipe 58–145 × 5–10 mm, central, white, cylindrical to tapering towards lamellae, with whitish bulbous to abrupt base; context white, thin, hollow. Annulus white, membranous, moveable, superior, usually sliding to stipe's base with maturity. Taste and odour are indistinct. Spore print black.

Habit and habitat: Growing solitary to scattered, on the ground or drying dung of buffalo.
Localities: Bhainskot, Khirsu forest, Kandoliya forest (Pauri Garhwal district).
Medicinal properties: Different extracts of *C. comatus* exhibit antioxidant, anti-inflammatory, antidiabetic, anticancer, anti-obesity, antimicrobial, anti-nematode, antifungal, and antiviral

FIGURE 16.5 (a) *Grifola frondosa*, (b) *Calocera viscosa*, (c) *Hydnum repandum*, (d) *Lactifluus piperatus*, (e) *Hericium erinaceus*.

activities. Natural antioxidant components, like ascorbic acid, β-carotene, lycopene, flavonoids, terpenoids, and saponins have been found in *C. comatus* extracts. According to Ozalp et al. (2014), a polysaccharide extract of *C. comatus* can help in curing liver damage caused by alcohol consumption (Kalaw and Albinto 2014; Nowakowski et al. 2020; Ratnaningtyas et al. 2022; Sánchez 2017; Yu et al. 2009; Zhou and Han 2008).

16.2.8 *Cordyceps militaris* (L.) Fr., Observ. Mycol. (Havniae) 2: 317 (cancellans) (1818) Figure 16.4b

Stroma 20–60 × 1–4 mm, cylindrical to slightly clavate, tapering towards base, upper surface granular, covered with raised pores, orange to mandarin orange (6A7–6B8); lower surface glabrous, curved, salmon (^A4), base attached to a dead underground pupa; context hollow, orange to mandarin orange (6A7–6B8). Taste and odour are mild. Spore print not obtained.

Habit and habitat: Growing solitary to gregarious on dead insect larvae.
Localities: Chopta-Baniyakund (Rudraprayag district).
Medicinal properties: *Cordyceps militaris* is an entomopathogenic fungus that can be easily identified by its orange colour and granular surface. It exhibits various therapeutic properties like anti-oxidative, antidiabetic, anti-tyrosinase, anti-cholesterol, and antiproliferative activities.

Wong et al. (2011) purified a protease known as cordymin from *C. militaris*, which showed antiproliferative activity towards breast cancer cells (MCF-7) and also inhibited HIV-1 reverse transcriptase. It also shows mitochondrial protection by scavenging reactive oxygen species. Cordycepinan adenosine analogue was first discovered in *C. militaris* and has anticancer properties. A cytotoxic antifungal protease form *C. militaris* has been found to exhibit antifungal effects against *Fusarium oxysporum*, *Bipolaris maydis*, *Mycosphaerella arachidicola*, *Rhizoctonia solani* and *Candida albicans* (Jin et al. 2018; Mehra et al. 2017).

16.2.9 *Grifola frondosa* (Dicks.) Gray, Nat. Arr. Brit. Pl. (London) 1: 643 (1821) Figure 16.5a

Basidiomata 175–340 × 125–210 mm, stipitate, multipileate. Pileus 10–75 mm in diam., imbricate, flabelliform, undulating fronds, surface dry, glabrous to finely tomentose, azonate to faintly zonate, smooth or radiately rugose, yellowish brown to golden brown (5D5–5D7), light brown (6D8) with maturity; context milk-white (1A2), thin, up to 7 mm; margin undulate or curled, whitish. Hymenophore deccurent, with mostly angular pores, rarely irregular, 2–4 per mm, white to pale (1A2); tubes up to 3 mm long, white to pale (1A2). Stipe 55–95 × 12–20 mm, attached to wood, pale (1A2), whitish towards base; context solid, pale (1A2). Taste and odour are pleasant. Spore print pale (1A2).

Habit and habitat: Found growing solitarily on dead and cut logs or living stems of *Quercus* spp.
Localities: Chopta-Baniyakund (Rudraprayag district), Dhakuri (Bageshwar district).
Medicinal properties:*Grifola frondosa* is commonly known as 'hen of the woods', 'maitake' in Japanese, and 'hui-shu-hua' in Chinese. Fractions of *G. frondosa* show antitumour effects against various organ cancers. Particularly, the D-fraction of *G. frondosa* is the most effective in the inhibition of cancer. The MD-fraction, resulting from the purification of the D-fraction, shows more potentiated activity on the immune system and stronger inhibition of tumour growth. In addition to its anti-tumour effect, antihypertensive, antidiabetes, anti-obesity, and anti-hepatitis activities have been demonstrated in *G. frondosa* (Bhatt et al. 2018; Wu et al. 2021). Anti-inflammatory and immunomodulatory activities in astrocytes and microglia cells are shown by o-orsellinaldehyde, a natural compound present in *Grifola frondosa* (Tomas-Hernandez et al. 2021).

16.2.10 *Hericium erinaceus* (Bull.) Pers., Commentatio de Fungis Clavaeformibus: 27 (1797) Figure 16.5e

Basidiomata 55–95 × 40–65 mm, white to pale (1A2), stipitate, white to yellowish white base, bearing primary branches. Primary branches wide up to 10 mm, pale (1A2), covered with small spines (up to 2 mm in length) all around. Secondary branches arise from primary branches, wide up to 8 mm, smooth, and lack spines. Tertiary branches arise from secondary branches, wide up to 4 mm, covered with fertile spines. Spines dense, pendant, white, up to 25 mm in length. Context pale (1A2), unchanged with $FeSO_4$, turning yellowish with KOH and orange with guaiacol. Taste and odour are pleasant. Spore print pale (1A2).

Habit and habitat: Growing solitarily on dead wood of *Quercus* spp. in mixed forests.
Localities: Chopta-Baniyakund, Pabdhar (Rudraprayag district), Dudhatoli trek (Pauri Garhwal district).
Medicinal properties:*Hericium erinaceus*is is commonly called the 'lion's mane'. Various diterpenes and sesterpenes like hericenones and erinacines have been found in the sporocarp and fermenting mycelium of *H. erinaceus*. Erinacines show biological activities as stimulators of nerve growth factor synthesis and could be promising in the treatment of neurodegenerative disorders and peripheral neuropathy. Studies have shown the anti-oxidative effects of the standardized aqueous extract of *H. erinaceous* induced by L-luthioninesulfoximine (BSO) in fibroblasts from a Friedreich's ataxia (FRDA) patient (Lew et al. 2020; Li et al. 2018; Spelman et al. 2017).

Studies have revealed that polysaccharides from submerged mycelial cultures of *H. erinaceus* exhibit anticancer activities. Various cancers like esophageal, intestinal, pancreatic, and stomach cancers can also possibly be treated by administering preparations based on *H. erinaceus.* Cancer patients treated with *H. erinaceus* have reported fewer side effects than those associated with conventional cancer treatments. Methanol and aqueous extracts of *H. erinaceus* have demonstrated significant antihyperglycemic and anti-hyperlipidemic effects in instreptozotocin-induced diabetic rats. Ethanol and ethyl acetate extracts exhibit activity against *Helicobacter pylori* (Shang et al. 2013; Thongbai et al. 2015).

16.2.11 *Hydnum repandum* L., Sp. Pl. 2: 1178 (1753) Figure 16.5c

Pileus 25–95 mm in diam., planno-convex, applanate with a central depression with age, surface dry, glabrous to slightly scaly, salmon (6A4) to Persian orange (6A7), or yellowish white (1A2); margin inrolled when young, becoming wavy to undulate with age, reddish golden (6C7) when bruised; context white, fleshy, firm, thick (up to 3 mm), paler when exposed. Hymenophore decurrent, concolourous to pileus, spines up to 6 mm long. Stipe 25–75 × 8–20 mm, central to excentric, equal or tapering towards apex, surface dry, glabrous, pale (1A2), reddish golden (6C7) when bruised. Taste indistinct. Odour pleasant or fruity. Spore print white.

Habit and habitat: growing scattered to gregariously on the ground under *C. deodara* and *Quercus* spp. in temperate to subalpine forests.

Localities: Chaubatta, Adwani, Teka, Nagdev-Jhandidhar, Lansdowne (Pauri Garhwal district).

Medicinal properties: *Hydnum repandum* is also called a 'wood hedgehog mushroom' due to its spiny hymenophore. It contains various secondary metabolites which exhibit antioxidant, antimicrobial, cytotoxic, antiproliferative, anti-inflammatory, genotoxic, and protective activity against chemotherapeutics. Repandiol, a diepoxide isolated from *H. repandum*, shows cytotoxic activity against various cancer cells. (Thu et al. 2020; Tubić et al. 2019; Zeb and Lee 2021).

16.2.12 *Lactifluus piperatus* (L.: Fr.) Kuntze, Revis. Gen. Pl. 2: 857. 1891 Figure 16.5d

Pileus 45–100 mm diam., convex to plano-convex with a depressed centre, sometimes infundibuliform in age; surface dry, smooth to wrinkled, sometimes areolate at maturity; white to yellowish white (2A2) with light yellow (3A4) patches or brownish yellow (5C7–8) when areolate; margin incurved when young, decurved to uplifted at maturity. Lamellae adnate to sub-decurrent, crowded, forked, white to yellowish white (2A2), light yellow (3A4) on bruising, lamellulae numerous. Stipe 35–110 × 20–42 mm, cylindrical, tapering downward, slightly wrinkled, white to yellowish white (2A2). Context firm, up to 11 mm at the disc, white, turning pale yellow (2A3) on exposure to air; taste mild at first, becoming very acrid; odour unpleasant. Latex abundant, white, unchanging to slightly yellowish white (2A2); taste acrid. Spore print white.

Habit and habitat: Gregarious to scattered under *Quercus leucotrichophora*, *Myrica esculenta*, and *Rhododendron arboreum* in mixed forests and under *Pinus roxburghii* in temperate coniferous forests.

Localities: Khirsu, Chaubatta, Phedkhal (Pauri Garhwal district); Badiyargarh (Tehri Garhwal district); Jasoli, Makku (Rudraprayag district).

Medicinal properties: *Lactifluus piperatus* is a striking milk-cap mushroom with whitish fruiting bodies and white, peppery latex. Fruiting bodies of this milk cap possess various bioactive compounds including phenolic, flavonoidic, ascorbic acid, β-carotene, and lycopene, which are responsible for the anti-oxidative activity of this fungus. The antioxidant compounds from *L. piperatus* can be used for different chronic diseases related to oxidative stress, which can be the reason for various ailments like cancer, Parkinson's disease, Alzheimer's disease, heart failure, myocardial infarction, schizophrenia, bipolar disorder, sickle cell disease, and chronic fatigue disease. *Lactifluus piperatus* also shows neuroprotective, genoprotective, and cytotoxic activities against cancer cells. It is also used in traditional medicine systems in the treatment of urinary tract diseases due to its positive effects on urine formation regulation. *Lactifluus piperatus* is also considered to have beneficial effects of joint and muscle relaxation. Methanol extracts of *L. piperatus* exhibits antimicrobial activity against *Escherichia coli*, *Proteus vulgaris*, and *Mycobacterium smegmatis*. Anti-microbial activity of *L. piperatus* is similar to that of antibiotics by inhibition of cell wall and protein synthesis or nucleic acid synthesis (Barros et al. 2007; Dulger et al. 2002; Joshi et al. 2013; Kosanić et al. 2020).

16.2.13 *Schizophyllum commune* Fr. [as 'Schizophyllus communis'], Observ. Mycol. (Havniae) 1: 103 (1815) Figure 16.4c

Pileus 8–35 mm in diam, fan-shaped bracket to shell-shaped, sessile, surface dry, white to grey (3B2), covered with dense, tiny fibrills, fused at the base with other brackets; context thin, leathery, white to grey (3B2). Hymenophore consists of longitudinally split lamellae, distant, pale red to dull red (9A3–9B3). Taste and odour are not distinct. Spore print white.

Habit and habitat: Growing gregarious to caespitose on dead logs or decaying wood.
Localities: Pauri, Khirsu, Phedkhal (Pauri Garhwal district).
Medicinal properties:*Schizophyllum communae* is generally called the 'split-gilled mushroom' due to the split hymenophore. It is a pathogenic fungus that is found growing on the wood of different trees, but due to its nutritional and medicinal potential, it is widely cultivated. It exhibits antidiabetic, anti-candida, anti-tumour, and antiviral properties. Moreover, various bioactivities of the split-gilled mushroom are due to schizophyllan, a β 1–3, β 1–6D-glucan. It also activates macrophages, which results in the augmentation of T-cell activities and increases the sensitivity of cytotoxic LAK and NK cells to Il-2. *Schizophyllum commune* extracts have been found to exhibit antibacterial activity against Gram-positive bacteria (Bhatt et al. 2018; Mirfat et al. 2014).

16.2.14 *Sparassis latifolia* Y.C. Dai and Zheng Wang, Mycologia 98: 586 (2006) Figure 16.4d

Basidiomata 250–310 mm in diam., stipitate. Flabellae up to 9–1 mm, azonate, dissected, abundant, arising from the common central mass, white to yellowish white (2A2) when young, brownish with maturity, cinnamon brown (6D6) when dry; margin wavy to broadly rugulose; context leathery, white to yellowish white (2A2). Stipe 115–145 × 10–12 mm, white to yellowish white (2A2). Odour pleasant. Taste indistinctive. Spore print not obtained.

Habit and habitat: Growing solitarily on the base of *Cedrus deodara*.
Localities: Abbot Mount (Champawat district).
Medicinal properties:*Sparassis latifolia* is commonly known as the 'cauliflower mushroom' and it is a prized edible species. Polysaccharides present in *S. latifolia* render their therapeutic potential. *Sparassis latifolia* β-glucan has immunoregulatory, anti-tumour, anti-inflammatory, and pro-hematopoietic activities. In addition, *S. latifolia* polysaccharides show reducing ability and DPPH-, -OH-, O_{2-}-, and free radical-scavenging activities. Studies suggest that polysaccharides of this mushroom can also promote the proliferation of macrophages (Wang et al. 2019; Zhang et al. 2022).

16.2.15 *Tremella fuciformis* Berk., Hooker's J. Bot. Kew Gard. Misc. 8: 277 (1856) Figure 16.4e

Basidiomata up to 100 × 65 mm in diam., foliose, gelatinous, translucent white to white, composed of thin, erect, branching fronds; surface glabrous, shiny; margin crenate, rarely entire or undulate. Taste and odour are not distinct. Spore print not obtained.

Habit and habitat: Growing solitary to scattered on the cut stem and dead logs of *Quercus* and *Abies* spp.

Localities: Duggalbitta, Tungnath trek, Jakhdhar (Rudraprayag district).

Medicinal properties: *Tremella fuciformis* has a long history of use in food and medicine in China and Japan. The fruiting body is rich in dietary fibre, protein, minerals, antioxidants, and vitamin D. In China and Japan, it has traditionally been used as a potent Jing and Chi (Qi) tonic. It exhibits antidiabetic, anti-tumour, anti-oxidation, anti-aging, immunomodulatory, and anti-inflammatory activities. *Tremella fuciformis* is also reported to enhance cognition and prevent neurodegenerative disorders. D-mannose present in *T. fuciformis* polysaccharides has demonstrated potential effects on the immune system. Moreover, it acts against metabolic syndromes by inhibiting the attachment of ST-10 *Salmonella typhimurium* to intestinal cells (Ban et al. 2018; Hu et al. 2016; Mohamad et al. 2020; Qin et al. 2022; Yui et al. 1995).

16.3 CONCLUSIONS AND FUTURE PROSPECTS

It is evident that wild mushrooms are rich sources of essential elements and macromolecules, and that these natural resources should be used as food or supplements and in medicine. With the increasing global population and shrinkage of agricultural land, the food crisis in many regions around the globe is inevitable. In such a scenario, mushroom cultivation, which is a cost-effective venture and requires only a few basic facilities, should be encouraged, and wild mushrooms which have rich nutritional values, should also be introduced into cultivation. Moreover, mushrooms must be utilized for their rich therapeutic value. The history of mushroom cultivation and use is quite long in some areas of the world, but many wild mushrooms are not exploited in terms of their spectra of biomolecules and bioactivity. More intense research is needed to explore the macrofungal diversity in the wild to develop their identification and utilization.

REFERENCES

Amono, K., H. Katayama, A. Saito, A. Ando & Y. Nagata. 2012. *Aleuria aurantia* lectin exhibits antifungal activity against *Mucorracemosus*. *Bioscience, Biotechnology, and Biochemistry* 76(5): 967–970.

An, S., W. Lu, Y. Zhang, Q. Yuan & D. Wang. 2017. Pharmacological basis for use of *Armillariamellea* polysaccharides in Alzheimer's disease: Antiapoptosis and antioxidation. *Oxidative Medicine and Cellular Longevity* 2017: 1–11.

Arnone, A., R. Cardillo & G. Nasini. 1986. Structures of melleolides B-D, three antibacterial sesquiterpenoids from *Armillaria mellea*. *Phytochemistry* 25: 471–474.

Ata, A. 2009. Novel bioactive natural products from marine and terrestrial sources. In: Şener, B. (eds.), *Innovations in Chemical Biology*. Springer, Dordrecht.

Ban, S., S.L. Lee, H.S. Jeong, S.M. Lim, S. Park, Y.S. Hong & J.E. Kim. 2018. Efficacy and safety of tremella fuciformis in individuals with subjective cognitive impairment: A randomized controlled trial. *Journal of Medicinal Food*: 1–8.

Barros, L., P. Baptista & I.C.F.R. Ferreira. 2007. Effect of *Lactarius piperatus* fruiting body maturity stage on antioxidant activity measured by several biochemical assays. *Food and Chemical Toxicology* 45: 1731–1737.

Bhambri, A., M. Srivastava, V.G. Mahale, S. Mahale & S.K. Karn. 2022. Mushrooms as potential sources of active metabolites and medicines. *Frontiers in Microbiology* 13: 837266.

Bhatt, R.P., U. Singh & S.L. Stephenson. 2016. Wild edible mushrooms from high elevations in the Garhwal Himalaya–I. *Current Research in Environmental & Applied Mycology* 6(2): 118–131.

Bhatt, R.P., U. Singh & P. Uniyal. 2018. Healing mushrooms of Uttarakhand Himalaya, India. *Current Research in Environmental & Applied Mycology* 8(1): 1–23.

Biswas, G., S. Nandi, D. Kuila & K. Acharya. 2017. A comprehensive review on food and medicinal prospects of *Astraeus hygrometricus*. *Pharmacognosy Journal* 9(6): 799–806.

Champion, H.G. & S.K. Seth. 1968. *A Revised Survey of the Forest Types of India*. Government of India Publications, New Delhi.

Chang, S.T. & P.G. Miles. 1992. Mushrooms biology a new discipline. *Mycologist* 6: 64–65.

Donnelly, D.M.X., F. Abe, D. Coveney, N. Fukuda & J. O'Reilly. 1985. Antibacterial sesquiterpene aryl ester from *Armillaria mellea*. *Journal of Natural Products* 48: 10–16.

Dulger, B., F. Yilmaz & F. Gucin. 2002. Antimicrobial activity of some *Lactarius* species. *Pharmaceutical Biology* 40(4): 304–306.

Elkhateeb, W.A. & G.M. Daba. 2021. Highlights on the golden mushroom *Cantharellus cibarius* and unique Shaggy ink cap mushroom *Coprinus comatus* and smoky bracket mushroom *Bjerkandera adusta* ecology and biological activities. *Journal of Mycology & Mycological Sciences* 4(2): 000143.

Gao, L.W., W.Y. Li, Y.L. Zhao & J.W. Wang. 2009. The cultivation, bioactive components and pharmacological effects of *Armillaria mellea*. *African Journal of Biotechnology* 8(25): 7383–7390.

Genevieve, L., C. Pierre-Luc, G.T. Roxanne, M. Amélie, B. Danny, M. Vincent & G. Hugo. 2019. Estimation of fungal diversity and identification of major abiotic drivers influencing fungal richness and communities in northern temperate and boreal quebec forests. *Forests* 10: 1096.

Hu, X., Y. Shi, P. Zhang, M. Miao, T. Zhang & B. Jiang. 2016. D-Mannose: properties, production, and applications: an overview. *Comprehensive Reviews in Food Science and Food Safety* 15: 773–785. http://doi.org/10.1111/1541-4337.12211

Jin, Y., X. Meng, Z. Qiu, Y. Su, P. Yu & P. Quc. 2018. Anti-tumor and anti-metastatic roles of cordycepin, one bioactive compound of *Cordyceps militaris*. *Saudi Journal of Biological Sciences* 25: 991–995.

Joshi, S., R.P. Bhatt & S.L. Stephenson. 2012. The current status of the family Russulaceae in the Uttarakhand Himalaya, India. *Mycosphere* 3(4): 486–501.

Joshi, S., M.P. Vishwakarma, R. Mahar & R.P. Bhatt. 2013. Medicinally important and edible species of genus *Lactarius* from Garhwal Himalaya, India. *Mycosphere* 4(4): 714–720.

Kalaw, S.P. & R.F. Albinto. 2014. Functional activities of Philippine wild strain of *Coprinus comatus* (O. F. Müll.: Fr.) Pers and *Pleurotus cystidiosus* O. K. Miller grown on rice straw based substrate formulation. *Mycosphere* 5(5): 646–655.

Khalili, M., M.A. Ebrahimzadeh, F. Omrani & M. Karami. 2014. Antihypoxic activities of the golden chanterelle mushroom, *Cantharellus cibarius* (Higher Basidiomycetes). *International Journal of Medicinal Mushrooms* 16(4): 339–344.

Kim, S., Y. Choi & K. Kim. 2022. Coacervate mediated novel pancreatic cancer drug *Aleuria aurantia* lectin delivery for augmented anticancer therapy. *Biomaterials Research* 26: 35.

Kornerup, A. & J.H. Wanscher. 1978. *Methuen Handbook of Color*. 3rd ed. Eyre Methuen Ltd., London.

Kosanić, M., N. Petrović, O.M. Djordjević, D. Grujičić, D. Tubić, A. Marković & T. Stanojković. 2020. The health promoting effects of the fruiting bodies extract of the peppery milk cap mushroom *Lactarius piperatus* (Agaricomycetes) from Serbia. *International Journal of Medicinal Mushrooms* 22(4): 347–357.

Lemieszek, M.K., F.M. Nunes, C. Cardoso, G. Marques & W. Rzeski. 2018. Neuroprotective properties of *Cantharellus cibarius* polysaccharide fractions in different in vitro models of neurodegeneration. *Carbohydrate Polymers* 197: 598–607.

Lew, S.Y., Y.Y. Yow, L.W. Lim & K.H. Wong. 2020. Antioxidant-mediated protective role of *Hericium erinaceus* (Bull.: Fr.) Pers. against oxidative damage in fibroblasts from Friedreich's ataxia patient. *Food Science and Technology* 40(1): 264–272.

Li, I.C., L.Y. Lee, T.T. Tzeng, W.P. Chen, Y.P. Chen, Y.J. Shiao & C.C. Chen. 2018. Neurohealth Properties of *Hericium erinaceus* Mycelia Enriched with Erinacines. *Behavioural Neurology* 2018: 10.

Lin, Z.B., D.M. Zhang & D. Xia. 1988. Effects of *Armillaria mellea* polysaccharide on immunological function in mice. *Chinese Pharmacological Bulletin* 4: 93–96.

Mallick, S., A. Dutta, A. Chaudhuri, D. Mukherjee, S. Dey, S. Halder, J. Ghosh, D. Mukherjee, S.S. Sultana, G. Biswas, T.K. Lai, P. Patra, I. Sarkar, S. Chakraborty, B. Saha, K. Acharya C. & Pal. 2016. Successful therapy of murine visceral leishmaniasis with astrakurkurone, a triterpene isolated from the mushroom *Astraeus hygrometricus*, involves the induction of protective cell-mediated immunity and TLR9. *Antimicrob Agents Chemother* 60: 2696–2708.

Mallick, S.K., S. Maiti, S.K. Bhutia & T.K. Maiti. 2010. Antitumor properties of a heteroglucan isolated from *Astraeus hygrometricus* on Dalton's lymphoma bearing mouse. *Food and Chemical Toxicology* 48: 2115–2121.

Mehra, A., K.U. Zaidi, A. Mani & V. Thawani. 2017. The health benefits of *Cordyceps militaris*—a review. *Kavaka* 48(1): 27–32.

Mirfat, A.H.S., A. Noorlidah & S. Vikineswary. 2014. Antimicrobial activities of split gill mushroom *Schizophyllum commune* Fr. *American Journal of Research Communication* 2(7): 113–124.

Mohamad, H.S., W. Sun, H. Shen & Q. Cheng. 2020. Chemical compounds and health benefits of *Tremella*, a valued mushroom as both cuisine and medicine in ancient China and modern era. *Amazonian Journal of Plant Research* 4(3): 692–697.

Monkai, J., K.D. Hyde, J. Xu & P.E. Mortimer. 2017. Diversity and ecology of soil fungal communities in rubberplantations. *Fungal Biology Reviews* 31: 1–11.

Muszyŋska, B. & K. Sułkowska-Ziaja. 2012a. Analysis of indole compounds in fruiting bodies and in mycelia from in vitro cultures of *Calocera viscosa* (Basidiomycota). *Acta Mycologica* 47(1): 57–64.

Muszyŋska, B., K. Sułkowska-Ziaja & H. Ekiert. 2012b. An antioxidant in fruiting bodies and in mycelia from in vitro cultures of *Calocera viscosa* (Basidiomycota) Preliminary results. *Acta Poloniae Pharmaceutica-Drug Research* 69: 135–138.

Nowakowski, P., S.K. Naliwajko, R. Markiewicz-Zukowska, M.H. Borawska & K. Socha. 2020. The two faces of *Coprinus comatus*—Functional properties and potential hazards. *Phytotherapy Research* 34: 2932–2944.

Ozalp, F.O., M. Canbek, M. Yamac, G. Kanbak, L.J. Van Griensven, M. Uyanoglu, H. Senturk, K. Kartkaya & A. Oglakci. 2014. Consumption of *Coprinus comatus* polysaccharide extract causes recovery of alcoholic liver damage in rats. *Pharmaceutical Biology* 52(8): 994–1002.

Pavithra, M., K.R. Sridhar, A.A. Greeshma & K.T. Yokotani 2016. Bioactive potential of the wild mushroom *Astraeus hygrometricus* in South-west India. *Mycology* 7(4): 191–202. http://doi.org/10.1080/21501203.2016.1260663

Qin, L., G. Su, C. Wu, Q. Zhou, X. Peng, L. Hu, Y. Liu, R. Wang, Q. Xu, Z. Fang, Y. Lin, S. Xu, B. Fang, J. Li, D. Wu & L. Chi. 2022. Effects of *Tremella fuciformis* extract on growth performance, biochemical and immunological parameters of weaned piglets challenged with lipopolysaccharide. *Animal Production Science* 62(5): 462–469. http://doi.org/10.1071/AN20425

Ratnaningtyas, N.I., E.N. Hernayanti, N. Ekowati & F. Husen. 2022. Ethanol extract of the mushroom *Coprinus comatus* exhibits antidiabetic and antioxidant activities in streptozotocin-induced diabetic rats. *Pharmaceutical Biology* 60(1): 1126–1136.

Sánchez, C. 2016. Reactive oxygen species and antioxidant properties from mushrooms. *Synthetic and Systems Biotechnology* 2(1): 13–22.

Sati, V.P. & Bandooni, S. K. 2018. Forests of Uttarakhand: Diversity, distribution, use pattern and conservation. *ENVIS Bulletin Himalayan Ecology* 26: 21–27.

Selem, E., Y. Alp, S. Sensoy, Y. Uzun, S. Cavusoglu, N. Karatas, S. Ercisli, N. Yilmaz, H. Ekiert, H.O. Elansary & A. Szopa. 2021. Biochemical and morphological characteristics of some macrofungi grown naturally. *Journal of Fungi* 7(851).

Semwal, K.C., S.L. Stephenson, V.K. Bhatt & R.P. Bhatt. 2014. Edible mushrooms of the Northwestern Himalaya, India: A study of indigenous knowledge, distribution and diversity. *Mycosphere* 5(3): 440–461.

Shang, X., Q. Tan, R. Liu, K. Yu, P. Li & G.P. Zhao. 2013. In vitro anti-*Helicobacter pylori* effects of medicinal mushroom extracts, with special emphasis on the Lion's Mane mushroom, *Hericium erinaceus* (higher Basidiomycetes). *International Journal of Medicinal Mushrooms* 15: 165–174.

Singh, U., R.P. Bhatt, S.L. Stephenson, P. Uniyal & T. Mehmood. 2017. Wild edible mushrooms from high elevations in the Garhwal Himalaya—II. *Current Research in Environmental & Applied Mycology* 7(3): 208–226.

Spelman, K., E. Sutherland & A. Bagade. 2017. Neurological Activity of Lion's Mane (*Hericium erinaceus*). *Journal of Restorative Medicine* 6: 19.

Stamets, P. 2000. *Growing Gourmet and Medicinal Mushrooms*. Ten Speed Press, Berkeley and Toronto, p. 339.

Strapáč, I., M. Baranová, M. Smrčová & Z. Bedlovičová. 2016. Antioxidant activity of honey mushrooms (*Armillaria mellea*). *Folia Veterinaria* 60(4): 37–41.

Thongbai, B., S. Rapior, K.D. Hyde, K. Wittstein & M. Stadler. 2015. *Hericium erinaceus*, an amazing medicinal mushroom. *Mycological Progress* 14: 1–23.

Thu, Z.W., K.K. Myo, H.T. Aung, M. Clericuzio, C. Armijos & G. Vidari. 2020. Bioactive phytochemical constituents of wild edible mushrooms from Southeast Asia. *Molecules* 25: 1–38.

Tomas-Hernandez, S., J. Blanco, S. Garcia-Vallvé, G. Pujadas, M.J. Ojeda-Montes, A. Gimeno, L. Arola, L. Minghetti, R. Beltrán-Debón & M. Mulero. 2021. Anti inflammatory and immunomodulatory effects of the *Grifola frondosa* natural compoundo orsellinaldehyde on LPS-challenged murine primary glial cells. Roles of NF-κβ and MAPK. *Pharmaceutics* 13: 806.

Tubić, J., D. Grujičić, M.R. Jakovljević, B. Ranković, M. Kosanić, T. Stanojković, A. Ćirić & O. Milošević-Djordjević. 2019. Investigation of biological activities and secondary metabolites of *Hydnum repandum* acetone extract. *Farmacia* 67(1): 174–183.

Wang, M.H., Z.Q. Hao, M.C. Chang, J.L. Meng, J.Y. Liu & C.P. Feng. 2019. Characterization, antioxidant and immunity activities of *Sparassis latifolia* polysaccharides. *Mycosystema* 38(5): 707–716.

Wang, X.Q., J.J. Yin, F.W. Meng, Y.H. Peng & F. Lin 2007. 65 cases of clinical observation of applying compound *Armillari amellea* tablets invertebro-basilar. *Chinese Journal of Clinical Healthcare* 10: 60–62.

Willis, K.J. (ed.) 2018. *State of the World's Fungi 2018. Report*. Royal Botanic Gardens, Kew.

Wong, J.H., Ng, T.B., Sze, S.C., Zhang, K.Y., Li, Q. & Lu, X. 2011. Cordymin, an antifungal peptide from the medicinal fungus *Cordyceps militaris. Phytomedicine* 18(5): 387–392.

Wu, J.-Y., K.-C. Siu & P. Geng. 2021. Bioactive ingredients and medicinal values of *Grifola frondosa* (Maitake). *Foods* 10: 95.

Yu, J., P. Cui, W. Zeng, X. Xie, W. Liang, G. Lin & L. Zeng. 2009. Protective effect of selenium polysaccharides from the mycelia of *Coprinus comatus* on alloxan-induced oxidative stress in mice. *Food Chemistry* 117: 42–47.

Yui, T., K. Ogawa, M. Kakuta & A. Misaki. 1995. Chain conformation of a glucurono-xylo mannan isolated from fruit body of *Tremella fuciformis* Berk. *Journal of Carbohydrate Chemistry* 14: 255–263. http://doi.org/10.1080/07328309508002068

Zahid, S., C.C. Udenigwe, A. Ata, M.O. Eze, E.P. Segstro & P. Holloway. 2006. New bioactive natural products from *Coprinus micaceus. Natural Product Research* 20(14): 1283–1289.

Zeb, M. & C.H. Lee. 2021. Medicinal Properties and Bioactive Compounds from Wild Mushrooms Native to North America. *Molecules* 26: 251.

Zhang, M., B.Y. Lou, Y.J. Zhang, S.J. Mei, L.Y. Gao & W.J. Chen. 2022. Preparation and characterization of *Sparassis latifolia* β-glucan microcapsules. *Open Chemistry* 20: 351–360.

Zhou, G. & C. Han. 2008. The co-effect of vanadium and fermented mushroom of *Coprinus comatus* on glycemic metabolism. *Biological Trace Element Research* 124(1): 20–27.

Index

Note: Page locators in **bold** indicate a table and page locators in *italics* indicate a figure.

D

G

H

N

O

P

Q

R

U

V

W